Gulf of Mexico Origin, Waters, and Biota

TEXAS A&M UNIVERSITY CORPUS CHRISTI | HARTE RESEARCH INSTITUTE FOR GULF OF MEXICO STUDIES

Harte Research Institute for Gulf of Mexico Studies Series, Sponsored by Texas A&M University–Corpus Christi

Larry McKinney, General Editor
John W. Tunnell Jr., Founding Editor

Gulf of Mexico Origin, Waters, and Biota

Volume 5, Chemical Oceanography

Edited by
Thomas S. Bianchi

Texas A&M University Press
College Station

First edition

This paper meets the requirements of ANSI/NISO Z39.48-1992
(Permanence of Paper).
Binding materials have been chosen for durability.
Manufactured in China through FCI Print Group.

∞

Library of Congress Cataloging-in-Publication Data

Gulf of Mexico origin, waters, and biota / [edited by John W. Tunnell Jr., Darryl L.
Felder, and Sylvia A. Earle] — 1st ed.
v. cm — (Harte Research Institute for Gulf of Mexico Studies Series)
Includes indexes.
Taken from the Harte Research Institute for Gulf of Mexico Studies website:
Gulf of Mexico origin, waters, and biota, is an updated and enlarged version of the
Gulf of Mexico: its origin, waters, and marine life, first published by US Fish and
Wildlife Service in Fishery bulletin, v. 89, 1954. Contents: V. 1. Biodiversity / edited
by Darryl L. Felder and David K. Camp ISBN-13: 978-1-60344-094-3 (cloth : alk. paper)
ISBN-10: 1-60344-094-1 (cloth : alk. paper)
1. Mexico, Gulf of. 2. Marine biology—Mexico, Gulf of. 3. Geology—Mexico,
Gulf of. 4. Oceanography—Mexico, Gulf of. I. Tunnell, John Wesley II. Felder, Darryl L.
III. Earle, Sylvia A., 1935– IV. Camp, David K. V. Series.
QH92.3.G834 2009
578.77'364—dc22

2008025312

Vol. 2, Ocean and Coastal Economy (2008)
Edited by James C. Cato
ISBN-13: 978-1-60344-086-8 (cloth : alk. paper)
ISBN-13: 978-1-60344-270-1 (ebook)

Vol. 3, Geology (2011)
Edited by Noreen A. Buster and Charles W. Holmes
ISBN-13: 978-1-60344-290-9 (cloth : alk. paper)
ISBN-13: 978-1-60344-293-0 (ebook)

Vol. 4, Ecosystem-Based Management (2013)
Edited by John W. Day and Alejandro Yáñez-Arancibia
ISBN-13: 978-1-60344-765-2 (cloth : alk. paper)
ISBN-13: 978-1-60344-776-8 (ebook)

Vol. 5, Chemical Oceanography (2019)
Edited by Thomas S. Bianchi
ISBN-13: 978-1-62349-774-3 (cloth : alk. paper)
ISBN-13: 978-1-62349-775-0 (ebook)

I thank my wife and son, Jo Ann and Christopher, for their unending support and patience through the years; they have always been my core inspiration. I also have to thank my canine buddy, Sir Felix of Cobblefield, for providing companionship while proofreading.

Thus the quantity of the different elements in sea water is not proportional to the quantity of elements which river water pours into the sea, but inversely proportional to the facility with which the elements in sea water are made insoluble by general chemical or organo-chemical actions in the sea.—**G. Forchammer, 1865**

In Memoriam

This book would not have been possible without the vision and guidance of Wes Tunnell. Wes was a champion of Gulf research for many years. We will miss him for his great wisdom about the Gulf of Mexico and, more important, for his unique qualities of caring and passion as a human being.

Contents

Preface

The Gulf of Mexico (GOM), located roughly between 18.2° to 30.4° north latitude and 81.0° to 97.9° west longitude, is a marginal sea that covers an area of more than 1.5×10^6 km^2 and reaches depths in excess of 4000 m over its abyssal plain. The basin of the GOM formed in the Late Triassic during rifting activity when the North American Plate began to drift from the African and South American Plates. The GOM is a particularly important body of water, since many of the hurricanes that reach landfall in the United States and Mexico pass through this dynamic system. It is bordered by very different estuarine and shelf systems, with more carbonate systems with significant groundwater inputs to the east and south, a river-dominated system in the north where the largest river system in the United States (Mississippi-Atchafalaya) empties in its coastal waters, and a river-starved coastal system to the west. The Gulf is characterized as having many warm-core eddies that spin off from the Florida Loop Current, which can entrain waters from the Caribbean that in many cases travel as far as to the northwestern Gulf and Texas shelf regions. These eddies have been shown to have significant effects on the chemistry and biology of Gulf waters as they are transported across the GOM. The northern GOM is one of the most productive US fisheries, at times only second to that in Alaska, and is home to many whale and shark species.

The GOM is perhaps most well known around the world for its oil and gas production. This notable feature became even more notable, and even infamous, when the Deepwater Horizon (DWH) explosion occurred in April 2010.

While this is not the main reason for writing this book, it is important to note that when researchers began to look for background data on the chemistry of the GOM waters, many people were surprised to find out how little was published in the peer-reviewed literature. One reason for this is a significant amount of shelf and deepwater research in the GOM over the past 20 to 30 years has been funded by Minerals Management Service, now the Bureau of Ocean Energy Management, Regulation, and Enforcement (BOEMRE), where typically technical reports are delivered as a final product.

The main goal of this book is to cover the basic chemical dynamics of this important marginal sea that is situated within the boundaries of a dynamic and changing semitropical region. We are already seeing some of the consequences of climate change on the migration of mangroves from the south to the northern reaches of the GOM, the increased frequency of harmful algal blooms whose causes in many cases remain enigmatic, complex interaction with large river plumes that in some cases come in contact with migrating warm-core eddies, and deep-sea communities that are linked with cold and warm seeps. The chemistry of the GOM is also particularly interesting in the wake of the DWH accident because it has a background signal of oil and gas from natural seeps that have existed for millions of years. This book is driven by a dynamic team of authors who represent the best and most knowledgeable in the field of chemical oceanography from the United States and Mexico.

Texas A&M University Press has published four other volumes in this series on the GOM: *Gulf of Mexico: Origin, Waters, and Biota,* Volume 1, *Biodiversity* (2009); Volume 2, *Ocean and Coastal Economy* (2009); Volume 3, *Geology* (2011); and Volume 4, *Ecosystem-Based Management* (2013). This book is to our knowledge the first of its kind that provides a solid background on the chemistry of sediments and water-column processes as they relate to natural and anthropogenic linkages for the entire GOM. We believe that the coverage of foundational chemistry topics in this book on this important marginal sea, should have long-lasting relevance and importance for basic science, management, and climate change issues for this region.

Acknowledgments

Special thanks go to the authors of each chapter, who supported the overall goal of the project and provided a comprehensive assessment of chemical oceanographic work in the Gulf of Mexico in their respective fields. In particular, thanks go to those lead authors who joined at a later date because of other authors having to drop out, and they still made the deadlines. I am very grateful for the time and energy put forth by each of the external reviewers of each chapter, listed here in alphabetical order: Jay A. Brandes, David J. Burdige, Robert Byrne, John M. Jaeger, William M. Landing, Ian R. MacDonald, Alfonso Mucci, Lisa H. Nowell, James L. Pinckney, Gilbert Rowe, Alan M. Shiller, and Kevin M. Yeager. I am also very grateful to Jo Ann M. Bianchi for helping me to proofread all chapters carefully. Thanks go to Elizabeth Sargent from Georgia Southern University for allowing me to use her image of the CTD for the book cover. Finally, I also thank the editorial staff at Texas A&M University Press; Stacy Eisenstark, who was fantastic to work with; and the late Wes Tunnell, who with his dedicated and relentless effort to get this topic finally covered in the Gulf of Mexico series did so with his usual charm and patience, even when dealing with personal health problems.

Gulf of Mexico Origin, Waters, and Biota

1

Introduction to the Gulf of Mexico

Humans and Their Impact

Thomas S. Bianchi and Elise Morrison

The Gulf of Mexico (GOM; Fig. 1.1) is one of the largest marginal seas in the world and supports a population of approximately 13.9 million along its shores (Wilson and Fischetti 2010) and many more far beyond by supplying moisture to the bread basket of North America. The GOM, located between 18.2° to 30.4° north latitude and 81.0° to 97.9° west longitude is a marginal sea that covers an area of more than 1.5×10^6 km^2 and reaches depths in excess of 4000 m over its abyssal plain (Schroeder and Wiseman 1999). The east and south of the GOM is bordered by carbonate systems with significant groundwater inputs, the north is a river-dominated system where the largest river system in the United States (Mississippi-Atchafalya) empties into its coastal waters, and the west is characterized by river-starved coastal systems (Bianchi et al. 1999). The GOM has many warm-core eddies that spin off from the Florida Loop Current, which can entrain waters from the Caribbean that may travel to the northwestern Gulf and Texas shelf regions. These eddies bring life-supporting

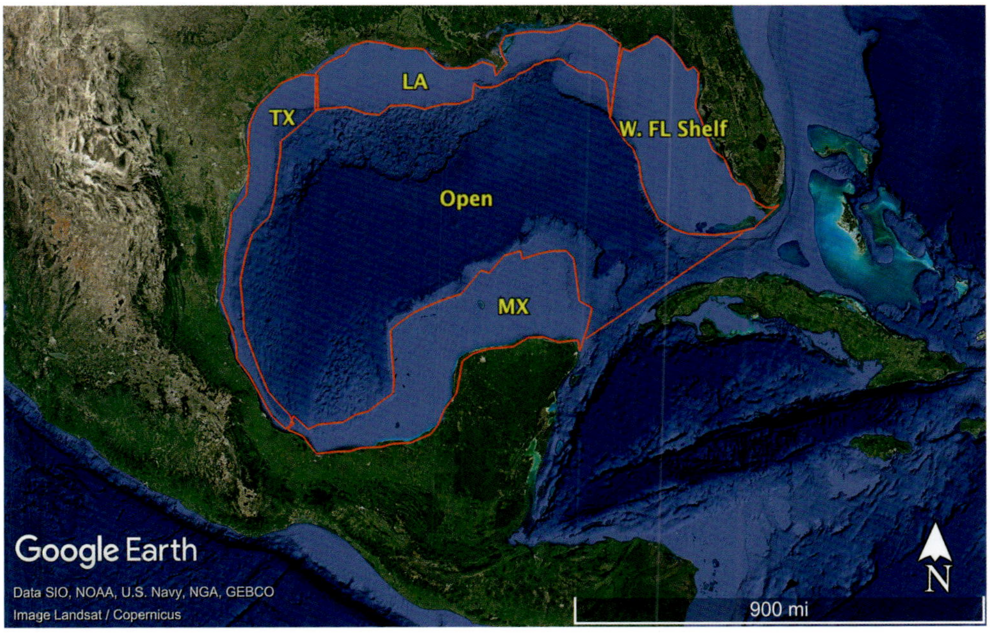

Figure 1.1. The 5 regions of the Gulf of Mexico (modified from Benway and Coble 2014). LA = Louisiana shelf; W. FL Shelf = west Florida shelf; MX = Mexican shelf; TX = Texas shelf; Open = Open Gulf.

nutrients from deep water where they impinge on the shelf, causing the northern GOM to be one of the most productive US fisheries, at times only second to that in Alaska. In addition to supporting regional and national economies, circulation dynamics in the GOM can influence regional storm events, as many of the hurricanes that reach landfall in the United States and Mexico pass through this dynamic system.

In a report from the US Gulf of Mexico Carbon Cycle Synthesis Workshop, the GOM was subdivided into 5 regions based on differing inputs, distinctive physical forcing, and ensuing biogeochemical characteristics and processes (see Fig. 1.1; Lohrenz et al. 2014). The eastern GOM is designated the west Florida shelf (WFS) and is influenced by upwelling, river discharge, and groundwater influx due to its karst geology. The northern GOM is the Louisiana shelf (LA), a highly river-dominated region that receives major freshwater discharge from the Mississippi-Atchafalaya River system. The western GOM, represented largely by the Texas shelf (TX), is dominated by upwelling and eddies shed from the Loop Current (LC). The Mexican shelf (MX) to the south is influenced by upwelling, groundwater, and Usumacinta-Grijalva River discharge. The open Gulf is a deep, semi-enclosed oligotrophic silled basin with an energetic surface circulation strongly connected to the Caribbean Sea and Atlantic Ocean.

The Gulf is bounded by the United States, Mexico, and Cuba and has been highly impacted by humans. Oil and gas production in the GOM has been important since the 1960s but became notable, and even infamous, when the Deepwater Horizon (DWH) explosion occurred on April 20, 2010. On that day, the DWH MC252 oil spill released 4.4 million barrels (6.4×10^8 L, 20%) of light south Louisiana crude oil (Crone and Tolstoy 2010) into the northern GOM. An estimated 205,000 metric tons of methane (CH_4) were also released (Lehr et al. 2010) and consumed by methanotrophs at some of the highest methane oxidation rates ever observed (Kessler et al. 2011). The release of oil at the DWH spanned 84 days with the most affected coastal areas in the northern GOM, where heavily oiled marsh sediments were found months after the spill (Bianchi et al. 2011; Lin and Mendelssohn 2012; Sabourin et al. 2012; Rosenheim et al. 2016). However, when researchers needed background data on the chemistry of the GOM waters, very few data were available in peer-reviewed literature because the majority of shelf and deepwater research in the Gulf over the past 20 to 30 years was funded by the Minerals Management Service, now the Bureau of Ocean Energy Management, Regulation, and Enforcement, where technical reports rather than peer-reviewed literature are delivered as a final product.

In addition to oil and gas production, other human interactions within and around the GOM have contributed to environmental conditions of the basin over the last century. Land-use change in North America, including the rapid growth of farming and industry as well as damming and flood control in the Mississippi River system, has changed the transport of sediment and dissolved constituents to the GOM. Sediment loads in the Mississippi have generally decreased to ~36% with proliferation of dams and other flood control structures (Meade and Moody 2010); however, our baseline from such measurements may have been affected by deforestation during the initiation of widespread agriculture in the basin. During the last century, the amount of particulate organic carbon transported into the northern GOM by the Mississippi-Atchafalaya River system has decreased despite higher concentrations of organic material in the particulate load (Bianchi et al. 2015).

This chapter provides an overview of the geologic history, water circulation, and biological oceanography of the GOM and how they relate to the chemical oceanography of the system. The main goal of this book is to provide an overview of the basic chemical dynamics of this important marginal sea that is situated within the boundaries of a dynamic and changing semitropical region.

Geologic History of the Gulf of Mexico

Delores M. Robinson and Brad E. Rosenheim

Tectonic

The Gulf of Mexico began forming in Late Triassic time (~240–210 Ma [mega-annum]; Pindell 1985; Salvador 1987; Marton and Buffler 1994; Kneller and Johnson 2011; Hudec et al. 2013) from northwest-southeast-directed extension, which separated Laurentia, the North American Plate, from Gondwana, the African and South American Plates. This breakup produced extensive networks of fractures and topographic depressions (grabens and half-grabens) that extend from the Canadian Atlantic offshore into eastern Mexico and northwestern South America (Bartok 1993). A second generation of rifting or a continuation of the first generation of rifting began in Middle Jurassic time with northeast-southwest-directed extension and continued into Late Jurassic time (~163–152 Ma; Stern and Dickinson 2010). During this rifting, the Yucatan block rotated counterclockwise to its present location along the southern margin of the GOM (Buffler and Sawyer 1985; Marton and Buffler 1994; Pindell and Kennan 2001; Pindell 1985; Salvador 1991). As the Yucatan rotated, the proto-GOM opened as continental crust stretched and thinned. The influx of seawater into the accommodation space facilitated the deposition of Middle Jurassic–aged (Callovian, 165 Ma) Louann salt (Bird et al. 2005; Marton and Buffler 1994; Salvador 1987).

The GOM contains a paucity of magnetic anomaly data that can be linked with its spreading history because it is covered by 9–18 km of sedimentary rock, including salt (e.g., Sawyer et al. 1991), which hinders exploration of the basement rock (Christeson et al. 2014; Stern and Dickinson 2010). However, high-resolution marine gravity models show oceanic crust, extinct mid-ocean ridges, and transform faults (plate boundaries with horizontal motion) in the deepest parts of the basin (Sandwell et al. 2014) and provide new interpretations of seismic data (e.g., Snedden et al. 2014). There are numerous models of the geologic opening of the GOM. Previous models included a transform fault through Florida, connecting the Central Atlantic and GOM spreading systems, but these models are not supported by the presence of an extinct mid-ocean ridge in the GOM (Dietz et al. 1971; Pindell and Dewey 1982; Klitgord et al. 1984; Pindell 1985). Marton and Buffler (1994) and Pindell and Kennan (2009) have models containing a transform fault during continental rifting in Early to Middle Jurassic time followed by seafloor spreading beginning in Late Jurassic time. However, other models proposed by Kneller and Johnson (2011) and MacRae and Watkins (1996) include a transtensional margin between the Yucatan and Florida blocks. Previous work by other authors has also invoked simple shear with a south-dipping detachment under the Yucatan (Marton and Buffler 1993) or a northeast-dipping detachment under Florida (Pindell et al. 2015) in their models. Stern and Dickinson (2010) place a north-dipping subduction zone, south of what is now northeastern Mexico, and open the GOM as a back-arc basin (a basin with seafloor spreading above a subduction zone) from rollback of the subducting slab in their model.

Regardless of the model invoked for the opening of the GOM, we know that organized seafloor spreading began in Late Jurassic time (Marton and Buffler 1994; Pindell and Kennan 2009; Kneller and Johnson 2011; Hudec et al. 2013), with the GOM rapidly subsiding and sediments depositing into the accommodation space. Seafloor spreading ended in Early Cretaceous time (~142–135 Ma) (Hudec et al. 2013; Kneller and Johnson 2011; Marton and Buffler 1994; Pindell and Kennan 2009; Snedden et al. 2014).

During Late Jurassic time (Oxfordian, ~163–157 Ma; Hudec et al. 2013), the Norphlet Formation was the first clastic sedimentary rock deposited into the rapidly opening basin and onto the flanks of the basin after the Louann salt (Fig. 1.2). The majority of the Norphlet Formation is an eolian sandstone and an important hydrocarbon res-

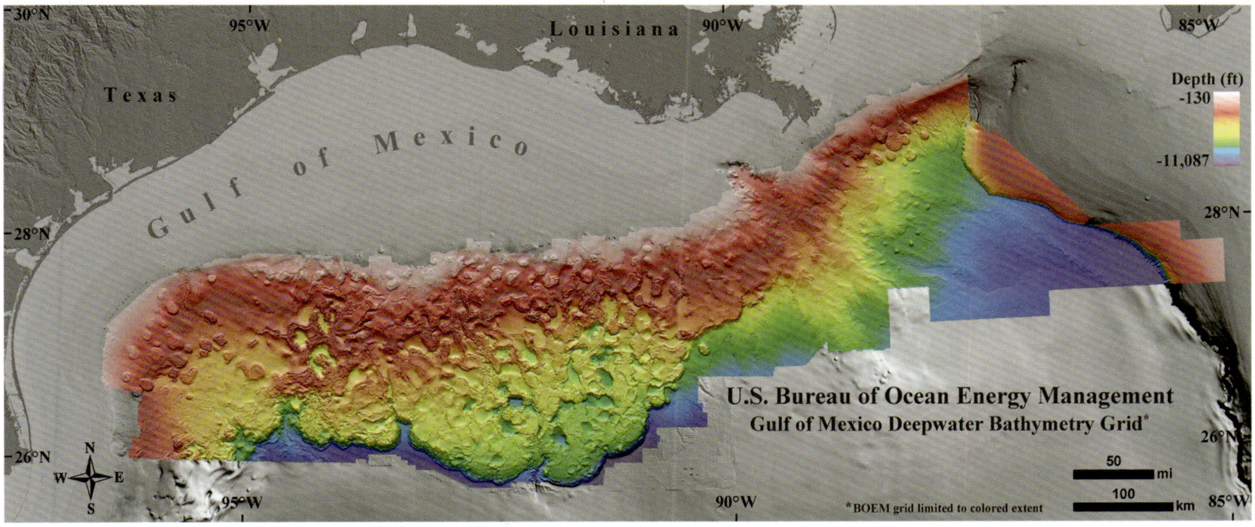

Figure 1.2. High-resolution bathymetry of the Louann salt basins and other slope features of the northern Gulf of Mexico (Kramer and Shedd 2017). The raw data for this map are available at https://www.boem.gov/Gulf-of-Mexico-Deepwater-Bathymetry/. The colored area of the map represents new bathymetric data within the Bureau of Ocean Energy Management (BOEM) grid.

ervoir. Eolian transport directions interpreted from dip log analyses are south-directed in southern Alabama and west- to northwest-directed in western Florida (Hunt et al. 2017). Provenance analyses (i.e., analyses investigating the origin of sediments) also indicate that river systems flowed southward from the Appalachian Mountains and eastward from Florida (Lovell 2010; Weislogel et al. 2015). Deposition of the Norphlet Formation was ended by a marine transgression (a rise in sea level) with the Smackover Formation, another prolific reservoir, then deposited on a carbonate ramp (Ahr 1973; Mancini and Benson 1980). Sediments continued to be deposited in the Late Jurassic and Cretaceous Periods and the Cenozoic Era. Because the 1–5 km thick Louann salt was at the bottom of the thick sedimentary section, the salt became mobile as the Norphlet Formation and subsequent sedimentary rocks were deposited on top (e.g., Hudec et al. 2013). This mobile salt has deformed into diapirs, domes of salt that are squeezed up through layers of sedimentary rock by the weight of the overlying rock, and allochthonous salt sheets and canopies (Fig. 1.2). The formation of salt diapirs, and the subsequent collapse of some of them in the northern Gulf, has resulted in hydrocarbon traps with organic material derived from the overlying sedimentary rocks.

Sedimentary Regimes

The Mississippi-Atchafalaya River system is the major source of sediment to the shelf of the modern GOM where sedimentary rocks are as thick as 15 km (Stern and Dickinson 2010). Other smaller sources of terrigenous, clastic sediments include the Rio Grande, the Brazos, and numerous smaller drainage systems in Mexico north of the Cenote provinces of the Yucatan Peninsula. The Rio Grande drains the valley formed by the Border rift system, an interarc rift system (rifts where there is no seafloor spreading) that was possibly congenital with the hypothesized back-arc basin of Stern and Dickinson (2010). Together, the Brazos and Rio Grande account for only 4.7% of the total suspended solids of the Mississippi despite draining a land area approximately 30% as large as the Mississippi River drainage basin (Milliman and Farnsworth 2013). In fact, all rivers in North America currently draining to the Gulf deliver only about 25% of the total suspended solids that the Mississippi does annually (n = 39, excluding the Mississippi River), even though the combined land area of these drainages is equivalent to 58% of the land area that the Mississippi drains (Milliman and Farnsworth 2013). The Mississippi River drainage served as an early conduit for the outlet of glacial and pro-glacial lake waters that formed on the southern extent of the Laurentide ice sheet during the last deglacial period (Broecker et al. 1989; Flower and Kennett 1990; Clark et al. 2001; Flower et al. 2004) and likely prior deglacial periods of the Pleistocene.

The 15 km of sedimentary rock in the semi-enclosed GOM play a primary role in the chemical oceanography of the present-day basin. Thicknesses range from 15 km in the northwestern GOM to ~1 km in parts of the basin that are farther from major riverine inputs (Stern and Dickinson 2010). Terrigenous sediment inputs are associated with inorganic nutrients, and nutrients delivered to the surface waters of the GOM fuel primary productivity that is the foundation of both a rich pelagic ecosystem and an active biological pump of marine sediment to the seafloor. Clastic sedimentary rocks in the GOM contain large amounts of organic matter, especially in the northwestern margin of the basin, that have been converted to hydrocarbon through diagenetic and catagenic processes. Hydrocarbons produced from buried organic material in GOM sedimentary rock are not only trapped in salt diapir structures producing large reserves but are also present as numerous cold-water seeps.

Soft, muddy sediments cover about 95% of the seafloor in the GOM (Fisher et al. 2016). However, microbial communities that metabolize oil and gas at numerous cold-water hydrocarbon seeps favor deposition of carbonates and can form hard-bottom habitats based on chemosynthesis (Fisher et al. 2016). Active seeps are numerous and widely distributed in the GOM (Roberts et al. 2007; Cordes et al. 2009; MacDonald et al. 2015; Fisher et al. 2016), and there are numerous areas with buried carbonate deposition features that are no longer actively releasing hydrocarbons. These features, resulting from underlying geologic conditions, support unique ecosystems and contribute to the chemistry of the deep GOM.

Human activities related to oil exploration have resulted in pollution events along the continental shelf and, most recently, into the deeper waters of the GOM. None of these pollution events rival the well blowout of the Deepwater Horizon in 2010, which released 4.6–6.0 $\times 10^{11}$ g of petrocarbon into the deep GOM, which had implications for the chemistry and biology of the GOM. The presence of microbial communities related to the cold seeps of the GOM resulted in some initial drawdown of

O_2 (Du and Kessler 2012), sinking particles of oil snow (Passow et al. 2012) were deposited on the seafloor (Chanton et al. 2015; Valentine et al. 2014), and reduced benthos biodiversity was noted (Schwing et al. 2015). It remains to be seen whether this signature of human activity will be preserved in the GOM geologic record, although it is present in the water column 4 years after the blowout event (Walker et al. 2017).

Circulation in the Gulf of Mexico

Ruoying He and Jennifer Warrillow

Introduction

The GOM is a semi-enclosed sea with a maximum depth around 4000 m. It opens to the Caribbean Sea through the Yucatan Channel and to the Atlantic Ocean through the Straits of Florida. The continental shelf of the GOM is broadest along the west coast of Florida. The narrowest portions are along the east coast of Mexico and south of the Mississippi River delta, and several major North American rivers flow into the Gulf, the largest of which is the Mississippi River.

The Loop Current and its rings directly or indirectly affect almost every aspect of oceanography in the Gulf. The LC originates at the Yucatan Channel, through which approximately 23–27 Sv transport passes (1 Sv = 10^6 m^3 s^{-1}; Johns et al. 2002). The LC episodically sheds warm-core rings at intervals of approximately 3 to 18 months (Leben 2005). Tides are generally weak in the Gulf, but strong tropical and extratropical storms often pass through the region, acting as strong synoptic forcing agents for water transport, mixing, and entrainment in the regional ocean (Schmitz et al. 2005). Together, these turbulent flows have an impact on the region's biology, chemistry, climate, circulation, material transport, and human activities.

Ever-increasing human activities, such as shoreline development, changes in land-use practices, and the resulting increase in pollutant and nutrient/carbon inputs, continue to threaten the well-being of the marine environment and ecosystems in the GOM. These impacts result in issues such as coastal eutrophication, recurring hypoxia (a.k.a. the dead zone), and coastal ocean acidification on the Louisiana-Texas shelf (Cai et al. 2011). The coastal regions of the GOM are also periodically threatened by surges induced by storms, which have exhibited a statistically significant trend of increasing in their frequency over the past 90 years (Grinsted et al. 2013).

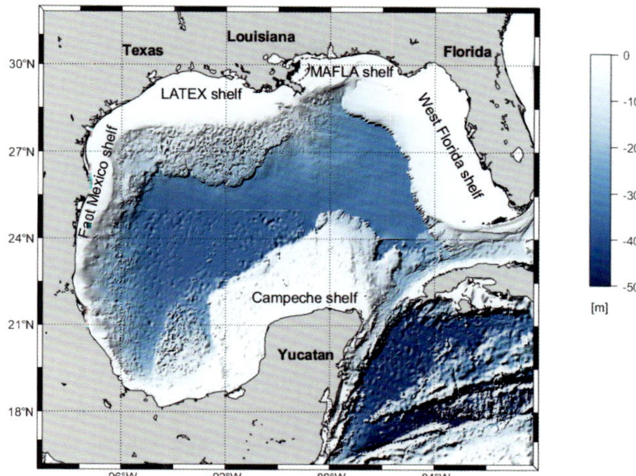

Figure 1.3. Circulation in the Gulf of Mexico and the western North Atlantic. The Caribbean Current, made up of northern and southern Atlantic water, flows into the Gulf of Mexico through the Yucatan Channel. It becomes the Loop Current, which may shed eddies, then exits the Gulf through the Florida Straits. It then enters the Atlantic as the Gulf Stream.

Circulation

Seawater in the GOM originates in the Atlantic Ocean, enters the Caribbean Sea, and then flows into the Gulf through the Yucatan Channel (Fig. 1.3). The net inflow of seawater into the Caribbean and then into the Gulf is ~45% of South Atlantic origin and ~55% of North Atlantic (Schmitz and Richardson 1991). At lower latitudes, Caribbean water experiences higher evaporation and solar heat contribution, resulting in high salinity and temperature inflow to the Gulf. The flow then exits primarily through the Florida Straits, with some deep counterflow south through the Yucatan Channel as well (Sturges 2005; Sturges and Kenyon 2008).

The upper-layer water-mass flux in the Gulf is to the west, driven by wind stress and the Coriolis force. This mass flux is driven downward by Ekman pumping because the western edge of the Gulf is closed. This downward flux exits the Gulf by deep flow through the Yucatan Channel counterflow and perhaps through the Straits of Florida (Sturges 2005; Sturges and Kenyon 2008).

The Loop Current

The LC dominates circulation in the GOM. As mentioned earlier, this fast, wide, deep-reaching flow is a segment of the Atlantic's western boundary current, carrying ca. 23–28 Sv (Johns et al. 2002) of warm, salty Caribbean water through the Gulf and into the Atlantic's Gulf Stream. The LC reaches peak speeds of 1.5 to 1.8 m s^{-1} (Oey et al. 2005) and can extend to approximately 1000 m deep, depending on topography. As the LC expands north and west into the GOM, it also displaces some Gulf water back into the Caribbean Sea through the Yucatan Channel at depths below ~800 m (Bunge et al. 2002). The extent to which the LC penetrates the Gulf varies irregularly. It may reach as far as 92°W and 27°N (Alvera-Azcárate et al. 2009) or race through the Florida Straits, barely intruding into the Gulf. Intrusions far into the Gulf advect Mississippi-Atchafalaya River water and its dissolved and particulate constituents (e.g., dissolved chemical compounds and organic and inorganic particulates) more quickly to the Florida Straits.

The LC periodically sheds eddies (also called rings), which swirl across the GOM generally westward. The LC may either (1) pinch off into warm-core, anticyclonic eddies of 100–300 km diameter; (2) generate small (<200 km), cold-core, cyclonic eddies that travel along the LC's edge (Huang et al. 2013); (3) shed then reattach an eddy; or (4) extend and retreat without shedding any eddies (Alvera-Azcárate et al. 2009; Zeng, Li, He, and Yen 2015). The position of the LC typically follows a process of (1) extension, whereby the LC intrudes farther into the Gulf; (2) shedding of an eddy by pinching off a loop to form a ring; (3) retraction, where the LC extends less into the Gulf; followed by (4) a normal state, where the LC reaches into the Gulf to about 25.5°N and 86°W (Zeng, Li, and He 2015). Intervals between eddy shedding vary from a few weeks up to 18 months without any clear annual pattern (Leben 2005). When a large eddy is shed, a longer period (mean = 12.63 months) follows without eddy shedding because when the LC sheds a large volume, it retreats far to the south and requires more time to grow to shedding size. Inversely, the shedding of a small-volume eddy takes less time (mean = 3.70 months) for the LC to recover and be able to shed another eddy. Further, a relationship exists between the distance the LC intrudes into the GOM and the size of the shed eddy. Extremely large intrusions into the GOM (to 28°N) result in smaller eddies being shed, while more usual intrusions (to 24°N) result in the shedding of larger eddies (Alvera-Azcárate et al. 2009).

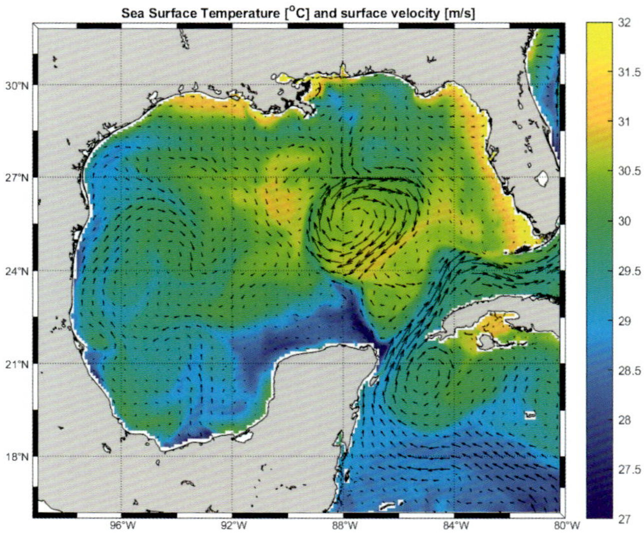

Figure 1.4. Regions and bathymetry of the Gulf of Mexico.

Warm-core eddies have a significant impact on the entire GOM as they carry Caribbean water westward across the Gulf for months to 1 year (translation speed of 2.16–5.18 km d^{-1}), then dissipate upon impact with the Gulf's west coast (Candela et al. 2002; Oey et al. 2005; Nguyen et al. 2015). While most rings migrate westward, they may drift northward or southward or reattach to the LC. Bottom topography, specifically the Campeche Bank (Fig. 1.4), can increase their westward velocity (Chérubin et al. 2006). These turbulent flows reach about 1000 m deep with swirl speeds of 1.8–2 m s^{-1} and impact mixing, material and larval transport, air-sea interactions, and human activities such as shipping and the thousands of oil rigs on the Texas-Louisiana shelf (Zeng, Li, and He 2015).

A wide range of rings, in addition to the anticyclonic warm rings, also form as a result of the LC. For example, smaller (<200 km), cyclonic, cold Loop Current frontal eddies (LCFEs) move downstream along the outside of the LC (Huang et al. 2013). Due to their relatively small size and weak thermal contrast, LCFEs are not easily detected. Recent advances in computing techniques such as self-organizing maps (Zeng, Li, He, and Yin 2015) and neural networks (Zeng, Li, and He 2015) have enabled the prediction of the periodic but highly irregular eddy shedding process. For example, by forecasting sea-surface height in the Gulf, scientists can predict LC variation and eddy shedding up to 6 weeks in advance. Further, eddy shedding is positively correlated with wind-stress curl in the Caribbean Sea (upstream of the Gulf; Zeng, Li, He, and Yin 2015).

Upper-Layer Circulation and Deep Circulation in the Open Gulf

Upper-layer (<800 m deep) circulation in the open waters of the Gulf (i.e., water-column depth >200 m) is driven by wind, the LC, and its shed eddies. Studies of drifters (trackable passive floats) revealed an elongated anticyclonic gyre in the western Gulf created by both wind-stress curl and shed eddies from the LC (Sturges 1993). In the Bay of Campeche, a wind-induced cold-core cyclonic flow persists at a depth of 1000 m, with seasonal variability in strength (Perez-Brunius et al. 2013; Vázquez de la Cerda et al. 2005).

At its center, the Gulf reaches more than 4300 m deep (Ellis and Dean 2012). While circulation in the upper layers is anticyclonic, the deep (>1000 m) circulation is cyclonic in both the eastern and western Gulf (1–2 cm s^{-1} at 2000 m) (DeHaan and Sturges 2005). Intermittent intensified currents (10–40 cm s^{-1}) lasting 10–30 days have been observed by moored current meters in the deep Gulf (Kolodziejczyk et al. 2012). These intensified currents occur at the boundary between flows rotating in opposite directions, which suggests that eddies in the upper layer transmit energy to the deeper layers.

Coastal (Shelf) Circulation

The continental shelf around the GOM, defined by the 200 m isobath, is narrow on the western edge along the coast of Mexico, wide on the Louisiana-Texas (LaTex) shelf, and very wide on the WFS and Campeche Bank (see Fig. 1.4). Both deep-ocean and local forcing drive shelf circulation. Surface momentum, buoyancy fluxes, and freshwater runoff from land contribute to local circulation, while deep-ocean forcing impacts the shelf through transmission of momentum and buoyancy fluxes to shallow water (He and Weisberg 2002a; Weisberg, He, Liu, et al. 2005). Wind-stress forcing mainly dominates inner shelves, with a seasonal cycle in intensity and direction that reverses over the year. The shallowness of the shelf causes bottom friction to be more important than in the deep ocean. The change in depth also inhibits exchanges between the shelf and deep ocean because of the conservation of potential vorticity; thus, current variability along-shelf is much greater than cross-shelf (He and Weisberg 2002a; Zavala-Hidalgo et al. 2014).

The shelf regions of the Gulf vary in their responses to these forcings based on regional topography, river systems, and boundary currents. In the western GOM, the Tamaulipas-Veracruz shelf has down-coast circulation from fall through spring, then switches to up-coast circulation in summer. In spring and summer, the up-coast current meets a down-coast current at southern Texas, resulting in offshore transport. In fall and winter, this meeting and offshore transport reach the southern Bay of Campeche (Zavala-Hidalgo et al. 2003). The LaTex shelf has cyclonic circulation except during summer, with mean southwestward along-coast flows with notable current reversals and seasonal variation. In fall and winter, cool, fresh Mississippi River outflow advected westward to the eastern GOM causes along-shelf temperature fronts and inversions at the outer shelf (He and Weisberg 2002a; Zavala-Hidalgo et al. 2003; Walker 2005). On the WFS, seasonal reversals of inner- and outer-shelf currents occur, and LC variation can influence shelf circulation (He and Weisberg 2002b; Weisberg and He 2003; Weisberg et al. 2005). The WFS (during winter) and Campeche Bank (during summer) are most notably impacted by upwelling of cold, nutrient-rich, deep-ocean water. Because of seasonal changes in wind direction, downwelling occurs at the opposite time of year from upwelling at any location (Zavala-Hidalgo et al. 2014). The LC impacts the WFS by extending to the shelf or by shedding an eddy that impacts the shelf (Weisberg and He 2003).

Tides

Barotropic tides are caused by the gravitational pull of the moon and sun on the oceans and have no vertical structure. Most of the Gulf Coast experiences tides of <0.5 m amplitude. Diurnal tides (1 high and 1 low tide each day) dominate in the Gulf because of near resonance of the Gulf with diurnal tide forcing, with the exception of Apalachicola Bay and the WFS, which experience semidiurnal tides (2 highs and 2 lows per day; He and Weisberg 2002a; Kantha 2005; Ellis and Dean 2012). In the deep Gulf, barotropic tidal currents have been measured ~2 cm s^{-1}, which is within the observational error range of the recording instruments, and barotropic tides in the deep Gulf are considered negligible (Kantha 2005).

Baroclinic, or internal, tides arise from the interaction of barotropic tides in a stratified ocean with ocean-bottom topography. The generation of baroclinic waves results from baroclinic wave drag and converts energy from barotropic tides into baroclinic waves (Simmons et al. 2004). These internal waves contribute to mixing through the water column. In the narrow channels of the Yucatan and Florida Straits, baroclinic tides impact the variability of energy and volume transport, which causes substantial forcing of observed circulation variability in the Gulf.

Baroclinic tides may also have an impact on mixing near the LC (Souza et al. 2016).

Meteorological Forcing

Wind forcing in the Gulf drives surface fluxes, generates waves, influences basin-scale circulation, and impacts mixed-layer depth. Winds over the Gulf include the intense forcing of tropical cyclones (hurricanes and tropical storms) and of winter cold fronts, in addition to weaker, but still important seasonal winds (Zavala-Hidalgo et al. 2014). These winds have seasonal variability in strength and direction and interact with the temperature and topography of the surrounding land. Wind stress over the entire GOM is generally westward. Time-averaged wind-stress curl is negative (anticyclonic) north of 23°N and positive south of that, with the coverage of negative wind-stress curl increasing in spring and summer, then decreasing in fall and winter (Velasco and Winant 1996). Wind stress creates along-shore currents on the inner continental shelves. The direction of wind, and therefore currents, changes and even reverses seasonally. Interannual variability in cross-shelf transport can be explained by variation in convergence of along-coast wind stress in the southern Bay of Campeche and on the LaTex shelf (Martínez-López and Zavala-Hidalgo 2009).

The mixed layer, in which properties (e.g., temperature, salinity, density) are vertically well mixed, is maintained by turbulence at the sea surface due to wind forcing and evaporation or heat loss. The addition of heat and precipitation decreases this mixing (Talley 2011). In the Gulf, mixed-layer depth reaches a month-averaged maximum of ~75 m in February and a minimum of ~18 m in July (Mendoza et al. 2005).

Tropical cyclones (TCs; including tropical depressions, tropical storms, and hurricanes) regularly pass through the Gulf, causing significant change to the regular patterns of wind stress, upper-layer mixing, ocean heat content, and circulation. The cyclonic rotation of a TC's intensified wind field causes mass flux of water outward from the eye. This flux creates a void, which is filled by water from the depths below the eye (upwelling). This upwelling increases nutrient availability in the upper layer and decreases surface temperature 1–6 °C (Shay 2010). For example, Hurricane Ivan in 2004 displaced isotherms upward by 50–65 m (Walker et al. 2005). The outward flux of water creates an Ekman spiral in the water column from the surface to below the mixed layer. This flow and resulting shear stress cause entrainment mixing of the warm upper layer and cooler stratified layers (Fig. 1.5).

Heat content in the upper ocean intensifies the strength of TCs by transmitting heat energy and moisture to the atmosphere. As a result, TC passage cools sea-surface temperature (SST). The decrease in SST behind the TC's eye is called the cold wake and results in a negative feedback to TC intensity. As SST decreases, heat and water fluxes from the ocean to the atmosphere decrease, and SST cooling and mixed-layer deepening cause a decline in TC intensity (Shay 2010). If a TC passes over the LC or a warm-core eddy in the Gulf, more heat energy is available to the TC, causing it to intensify, as in the case of Hurricanes Isaac (Jaimes and Shay 2015), Katrina,

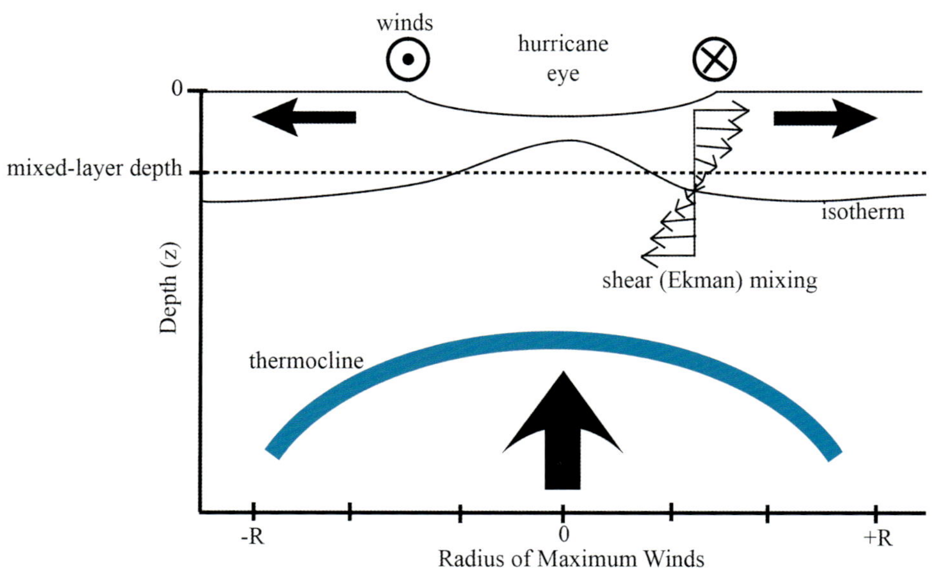

Figure 1.5. Cross-sectional view of the ocean under the eye of a Northern Hemisphere hurricane. Arrows indicate water flow. Winds are into the page on the right side of the eye (*circle with an x*) and out of the page on the left side (*circle with a dot*). The Ekman spiral due to wind shear and the vertical displacement of an isotherm are illustrated.

Rita, Wilma (Mainelli et al. 2008), and Opal (Shay et al. 2000), among many others. Conversely, passing over a cold-core anticyclonic eddy decreases the intensity of a TC, as did Hurricanes Ivan and Rita just before landfall (Mainelli et al. 2008).

In terms of heat-flux exchange, the GOM gains heat in March to September and has peak heat gain in May to July. The Gulf also receives significant heat from the Caribbean Sea via the LC. Heat fluxes are responsible for the development (when heat flux is out of the ocean) or reduction (when heat enters the ocean) of the mixed layer. Through different rates of heating/cooling of shallow/deep water, heat flux also develops cross-shelf pressure gradients, which in turn produce cross-shelf flow (He and Weisberg 2002b; Zavala-Hidalgo et al. 2002). Models show that steady westerly wind increases heat in the upper Gulf by increasing shelf currents, which leads to increased downwelling in the western GOM. Increased heat in the upper Gulf can also be due to a larger LC with deeper-reaching rings, which transport heat across the central GOM (Chang and Oey 2010).

Connectivity of Circulation, Material Transport, and Dispersion

Circulation patterns in the GOM impact the movement of material properties (e.g., larvae, sediments, pollutants, plankton, and oil spills) and ocean-state variables (e.g., salinity, heat, density, and vorticity). Intense, discrete events such as TCs add to the mixing and transport environment. Material transport occurs by advection (the flow of water) and by entrainment (the transport of material or mass from a less turbulent medium into a more turbulent medium). Entrainment occurs at an interface of 2 layers when shear at the interface creates internal waves that break, mixing material and properties from the less turbulent layer into the more turbulent layer. For example, entrainment due to shear from tropical cyclones causes water from below the thermocline to mix with upper-layer water (Shay 2010).

As discussed previously, wind and the LC have the greatest influence on currents and therefore on material transport. Wind affects circulation on the continental shelves by forcing currents there, which impacts the extent and direction of river plumes such as the Mississippi-Atchafalaya River system—a major source of freshwater and sediment and the chemicals and pollutants they contain (Ellis and Dean 2012). Wind-induced circulation can change the Gulf from a sink to a source of CO_2 to the atmosphere and influences the availability of dissolved oxygen for primary production (Huang et al. 2015; Xue et al. 2016). At times, river discharge can be great enough that it persists with only partial mixing. For example, a coincidence of conditions in 1993 maintained a low-salinity lens (a practical salinity of 5 below normal) from the Mississippi River (north-central GOM) to the open waters of the Gulf to the shallow Florida shelf (eastern GOM) (Gilbert et al. 1996).

Intereddy advection and the translation of eddies westward allow for the movement of material across the Gulf. Ocean color, as detected by satellites, can track chlorophyll-*a* in phytoplankton. Cyclonic and anticyclonic eddies, whose interacting margins can have higher flow rates, can transport shelf water seaward at least 200 km and concentrate particles (including phytoplankton) in the flow confluence between the eddy pairs (Biggs and Müller-Karger 1994). The northern wall of the LC can serve as the anticyclonic member of an eddy pair, which can advect chlorophyll from the south Florida shelf to the eastern Campeche shelf (Toner et al. 2003). The same authors also report a chlorophyll-*a* plume spun out between a large anticyclone and several adjacent cyclones from the Louisiana shelf break across the Gulf to the Yucatan Peninsula.

The occasional chaos of tropical cyclones increases vertical mixing both by causing upwelling in the open ocean and by increasing the depth of the mixed layer, as described previously. The significance of deeper water rising to the surface is evident in the regular occurrence of phytoplankton blooms in the wake of TCs (Shi and Wang 2007; Gierach and Subrahmanyam 2008; Zhao and He 2012; Shropshire et al. 2016). The influence of TCs on horizontal mixing has also been shown to affect fish distribution. For example, adult red snapper were found 2 orders of magnitude farther from their initial capture site when a hurricane occurred, relative to their movement in the absence of a hurricane (Johnson et al. 2009).

Currents can also influence the movement of particulates in the GOM. Particulates are transported when fluid velocity has enough force to move the particles. The threshold velocity depends on particle size, slope, and particle density. Particle size and composition also affect dispersion, as modeled by North, Adams, Schlag, et al. (2011) and North, Adams, Thessen, et al. (2015) for droplets from an oil spill. Small and/or less buoyant particles may be trapped in density-stratified water layers to such an extent that a subsurface plume may form. Additionally, the decay of particulates, such as pollutants, over time changes the particle size and composition, and thus particle distribution.

Coupled ocean-circulation models have been used to model the dispersal of larvae, which can be viewed as particles with some swimming ability. These models indicate that oceanographic features, such as point of origin and entrainment into currents, determine the destination of holoplankton (or any particulate) far more than meroplankton (larval) behavior (e.g., Qian et al. 2015). Larval dispersion may be unidirectional on parts of the shelf due to current structure. On the LaTex shelf, westward dispersal can occur closer to shore, while eastward dispersal can occur offshore near the slope, where ichthyoplankton have much lower survival rates (Johnson et al. 2009). Larval dispersal distance also varies with the depth at which they travel. For example, larvae in the upper water column move farther than those below the thermocline with the same drift duration (Young et al. 2012).

In conjunction with in situ observations, computer models for the prediction of GOM circulation and material transport have improved in recent years (He and Weisberg 2002a, 2002b, 2003; Oey et al. 2005; Zambon 2009a, 2009b; Warner et al. 2010; Olabarrieta et al. 2012; Putman and He 2013; Zambon et al. 2014; Xue et al. 2015; Xue et al. 2016). These models combine data from satellites, moorings, tide gauges, and floats to predict Gulf conditions and fill gaps in the data, making valuable contributions to improve the fundamental understanding of connectivity, transport, and dispersion in the GOM.

Acknowledgments

Research support provided by Gulf of Mexico Research Initiative/GISR through grant 02-S130202, NOAA grant NA11NOS0120033, NASA grants NNX10AU06G, NNX12AP84G, and NNX13AD80G are much appreciated.

Biological Oceanography

Gary Hitchcock, Sumit Chakraborty, and Kendra Daly

Introduction

The pelagic food web plays an important role in the exchange of CO_2 between the atmosphere and the surface ocean and the subsequent export of particulate and dissolved carbon to the deep ocean (Longhurst and Harrison 1989). This suite of biologically driven processes is termed the "biological pump." Biota influence the distribution of elements primarily through the production and regeneration of organic matter. Phytoplankton take up CO_2 through photosynthesis, thereby reducing pCO_2 in the surface layer, which leads to enhanced CO_2 flux from the atmosphere into the surface ocean. About two-thirds of the carbon flux into deep water is due to the biological pump, with the remaining amount due to the solubility pump (Passow and Carlson 2012). In the open waters of the GOM (see Fig. 1.1), the vertical distribution of the "nutrient-like" elements required to sustain life, such as carbon, nitrogen, and phosphorus, reflect their incorporation into primary producers in the surface waters and subsequent recycling at depths below the euphotic zone (Parks 2015). Additionally, the biota influence the vertical distributions of elements such as nickel, aluminum, and zinc through the production and subsequent sinking of particles that sorb, or scavenge, trace elements and transport them to depth via the action of the biological pump (Donat and Bruland 1995). Primary producers in the GOM range in size from the unicellular phytoplankton that dominate the pelagic communities, to sea grasses, mangroves, and salt-marsh communities of coastal ecosystems. Irrespective of size, all photoautotrophs create organic matter through a 2-step sequence of light-dependent processes converting energy from absorbed photons to chemical energy and subsequent light-independent reactions that convert the inorganic substrates to sugars and additional organic compounds.

The coastal ecosystems of the GOM include the shoreline habitats of the west Florida, Louisiana, Texas, and Mexico regions designated in Figure 1.1. These coastal ecosystems are dominated by mangroves in the southern latitudes and salt marshes in temperate latitudes, each of which has exceptionally high net productivity rates on the order of 3–5 gC m^{-2} d^{-1} (Duarte and Cebrián 1996). Few animals directly graze on mangroves or salt-marsh plants, so these communities contribute detrital carbon that sustains detrital-based food chains at rates in excess of 1 kg C m^{-2} yr^{-1}. Sea grasses also contribute to the high productivity of the Gulf coastal ecosystems but are a direct food source for vertebrate herbivores such as manatees, sea turtles, and fish, unlike mangroves and *Spartina* spp. in salt marshes (Ray et al. 2014). Collectively the mangroves and salt-marsh and sea-grass communities store large quantities of organic carbon, often called blue carbon, in their sediments on an aerial (m^{-2}) basis that exceeds that in temperate forests (McLeod et al. 2011). Mangrove and salt marshes also function as nutrient "traps" that sequester autochthonous and allochthonous nitrogen in sediments (Morell and Corredor 1993). There has been an observed northward expansion of mangroves that may reflect a warming of the coastal waters in the northern GOM that is referred to as the "tropicalization" of coastal habitats. This could lead to significant changes in the distribution of plant communities and their associated ecosystem services, such as carbon sequestration and nutrient cycling at the interface of the land and the Gulf (Hyndes et al. 2016).

Pelagic Primary Productivity in the Gulf of Mexico

The north-central and northeastern shelf of the GOM, referred to as the LA shelf (see Fig. 1.1), is influenced by large river systems, such as the Mississippi-Atchafalaya, Mobile, and Apalachicola, and numerous estuaries and bays. Annual average productivity in the LA shelf is ~400 gC m^{-2} yr^{-1} (Lohrenz et al. 2014). Freshwater-mediated nutrient supply and the variability of the light attenuation coefficient (Kd m^{-1}) are key variables that regulate primary production in this region. Productivity may be light limited in low-salinity waters within river plumes, while estuaries and bays shift to predominantly nutrient-limited conditions outside the plume-influenced areas of the shelf (Dortch and Whitledge 1992; Smith and Hitchcock 1994; Lohrenz et al. 1999; Rabalais et al. 2002; Sylvan et al. 2006; Dagg et al. 2008). High productivity in the LA shelf has been implicated in recurrent episodes of hypoxia in bottom waters off the shelf (Rabalais and Turner 2001; Rabalais et al. 2002). Additionally, the dynamics of the Loop Current and its occasional penetration into the northern and eastern GOM (Wiseman and Sturges 1999) can also impact the primary productivity of the region.

High productivity rates in the west Florida shelf are observed in association with rivers, estuaries, and bays with an average annual production of ~475 gC m^{-2} yr^{-1} (Lohrenz et al. 2014). Blooms of the dinoflagellate *Karenia brevis* occur off the WFS, and rates of daily primary production vary widely between regions and between bloom and nonbloom periods. Primary production ranges between 1.6 and 2.88 gC m^{-2} d^{-1} in typical bloom conditions (Lohrenz, unpublished data), while under nonbloom periods primary production can range from 0.09 to 0.5 gC m^{-2} d^{-1} (Vargo et al. 1987). On the Texas shelf, in the northwestern GOM, annual primary production rates are lower, averaging ~120 gC m^{-2} yr^{-1} (Chen et al. 2000; Lohrenz et al. 2014).

Primary productivity has been shown to vary with the spatial distribution of nutrients and can be related to seasonal wind forcing, river discharge, and associated coastal circulation (Chen et al. 2000; Chen et al. 1997, 2002). Interactions between LC eddies and the continental margin of the Texas shelf have been shown to influence rates of primary production (Biggs 1992) and chlorophyll concentrations (Biggs and Müller-Karger 1994). Upwelling at the periphery of anticyclonic rings and at the center of cyclonic rings may also lead to increased vertical inputs of nutrients, subsequently stimulating rates of primary production in this region (Müller-Karger 2000; Weisberg et al. 2004).

There is a paucity of data available for the primary production of continental shelf waters in the southern GOM (Mexican shelf). El-Sayed and Turner (1977) reported values of 0.29–0.32 gC m^{-2} d^{-1} for daily productivity in continental shelf waters of the Bay of Campeche and annual productivity rates of 100–120 gC m^{-2} yr^{-1}. Instabilities associated with the LC may influence primary production on the Campeche Bank, off the Yucatan Peninsula (Wiseman and Sturges 1999). Regions of apparent upwelling have been observed on the Campeche Bank and the western part of the Gulf of Campeche and have also been reported for the Yucatan shelf by Merino (1997). Satellite algorithm-derived estimates report that productivity in the Yucatan shelf ranged from 1.18 to 1.22 gC m^{-2} d^{-1}(Hidalgo-González et al. 2005).

Primary production in the open GOM is controlled mainly by seasonal variations in mixed-layer depth, sea-surface temperature, winds, and large-scale circulation dominated by the LC. These processes likely affect the water-column structure of primary production by exerting influence on irradiance and nutrient availability. Numerical simulation (Walsh et al. 1989) and analysis of satellite imagery (Müller- Karger et al. 1991; Müller-Karger et al. 2015) infer a temporal pattern characterized by high net primary production in winter months and low values during late spring and summer.

Role of Secondary Producers in the Gulf of Mexico

Zooplankton play an important role in biogeochemical cycles in the ocean, as a source of particulate and dissolved materials and a sink for particulate material (Longhurst and Harrison 1989). They are active agents of organic matter transfer by grazing on phytoplankton and, in turn, are eaten by higher-trophic-level predators. They contribute to nutrient regeneration through the excretion of ammonia, urea, amino acids, phosphate, and dissolved organic carbon (LeBorgne 1986). In addition, zooplankton are an essential component of the biological pump by transferring particulate and dissolved organic matter produced in surface waters to deep water via several mechanisms, including sinking fecal material, exuviae, dead organisms, and active vertical migration of zooplankton (Longhurst and Harrison 1989). Microzooplankton are an important component of the microbial loop (Azam et al. 1983) and an essential component of the diet of mesozooplankton and larval fish (Gifford 1991; Liu et al. 2005). The microbial loop is a trophic pathway in which dissolved organic carbon is incorporated into bacterial biomass and then transferred to higher trophic levels through flagellates and other protozoans (Fenchel 2008). The microbial loop is a dominant pathway in offshore oligotrophic waters, where it can be a sink for fixed carbon, in contrast to a link to higher trophic levels. However, Ortell and Ortmann (2014) reported that the microbial loop supported a link to higher trophic levels across an estuarine-offshore gradient in the northern GOM.

Composition and Spatial Distribution of Secondary Producers

The composition and spatial distribution of microzooplankton in the GOM are poorly understood. Several studies have shown that microzooplankton, such as aloricate ciliates, tintinnids, heterotrophic dinoflagellates, and small metazoans (e.g., copepod nauplii), are abundant in estuarine and coastal regions (Buskey 1993; Putland and Iverson 2007). Ciliate concentrations in a Texas estuary ranged from 5 to 400 × 10^3 ciliates L^{-1}, while mesozooplankton ranged from 6100 during the day to 10,100 m^{-3} at night (Buskey 1993). Putland and Iverson (2007) reported maximum abundances in summer in Apalachicola Bay as 26 × 10^3 cells L^{-1} for ciliates, 4 × 10^6 cells L^{-1} for heterotrophic flagellates, and 1 × 10^4 cells L^{-1} for heterotrophic dinoflagellates (>20 μm), with lower abundances of all microzooplankton in winter.

The composition and spatial distribution of mesozooplankton species were reviewed by Iverson and Hopkins (1981) and Biggs and Ressler (2001). Most deepwater species are typical of those reported from the western tropical Atlantic and Caribbean Sea, as the LC provides a conduit linking these geographic provinces. Many of the estuarine and continental shelf zooplankton, including species of jellyfish, copepods, ctenophores, and larvaceans, have very broad distributions and are commonly observed all along the GOM and Atlantic coastlines (Johnson and Allen 2012). In general, the dominant GOM zooplankton taxa are copepods and chaetognaths, with euphausiids and micronekton, such as decapod shrimp and fishes, having higher abundances in off-shelf waters. Episodically, gelatinous zooplankton, ostracods, and larvaceans also may be very abundant.

In general, mesozooplankton have higher abundances nearshore and decrease across the continental shelf. The highest reported abundances occur in the northeastern GOM, in productive regions of high river outflow, especially in the vicinity of the Mississippi, Atchafalaya, and Mobile River systems and downstream river plumes. Examples of zooplankton abundances on the productive Alabama shelf and the more oligotrophic west Florida shelf are provided by Carassou et al. (2015) and Lester et al. (2008), respectively. Zooplankton abundance and biomass are relatively low off-shelf and typical of oligotrophic waters (Hopkins 1982). Although zooplankton abundances are low in offshore waters, there are hot spots of higher-standing stocks associated with upwelling at the edge of the LC and cyclonic and anticyclonic hydrographic features, with the highest-standing stock in cyclonic cold-core eddies. There are also seasonal changes in abundance, with lowest zooplankton densities in winter and higher densities in spring and summer, concomitant with an increase in phytoplankton. The large hypoxic region that occurs during summer on the Louisiana shelf can also influence zooplankton abundance and distribution in the GOM. While bottom-water hypoxia does not appear to affect the total zooplankton biomass, and zooplankton vertical distributions were not correlated with oxygen levels, smaller zooplankton were more abundant near the surface and <40% of the zooplankton collected in hypoxic bottom waters were alive (Kimmel et al. 2009; Roman et al. 2012). Harmful algal blooms in the GOM also impact zooplankton by reducing the dominant species and changing species assemblages (Lester et al. 2008). Human activities may be changing zooplankton abundances over time, which would affect biochemical cycles. For example, Walsh et al. (2011) reported a 10-fold decline in zooplankton on the west Florida shelf between 1973 and 1993, as the result of a trophic cascade due to overfishing. In contrast, Graham (2001) reported an increase in jellyfish during summer and fall on the Alabama and Louisiana shelves between 1985 and 1997.

Role of Zooplankton in Biogeochemical Cycles

Zooplankton influence biogeochemical cycles through grazing, respiration, nitrogen excretion, and the release of particulate and dissolved matter. Microzooplankton have short turnover times, typically ranging from a few hours to a few days (Fahnenstiel et al. 1995) and relatively high metabolic rates due to their small size. Putland and Iverson (2007) observed microzooplankton grazing rates on phytoplankton in a GOM estuary and nearshore waters to be 10 times greater than that attributed to mesozooplankton, which can sometimes act to keep net phytoplankton growth near zero (Strom and Strom 1996; Juhl and Murrell 2005). Tropical and subtropical mesozooplankton have greater specific respiration rates than zooplankton in higher latitudes due to warm water temperature and smaller body size, as reviewed in Hernández-León and Ikeda (2005). These authors estimated that mesozooplankton carbon respiration rates were on average 97.52 mg C m^{-2} d^{-1} (at 0–200 m) between 20°N and 50°N. Microzooplankton excretion rates were estimated to supply 20–39% of the phytoplankton nitrogen demand between Mobile Bay and nearshore shelf stations (Lehrter et al. 1999). Donnelly and Torres (1988) reported respiration rates for many deepwater taxa in the eastern GOM. Diel vertical migration by zooplankton and fish provides an active transport of carbon and nitrogen from surface waters, where they feed at night, to mesopelagic depths, which they inhabit during the day (Longhurst et al. 1990). At GOM offshore locations >50% of the zooplankton biomass occurs in the upper 200 m (Hopkins 1982), and a significant proportion of the zooplankton and micronekton undergo diel vertical migration (D'Elia et al. 2016). Migrators include copepods, euphausiids, pelagic shrimp, cephalopods, and myctophid fish (Hopkins 1982). Respiration by diel migrants may account for 13–58% of the total particulate organic carbon flux to the deep sea in subtropical waters (Longhurst et al. 1990).

Turner (2002) noted that micro- and mesozooplankton and fish fecal pellets contribute to vertical particle flux in the GOM. Pellet sinking rates range from <10 to hundreds of meters per day. However, many pellets are likely recycled in the water column through microbial remineralization or reingestion by other zooplankton. Lipid, fatty acid, and amino acid composition of pellets, and thus their value as a food source, varies within pellets depending on the food ingested and pellet age (reviewed in Turner 2002). The ingestion of lithogenic particles by copepods and larvaceans in the Mississippi River plume likely increases the density and sinking rates of those pellets (Turner 1984; Dagg et al. 1996). Larvaceans may episodically be important contributors to vertical carbon flux because they have high production rates in coastal waters. Their discarded gelatinous houses and fecal pellets often provide more than 1 g m^{-2} d^{-1} of organic carbon (Dagg et al. 2008).

Eutrophication and hypoxia promote large blooms of gelatinous zooplankton, such as scyphozoan jellyfish, in the northern GOM (Miller and Graham 2012). Harvesting forage fish also enhances the productivity of these jellyfish, which decreases the amount of carbon being transferred to upper trophic levels (Robinson et al. 2015). Gelatinous zooplankton, such as ctenophores and scyphomedusae, may release substantial amounts of highly labile C-rich DOM (dissolved organic matter), relative to nitrogen (25.6C:1N), which is rapidly metabolized by bacterioplankton at uptake rates 2 to 6 times that of bulk DOM pools (Condon et al. 2011), thereby contributing to the microbial loop. In addition, mass falls of large jellyfish significantly contribute to the vertical carbon flux and act as a source of food for benthic communities (Sweetman et al. 2014).

Gulf of Mexico Benthic Communities

Benthic ecosystems of the GOM continental shelf vary in species composition from temperate communities along the northern shelf to subtropical and tropical communities in the south (Rabalais et al. 1999). Rowe (2017) has recently compiled general trends in shelf biota. Biomass consistently decreases from onshore to offshore because of higher terrestrially-derived nutrient inputs and productivity in coastal waters. In general, the macrofauna are dominated by annelid polychaete worms, with amphipods and bivalve mollusks of secondary importance (Rowe 2017). The bottom waters west of the Mississippi Delta have a distinct benthic community influenced by seasonal hypoxic to anoxic conditions in summer, with distinct assemblages of polychaetes and nematodes present with sulfur-oxidizing bacteria (*Beggiatoa* spp.) on the sediment surface (Baustian and Rabalais 2009). In the southern GOM the benthic communities are more subtropical to tropical with coral reefs associated with hermatypic corals, such as on Campeche Bank near the Yucatan Peninsula. Coral communities do exist in the northern

GOM, although they are not as prevalent as in the south. The East and West Flower Garden Banks, south of Galveston, Texas, are among the most extensive coral reefs in the northern GOM (Rezak et al. 1985) and are designated as a National Marine Sanctuary.

Mid-water to deep-sea benthic populations in the GOM display characteristics common to most oceanic communities, with a decrease in macrofauna and megafauna as depth increases. Additionally, there are spatial gradients in biomass and species occurrence that reflect the flux of sinking particulate organic matter from the overlying surface waters (Carney 2005; Wei et al. 2010). Sediment metazoan biomass in the Gulf is dominated by nematodes, as are the majority of oceanic populations (Danovaro et al. 2008). Interest in the community structure of deepwater benthic communities has increased as it has become apparent that the communities have critical roles in stabilization of deep-sea sediments, as well as organic matter recycling and nutrient regeneration.

Increasing numbers of observations over the past 2 decades indicate that deep-sea communities are characterized by exceptionally high biodiversity in the GOM, as well as globally (Haedrich et al. 2008). Many recent benthic studies in the GOM seek to explain, or better understand, the factors that contribute to the observed peak in biodiversity at mid-depths along the slope. In northern Gulf waters, both megafauna (e.g., Porifera, echinoderms, fishes) and macrofauna (e.g., amphipods, echinoderms) benthic species distributions have been related to both abiotic and biological factors, including depth, temperature, and notably, the flux of particulate organic matter from overlying surface waters (Haedrich et al. 2008). In general, a suite of 18 parameters, both abiotic and biotic, have been found to group into food-related, habitat-related, pollution-related, and location-related categories that are linked to species richness. However, no single factor can reliably account for observed patterns within a given species. In this sense, the benthic fauna of the northern GOM reflects patterns of species richness observed in other ocean basins (Snelgrove and Smith 2002).

Chemosynthetic ecosystems occur on the GOM slope associated with hydrocarbon seeps above salt diapirs. These ecosystems are dominated by 2 groups of symbiont-containing megafauna: vestimentiferan polychaetes and bathymodiolin mussels. The initial discovery of these "hydrothermal vent-like" communities was on the Florida Escarpment (Paull et al. 1984). Subsequent surveys have identified more than 20 locations of these communities along the northern GOM slope at depths of a few hundred meters to >2000 m (Cordes et al. 2009). Methane and sulfide serve as the two primary energy sources that support chemosynthetic endosymbionts in tubeworms and mussels (MacDonald et al. 1989).

The cold-seep megafauna are found primarily on carbonate substrates, which form in seep regions characterized by epibenthic waters with high alkalinity and carbonate precipitation. The bathymodiolin mussels frequently contain methanotropic bacteria and are associated with high methane concentrations in the surrounding epibenthic waters (Fisher et al. 1987). The vestimentiferan tubeworm–dominated communities occur in regions with high hydrocarbon concentrations, where the tubeworms rely on sulfide-oxidizing endosymbionts (Fisher 1990). As the age of seep communities increases, the rate of release of hydrocarbons and sulfide decreases. This change in seepage coincides with a shift in the epibenthic community from mussels and tubeworms to scerlactinian corals that lack photosynthetic symbionts and are predominantly solitary (Cairns et al. 1993). The deepwater corals are filter feeders and, in contrast to vestiminiferans and mussels, are dependent on sinking organic matter from the surface ocean (Prouty et al. 2011).

Gulf of Mexico Fisheries

In addition to supporting unique benthic communities, the GOM supports valuable fisheries and provides more than 100 species of finfish and shellfish that contribute to both commercial and recreational harvests. Statistics for US fisheries are regularly updated by the National Marine Fisheries Service of the National Oceanic and Atmospheric Administration. An assessment for 2014, for example, indicated that the Gulf provided about 546,000 metric tons (MT) of seafood in 2014, which make up approximately 15% of the total marine landings for the United States (National Marine Fisheries Service 2015). Commercial catch of menhaden (*Brevoortia patronus*) is the largest pelagic finfish fishery in terms of dollar value in the GOM. Annually, about 544,000 MT of menhaden are converted to fish meal and fish oil, and the commercial value of this fishery is approximately $125 million. The Gulf also provides about 68,000 MT of oysters, or half the national landings, and the largest shrimp harvest at 40,000 MT. The total dollar value of commercial fisheries in the GOM is about $800 to $900 million annually, on the order of 15% to 20% of total US landing values. Recreational fisheries are also an important component of fish catch in the Gulf. Fishing trips in the GOM constitute more than 30% of all trips in US coastal waters because of the favorable climate in the region. The most common species caught by recreational fishers include spotted sea trout, gray snapper, red drum, and blue runners, which are different from the main target species of commercial fleets. Recreational fishers also release up to 50% of caught fish, which can contribute to reduced mortality for some species.

The fisheries in the northern GOM are potentially influenced by nutrient flux delivered to coastal waters from the Mississippi River and the associated summer hypoxia. As the water column west of the Mississippi Delta stratifies during the summer, when seasonal winds are weak, the decay of sinking organic matter from the overlying surface layer results in hypoxic conditions in bottom waters. Since pelagic species have the potential to escape low-oxygen regions, the pelagic adults are less affected than benthic or demersal biota. Early life stages of pelagic species are less mobile than adults and are, therefore, also more likely to be affected by low oxygen (Breitburg et al. 1997). Fish eggs and larvae have also been found to be highly susceptible to low oxygen concentrations (Rombough 1988). Nevertheless, Chesney and Baltz (2001) assessed the effects of hypoxia on fisheries and concluded that 12 fish stocks may be "buffered" from any major reduction attributed to seasonal hypoxia.

As can be seen throughout this chapter, the geologic, physical, and biological characteristics of the GOM can influence the chemistry of this valuable and dynamic marginal sea. However, threats to the GOM mentioned in this chapter, such as eutrophication, acidification, and land-use change, continue to impact many aspects of the GOM system in numerous ways.

Acknowledgments

The authors thank the 2 anonymous reviewers who provided comments and suggestions that greatly improved this section. We also thank Tom Bianchi for his dedication and guidance in completing this book.

References

Ahr, W. M. 1973. The carbonate ramp: An alternative to the shelf model. *Gulf Coast Association of Geological Societies Transactions* 23:221–225.

Alvera-Azcárate, A., A. Barth, and R. H. Weisberg. 2009. The surface circulation of the Caribbean Sea and the Gulf of Mexico as inferred from satellite altimetry. *Journal of Physical Oceanography* 39(3):640–657.

Azam, F., T. Fenchel, J. G. Field, J. S. Grey, L. A. Meyer-Reil, and F. Thingstad. 1983. The ecological role of water-column microbes in the sea. *Marine Ecology Progress Series* 10:257–263.

Bartok, P. 1993. Prebreakup geology of the Gulf of Mexico-Caribbean: Its relation to Triassic and Jurassic rift systems of the region. *Tectonics* 12(2):441–459.

Baustian, M., and N. N. Rabalais. 2009. Seasonal composition of benthic macrofauna exposed to hypoxia in the northern Gulf of Mexico. *Estuaries and Coasts* 32:975–983.

Benway, H. M., and P. G. Coble (eds.). 2014. *Report of the US Gulf of Mexico Carbon Cycle Synthesis Workshop*. Woods Hole, MA: Ocean Carbon and Biogeochemistry Program and North American Carbon Program. 67 pp.

Bianchi, T. S., V. Galy, B. E. Rosenheim, M. Shields, X. Cui, and P. van Metre. 2015. Paleoreconstruction of organic carbon inputs to an oxbow lake in the Mississippi River watershed: Effects of dam construction and land use change on regional inputs. *Geophysical Research Letters* 42(19):7983–7991.

Bianchi, T. S., J. R. Pennock, and R. R. Twilley. 1999. *Biogeochemistry of the Gulf of Mexico Estuaries*. Hoboken, NJ: John Wiley and Sons. 448 pp.

Bianchi, T. S., L. A. Wysocki, K. M. Schreiner, D. R. Corbett, and A. S. Kolker. 2011. Sources of terrestrial organic carbon in the Mississippi plume region: Evidence for the importance of coastal marsh inputs. *Aquatic Geochemistry* 17(4–5):431–456.

Biggs, D. C. 1992. Nutrients, plankton and productivity in a warm-core ring in the western Gulf of Mexico. *Journal of Geophysical Research* 97:2143–2154.

Biggs D. C., and F. E. Müller-Karger. 1994. Ship and satellite observations of chlorophyll stocks in interacting cyclone-anticyclone pairs in the western Gulf of Mexico. *Journal of Geophysical Research* 99:7371–7384.

Biggs, D. C., and P. H. Ressler. 2001. Distribution and abundance of phytoplankton, zooplankton, ichthyoplankton, and micronekton in the deepwater Gulf of Mexico. *Gulf of Mexico Science* 1:7–29.

Bird, D. E., S. A. Hall, and J. F. Casey. 2005. Gulf of Mexico tectonic history: Hotspot tracks, crustal boundaries, and early salt distribution. *AAPG Bulletin* 89(3):311–328.

Breitburg, D. L., M. T. Loher, C. A. Pacey, and A. Gerstein. 1997. Varying effects of low dissolved oxygen on trophic interactions in an estuarine food web. *Ecological Monographs* 67:489–507.

Broecker, W. S., J. P. Kennett, B. P. Flower, J. T. Teller, S. Trumbore, G. Bonani, and W. Rolfi. 1989. Routing of meltwater from the Laurentide ice sheet during the Younger Dryas cold episode. *Nature* 341(6240):318–321.

Buffler, R. T., and D. S. Sawyer. 1985. Distribution of crust and early history, Gulf of Mexico basin. *Gulf Coast Association of Geological Societies Transactions* 35:333–344.

Bunge, L., J. Ochoa, A. Badan, J. Candela, and J. Sheinbaum. 2002. Deep flows in the Yucatan Channel and their relation to changes in the Loop Current extension. *Journal of Geophysical Research: Oceans* 107(C12):26–27.

Buskey, E. J. 1993. Annual pattern of micro- and mesozooplankton abundance and biomass in a subtropical estuary. *Journal of Plankton Research* 15:907–924.

Cai, W. J., X. Hu, W. J. Huang, M. C. Murrell, J. C. Lehrter, S. E. Lohrenz, W. C. Chou, et al. 2011. Acidification of subsurface coastal waters enhanced by eutrophication. *Nature Geoscience* 4(11):766–770.

Cairns, S. D., D. M. Opresko, T. S. Hopkins, and W. W. Schroeder. 1993. New records of deep-water Cnidaria (Scleractinia and Antipatharia) from the Gulf of Mexico. *Northeastern Gulf Science* 13:1–11.

Candela, J., J. Sheinbaum, J. Ochoa, A. Badan, and R. Leben. 2002. The potential vorticity flux through the Yucatan Channel and the Loop Current in the Gulf of Mexico. *Geophysical Research Letters* 29(22):1–4.

Carassou, L., F. J. Hernandez, and W. M. Graham. 2015. Corrigendum: Change and recovery of coastal mesozooplankton community structure during the Deepwater Horizon oil spill. *Environmental Research Letters* 10:049501.

Carney, R. S. 2005. Zonation of deep-sea biota on continental shelves. *Oceanography and Marine Biology Annual Review* 43:211–279.

Chang, Y. L., and L. Y. Oey. 2010. Eddy and wind-forced heat transports in the Gulf of Mexico. *Journal of Physical Oceanography* 40(12):2728–2742.

Chanton, J., T. Zhao, B. E. Rosenheim, S. Joye, S. Bosman, C. Brunner, K. M. Yeager, A. R. Diercks, and D. Hollander. 2015. Using natural abundance radiocarbon to trace the flux of petrocarbon to the seafloor following the Deepwater Horizon oil spill. *Environmental Science & Technology* 49(2):847–854.

Chen, C., D. A. Wiesenburg, and L. Xie. 1997. Influence of river discharge on biological production in the inner shelf: A coupled biological and physical model of the Louisiana-Texas shelf. *Journal of Marine Research* 55:293–320.

Chen, X., S. E. Lohrenz, and D. A. Wiesenburg. 2000. Distribution and controlling mechanisms of primary production on the Louisiana-Texas continental shelf. *Journal of Marine Systems* 25:179–207.

Chérubin, L. M., Y. Morel, and E. P. Chassignet. 2006. Loop Current ring shedding: The formation of cyclones and the effect of topography. *Journal of Physical Oceanography* 36(4):569–591.

Chesney, E. J., and D. Baltz. 2001. The effects of hypoxia on the northern Gulf of Mexico coastal ecosystem: A fisheries perspective. Pp. 321–354 in N. N. Rabalais and R. E. Turner (eds.), *Coastal Hypoxia: Consequences for Living Resources and Ecosystems*. Washington, DC: American Geophysical Union.

Christeson, G. L., H. J. A. Van Avendonk, I. O. Norton, J. W. Snedden, D. R. Eddy, G. D. Karner, and C. A. Johnson. 2014. Deep crustal structure in the eastern Gulf of Mexico. *Journal of Geophysical Research: Solid Earth* 119(9):6782–6801.

Clark, P. U., S. J. Marshall, G. K. C. Clarke, S. W. Hostetler, J. M. Licciardi, and J. T. Teller. 2001. Freshwater forcing of abrupt climate change during the last glaciation. *Science* 293(5528):283–287.

Condon, R. H., D. K. Steinberg, P. A. del Giorgio, T. C. Bouvier, D. A. Bronk, W. M. Graham, and H. W. Ducklow. 2011. Jellyfish blooms result in a major microbial respiratory sink of carbon in marine systems. *Proceedings of the National Academy Sciences of the USA* 108:10225–10230.

Cordes, E. E., D. C. Bergquist, and C. R. Fisher. 2009. Macroecology of Gulf of Mexico cold seeps. *Annual Review of Marine Science* 1(1):143–168.

Crone, T. J., and M. Tolstoy. 2010. Magnitude of the 2010 Gulf of Mexico oil leak. *Science* 330(6004):634.

Dagg, M. J., T. S. Bianchi, B. McKee, and R. Powell. 2008. Fates of dissolved and particulate materials from the Mississippi River immediately after discharge into the northern Gulf of Mexico, USA, during a period of low wind stress. *Continental Shelf Research* 28:1443–1450.

Dagg, M. J., E. P. Green, B. A. McKee, and P. B. Ortner. 1996. Biological removal of fine-grained lithogenic particles from a large river plume. *Journal of Marine Research* 54:149–160.

Dagg, M. J., R. Sato, H. Liu, T. S. Bianchi, R. Green, and R. Powell. 2008. Microbial food web contributions to bottom water hypoxia in the northern Gulf of Mexico. *Continental Shelf Research* 28:1127–1137.

Danovaro, R., C. Gambi, N. Lampidariou, and A. Tselepides. 2008. Deep-sea nematode diversity in the Mediterranean basin: Testing for longitudinal, bathymetric and energetic gradients. *Ecography* 31:231–244.

D'Elia, M., J. D. Warren, I. Rodriguez-Pinto, T. T. Sutton, A. Cook, and K. M. Boswell. 2016. Diel variation in the vertical distribution of deep-water scattering layers in the Gulf of Mexico. *Deep Sea Research Part I: Oceanographic Research Papers* 115:91–102.

DeHaan, C. J., and W. Sturges. 2005. Deep cyclonic circulation in the Gulf of Mexico. *Journal of Physical Oceanography* 35(10):1801–1812.

Dietz, R. S., J. C. Holden, and W. P. Sproll. 1971. Geotectonic evolution and subsidence of Bahama platform. *Geological Society of America Bulletin* 82(3):811–814.

Donat, J. R., and K. Bruland. 1995. Trace metals in the ocean. Pp. 247–281 in B. Salbu and E. Steinnes (eds.), *Trace Metals in Natural Waters*. Boca Raton, FL: CRC Press.

Donnelly, J., and J. J. Torres. 1988. Oxygen consumption of midwater fishes and crustaceans from the eastern Gulf of Mexico. *Marine Biology* 97:483–494.

Dortch, Q., and T. E. Whitledge. 1992. Does nitrogen or silicon limit phytoplankton production in the Mississippi River plume and nearby regions? *Continental Shelf Research* 12:1293–1309.

Du, M., and J. D. Kessler. 2012. Assessment of the spatial and temporal variability of bulk hydrocarbon respiration following the Deepwater Horizon oil spill. *Environmental Science & Technology* 46(19):10499–10507.

Duarte, C. M., and J. Cebrián. 1996. The fate of marine autotrophic production. *Limnology and Oceanography* 41:1758–1766.

Ellis, J. T., and B. J. Dean. 2012. Gulf of Mexico processes. *Journal of Coastal Research* 60:6–13.

El-Sayed S. Z., and J. T. Turner. 1977. Productivity of the Antarctic and tropical/subtropical regions: A comparative study. Pp. 463–503 in M. J. Dunbar (ed.), *Polar Oceans*. Montreal: McGill University.

Fahnenstiel, G. L., M. J. McCormick, G. A. Lang, D. G. Redale, S. E. Lohrenz, M. Markowitz, B. Wagoner, and H. J. Carrick. 1995. Taxon-specific growth and loss rates for dominant phytoplankton populations from the northern Gulf of Mexico. *Marine Ecology Progress Series* 117:229–239.

Fenchel, T. 2008. The microbial loop—25 years later. *Journal of Experimental Marine Biology and Ecology* 366:99–103.

Fisher, C. R. 1990. Chemoautotrophic and methanotropic symbioses in the marine invertebrates. *Reviews in Aquatic Science* 2:399–436.

Fisher, C. R., J. J. Childress, R. S. Oremland, and R. R. Bidigare. 1987. The importance of methane and thiosulfate in the metabolism of bacterial symbionts of two deep-sea mussels. *Marine Biology* 96:59–71.

Fisher, C. R., P. Montagna, and T. Sutton. 2016. How did the Deepwater Horizon oil spill impact deep-sea ecosystems? *Oceanography* 29(3):182–195.

Flower, B. P., D. W. Hastings, H. W. Hill, and T. M. Quinn. 2004. Phasing of deglacial warming and Laurentide ice sheet meltwater in the Gulf of Mexico. *Geology* 32(7):597–600.

Flower, B. P., and J. P. Kennett. 1990. The Younger Dryas cool episode in the Gulf of Mexico. *Paleoceanography* 5(6):949–961.

Gierach, M. M., and B. Subrahmanyam. 2008. Biophysical responses of the upper ocean to major Gulf of Mexico hurricanes in 2005. *Journal of Geophysical Research: Oceans* 113(C4):1–11.

Gifford, D. J. 1991. The protozoan-metazoan trophic link in pelagic ecosystems. *Journal of Protozoology* 38:81–86.

Gilbert, P. S., T. N. Lee, and G. P. Podesta. 1996. Transport of anomalous low-salinity waters from the Mississippi River flood of 1993 to the Straits of Florida. *Continental Shelf Research* 16(8):1065–1085.

Graham, W. M. 2001. Numerical increases and distributional shifts of *Chrysaora quinquecirrha* (Desor) and *Aurelia aurita* (Linné) (Cnidaria: Scyphozoa) in the northern Gulf of Mexico. *Hydrobiologia* 451:97–111.

Grinsted, A., J. C. Moore, and S. Jevrejeva. 2013. Projected Atlantic hurricane surge threat from rising temperatures. *Proceedings of the National Academy of Sciences* 110(14):5369–5373.

Haedrich, R., A. Devine, and V. J. Kendall. 2008. Predictors of species richness in the deep benthic fauna of the northern Gulf of Mexico. *Deep Sea Research Part II: Tropical Studies in Oceanography* 55:2650–2656.

He, R., and R. H. Weisberg. 2002a. Tides on the West Florida Shelf. *Journal of Physical Oceanography* 32(12):3455–3473.

———. 2002b. West Florida shelf circulation and temperature budget for the 1999 spring transition. *Continental Shelf Research* 22(5):719–748.

———. 2003. A Loop Current intrusion case study on the west Florida shelf. *Journal of Physical Oceanography* 33:465–477.

Hernández-León, S., and T. Ikeda. 2005. A global assessment of mesozooplankton respiration in the ocean. *Journal of Plankton Research* 27:153–158.

Hidalgo-González, R. M., S. Alvarez-Borrigo, C. Fuentes-Yaco, and T. Platt. 2005. Satellite-derived total and new phytoplankton production in the Gulf of Mexico. *Indian Journal of Marine Sciences* 34:408–417.

Hopkins, T. L. 1982. The vertical distribution of zooplankton in the eastern Gulf of Mexico. *Deep Sea Research Part A: Oceanographic Research Papers* 29(9):1069–1083.

Huang, H-J., N. D. Walker, Y. Hsueh, Y. Chao, and R. R. Leben. 2013. An analysis of Loop Current frontal eddies in a ⅙ Atlantic Ocean model simulation. *Journal of Physical Oceanography* 43(9):1924–1939.

Huang, W-J., W-J. Cai, Y. Wang, X. Hu, B. Chen, S. E. Lohrenz, S. Chakraborty, R. He, J. Brandes, and C. S. Hopkinson. 2015. The response of inorganic carbon distributions and dynamics to upwelling-favorable winds on the northern Gulf of Mexico during summer. *Continental Shelf Research* 111:211–222.

Hudec, M., R. Martin, P. A. Jackson, and F. J. Peel. 2013. Influence of deep Louann structure on the evolution of the northern Gulf of Mexico. *AAPG Bulletin* 97(10):1711–1735.

Hudec, M. R., I. O. Norton, M. P. A. Jackson, and F. J. Peel. 2013. Jurassic evolution of the Gulf of Mexico salt basin. *AAPG Bulletin* 97(10):1683–1710.

Hunt, B. D. Robinson, A. Weislogel, and R. C. Ewing. 2017. Sediment source regions and paleotransport of the Upper Jurassic Norphlet Formation, eastern Gulf of Mexico. *AAPG Bulletin* 101(9):1519–1542.

Hyndes, G. A., K. L. Heck Jr., A. Verges, E. S. Harvey, G. A. Kendrick, P. S. Laverny, K. McMahon, et al. 2016. Accelerating tropicalization and the transformation of temperate seagrass meadows. *Bioscience* 66:938–948.

Iverson, R. L., and T. L. Hopkins. 1981. A summary of knowledge of plankton production in the Gulf of Mexico: Recent phytoplankton and zooplankton research. Pp. 147–211 in *Proceedings of a Symposium on Environmental Research Needs in the Gulf of Mexico (GOMEX)*, vol. 2A. Miami, FL: Atlantic Oceanographic and Meteorological Laboratories, NOAA.

Jaimes, B., and L. K. Shay. 2015. Enhanced wind-driven downwelling flow in warm oceanic eddy features during the intensification of Tropical Cyclone Isaac (2012): Observations and theory. *Journal of Physical Oceanography* 45(6):1667–1689.

Johns, W. E., T. L. Townsend, D. M. Fratantoni, and W. D. Wilson. 2002. On the Atlantic inflow to the Caribbean Sea. *Deep Sea Research Part I: Oceanographic Research Papers* 49(2):211–243.

Johnson, D. R., H. M. Perry, J. Lyczkowski-Shultz, and D. Hanisko. 2009. Red snapper larval transport in the northern Gulf of Mexico. *Transactions of the American Fisheries Society* 138(3):458–470.

Johnson, W. S., and D. M. Allen. 2012. *Zooplankton of the Atlantic and Gulf Coasts: A Guide to Their Identification and Ecology*. Baltimore: Johns Hopkins University Press. 452 pp.

Juhl, A. R., and M. C. Murrell. 2005. Interactions between nutrients, phytoplankton, growth, and microzooplankton grazing in a Gulf of Mexico estuary. *Aquatic Microbial Ecology* 38:147–156.

Kantha, L. 2005. Barotropic tides in the Gulf of Mexico. Pp. 159–163 in W. Sturges and A. Lugo-Fernandez (eds.), *Circulation in the Gulf of Mexico: Observations and Models*. Washington, DC: American Geophysical Union.

Kessler, J. D., D. L. Valentine, M. C. Redmond, M. Du, E. W. Chan, S. D. Mendes, E. W. Quiroz, et al. 2011. A persistent oxygen anomaly reveals the fate of spilled methane in the deep Gulf of Mexico. *Science* 331(6015):312–315.

Kimmel, D. G., W. C. Boicourt, J. J. Pierson, M. R. Roman, and X. Zhang. 2009. A comparison of the mesozooplankton response to hypoxia in Chesapeake Bay and the northern Gulf of Mexico using the biomass size spectrum. *Journal of Experimental Marine Biology and Ecology* 381:S65–S73.

Klitgord, K. D., P. Popenoe, and H. Schouten. 1984. Florida: A Jurassic transform plate boundary. *Journal of Geophysical Research: Solid Earth* 89(B9):7753–7772.

Kneller, E. A., and C. A. Johnson. 2011. Plate kinematics of the Gulf of Mexico based on integrated observations from the Central and South Atlantic. *Gulf Coast Association of Geological Societies Transactions* 61:283–299.

Kolodziejczyk, N., J. Ochoa, J. Candela, and J. Sheinbaum. 2012. Observations of intermittent deep currents and eddies in the Gulf of Mexico. *Journal of Geophysical Research: Oceans* 117(C9):1–20.

Kramer, K. V., and W. W. Shedd. 2017. A 1.4-billion-pixel map of the Gulf of Mexico seafloor. *Eos* 98(8):16–21.

Le Borgne, R. 1986. The release of soluble end products of metabolism. Pp. 109–164 in E. D. S. Corner and S. C. M. O-Hara (eds.), *The Biological Chemistry of Marine Copepods*. Oxford: Clarendon Press.

Leben, R. R. 2005. Altimeter-derived Loop Current metrics. Pp. 181–201 in W. Sturges and A. Lugo-Fernandez (eds.), *Circulation in the Gulf of Mexico: Observations and Models*. Washington, DC: American Geophysical Union.

Lehr, B. S. Bristol, A. Possolo, X. Bai, J. Wang, C. Sellinger, A. Clites, and R. Assel. 2010. *Oil Budget Calculator Deepwater Horizon: Technical Documentation*. A report by the Federal Interagency Solutions Group, Oil Budget Calculator Science and Engineering Team. Durham: University of New Hampshire. 217 pp.

Lehrter, J. C., J. R. Pennock, and G. B. McManus. 1999. Microzooplankton grazing and nitrogen excretion across a surface estuarine-coastal interface. *Estuaries* 22:113–125.

Lester, K. M., C. A. Heil, M. B. Neely, D. N. Spence, S. Murasko, T. L. Hopkins, T. T. Sutton, et al. 2008. Zooplankton and *Karenia brevis* in the Gulf of Mexico. *Continental Shelf Research* 28:99–111.

Lin, Q., and I. A. Mendelssohn. 2012. Impacts and recovery of the Deepwater Horizon oil spill on vegetation structure and function of coastal salt marshes in the northern Gulf of Mexico. *Environmental Science & Technology* 46:3737–3743.

Liu, H. B., M. J. Dagg, C. J. Wu, and K. P. Chiang. 2005. Mesozooplankton consumption of microplankton in the Mississippi River plume, with special emphasis on planktonic ciliates. *Marine Ecology Progress Series* 286:133–144.

Lohrenz, S. E., S. Chakraborty, M. Huettel, J. Herrera Silveira, K. Gundersen, D. Redalje, J. Wiggert, B. E. Denton, and J. Lehrter. 2014. Primary production. Pp. 24–27 in H. M. Benway and P. G. Coble (eds.), *Report of the US Gulf of Mexico Carbon Cycle Synthesis Workshop*. Woods Hole, MA: Ocean Carbon and Biogeochemistry Program and North American Carbon Program.

Lohrenz, S. E., G. L. Fahnenstiel, D. G. Redalje, G. A. Lang, M. J. Dagg, T. E. Whitledge, and Q. Dortch. 1999. Nutrients, irradiance, and mixing as factors regulating primary production in coastal waters impacted by the Mississippi River plume. *Continental Shelf Research* 19:1113–1141.

Longhurst, A. R., A. W. Bedo, W. G. Harrison, E. J. H. Head, and D. D. Sameoto. 1990. Vertical flux of respiratory carbon by oceanic diel migrant biota. *Deep Sea Research Part A: Oceanographic Research Papers* 37(4):685–694.

Longhurst, A. R., and W. G. Harrison. 1989. The biological pump: Profiles of plankton production and consumption in the upper ocean. *Progress in Oceanography* 22:47–123.

Lovell, T. 2010. Detrital zircon U-Pb age constraints on the provenance of the Upper Jurassic Norphlet Formation, eastern Gulf of Mexico: Implications for paleogeography. *Gulf Coast Association of Geological Societies Transactions* 60:443–460.

MacDonald, I. R., G. S. Boland, J. S. Baker, J. M. Brooks, M. C. Kennicutt, and R. R. Bidigare. 1989. Gulf of Mexico hydrocarbon seep communities. II: Spatial distribution of seep organisms and hydrocarbons at Bush Hill. *Marine Biology* 101:235–247.

MacDonald, I. R., O. Garcia-Pineda, A. Beet, S. Daneshgar Asl, L. Feng, G. Graettinger, D. French-Mccay, et al. 2015. Natural and unnatural oil slicks in the Gulf of Mexico. *Journal of Geophysical Research: Oceans* 120(12):8364–8380.

MacRae, G., and J. S. Watkins. 1996. Desoto Canyon salt basin: Tectonic evolution and salt structural styles. *Structural Framework of the Northern Gulf of Mexico: Gulf Coast Association of Geological Societies Special Publication* 46:53–61.

Mainelli, M., M. DeMaria, L. K. Shay, and G. Goni. 2008. Application of oceanic heat content estimation to operational forecasting of recent Atlantic category 5 hurricanes. *Weather and Forecasting* 23(1):3–16.

Mancini, E. A., and D. J. Benson. 1980. Regional stratigraphy of Upper Jurassic Smackover carbonates of southwest Alabama. *Gulf Coast Association of Geological Societies Transactions* 56:551–563.

Martínez-López, B., and J. Zavala-Hidalgo. 2009. Seasonal and interannual variability of cross-shelf transports of chlorophyll in the Gulf of Mexico. *Journal of Marine Systems* 77(1):1–20.

Marton, G., and R. T. Buffler. 1993. Application of simple-shear model to the evolution of passive continental margins of the Gulf of Mexico basin. *Geology* 21:495–498.

———. 1994. Jurassic reconstruction of the Gulf of Mexico basin. *International Geology Review* 36(6):545–586.

McLeod, E., G. L. Chumura, S. Bouillon, R. Salm, M. Björk, C. M. Duarte, C. E. Lovelock, W. H. Schesinger, and B. R. Sillman. 2011. A blueprint for blue carbon: Towards an improved understanding of the role of vegetated coastal habitats in sequestering CO_2. *Frontiers in Ecology and the Environment* 9:552–560.

Meade, R. H., and J. A. Moody. 2010. Causes for the decline of suspended-sediment discharge in the Mississippi River system, 1940–2007. *Hydrological Processes* 24(1):35–49.

Mendoza, V. M., E. E. Villanueva, and J. Adem. 2005. On the annual cycle of the sea surface temperature and the mixed layer depth in the Gulf of México. *Atmósfera* 18(2):127–148.

Merino, M., 1997. Upwelling on the Yucatan shelf: Hydrographic evidence. *Journal of Marine Systems* 13:101–121.

Miller, M.-E. C., and W. M. Graham. 2012. Environmental evidence that seasonal hypoxia enhances survival and success of jellyfish polyps in the northern Gulf of Mexico. *Journal of Experimental Marine Biology and Ecology* 432–433:113–120.

Milliman, J. D., and K. L. Farnsworth. 2013. *River Discharge to the Coastal Ocean: A Global Synthesis*. Cambridge: Cambridge University Press. 382 pp.

Morell, J. M., and J. E. Corredor. 1993. Sediment nitrogen trapping in a mangrove lagoon. *Estuarine, Coastal and Shelf Science* 37:203–212.

Müller-Karger, F. E. 2000. The spring 1998 NEGOM cold water event: Remote sensing evidence for upwelling and for eastward advection of Mississippi Water (or: How an errant LC anticyclone took the NEGOM for a spin). *Gulf of Mexico* Science 1:55–67.

Müller-Karger, F. E, J. P. Smith, S. Werner, R. Chen, M. Roffer, Y. Liu, B. Muhling, D. Lindo-Atichati, and J. Lamkin. 2015. Natural variability of surface oceanographic conditions in the offshore Gulf of Mexico. *Progress in Oceanography* 134:54–76.

Müller-Karger F. E., J. J. Walsh, R. H. Evans, and M. B. Meyers. 1991. On the seasonal phytoplankton concentration and sea surface temperature cycles of the Gulf of Mexico as determined by satellites. *Journal of Geophysical Research* 96:12645–12665.

National Marine Fisheries Service. 2015. *Fisheries of the United States*, 2014. NOAA Current Fishery Statistics No. 2014. Silver Spring, MD: US Department of Commerce. 136 pp.

Nguyen, T. T., S. L. Morey, D. S. Dukhovskov, and E. P. Chassignet. 2015. Nonlocal impacts of the Loop Current on cross-slope near-bottom flow in the northeastern Gulf of Mexico. *Geophysical Research Letters* 42(8):2926–2933.

North, E. W., E. E. Adams, Z. Schlag, C. R. Sherwood, R. He, K. H. Hyun, and S. A. Socolofsky. 2011. Simulating oil droplet dispersal from the Deepwater Horizon spill with a Lagrangian approach. Pp. 217–226 in Y. Lui and A. McFadden (eds.), *Monitoring and Modeling the Deepwater Horizon Oil Spill: A Record-Breaking Enterprise*. Washington, DC: American Geophysical Union.

North, E. W., E. Adams, A. Thessen, Z. Schlag, R. He, S. A. Socolofsky, S. Masutani, and P. Scott. 2015. The influence of droplet size and biodegradation on the transport of subsurface oil droplets during the Deepwater Horizon spill: A model sensitivity study. *Environmental Research Letters* 10:12.

Oey, L-Y, T. Ezer, and H.-C. Lee. 2005. Loop Current, rings and related circulation in the Gulf of Mexico: A review of numerical models and future challenges. Pp. 31–56 in W. Sturges and A. Lugo-Fernandez (eds.), *Circulation in the Gulf of Mexico: Observations and Models*. Washington, DC: American Geophysical Union.

Olabarrieta, M., J. C. Warner, B. Armstrong, J. B. Zambon, and R. He. 2012. Ocean-atmosphere dynamics during Hurricane Ida and Nor'Ida: An application of the coupled ocean–atmosphere–wave–sediment transport (COAWST) modeling system. *Ocean Modelling* 43:112–137.

Ortell, N., and A. Ortmann. 2014. Interactions among members of the microbial loop in an estuary dominated by microzooplankton grazing. *Aquatic Microbial Ecology* 72:63–71.

Parks, A. A. 2015. Dissolved nutrient distributions in the Gulf of Mexico following the Deepwater Horizon oil spill. Master's thesis, University of South Florida. 42 pp. https://scholarcommons.usf.edu/etd/6010.

Passow, U., and C. A. Carlson. 2012. The biological pump in a high CO_2 world. *Marine Ecology Progress Series* 470:249–271.

Passow, U., K. Ziervogel, V. Asper, and A. Diercks. 2012. Marine snow formation in the aftermath of the Deepwater Horizon oil spill in the Gulf of Mexico. *Environmental Research Letters* 7(3):035301.

Paull, C. K., B. Hecker, R. Commeau, R. P. Freeman-Lynde, C. Neumann, W. P. Corso, S. Golubic, J. E. Hook, E. Sikes, and J. Curray. 1984. Biological communities at the Florida Escarpment resemble hydrothermal vent taxa. *Science* 226:964–967.

Perez-Brunius, P., P. García-Carrillo, J. Dubranna, J. Sheinbaum, and J. Candela. 2013. Direct observations of the upper layer circulation in the southern Gulf of Mexico. *Deep Sea Research Part II: Tropical Studies in Oceanography* 85:182–194.

Pindell, J. L. 1985. Alleghenian reconstruction and subsequent evolution of the Gulf of Mexico, Bahamas, and Proto-Caribbean. *Tectonics* 4(1):1–39.

Pindell, J., and J. F. Dewey. 1982. Permo-Triassic reconstruction of western Pangea and the evolution of the Gulf of Mexico/Caribbean region. *Tectonics* 1(2):179–211.

Pindell, J., and L. Kennan. 2001. Kinematic evolution of the Gulf of Mexico: Petroleum systems of deep-water basins. Paper presented at the GCSSEPM Foundation 21st Annual Bob F. Perkins Research Conference, December 2–5, Houston, Texas.

———. 2009. Tectonic evolution of the Gulf of Mexico, Caribbean and northern South America in the mantle reference frame: An update. *Geological Society, London, Special Publications* 328(1):1–55.

Pindell, J., B. Radovich, E. Haire, D. Howard, A. Goswami, G. Dinc, and B. Horn. 2015. Structure maps of the top-rift unconformity/oceanic crust and top Cretaceous surfaces, and the Oxfordian rift-drift reconstruction. *Gulf of Mexico: Gulf Coast Association of Geological Societies Transactions* 65:821–831.

Prouty, N. G., E. B. Roark, N. A. Buster, and S. W. Ross. 2011. Growth rate and age distribution of deep-sea black corals in the Gulf of Mexico. *Marine Ecology Progress Series* 423:101–115.

Putland, J. N., and R. L. Iverson. 2007. Microzooplankton: Major herbivores in an estuarine planktonic food web. *Marine Ecology Progress Series* 345:63–73.

Putman, N. F., and R. He. 2013. Tracking the long-distance dispersal of marine organisms: Sensitivity to ocean model resolution. *Journal of the Royal Society Interface* 10(81):20120979.

Qian, H., Y. Li, R. He, and D. Eggleston. 2015. Connectivity in the Intra-American Seas and implications for potential larval transport. *Coral Reefs* 34(2):403–417.

Rabalais, N. N., R. Carney, and E. Escobar-Briones. 1999. Overview of continental shelf benthic communities of the Gulf of Mexico. Pp. 211–240 in H. Kumpf, K. Steidinger, and K. Sherman (eds.), *The Gulf of Mexico Large Marine Ecosystem: Assessment, Sustainability and Management*. Malden, MA: Blackwell Science.

Rabalais, N. N., and R. E. Turner. 2001. Hypoxia in the northern Gulf of Mexico: Description, causes and change. Pp. 1–36 in N. N. Rabalais and R. E. Turner (eds.), *Coastal Hypoxia: Consequences for Living Resources and Ecosystems*. Washington, DC: American Geophysical Union.

Rabalais, N. N., R. E. Turner, and W. J. Wiseman Jr. 2002. Gulf of Mexico hypoxia, a.k.a. the dead zone. *Annual Review of Ecology and Systematics* 33:235–263.

Ray, B. R., M. W. Johnson, K. Cammarata, and D. L. Smee. 2014. Changes in seagrass species composition in northwestern Gulf of Mexico estuaries: Effects on associated seagrass fauna. *PLOS ONE* 9:1–12.

Rezak, R., T. Bright, and D. McGrail. 1985. *Reefs and Banks of the Northwestern Gulf of Mexico*. New York: Wiley. 259 pp.

Roberts, H. H., C. Fisher, B. Bernard, J. Brooks, M. Bright, R. Varney, E. Cordes, et al. 2007. Alvin explores the deep northern Gulf of Mexico slope. *Eos* 88(35):341–342.

Robinson, K. L., J. J. Ruzicka, F. J. Hernandez, W. M. Graham, M. B. Decker, R. D. Brodeur, and M. Sutor. 2015. Evaluating energy flows through jellyfish and Gulf menhaden (*Brevoortia patronus*) and the effects of fishing on the northern Gulf of Mexico ecosystem. *ICES Journal of Marine Science* 72:2301–2312.

Roman, M. R., J. J. Pierson, D. G. Kimmel, W. C. Boicourt, and X. Zhang. 2012. Impacts of hypoxia on zooplankton spatial distributions in the northern Gulf of Mexico. *Estuaries and Coasts* 35:1261–1269.

Rombough, P. J. 1998. Respiratory gas exchange, aerobic metabolism, and effects of hypoxia during early life. Pp 59–161 in W. S. Hoar and D. J. Randall (eds.), *The Physiology of Developing Fish*, vol. 11: *Fish Physiology*. New York: Academic Press.

Rosenheim, B. E., M. A. Pendergraft, G. C. Flowers, R. Carney, J. L. Sericano, R. M. Amer, J. Chanton, Z. Dincer, and T. L. Wade. 2016. Employing extant stable carbon isotope data in Gulf of Mexico sedimentary organic matter for oil spill studies. *Deep Sea Research Part II: Tropical Studies in Oceanography* 129:249–258.

Rowe, G. T. 2017. Offshore plankton and benthos of the Gulf of Mexico. Pp. 641–747 in C. Ward (ed.), *Habitats and Biota of the Gulf of Mexico: Before the Deepwater Horizon Oil Spill*. New York: Springer.

Sabourin, D. T., J. E. Silliman, and K. B. Strychar. 2012. Polycyclic aromatic hydrocarbon contents of coral and surface sediments off the South Texas coast of the Gulf of Mexico. *International Journal of Biology* 5(1):1.

Salvador, A. 1987. Late Triassic-Jurassic paleogeography and origin of Gulf of Mexico basin. *American Association of Petroleum Geologists Bulletin* 71(4):419–451.

———. 1991. Origin and development of the Gulf of Mexico basin. Pp. 389–443 in A. Salvador (ed.), *The Gulf of Mexico Basin*, vol. J: *The Geology of North America*. Boulder, CO: Geological Society of America.

Sandwell, D. T., R. D. Muller, W. H. F. Smith, E. Garcia, and R. Francis. 2014. New global marine gravity model from CryoSat-2 and Jason-1 reveals buried tectonic structure. *Science* 346(6205):65–67.

Sawyer, D. S., R. T. Buffler, and R. H. Pilger Jr. 1991. The crust under the Gulf of Mexico basin. Pp. 53–72 in A. Salvador (ed.), *The Gulf of Mexico Basin*, vol. J: *The Geology of North America*. Boulder, CO: Geological Society of America.

Schmitz, W. J., D. C. Biggs, A. Lugo-Fernandez, Y. Oey, and W. Sturges. 2005. A synopsis of the circulation in the Gulf of Mexico and on its continental margins. Pp. 11–29 in W. Sturges and A. Lugo-Fernandez (eds.), *Circulation in the Gulf of Mexico: Observations and Models*. Washington, DC: American Geophysical Union.

Schmitz, W. J., and P. L. Richardson. 1991. On the sources of the Florida Current. *Deep Sea Research Part I: Oceanographic Research Papers* 38:S379–S409.

Schroeder, W. W., and W. J. Wiseman. 1999. Geology and hydrodynamics of Gulf of Mexico estuaries. Pp. 3–28 in T. S. Bianchi, J. R. Pennock, and R. R. Twiley (eds.), *Biogeochemistry of Gulf of Mexico Estuaries*. Hoboken, NJ: John Wiley and Sons.

Schwing, P. T., I. C. Romero, G. R. Brooks, D. W. Hastings, R. A. Larson, and D. J. Hollander. 2015. A decline in benthic foraminifera following the Deepwater Horizon event in the northeastern Gulf of Mexico. *PLOS ONE* 10(3):e0120565.

Shay, L. K. 2010. Air-sea interactions in tropical cyclones. *Global Perspectives on Tropical Cyclones: From Science to Mitigation* 4:93–131.

Shay, L. K., G. J. Goni, and P. G. Black. 2000. Effects of a warm oceanic feature on Hurricane Opal. *Monthly Weather Review* 128(5):1366–1383.

Shi, W., and M. Wang. 2007. Observations of a Hurricane Katrina–induced phytoplankton bloom in the Gulf of Mexico. *Geophysical Research Letters* 34(11) L11607. doi:10.1029/2007GL029724.

Shropshire, T., Y. Li, and R. He. 2016. Storm impact on sea surface temperature and chlorophyll-*a* in the Gulf of Mexico and Sargasso Sea based on daily cloud-free satellite data reconstructions. *Geophysical Research Letters* 43(23):12199–12207.

Simmons, H. L., R. W. Hallberg, and B. K. Arbic. 2004. Internal wave generation in a global baroclinic tide model. *Deep Sea Research Part II: Tropical Studies in Oceanography* 51(25):3043–3068.

Smith, S. M., and G. L. Hitchcock. 1994. Nutrient enrichments and phytoplankton growth in the surface waters of the Louisiana Bight. *Estuaries* 17:740–753.

Snedden, J. W., I. O. Norton, G. L. Christeson, and J. C. Sanford. 2014. Interaction of deepwater deposition and a mid-ocean spreading center, eastern Gulf of Mexico basin, USA. *Gulf Coast Association of Geological Societies Transactions* 64:371–383.

Snelgrove, P. V. R., and C. R. Smith. 2002. A riot of species in an environmental calm: The paradox of the species-rich deep-sea floor. *Oceanography and Marine Biology Annual Review* 40:311–342.

Souza, J. M., J. Candela, and J. Ochoa. 2016. Internal tide generation in the Gulf of Mexico and its impact on the Yucatan Channel and Florida Strait exchanges. American Geophysical Union, Ocean Sciences Meeting, New Orleans, Louisiana, Abstract# P014H-2901.

Stern, R. J., and W. R. Dickinson. 2010. The Gulf of Mexico is a Jurassic backarc basin. *Geosphere* 6(6):739–754.

Strom, S. L., and M. W. Strom. 1996. Microplankton growth, grazing, and community structure in the northern Gulf of Mexico. *Marine Ecology Progress Series* 130:229–240.

Sturges, W. 1993. The annual cycle of the Western Boundary Current in the Gulf of Mexico. *Journal of Geophysical Research: Oceans* 98(C10):18053–18068.

———. 2005. Deep-water exchange between the Atlantic, Caribbean, and Gulf of Mexico. Pp. 263–278 in W. Sturges and A. Lugo-Fernandez (eds.), *Circulation in the Gulf of Mexico: Observations and Models*. Washington, DC: American Geophysical Union.

Sturges, W., and K. E. Kenyon. 2008. Mean flow in the Gulf of Mexico. *Journal of Physical Oceanography* 38(7):1501–1514.

Sweetman A. K., C. R. Smith, T. Dale, and D. O. B. Jones. 2014. Rapid scavenging of jellyfish carcasses reveals the importance of gelatinous material to deep-sea food webs. *Proceedings of the Royal Society* B281:20142210.

Sylvan, J. B., Q. Dortch, D. M. Nelson, A. F. Maier Brown, W. Morrison, and J. W. Ammerman. 2006. Phosphorus limits phytoplankton growth on the Louisiana shelf during the period of hypoxia formation. *Environmental Science & Technology* 40:7548–7553.

Talley, Lynne D. 2011. *Descriptive Physical Oceanography: An Introduction*. Cambridge, MA: Academic Press. 560 pp.

Toner, M., A. D. Kirwan, A. C. Poje, L. H. Kantha, F. E. Muller-Karger, and C. K. R. T. Jones. 2003. Chlorophyll dispersal by eddy-eddy interactions in the Gulf of Mexico. *Journal of Geophysical Research: Oceans* 108(C4) L11607. doi.org/10.1029/2002JC001499.

Turner, J. T. 1984. Zooplankton feeding ecology: Contents of fecal pellets of the copepods *Acartia tonsa* and *Labidocera aestiva* from continental shelf waters near the mouth of the Mississippi River. *Marine Ecology–Publicazioni della Stazione Zoologica di Napoli* 5:265–282.

———. 2002. Zooplankton fecal pellets, marine snow, and sinking phytoplankton blooms. *Aquatic Microbial Ecology* 27:57–102.

Valentine, D. L., G. B. Fisher, S. C. Bagby, R. K. Nelson, C. M. Reddy, S. P. Sylva, and M. A. Woo. 2014. Fallout plume of submerged oil from Deepwater Horizon. *Proceedings of the National Academy of Sciences* 111(45):15906–15911.

Vargo, G. A., K. L. Carder, W. W. Gregg, E. Shanley, C. Heil, K. A. Steidinger, and K. D. Haddad. 1987. The potential contribution by red tides to the west Florida shelf ecosystem. *Limnology and Oceanography* 32:762–767.

Vázquez de la Cerda, A. M., R. O. Reid, S. F. DiMarco, and A. E. Jochens. 2005. Bay of Campeche circulation: An update. Pp. 279–293 in W. Sturges and A. Lugo-Fernandez (eds.), *Circulation in the Gulf of Mexico: Observations and Models*. Washington, DC: American Geophysical Union.

Velasco, G. G., and C. D. Winant. 1996. Seasonal patterns of wind stress and wind stress curl over the Gulf of Mexico. *Journal of Geophysical Research: Oceans* 101(C8):18127–18140.

Walker, B. D., E. R. M. Druffel, J. Kolansinski, B. J. Roberts, X. Xu, and B. E. Rosenheim. 2017. Stable and radiocarbon isotopic composition of dissolved organic matter in the Gulf of Mexico. *Geophysical Research Letters* 44(16):8424–8434.

Walker, N. D. 2005. Wind and eddy-related shelf/slope circulation processes and coastal upwelling in the northwestern Gulf of Mexico. Pp. 295–313 in W. Sturges and A. Lugo-Fernandez (eds.), *Circulation in the Gulf of Mexico: Observations and Models*. Washington, DC: American Geophysical Union.

Walker, N. D., R. R. Leben, and S. Balasubramanian. 2005. Hurricane-forced upwelling and chlorophyll-a enhancement within cold-core cyclones in the Gulf of Mexico. *Geophysical Research Letters* 32(18) L18610. doi:10.1029/2005GL023716.

Walsh, J. J., D. A. Dieterle, M. B. Meyers, and F. E. Müller-Karger. 1989. Nitrogen exchange at the continental margin: A numerical study of the Gulf of Mexico. *Progress in Oceanography* 23:245–301.

Walsh, J. J., C. R. Tomas, K. A. Steidinger, J. M. Lenes, F. R. Chen, R. H. Weisberg, L. Zheng, J. H. Landsberg, G. A. Vargo, and C. A. Heil. 2011. Imprudent fishing harvest and consequent trophic cascades on the west Florida shelf over the last century: A harbinger of increased human deaths from paralytic shellfish poisoning along the southeastern United States, in response to oligotrophication? *Continental Shelf Research* 31:891–911.

Warner, J. C., B. Armstrong, R. He, and J. B. Zambon. 2010. Development of a coupled ocean–atmosphere–wave–sediment transport (COAWST) modeling system. *Ocean Modelling* 35(3):230–244.

Wei, C. L., G. T. Rowe, G. F. Hubbard, A. H. Scheltema, G. D. F. Wilson, I. Petrescu, J. M. Foster, et al. 2010. Bathymetric zonation of deep-sea macrofauna in relation to export of surface phytoplankton production. *Marine Ecology Progress Series* 399:1–14.

Weisberg, R. H., and R. He. 2003. Local and deep-ocean forcing contributions to anomalous water properties on the west Florida shelf. *Journal of Geophysical Research: Oceans* 108(C6), 3184. doi:10.1029/2002JC001407.

Weisberg, R. H., R. He, G. Kirkpatrick, F. Müller-Karger, and J. J. Walsh. 2004. Coastal ocean circulation influences on remotely sensed optical properties: A west Florida shelf case study. *Oceanography* 17:68–75.

Weisberg, R. H., Y. He, Y. G. Liu, and J. I. Virmani. 2005. West Florida shelf circulation on synoptic, seasonal, and interannual time scales. Pp. 325–347 in W. Sturges and A. Lugo-Fernandez (eds.), *Circulation in the Gulf of Mexico: Observations and Models*. Washington, DC: American Geophysical Union.

Weislogel, A. L., B. Hunt, A. Lisi, T. Lovell, and D. M. Robinson. 2015. Detrital zircon provenance of the eastern Gulf of Mexico subsurface: Constraints on Late Jurassic paleogeography and sediment dispersal of North America. *Geological Society of America Special Papers* 513:89–105.

Wilson, S. G., and T. R. Fischetti. 2010. Coastline Population *Trends in the United States* 1960 to 2008. Report no. P25-1139. Washington, DC: US Census Bureau. 28 pp.

Wiseman, W. J., Jr., and W. Sturges. 1999. Physical oceanography of the Gulf of Mexico: Processes that regulate its biology. Pp. 77–92 in H. Kumpf, K. Steidinger, and K. Sherman (eds.), *The Gulf of Mexico Large Marine Ecosystem*. Malden, MA: Blackwell Science.

Xue, Z., R. He, K. Fennel, W. J. Cai, S. Lohrenz, W. J. Huang, H. Tian, W. Ren, and Z. Zang. 2016. Modeling pCO_2 variability in the Gulf of Mexico. *Biogeosciences* 13(15):4359–4377.

Xue, Z., J. Zambon, Z. Yao, Y. Liu, and R. He. 2015. An integrated ocean circulation, wave, atmosphere, and marine ecosystem prediction system for the South Atlantic Bight and Gulf of Mexico. *Journal of Operational Oceanography* 8(1):80–91.

Young, C. M., R. He, R. B. Emlet, Y. Li, H. Qian, S. M. Arellano, A. Van Gaest, et al. 2012. Dispersal of deep-sea larvae from the Intra-American Seas: Simulations of trajectories using ocean models. *Integrative and Comparative Biology* 52(4):483–496.

Zambon, J. B. 2009a. An examination of tropical cyclone dynamics utilizing the 3-way coupled ocean atmosphere wave sediment transport (COAWST) model. Raleigh: North Carolina State University.

———. 2009b. Investigations of landfalling tropical storms using the coupled ocean atmosphere wave and sediment transport (COAWST) model. Raleigh: North Carolina State University.

Zambon, J. B., R. He, and J. C. Warner. 2014. Investigation of Hurricane Ivan using the coupled ocean–atmosphere–wave–sediment transport (COAWST) model. *Ocean Dynamics* 64(11):1535–1554.

Zavala-Hidalgo, J., S. L. Morey, and J. J. O'Brien. 2003. Seasonal circulation on the western shelf of the Gulf of Mexico using a high-resolution numerical model. *Journal of Geophysical Research: Oceans* 108(C12), 3389. doi:10.1029/2003JC001879.

Zavala-Hidalgo, J., A. Pares-Sierra, and J. Ochoa. 2002. Seasonal variability of the temperature and heat fluxes in the Gulf of Mexico. *Atmósfera* 15(2):81–104.

Zavala-Hidalgo, J., R. Romero-Centeno, A. Mateos-Jasso, S. L. Morey, and B. Martinez-Lopez. 2014. The response of the Gulf of Mexico to wind and heat flux forcing: What has been learned in recent years? *Atmósfera* 27(3):317–334.

Zeng, X., Y. Li, and R. He. 2015. Predictability of the Loop Current variation and eddy shedding process in the Gulf of Mexico using an artificial neural network approach. *Journal of Atmospheric and Oceanic Technology* 32(5):1098–1111.

Zeng, X., Y. Li, R. He, and Y. Yin. 2015. Clustering of Loop Current patterns based on the satellite-observed sea surface height and self-organizing map. *Remote Sensing Letters* 6(1):11–19.

Zhao, Y. and R. He. 2012. Cloud-free sea surface temperature and colour reconstruction for the Gulf of Mexico: 2003–2009. *Remote Sensing Letters* 3(8):697–706.

2

Physicochemical Properties of Gulf of Mexico Seawater

Andrew R. Margolin and Frank J. Millero

The Composition and Salinity of Seawater

Throughout the global ocean, the relative composition of the major constituents of seawater is consistent; therefore, Gulf of Mexico (GOM) waters can be treated the same as ocean waters. Table 2.1 provides a list of the major ion concentrations in seawater. Ion concentrations reported in the table are relative to the chlorinity of seawater, measured as the sum of halides that co-precipitate with silver chloride, reported in units g kg^{-1}.

Most of the earlier physicochemical properties of seawater were measured as a function of chlorinity. The sum of the components listed in Table 2.1 has been used to determine the reference salinity of seawater (i.e., $S_R = 35.16504$ g kg^{-1}; Millero et al. 2008), and the ratio of S_R to Cl is given by

$$S_R\ (g\ kg^{-1}) \approx 1.815069 * Cl. \quad (2.1)$$

The ratio of S_R to Cl in Equation 2.1 is based on recent measurements and is in reasonable agreement with the value of 1.8148 determined by Lyman and Fleming (1940), which is based on measurements made by Forchhammer (1865) and Dittmar (1884).

For more than half a century (1901–67), oceanographers determined the salinity (S) of ocean waters using the equations of Knudsen (1901):

$$S\ (‰) = 0.030 + 1.805 * Cl, \quad (2.2)$$

based on the measurement of residual salts following high-temperature dehydration (i.e., evaporation) of seawater. Unfortunately, evaporating seawater results in a loss of CO_2, HCl, and $B(OH)_3$, causing the salinity calculated from Equation 2.2 to be lower than S_R. Another issue with this equation is that it is based on measurements from samples of the Baltic Sea water, which has a salinity of 0.03‰ when Cl = 0‰. The relationship in Equation 2.2 was later updated for ocean waters by Cox et al. (1967)

$$S\ (‰) = 1.80655 * Cl. \quad (2.3)$$

Having an intercept of 0, it is an improved estimate of salinity for natural waters (Wooster et al. 1969). The slope in Equation 2.3 (1.80655) is the value for samples with S = 35‰ and Cl = 19.374‰ determined from Equation 2.2.

Using the relationship in Equation 2.3, the practical salinity scale of seawater (S_P, no units) was established by the Joint Panel for Oceanographic Tables and Standards (JPOTS), based on conductivity measurements made as a function of chlorinity (Lewis and Perkin 1978). Whereas the S-to-Cl ratio in Equation 2.3 used to determine S_P is an improvement from Equation 2.2, the ratio of 1.8148 determined by Lyman and Fleming (1940) would have made S_P closer to the actual salinity (i.e., closer to S_R in Equation 2.1); it is unfortunate that a S-to-Cl ratio of 1.8148 was not used for the calculation of S_P. Despite the slight discrepancy, S_P is valid for typical seawater samples

Table 2.1. Composition of the major ions in seawater

	$(g\ kg^{-1})\ Cl^{-1}$	$g\ kg^{-1}$	$g\ mol^{-1}$	$mol\ kg^{-1}\text{-}H_2O$
Cl^-	0.9989041	19.35271	35.4530	0.5657619
Na^+	0.5564924	10.78145	22.9898	0.4860573
SO_4^{2-}	0.1400000	2.71235	96.0626	0.0292642
Mg^{2+}	0.0662600	1.28372	24.3050	0.0547419
Ca^{2+}	0.0212700	0.41208	40.0780	0.0106566
K^+	0.0206000	0.3991	39.0983	0.0105796
HCO_3^-	0.0054100	0.10481	61.0168	0.0017803
Br^-	0.0034730	0.06728	79.9040	0.0008727
$B(OH)_3$	0.0010030	0.01944	61.8330	0.0003259
CO_3^{2-}	0.0007400	0.01434	60.0089	0.0002477
$B(OH)_4^-$	0.0004100	0.00795	78.8404	0.0001045
Sr^{2+}	0.0004100	0.00795	87.6200	0.0000940
F^-	0.0000670	0.0013	18.9984	0.0000709
CO_2	0.0000220	0.00042	44.0095	
OH^-	0.0000070	0.00014	17.0073	0.0000085
Total	1.8150685	35.16504		1.1605659
H_2O		964.83496	31.9980	0.5802830
Grand total		1000.00000		

Source: Adapted from Millero et al. (2008); and Millero (2013).
Note: Determined at a practical salinity (S_P) = 35; temperature = 25°C; fugacity of CO_2 = 333 μatm; total alkalinity (TA) = 2300 μmol kg^{-1}; and calculated pH = 8.096 (seawater scale). Values are representative of surface water.

with salinities that range from 2 to 42 (Millero 2013). However, for estuarine systems, such as those found in the regions of the GOM where S_P can be <2 (Bianchi et al. 1999), extensions to lower salinities are available (Hill et al. 1986). For samples with S_P >42, S_P can be determined by diluting a sample by weight and measuring the conductivity (Millero 2013).

Today, oceanographers conventionally use S_P derived from conductivity, temperature, and depth (CTD) sensor measurements (e.g., Sea-Bird Scientific SBE 911plus CTD), since measuring all of the ions listed in Table 2.1 to determine S_R would be impractical. While not conventionally used, S_R can be calculated from S_P using Equations 2.1 and 2.3 (Millero et al. 2008), providing a better estimate of the actual, or *absolute*, salinity:

$$S_R\ (g\ kg^{-1}) = (35.16504/35) * S_P. \tag{2.4}$$

Brewer and Bradshaw (1975) first suggested that conductivity measurements of seawater are affected by its composition, motivating the introduction of the absolute salinity of seawater (S_A) by Millero, Gonzalez, et al. (1976) and Millero, Lawson, et al. (1976) and defined as

$$S_A\ (g\ kg^{-1}) = S_R + \delta S_A, \tag{2.5}$$

where, δS_A accounts for variations in nutrient concentrations, including nitrate, soluble reactive phosphate, and silicic acid [$Si(OH)_4$], as well as total alkalinity (TA) and total dissolved inorganic carbon (DIC or TCO_2) concentrations (McDougall et al. 2009). Whereas the effects of nutrients, TA, and DIC on density and S_A are well understood, the effect of dissolved organic carbon remains uncertain (Woosley et al. 2014).

As a result of S_A accounting for subtle changes in seawater composition, Millero, Gonzalez, et al. (1976) and Millero, Lawson, et al. (1976) suggested that S_A should be used to obtain the most accurate calculation of the physicochemical properties of marine waters. They also suggested that the physicochemical properties of rivers and estuaries were equal to the values determined from seawater at the same S_A, making S_A the universal salinity for the study of most natural waters (Millero, Lawson, et al. 1976). More recently, Woosley et al. (2014) made direct measurements of the density of seawater from the world's oceans, demonstrating that Brewer and Bradshaw (1975) were correct, as conductivity variations are observed in response to subtle deviations in seawater composition.

The value of δS_A in Equation 2.5 has been estimated by comparing the measured density (ρ_{meas}) of seawater samples with the values calculated from the equation of state using S_P (ρ_{calc}) based on the following relationship:

$$\delta S_A\ (\mathrm{g\ kg^{-1}}) = \delta\rho/0.7519, \tag{2.6}$$

where, $\delta\rho = \rho_{meas} - \rho_{calc}$ (McDougall et al. 2012). The values of δS_A calculated in this manner are linearly related to $[Si(OH)_4]$ and TA (Woosley et al. 2014); McDougall et al. (2009) found the δS_A and $[Si(OH)_4]$ (in μmol kg^{-1}) relationship for the global ocean to be

$$\delta S_A\ (\mathrm{g\ kg^{-1}}) = 9.824 \times 10^{-5} * [Si(OH)_4], \tag{2.7}$$

which can be applied to GOM waters.

The highest δS_A values found in the global ocean are ~0.024 g kg^{-1} (equivalent to $[Si(OH)_4] \approx 240$ µmol kg^{-1}), found in the deep Bering Sea and Sea of Okhotsk in the far North Pacific (Key et al. 2015). Most deep waters have high concentrations of nutrients, TA, and DIC relative to surface waters, including those found in the GOM, which can notably affect the physicochemical properties of seawater. Since the deep waters of the GOM have relatively low levels of $[Si(OH)_4]$ (~25 µmol kg^{-1}), Equation 2.7 gives a δS_A of ~0.002 g kg^{-1}, approximately one-tenth of the δS_A found in the deep seas of the far North Pacific. Given the small δS_A found in GOM waters, δS_A does not typically have a large effect on the physicochemical properties of seawater within the basin.

Since the relative composition of GOM waters is the same as that found in the ocean, CTD measurements of GOM waters are reliable for determining S_P and for using S_P to calculate the physicochemical properties of seawater. Most of the physicochemical properties of seawater needed to calculate those found in GOM waters are tabulated in marine chemistry textbooks (Millero 2001, 2013). Another reliable resource is the 2010 Thermodynamic Equation of Seawater (TEOS-10), a computer program that can be used to calculate the thermodynamic properties of GOM waters. The TEOS-10 program is available for free online at www.teos-10. org (McDougall and Barker 2011). The TEOS-10 equations are also built into the free Ocean Data View (ODV) program (Schlitzer 2017), which is commonly used for plotting and interpreting oceanographic data sets.

Physicochemical Properties of Gulf of Mexico Waters

The distribution of the physically and biologically important components of GOM waters, including oxygen and nutrient concentrations, are given in Figures 2.1 to 2.7, based on data from the second Gulf of Mexico East Coast Carbon cruise (GOMECC-2; see http://www.aoml.noaa.gov/ocd/gcc/GOMECC2/), the *World Ocean Atlas 2013* (*WOA13*; Garcia et al. 2013; Locarnini et al. 2013; Zweng et al. 2013), and the World Ocean Database (WOD13; Boyer et al. 2013). Such information can be used to examine changes in the physical and chemical properties of the GOM as a function of location and depth.

Practical Salinity

A profile based on all S_P data from the GOM acquired during GOMECC-2 and a section of S_P that spans the basin, utilizing all available data from WOD13, are shown in Figures 2.1a and 2.2. In the western portion of the basin (i.e., to the west of the Yucatan Peninsula), S_P is highest at the surface and gradually decreases with depth to a minimum at ~600–1100 m, which corresponds to Antarctic Intermediate Water (AAIW). However, the lower portion of AAIW (~800–1100 m) is commonly referred to as Caribbean Midwater (CMW) (Morrison and Merrell 1983), having an S_P <34.95 that distinguishes it from the overlying portion of AAIW. In the eastern basin, there is a subsurface S_P maximum (S_P >36.5) at ~100–250 m, corresponding to Subtropical Underwater (STUW), which is also referred to as either Caribbean, Loop Current, or Yucatan water (Shiller 1999). As a result of the Loop Current and STUW, waters with S_P values >35 pene-

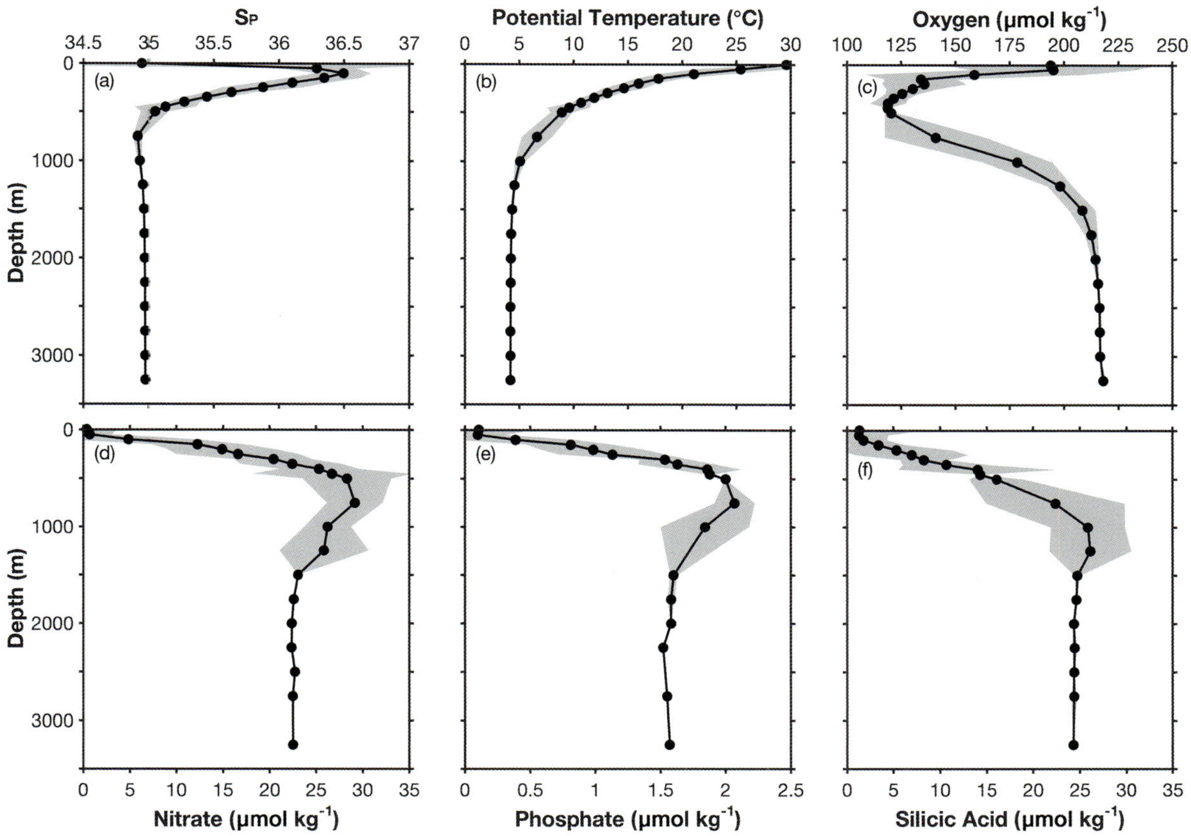

Figure 2.1a–f. Profiles of (*a*) S_P, (*b*) potential temperature, (*c*) oxygen, (*d*) nitrate, (*e*) soluble reactive phosphate, and (*f*) silicic acid. Mean values from GOMECC-2 data are plotted with shading representing 2 standard deviations, displayed in 50 m increments in the upper 500 m and 250 m increments below, following Margolin (2017).

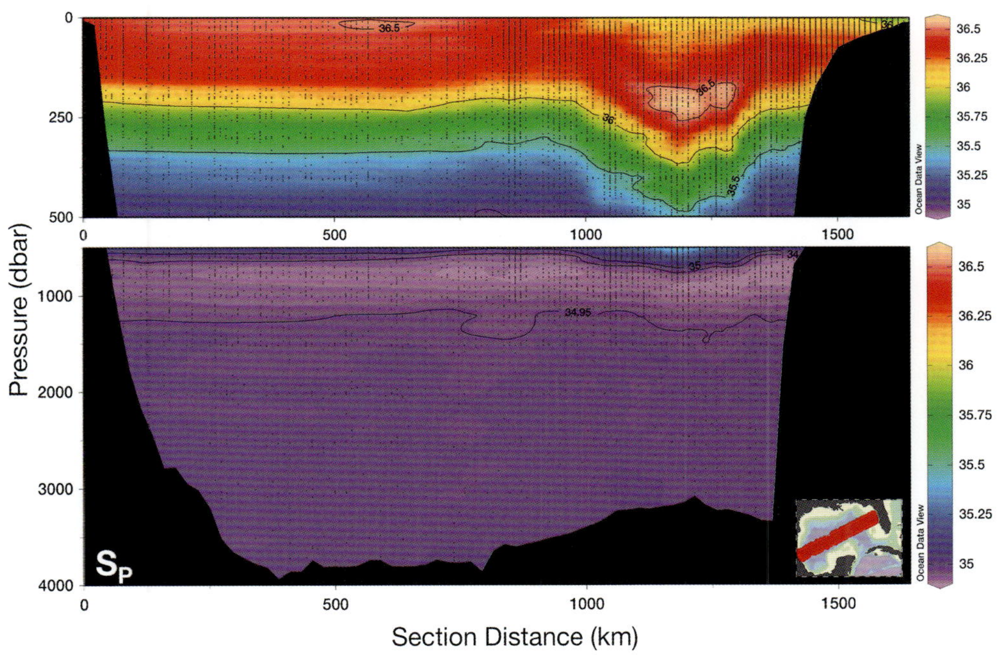

Figure 2.2. A basin-wide section of S_P, from WOD13, plotted using Ocean Data View (ODV). Upper panel is from 0 to 500 m, and lower panel is from 500 to 4000 m. Color maps of salinity (S_P, Practical Salinity Scale, no units) showing a range from 34.9 to 36.6, with contour lines every 0.5 from 35 to 37.5, and including 34.95.

trate to greater depths in the eastern basin than in the western basin, overlying the AAIW. Below the CMW, at depths greater than ~1100 m, the S_P of the GOM's deep water increases slightly from 34.95, having S_P values of ~34.96–34.98 below ~1500 m. The most saline deep GOM waters correspond to Upper North Atlantic Deep Water (UNADW), which enters the GOM via the Caribbean's Yucatan Basin. The Yucatan Basin's UNADW is the deepest water that enters the GOM over the Yucatan Channel sill, sinking as it enters the GOM because of its relatively high density.

Surface S_P values in the GOM range from as low as 0 in the Mississippi River delta, while coastal waters in the northern GOM have S_P values of ~25–30 (Fig. 2.3). Waters with S_P values <35 are mostly confined to the northern GOM in the summer, while they extend farther from the Mississippi River delta in the other seasons. The surface waters in the interior of the GOM, as well as the coastal waters south of ~25°N and along Florida's western coast, have S_P values of ~35–37, varying within that range throughout the year. The most saline waters are typically found in the Bay of Campeche and to its north, having S_P values >36.5; this feature is most pronounced in the summer. Seasonal variations in the Loop Current are not captured well by S_P, tending to have perennial S_P values of 36–36.5, like much of the basin.

To complement the *WOA13* climatologies of S_P, sea-surface S_P (SSS) values have recently been derived from measurements using radiometers and a scatterometer made onboard the Aquarius satellite (Lagerloef et al. 2015). The Aquarius satellite was launched in June 2011, suffering a power supply failure in June 2015 that caused the mission to end, resulting in a 3.75-year SSS data set that allowed for an understanding of S_P dynamics throughout the global ocean. However, the relatively coarse (1° × 1°) geographic resolution of Aquarius, which is not ideal for understanding small basins like the GOM, diminishes the utility of Aquarius measurements for the Gulf. Despite this shortcoming, Aquarius SSS measurements have been utilized to capture some of the dynam-

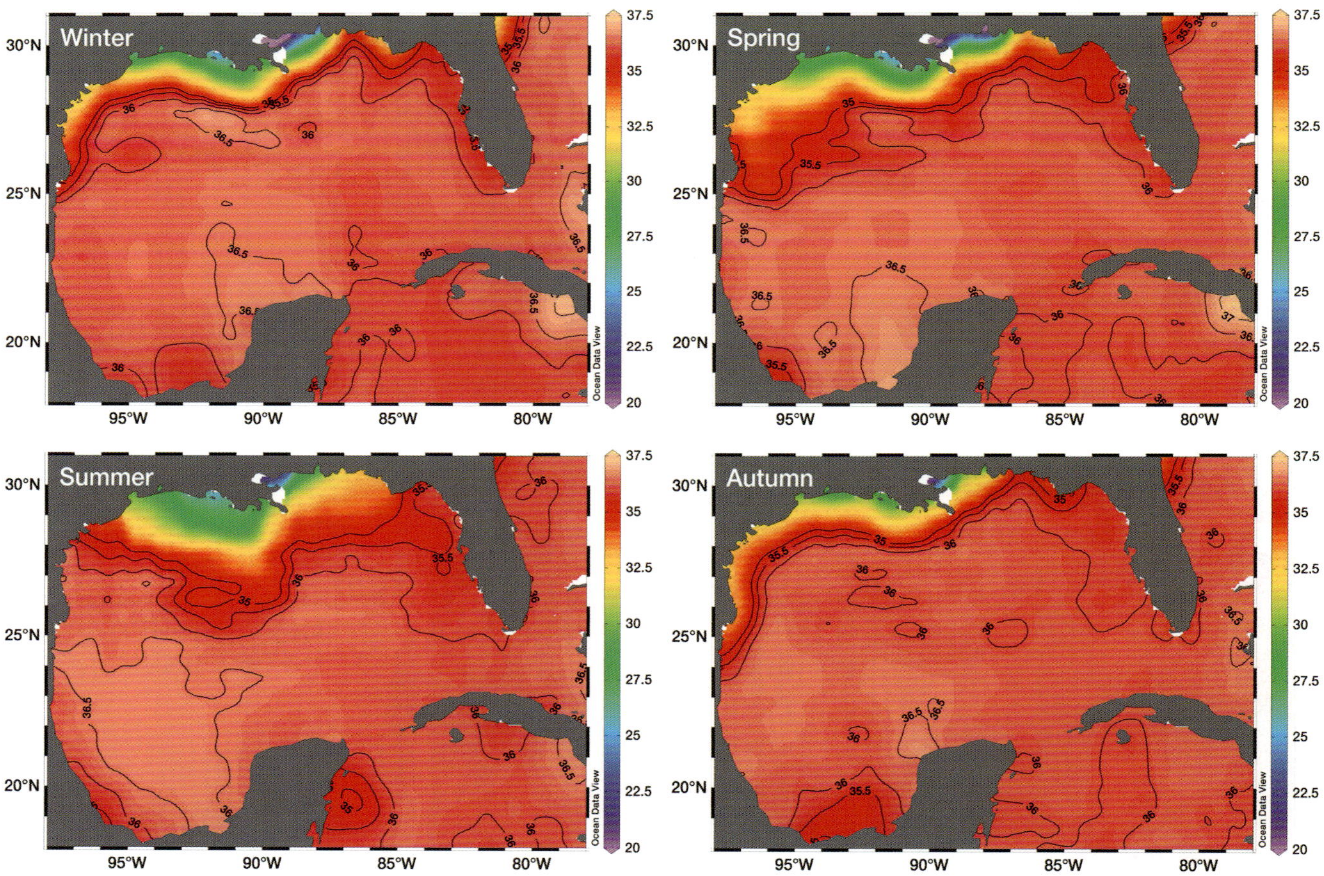

Figure 2.3. Seasonal surface salinities' (S_P) climatologies in winter, spring, summer, and autumn, from WOA13, plotted using ODV. Color maps range from 20 to 37.5, with contour lines every 0.5 from 35 to 37.5.

ics in the GOM, such as flooding events in the northern Gulf (Gierach et al. 2013). The SSS data are easily accessible via the National Aeronautics and Space Administration's (NASA) Ocean Color website (https://oceancolor.gsfc.nasa.gov).

Temperature

All temperature measurements reported since January 1990 should be on the International Temperature Scale (ITS-90) in °C, which was adopted by the International Committee for Weights and Measures in 1989 (McDougall and Barker 2011). Prior to ITS-90, the International Practical Temperature Scale (IPTS) was used, which was initially established in 1948 (IPTS-48) and later amended in 1968 (IPTS-68) (Barber 1969). If older data are utilized, conversion from IPTS-68 to ITS-90 can be done based on the following equation:

$$t_{90}\ (°C) = t_{68}/1.00024, \tag{2.8}$$

where t_{90} and t_{68} are temperature according to ITS-90 and IPTS-68, respectively (Saunders 1990). Whereas it is unlikely to obtain GOM temperature data from before 1948, IPTS-48 can be converted to ITS-90 based on the following equation:

$$t_{90}\ (°C) = [t_{48} - 4.4 \times 10^{-6} * t_{48}\ (100 - t_{48})/1.00024, \tag{2.9}$$

with t_{48} being the temperature according to IPTS-48 (Fofonoff and Bryden 1975; Saunders 1990).

In the interior of the GOM, as in any basin, in situ temperature measurements are affected by the increased pressure, notably at depths greater than ~1000 m, causing measured (in situ) temperatures to be warmer than if they were to be measured at atmospheric pressure (Millero 2013). For example, if seawater from the bottom of the GOM were to be brought to the surface adiabatically, the decrease in pressure would cause the water to cool by ~0.5 °C, while surface water brought to the bottom would conversely warm by ~0.5 °C because of pressure. To address these pressure-induced variations in temperature, potential temperature at a reference pressure of 0 dbar (θ) is commonly used in oceanographic sections, which is the temperature of the water if it moved adiabatically relative to its reference pressure, shown as a profile (Fig. 2.1b) and section spanning across the GOM (Fig. 2.4).

In the upper ~50 m, θ is 24–28 °C (Figs. 2.1b and 2.4), while the warmer waters of the Loop Current are distinguished by θ >28 °C. The Loop Current enters the GOM over the relatively deep Yucatan Channel sill at a depth of 2040 m, while it exits the basin via the Straits of Florida, having a restricting sill depth of 740 m (Rivas et al. 2005). In addition to the Loop Current bringing warm waters through the GOM at the surface, its influence can be observed at depths of ~350–500 m where θ exceeds 12 °C. At these depths, water originates from the Sargasso Sea with a temperature of 18 °C and is commonly referred

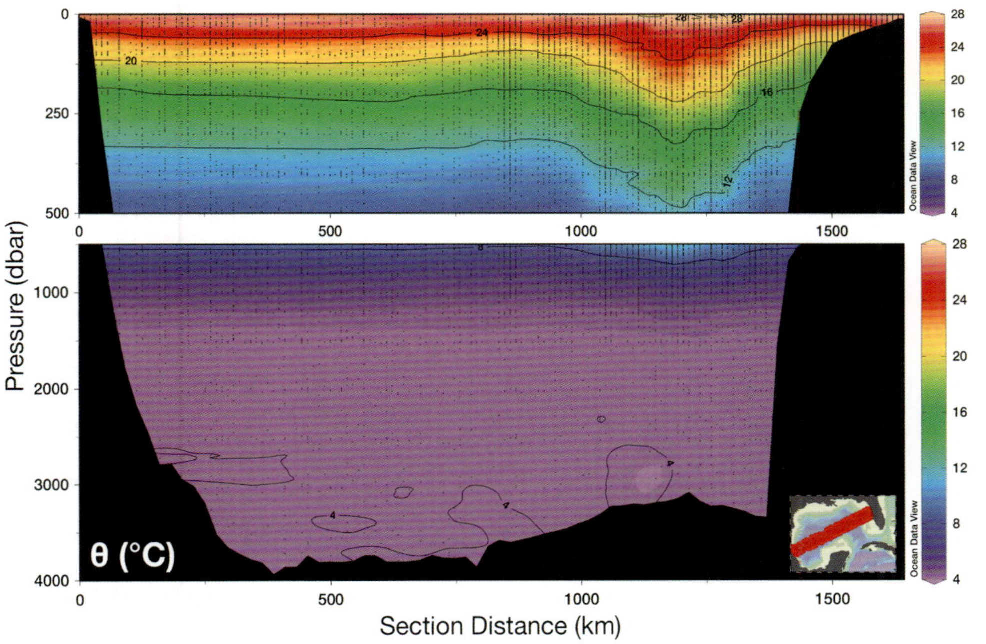

Figure 2.4. A basin-wide section of potential temperature at a reference pressure of 0 dbar (θ), from WOD13, plotted using ODV. Upper panel is from 0 to 500 m, and lower panel is from 500 to 4000 m. Color maps range from 4° C to 28° C, with contour lines every 4° C.

to as "18° Water" or Subtropical Mode Water (STMW) (Worthington 1959; Talley et al. 2011). Below 500 m, the influence of the Loop Current is still made apparent by waters having θ values warmer than 8 °C. The warm waters that characterize the Loop Current and inflowing Caribbean deep waters have been observed at depths of ~2000 m near the Yucatan Channel (DeHaan and Sturges 2008), although the deepest penetration of the warm Loop Current waters into the interior of the GOM is typically ~1000 m. The GOM's thermocline (i.e., depths where temperature decreases most rapidly) extends to a depth of ~1000 m through most of the basin, but the changes below 500 m are small, especially in the western basin where the influence of the Loop Current is minimal, if not absent. The GOM deep waters decrease from a θ of ~5 °C at a depth of ~1000 m to a minimum θ of ~4 °C below 1800 m. Johnson and Purkey (2009) identified subtle warming in deep basins of the Caribbean since the 1970s, which may also be occurring in the deep GOM but has not yet been investigated.

In the GOM surface waters, temperatures range from a minimum of ~15 °C in the winter to a maximum of ~30 °C in the summer (Fig. 2.5). The warmest waters tend to be in the southern basin, either in the Bay of Campeche to the west of the Yucatan Peninsula or to the north of the Yucatan Channel where the Loop Current brings warm, tropical waters from the Caribbean into this mostly subtropical basin. The coldest waters are usually in the northern GOM along the coast of the southern United States, since these waters are farthest from the equator. Nevertheless, all surface waters tend to have similar temperatures in the summer, ~28–29 °C across the entire basin.

In winter (January–March), the Loop Current brings warm waters through the Yucatan Channel into the GOM that feed into the Florida Loop Current, which exits the basin through the Straits of Florida—both currents are apparent by surface temperatures of 24–26 °C (Fig. 2.5). The Bay of Campeche also has temperatures >24 °C, but they are not as pronounced as the warm Loop and Florida

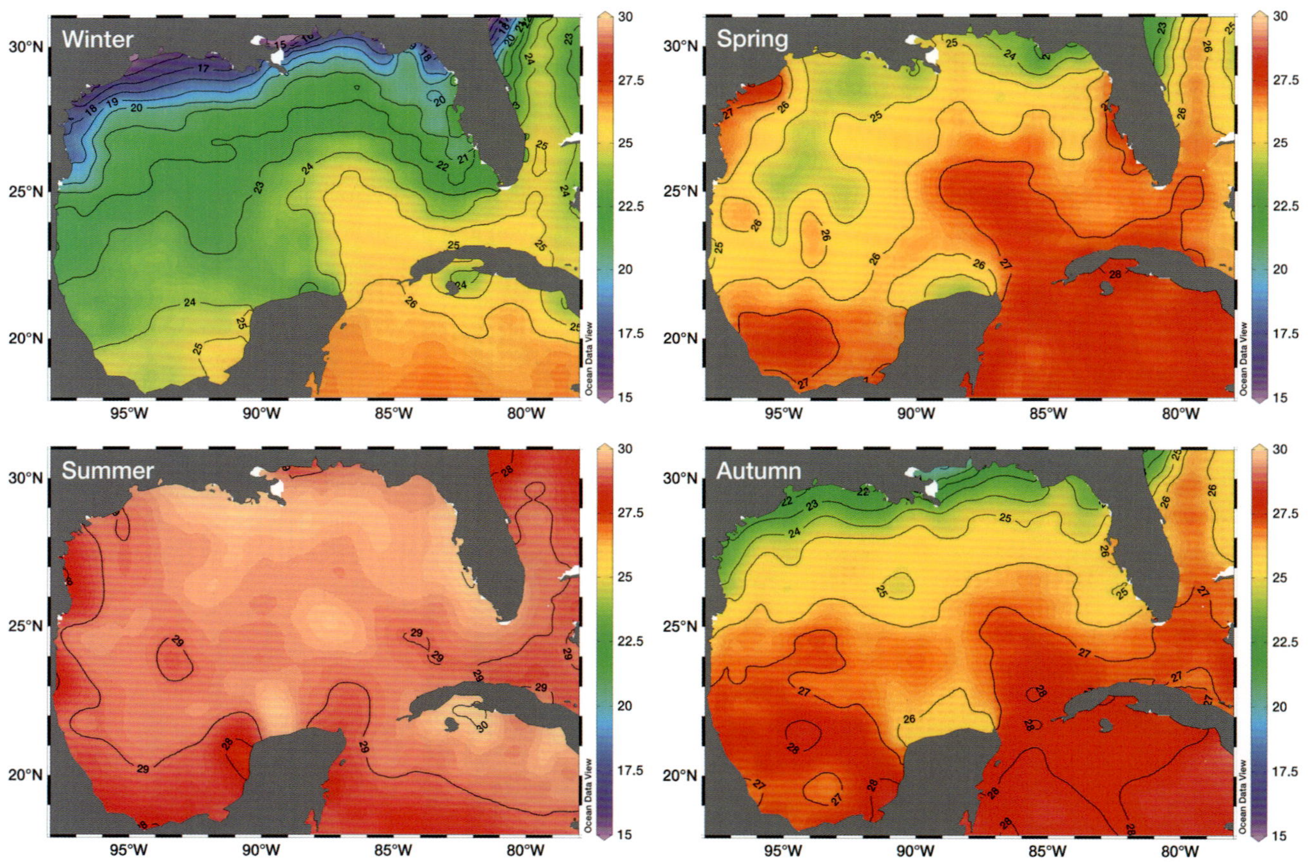

Figure 2.5. Seasonal surface-temperature climatologies in winter, spring, summer, and autumn, from WOA13, plotted using ODV. Color maps range from 15 °C to 30 °C, with contour lines every 1 °C.

Loop Current waters. The coldest waters follow the coast of the United States, having temperatures <20 °C that extend south to ~26°N near the Rio Grande delta on the GOM's western side and Tampa Bay on its eastern side. In the interior of the basin, the majority of the GOM's surface waters have temperatures of 20–24 °C.

In spring (April–June) and autumn (October–December), the Loop and Florida Loop Currents are still distinct from GOM surface waters, with surface temperatures >27 °C (Fig. 2.5). As the Florida Loop Current exits the basin through the Straits of Florida, it becomes slightly cooler, with temperatures of 26–27 °C. Like the Loop Current waters, the Bay of Campeche has surface temperatures that exceed 27 °C in the spring and autumn, while temperatures are also >27 °C along the coast of Texas in the spring. The warm Bay of Campeche surface waters extend to higher latitudes in the autumn than in the spring, reaching a maximum latitude of ~25°N. The temperatures of the remaining surface waters during these seasons are between 24 °C and ~27 °C, with exceptions along Florida's Panhandle in the spring and along the coastal United States in the autumn where temperatures are lower (Fig. 2.5).

In summer (July–September), the warm surface temperatures that distinguish the Loop Current from the interior GOM waters are lost, since surface waters for the entire basin are >28 °C (Fig. 2.5). The Mexican coastal waters, along with southern Texas waters, have temperatures <29 °C, while the remaining waters have temperatures of 29–30 °C. The waters that leave the GOM with the Florida Loop Current in the summer are slightly warmer than the Atlantic surface waters at temperatures >28 °C.

In addition to the *WOA13* climatologies as an important resource for understanding seasonal variations in temperature at the sea surface, some satellites determine SST based on infrared measurements, which can be utilized to understand GOM temperature variations on short time scales (i.e., daily, weekly) and fairly small spatial (4 km^2 resolution) scales. Satellite SST measurements began in the late 1960s, providing more accurate and reliable temperature measurements than conventional ship-based measurements (Krishna Rao et al. 1972; Kilpatrick et al. 2015). Currently, SST is determined by the Terra and Aqua satellites, which are equipped with the Moderate Resolution Imaging Spectroradiometer (MODIS) measurements that began in 2000 and 2002, respectively. The Suomi National Polar-Orbiting Partnership (SNPP) spacecraft, equipped with the Visible Infrared Imaging Radiometer Suite (VIIRS) and having begun measurements more recently in 2012, also determines SST. Like SSS, the SST data are easily accessible via NASA's Ocean Color website.

Density

The density (ρ) of seawater is primarily a function of temperature, S_P, and pressure. Colder and saltier waters have a greater ρ than warmer and fresher waters, and increasing pressure also increases the ρ. Given that temperature is a determinant factor in ρ, the densest waters in the world ocean tend to be formed at high latitudes, such as in the far North Atlantic and Arctic and Antarctic seas. The waters formed in these regions tend to be subsequently subducted underneath the warmer upper layers found at lower latitudes. Largely controlled by temperature, the densest water in any basin is found at its bottom or on features such as shelves, with ρ decreasing toward the surface.

Dilution of seawater caused by river runoff, precipitation, and melting ice decreases S_P and ρ, while evaporation and brine rejection increase S_P, and thus also increase ρ. The ρ of pure water is 1000 kg m^{-3} at 4 °C and 1 atm. Seawater is denser due to its solute content, typically having a ρ range of ~1020–1029 kg m^{-3} (Key et al. 2015). Given that the ρ range of seawater is 2 orders of magnitude smaller than the overall ρ of pure water, oceanographers conventionally express the density of seawater relative to that of pure water. This difference, referred to as the density anomaly (σ), is calculated as follows:

$$\sigma\,(\text{kg m}^{-3}) = \rho - 1000, \qquad (2.10)$$

resulting in typical marine waters having a σ range of ~20–30 kg m^{-3}.

To account for the effect that pressure (or depth) has on temperature, and thus σ, the potential density anomaly (σ_θ) at a reference pressure of 0 dbar is considered. As depth increases in the water column, so does pressure as the weight of the overlying water increases, and for every 10 m of depth increase there is an associated increase in pressure of ~1 bar (i.e., 1 m $\approx$ 1 dbar). As pressure increases, the parcel of water under consideration is compressed, which results in an increase in ρ. Differences between σ and σ_θ for typical GOM waters are as high as 0.02 kg m^{-3} at depths of ~3800 m, while the deepest trenches of the ocean can have σ and σ_θ differences as large as ~0.1 kg m^{-3} (Millero 2010, 2013).

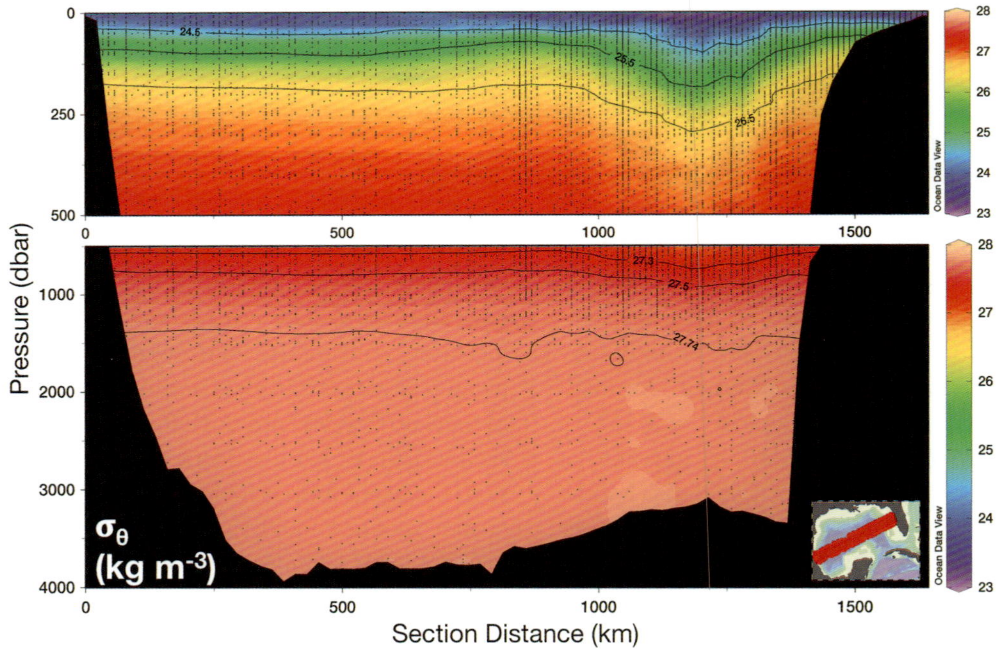

Figure 2.6. A basin-wide section of σ_θ, calculated from WOD13 temperature and S_P data using TEOS-10, plotted using ODV. Upper panel is from 0 to 500 m, and lower panel is from 500 to 4000 m. Color maps range from 23 to 28 kg m^{-3}, with contour lines representing surface water (<24.5 kg m^{-3}), STUW (~25.5 kg m^{-3}), GMMW (26.5–27.3 kg m^{-3}), AAIW (27.3–27.5 kg m^{-3}), CMW (27.5–27.74 kg m^{-3}), and UNADW (>27.74 kg m^{-3}).

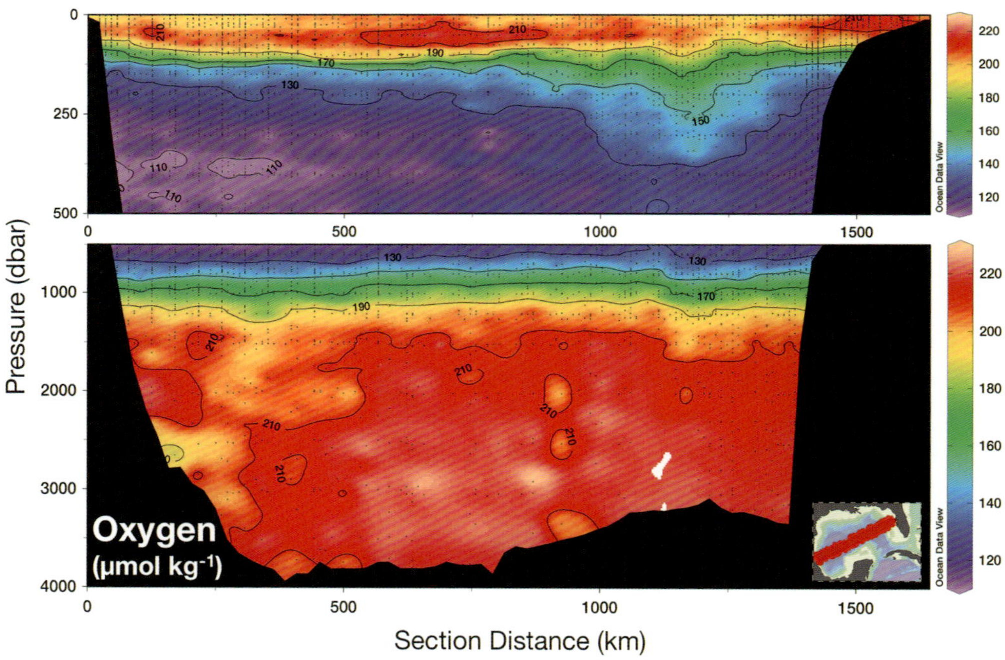

Figure 2.7. A basin-wide section of oxygen, from WOD13, plotted using ODV. Upper panel is from 0 to 500 m, and lower panel is from 500 to 4000 m. Color maps range from 110 to 230 μmol kg^{-1}, with contour lines every 20 μmol kg^{-1} from 110 to 210 μmol kg^{-1}.

In oceanography, σ_θ is commonly used to describe hydrographic layers in the water column, complementing layer distinctions on the basis of S_P and θ. A cross-basin σ_θ section for the GOM is shown in Figure 2.6. For GOM surface waters, $\sigma_\theta < 24.5$ kg m^{-3} (Schott et al. 1998; Smith 2010), while $\sigma_\theta \approx 25.5$ kg m^{-3} in the underlying STUW between ~100 and 250 m (Talley et al. 2011). The STMW has a $\sigma_\theta = 26.5$ kg m^{-3} between ~200 and 400 m, while Subpolar Mode Water is slightly denser ($\sigma_\theta = 27.2$ kg m^{-3}) (Shiller 1999; Talley et al. 2011); here, these two water masses are considered together, collectively referred to as Gulf of Mexico Mode Water (GMMW) between $\sigma_\theta = 26.5$ and 27.3 kg m^{-3}. Directly below the GMMW is AAIW, ranging from $\sigma_\theta = 27.3$ to 27.74 kg m^{-3} near ~600–1100 m (Talley et al. 2011), with its lower portion ($\sigma_\theta = 27.5$–27.74 kg m^{-3} near ~800–1100 m) referred to as CMW at ~800–1100 m (Morrison et al. 1983; Shiller 1999). Like S_P, greater σ_θ water penetrates to greater depths in the eastern basin than in the western basin as a result of the Loop Current and deeper STUW. Below the CMW at depths >1400 m are the densest deep GOM waters, which correspond to UNADW that enters the GOM via the Caribbean's Yucatan Basin. The Yucatan Basin's UNADW is the deepest water carried over the Yucatan Channel sill, and it sinks as it enters the GOM because of its relatively high density.

Oxygen

Oxygen concentrations are high near the sea surface, where they exceed 190 μmol kg^{-1} (Figs. 2.1c and 2.7) because of air-sea gas exchange driving surface waters to saturation, and sustained high primary productivity causes these waters to become supersaturated. Nevertheless, the highest upper-water-column concentrations (~210 μmol kg^{-1}) are typically found at a depth of ~50 m, corresponding to the subsurface chlorophyll maximum near the base of the euphotic zone. Below this oxygen maximum in the western basin, concentrations decrease with depth to <130 μmol kg^{-1} between depths of ~200 and 650 m, and as low as ~110 μmol kg^{-1} at the minimum's core. Below the oxygen minimum at ~500 m, concentrations steadily increase to ~200 μmol kg^{-1} at a depth of ~1250 m. For basin waters below ~1500 m, oxygen concentrations are generally >210 μmol kg^{-1} but are slightly lower than this in the western basin.

Nutrients

Nutrient concentrations are lowest, if not undetectable, at either the surface or the subsurface chlorophyll maximum at a depth of ~50 m (Fig. 2.1d–f), where nutrients are utilized by phytoplankton. Below the euphotic zone, nutrient concentrations increase with depth, with nitrate and soluble reactive phosphate (SRP, or simply "phosphate") maxima found at the same intermediate depths and a silicic acid maximum observed slightly deeper. At depths greater than ~1500 m, nutrient concentrations are nearly constant. A more extensive consideration of nutrient distributions and dynamics is presented in Chapter 3.

The maximum concentrations of nitrate (~28 μmol kg^{-1}) and phosphate (~2.2 μmol kg^{-1}) in the GOM are found in the AAIW at depths of ~600–800 m (Fig. 2.1d–e; Morrison et al. 1983). Below this nutrient maximum, nitrate and phosphate concentrations decrease across the CMW by ~5.5 and ~0.7 μmol kg^{-1}, respectively, having concentrations of ~22.5 and ~1.5 μmol kg^{-1} in the UNADW below ~1500 m. In the GOM's deep waters, the nitrate-to-phosphate ratio is ~13–14, similar to the ratio found in oceanic deep waters (Anderson and Sarmiento 1994).

Similar to nitrate and phosphate, silicic acid increases with depth in the upper water column (Fig. 2.1f) but reaches its maximum (~24.5 μmol kg^{-1}) at slightly greater depths (~950 m), corresponding to the CMW (Morrison et al. 1983). Unlike nitrate and phosphate, silicic acid concentrations do not change notably below their intermediate depth maximum, which is ~24.5 μmol kg^{-1} throughout the bottom ~2000 m of the water column. In the deep waters of the western basin (i.e., west of the Yucatan Peninsula), silicic acid concentrations are ~3–4 μmol kg^{-1} higher than in eastern basin deep waters, indicating that the western basin has a longer deepwater residence time than the eastern basin (Carder et al. 1977).

The Carbonate System in Seawater

Definitions

The 5 measurable carbonate system parameters are pH, DIC, TA, carbonate ion concentrations, and either the CO_2 partial pressure (pCO_2) or fugacity (fCO_2). The latter two are quantitatively very similar, with fugacity the preferred thermodynamic parameter. If any 2 of these carbonate system parameters are measured, all other carbonate system parameters can be calculated based on thermodynamic relationships that have been experimentally determined and incorporated into computer programs such as C02SYS (Lewis and Wallace 1998; Pierrot et al. 2006; van Heuven et al. 2011). An overview of the carbonate system in seawater is provided here, based on the 4 commonly measured parameters, pH, TA, DIC, and pCO_2/fCO_2. Since methane is not a part of the carbonate system, it is not discussed in this chapter; however it is considered in Chapters 4 and 7 in relation to trace metals and sediments, respectively. For in-depth descriptions of the marine carbonate system, there are numerous excellent resources for understanding the carbonate system and marine carbon cycle (e.g., Zeebe and Wolf-Gladrow 2001; Emerson and Hedges 2008; Millero 2013).

The 4 dissolved species of inorganic carbon in seawater are bicarbonate (HCO_3^-), carbonate (CO_3^{2-}), aqueous carbon dioxide ($CO_{2(aq)}$), and carbonic acid (H_2CO_3), with the latter two often being combined and denoted as CO_2^* because H_2CO_3 concentrations are ~0.1% of $CO_{2(aq)}$ (Zeebe and Wolf-Gladrow 2001). The relative abundances of these carbon species vary as a function of pH, which is illustrated in Figure 2.8. As more CO_2 is introduced to seawater, pH will decrease, along with $[CO_3^{2}]$ relative to that CO_2, making seawater less favorable to the proliferation of healthy populations of calcifying organisms. For seawater with $S_P = 35$ and t = 25 °C, HCO_3^- is the dominant species with the natural range of pH conditions. For pH >~7.4, CO_3^{2-} is the second most abundant inorganic carbon species. Thus, CO_2^* is the least abundant inorganic carbon species under most conditions, becoming more abundant than CO_3^{2-} under hypoxic conditions that are conducive to low pH. The pH values at which $[CO_2^*] = [HCO_3^-]$ and $[HCO_3^-] = [CO_3^{2-}]$ are $pK_1 \approx 5.85$ and $pK_2 \approx 8.9$, respectively; these values are logarithmic-scale dissociation constants of carbonic acid and bicarbonate at 25 °C.

The first commonly measured carbonate system parameter is pH, which is typically reported on either the total scale (pH_T) or seawater scale (pH_{SWS}). These scales are defined as

$$pH_T = -\log([H^+] + [HSO_4^-]) = -\log[H^+]_T \quad (2.11)$$

and

$$pH_{SWS} = -\log([H^+] + [HSO_4^-] + [HF]) = -\log[H^+]_{SWS}. \quad (2.12)$$

At $S_P = 35$ and t = 25 °C, pH_T is 0.01 lower than pH_{SWS}. Although 0.01 may appear to be a subtle difference, when considering pH in the context of climate change on decadal timescales, precisions of 0.001 or better are required. While the distinction between pH scales is important, for the sake of simplicity, pH is discussed here generally, without specifying between different scales. Since pH is affected by temperature, it is common for the temperature at which it was measured (typically 25 °C) to be reported. In the GOM, pH ranges from ~7.6 to 8.2 at 25 °C (Fig. 2.9a), decreasing from ~8.2 near the surface to a minimum of ~7.6 at a depth of ~500–750 m, corresponding to AAIW, while deep waters maintain a slightly higher pH of ~7.7. The pH range

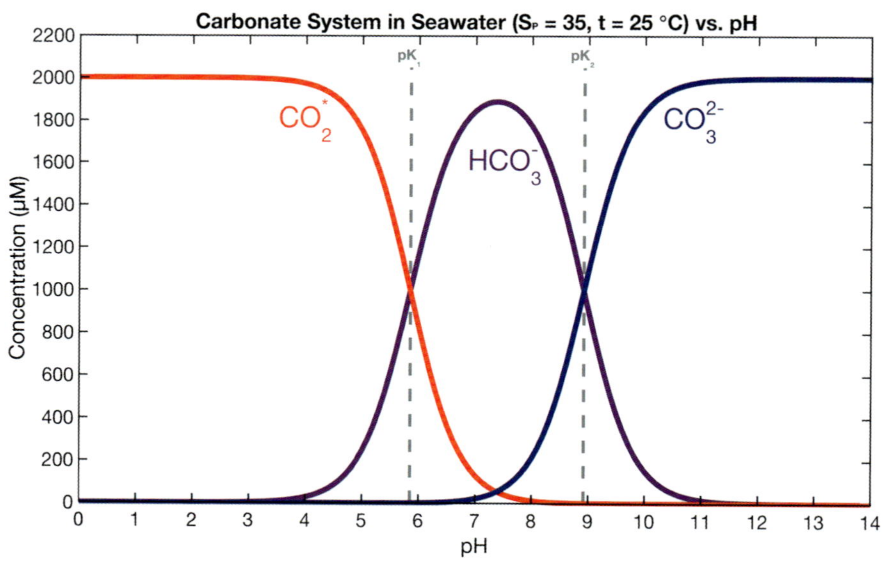

Figure 2.8. Bjerrum diagram representing the carbonate system in seawater at $S_P = 35$ and temperature = 25 °C. Colors represent CO_2^* (*red*), CO_3^{2-} (*blue*), and HCO_3^- (*purple*), while gray dashed lines represent the 2 dissociation constants for carbonic acid (pK_1 and pK_2).

in the GOM, and throughout most of the world ocean, lies between the values of pK_1 and pK_2.

The second commonly measured carbonate system parameter is DIC, which is represented in Figure 2.8 as the sum of the HCO_3^-, CO_3^{2-}, and CO_2^* concentrations. Since the pH of GOM seawater ranges from ~7.6 to 8.2 (Fig. 2.9a), speciation calculations suggest that its DIC comprises ~84–94% HCO_3^-, ~5–16% CO_3^{2-}, and <1–2% CO_2^*. However, the speciation of DIC can differ from calculated values, so measuring carbonate ion concentrations directly is ideal when considering DIC composition. In the GOM, DIC concentrations range from ~2000 μmol kg^{-1} near the surface to ~2200 μmol kg^{-1} at a depth of ~750 m (Fig. 2.9b), while concentrations are slightly lower in the deep basin (~2190 μmol kg^{-1}).

The third commonly measured carbonate system parameter is TA, which is the sum of the concentration of bases that accept hydrogen cations (H^+) when the sample is titrated with HCl and until HCO_3^- is converted to H_2CO_3. As defined by Dickson (1981), TA can be written as follows:

$$\begin{aligned} TA = {} & [HCO_3^-] + 2[CO_3^{2-}] + [B(OH)_4^-] \\ & + [OH^-] + [HPO_4^{2-}] + 2[PO_4^{3-}] - [SiO(OH)_3^-] \\ & + [HS^-] + 2[S^{2-}] + [NH_3] - [H^+] - [HSO_4^-] \\ & - [HF] - [H_3PO_4], \end{aligned} \tag{2.13}$$

while carbonate alkalinity (CA) is defined as

$$CA = [HCO_3^-] + 2[CO_3^{2-}]. \tag{2.14}$$

Of all the chemical species that make up TA, HCO_3^-, CO_3^{2-}, and $B(OH)_4^-$ are the most abundant, making them the most important constituents of TA. However, in anoxic waters, and perhaps hypoxic waters such as those found on the Louisiana shelf during eutrophication events, NH_3, HS^-, and S^{2-} can contribute notably to

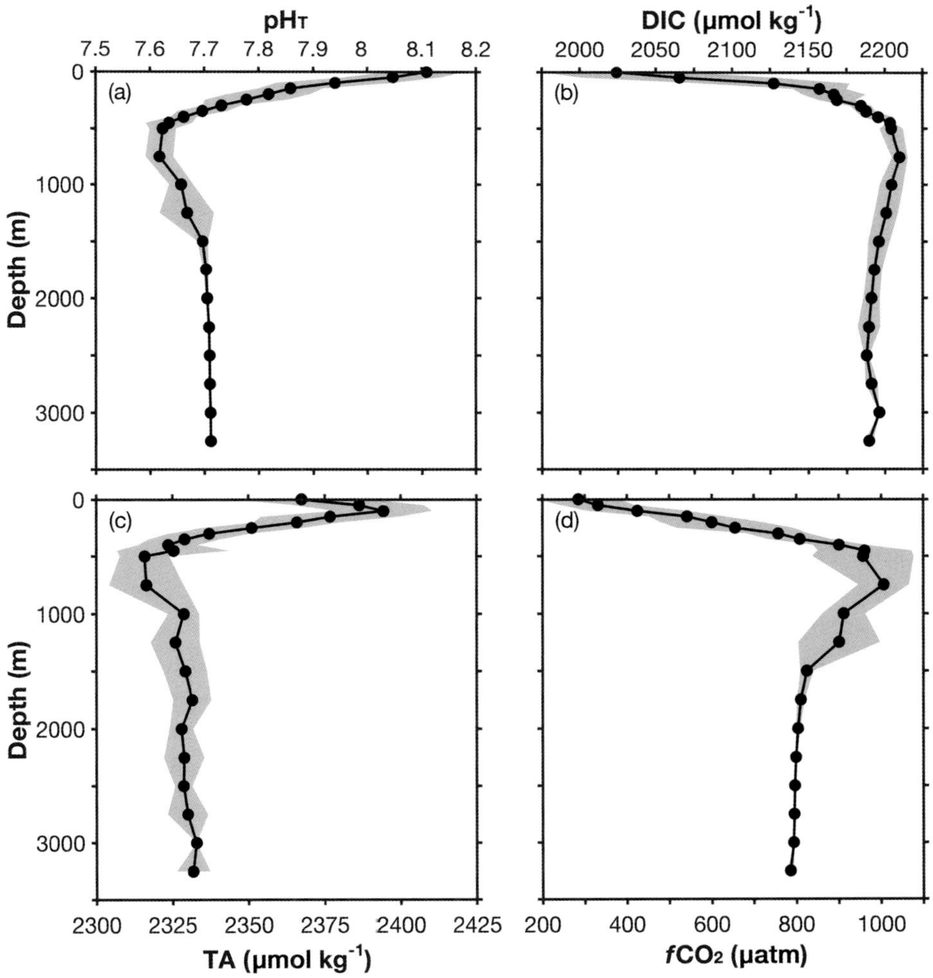

Figure 2.9a–d. Profiles of (*a*) pH_T, (*b*) DIC, (*c*) TA, and (*d*) fCO_2. Mean values from GOMECC-2 data are plotted with shading representing 2 standard deviations, displayed in 50 m increments in the upper 500 m and 250 m increments below, following Margolin (2017).

TA. In the GOM, TA concentrations range from ~2300 to 2425 μmol kg^{-1} (Fig. 2.9c), being highest just below the surface, with minimum values of 2305 μmol kg^{-1} at depths between ~500 and 750 m in the AAIW and slightly higher (~2330 μmol kg^{-1}) in the underlying CMW and UNADW.

The fourth commonly measured carbonate system parameter is pCO_2/fCO_2, with pCO_2 defined as the partial pressure of CO_2 in a gas headspace that is equilibrated with a seawater sample, which is typically reported in μatm at a temperature of 20 °C. The fCO_2 is similar to pCO_2 and can be calculated from either pCO_2 or CO_2^* plus the dependence of CO_2^* solubility on temperature and salinity. The numerical distinction between pCO_2 and fCO_2 is on the order of only ~1 μatm. The rate of CO_2 exchange across the air-sea interface is critically dependent on the difference between the fCO_2 of the atmosphere and the fCO_2 of surface water. Thus, fCO_2 provides valuable information on the air-sea exchange of this greenhouse gas. The fCO_2 value of surface waters tends to be close to equilibrium with the atmosphere; however, it may deviate from equilibrium because of variations in temperature, photosynthesis, and respiration, as well as localized point sources. In the GOM, the fCO_2 distribution generally includes values that are ~300 μatm at the surface, increasing to a maximum near ~1000 μatm between ~500 and 750 m and declining to ~800 μatm at depth (Fig. 2.9d).

Considering the 4 commonly measured carbonate system parameters, TA is similar to DIC in that both are concentration measurements and independent of temperature and pressure (pH and pCO_2/fCO_2 are not). Perhaps for this reason, DIC and TA are the 2 carbonate system parameters that were measured during the World Ocean Circulation Experiment (WOCE) in the 1990s. Today, the Global Ocean Ship-Based Hydrographic Investigations Program (GO-SHIP) usually includes measurements of at least 3 of the carbonate system parameters so the internal consistency of the measurements can be examined. The 3 most common carbonate system parameters measured in discrete samples are pH, DIC, and TA, while pCO_2/fCO_2 of surface waters is commonly measured via the automated underway system, which most research vessels are equipped with.

Application and Dynamics

Exchanges of CO_2 across the air-sea interface can occur by 2 mechanisms. The first, a physical process referred to as the solubility pump, occurs because of differences in fCO_2 between the atmosphere and the sea surface. The second is referred to as the biological pump, in which $CO_{2(aq)}$ is assimilated by phytoplankton for photosynthesis and organic matter production. The organic detritus from the fixed inorganic carbon eventually sinks below the mixed layer of the water column, decreasing the surface ocean fCO_2 and allowing more atmospheric CO_2 to cross the air-sea interface (Ducklow et al. 2001).

As the ocean absorbs more anthropogenic CO_2 from the atmosphere, the pH and carbonate ion concentration (CO_3^{2-}) of seawater will decrease in a process referred to as ocean acidification, which is problematic for calcifying organisms, such as coccolithophores, foraminifera, and reef corals (for a review of this process, see Doney et al. 2009). One method of quantifying the effect of ocean acidification on organisms is by determining the calcium carbonate ($CaCO_3$) saturation state (Ω). Saturation states are determined based on the product of the concentrations of Ca^{2+} and CO_3^{2-} ions relative to the stoichiometric solubility product (K_{sp}) of the mineral of interest. The latter is temperature, salinity, and pressure dependent, and Ω is obtained by the following equation:

$$\Omega = [Ca^{2+}][CO_3^{2-}]/K_{sp}^{*}. \tag{2.15}$$

The Ω can be determined for various forms of $CaCO_3$, including its 2 most common polymorphs, calcite and aragonite, the latter being a metastable form that is 50% more soluble than calcite. Thus, aragonitic organisms are more susceptible than calcitic organisms to the threat of ocean acidification. With respect to $CaCO_3$, $\Omega > 1$ indicates seawater supersaturation, $\Omega = 1$ indicates saturation, and $\Omega < 1$ indicates undersaturation and susceptibility to $CaCO_3$ dissolution. As the ocean acidifies, CO_3^{2-} and Ω decrease, and surface waters in some regions of the world ocean, particularly at high latitudes (Orr et al. 2005; Feely et al. 2009), will soon be progressively undersaturated; $CaCO_3$ dissolution is already occurring in some oceanic regions and having harmful effects on $CaCO_3$-secreting organisms (Feely et al. 2002; Doney et al. 2009). Like the 5 measurable carbonate system parameters, if 2 are measured, Ω can be calculated for either calcite or aragonite using computer programs such as C02SYS (Lewis and Wallace 1998; Pierrot et al. 2006; van Heuven et al. 2011).

Based on data from the 2012 GOMECC-2 cruise and its 2007 predecessor, GOMECC-1, Wanninkhof et al. (2015) calculated and compared Ω values for aragonite (Ω_{ar}) over the 5-year period. They identified slight decreases in Ω_{ar} in the surface mixed layer of the coastal waters; however, regional and small-scale variability overshadowed many of the changes caused by anthropogenic uptake. The carbonate system was also investigated in the northern GOM in reference to the health of scleractinian (i.e., stony) cold-water coral reefs, and Ω_{ar} of the regional waters was found to be near—yet above—the saturation threshold ($\Omega_{ar} > 1$) (Lunden et al. 2013; Georgian et al. 2016).

Wang et al. (2013) described the carbonate system in the GOM based on 3 cross sections in the US shelf waters and concluded that TA behaves approximately conservatively with respect to S_P, consistent with Yang et al. (2015). However, Cai et al. (2010) showed that alkalinity input from the Mississippi (TA ≈ 2400 μmol kg^{-1}) and Atchafalaya (TA ≈ 2000 μmol kg^{-1}) causes deviations in the otherwise linear TA-S_P relationship. Since TA is nearly conservative, it can be estimated from S_P, which Lee et al. (2006) did for the global ocean. Expanding on the work of Lee et al. (2006), Fine et al. (2017) applied a similar approach to model TA distributions throughout the global ocean by utilizing Aquarius SSS and National Oceanic and Atmospheric Administration (NOAA) SST data (Reynolds et al. 2002; Lagerloef et al. 2015), resulting in surface TA concentrations with highly resolved temporal (~daily) and spatial (1° × 1°) coverage. If a new Aquarius satellite were to be launched and measurement resolution was improved to ~4 km^2 (similar to MODIS SST), it would have great utility for future GOM studies that consider the marine carbonate system.

Wang et al. (2013) also considered in addition to TA the TA to DIC ratio, which showed dominant signs of biological CO_2 uptake in 2007. However, Wanninkhof et al. (2015) followed up on their work and found that much of the GOM's carbonate system variability results from evaporation and precipitation/river input rather than biological processes. Because of the important role of rivers on the carbonate system in the GOM, Cai et al. (2015) considered the Mississippi River discharge of inorganic carbon and found that it delivers 13.6 Tg C yr^{-1} as DIC to the interior of the GOM, which has increased as a result of anthropogenic-driven climate change in the 21st century (Ren et al. 2015).

The role of shelf waters, including the Mississippi River plume, as CO_2 sources/sinks to/from the atmosphere varies seasonally. In the spring and early summer, shelf waters tend to be net sinks for atmospheric CO_2 and sources in late summer, and they can vary between sources and sinks in the autumn and winter (Benway and Coble 2014; Huang et al. 2015). In general, river plumes act as CO_2 sources on the shelf, which are balanced by photosynthesis exceeding respiration in the highly productive coastal waters (Benway and Coble 2014). Preliminary results suggest that the GOM acts as a net sink for CO_2, sequestering ~3.6 Tg C from the atmosphere annually (Benway and Coble 2014). However, much of the CO_2 data in the GOM are limited to shelf waters in the northern basin, and more extensive sampling is required to adequately constrain the GOM's carbon budget, especially for the deep waters (Coble et al. 2010; Benway and Coble 2014). The deep waters of the GOM are assumed to contain little anthropogenic carbon due to the lack of active overturning circulation (Lee et al. 2011); however, few data exist that can be used to quantify the anthropogenic carbon in the deep basin.

Overall, understanding the carbonate system in the GOM is important for managing climate change, as variations to this system result in release of CO_2 into or stored from the atmosphere, respectively amplifying or dampening the greenhouse warming effect.

Dynamics on the Continental Shelf

The surface area of the GOM is ~1.6 million km^2, with roughly one-third of this area made up of waters that reside on the North American continental shelf. The Yucatan shelf and west Florida shelf are the most expansive shelf regions within the GOM. The basin's northern shelf to the south of the United States (including coastal waters of Texas, Louisiana, Mississippi, Alabama, and the Florida Panhandle) is broader than the basin's western shelf to the east of Mexico and Texas. The northern and western shelves of the basin both have gradual continental slopes, while the WFS and Campeche Bank to the northwest of the Yucatan Peninsula have relatively steep continental slopes. Understanding the dynamics on these contrasting shelf and slope environments is important for characterizing the physicochemical properties of GOM seawater, as these regions make up nearly half of the basin's surface area.

The Louisiana shelf is perhaps the most dynamic shelf system in the GOM due to massive amounts of freshwater that are delivered to this region via the Mississippi-Atchafalaya River system (MARS). The Mississippi River discharge alone represents nearly half of the river water that is released into the GOM. This discharge combines with that from the Atchafalaya, so MARS delivers >65% of the river water to the basin (see Chapter 6 for details on river discharge). The remaining ~35% of river discharge that enters the GOM is delivered by 40 or more Mexican and American rivers, the largest of which is Mexico's Usumacinta River, responsible for discharging ~6% of the total river water that enters the basin. Overall, rivers deliver ~1100 $km^3\ yr^{-1}$ of freshwater to the GOM, equivalent to all the seawater contained in the GOM's top ~0.7 m.

River discharge alters the physicochemical properties of seawater, such as near the Louisiana shelf where S_P seasonally varies because of fluctuations in river input from MARS (see Fig. 2.3). Discharge from MARS can also alter the physicochemical properties of seawater in the interior of the basin, such as occurred following a 2011 freshening event that led to SSS decreases of ~2 off the shelf as a result of surface currents transporting low-S_P waters away from the Louisiana shelf (Gierach et al. 2013). In the vicinity of the Louisiana shelf, S_P typically ranges from ~5 to 37, depending on sampling depth, location, and seasonal variations in river discharge (Cardona et al. 2016; Zhou et al. 2016).

In estuaries of the GOM, the S_P can reach as low as ~0, depending on proximity to the riverine sources as well as the size and number of sources. River water near MARS sampled by Zhou et al. (2016) had a S_P of 0.1, consistent with the typical salt concentrations found in rivers: <200 mg kg^{-1}, equivalent to <0.2 (Millero 2013). In river water, the ratios of SO_4^{2-}, Mg^{2+}, Ca^{2+}, K^+, and HCO_3^- to Cl^- are typically greater than in standard seawater listed in Table 2.1 (Millero 2013); however, this should not affect the calculation of S_P notably. Since the extension to lower S_P by Hill et al. (1986) is included in the TEOS-10 calculation of S_P from conductivity, estuarine and riverine samples with S_P <2 can be considered (McDougall and Barker 2011), allowing comparison with marine samples from the open GOM. Rather than use S_P calculated from CTD sensor software packages for samples with S_P <2, which is conventional for purely marine samples that are within the 2–42 range, it is recommended that S_P be manually calculated from conductivity measurements using TEOS-10.

The numerous estuaries in the GOM are discussed in detail by Bianchi et al. (1999), with a special focus on their biogeochemistry, as well as in this book, focusing on nutrients (Chapter 3), trace metals (Chapter 4), radioactive nuclides (Chapter 5), and organic matter (Chapter 6). In regard to the physicochemical properties of seawater, estuarine discharge generates unique and often dramatic gradients in, for example, S_P, which is affected by not only river runoff but also by tidal currents and local circulation patterns driven by coastal boundaries. Estuaries typically have lower S_P at the surface and higher values at depth. Nevertheless, this is not always the case. Pickard and Emery (1982) classically defined estuaries according to major types, including positive and negative estuaries, which are distinguished by their vertical salinity gradients (i.e., stratification). Relative to the neighboring sea, positive estuaries are fresher while negative estuaries are more saline. For the most part, GOM estuaries are positive, having fresher waters than the open GOM due to precipitation and river runoff exceeding evaporation. If an estuary is vertically mixed, its salinity will not vary with depth but will increase from the head of the river toward the mouth of the estuary (Fig. 2.10a), which is typical for estuaries that have a net seaward flow. Stratified estuaries with higher salinity at depth than near the surface have a net seaward flow in the upper layer and an inland flow in the lower layer; the salinity difference between the upper and lower layers is small for slightly stratified estuaries (Fig. 2.10b), while it is large for those that are highly stratified (Fig. 2.10c). The final type is the salt wedge estuary, which results from large rivers, such as the Mississippi, creating wedge-like salinity features (Fig. 2.10d).

In addition to altering the S_P and σ_θ on the shelves and within estuaries, rivers deliver nutrient-rich waters to the GOM, which affect the physicochemical properties of seawater, such as S_A. Nitrate, SRP, and silicic acid can be as high as ~250, 7, and 150 μM in the Mississippi (Turner et al. 2007), respectively, resulting in a δS_A of ~0.015 g kg^{-1} from silicic acid alone (McDougall et al. 2009). River waters also tend to be rich in organic matter, but the influence of dissolved organic carbon (DOC) on physicochemical properties of the waters is uncertain and requires further study (Woosley et al. 2014). As a result of MARS discharge, some of the highest nutrient concentrations in the GOM can be found on the Louisiana shelf, where their concentrations tend to correlate negatively with salinity (Hitchcock et al. 1997). Nevertheless, much of the nutrient load released into the GOM by MARS and other rivers is rapidly utilized by phytoplankton, which can lead to eutrophication and extensive hypoxic "dead zones" during summer stratification (Dagg and Breed 2003). Eutrophication in the northern GOM has recently occurred on unprecedented scales, as reported by NOAA, largely as a result of anthropogenic release of fertilizers into the Mississippi River watershed (Rabalais et al. 2002). For more details on anthropogenic impacts and pollutants in the GOM, see Chapters 1 and 8.

Physical circulation within the GOM can also influence the physicochemical properties of seawater, such as

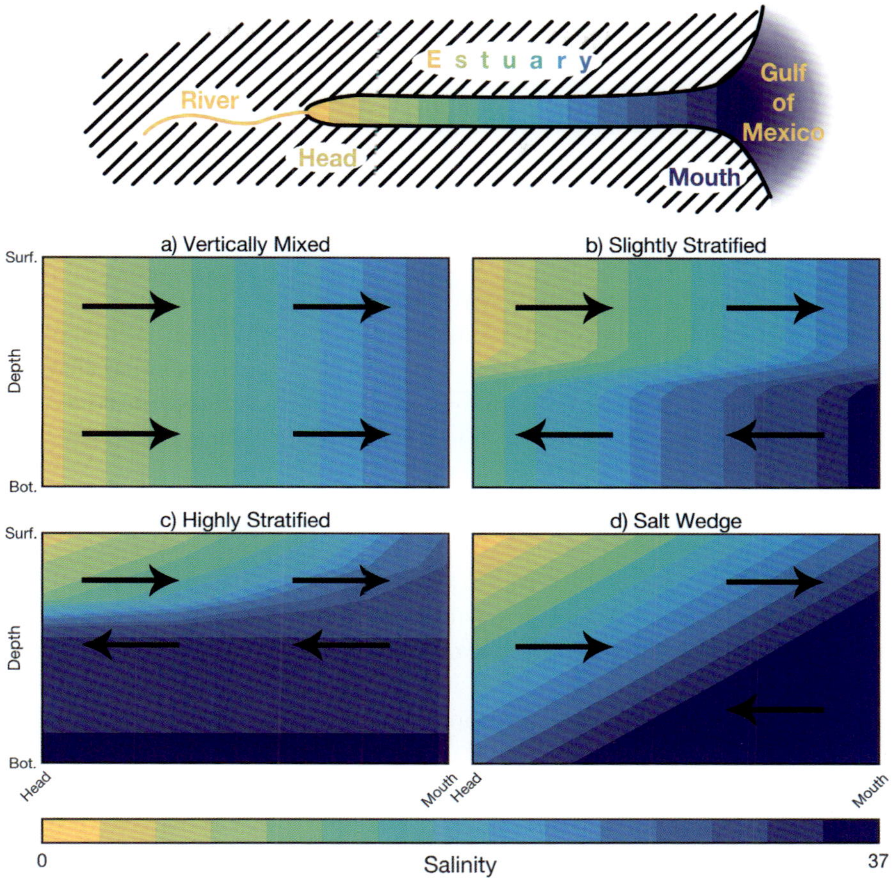

Figure 2.10a–d. Schematic map of a Gulf of Mexico estuary and salinity/depth sections for (*a*) vertically mixed, (*b*) slightly stratified, (*c*) highly stratified, and (*d*) salt-wedge estuaries (adapted from Millero 2013).

variations in the Loop Current, which frequently sheds anticyclonic eddies that redistribute warm waters, with similar properties as the Loop Current waters, throughout the surface of the GOM (Schmitz et al. 2005; see Fig. 2.5). Upwelling on the GOM's shelves, such as the Yucatan shelf, has also been observed, which increases the density of shelf waters, bringing colder and saltier waters near the surface of the GOM (Merino 1997). Another physical phenomenon that can have local impacts on the physicochemical properties of GOM water is the seasonal occurrence of tropical storms and hurricanes, which often make their way into the basin. Between 1987 and 2016, 61 tropical storms and 44 hurricanes have developed or passed through the GOM (see NOAA, https://coast.noaa.gov/hurricanes/), averaging ~2 tropical storms and ~1.5 hurricanes per year. Grodsky et al. (2012) observed temporary changes in SSS near the Amazon plume based on Aquarius measurements, resulting from the wake of Hurricane Katia in 2011; surely similar alterations to S_P, as well as temperature and other physicochemical properties of seawater, occur in the GOM as a result of hurricanes.

Brine Pools

Processes that occur near the surface of the GOM generally cause the S_P to vary between ~0 and 37 (e.g., river discharge, precipitation, storms), which is well within the extended S_P scale defined between 0 and 42. In contrast, the S_P ranges found in some environments exceed 42, such as in brine pools, which are hypersaline. Brine pools can be found at the bottom of the GOM in relatively small pools and canyons, with salinity values that exceed 42 (i.e., off the S_P scale). Given these extremely high S_P values, salinity values will be referred to as salinity, in g kg^{-1} or simply in ‰. Shokes et al. (1977) identified a 400 km^2 brine pool in the northern GOM. The bottom 200 m of the pool contain water with a salinity of ~250‰ (i.e., ~25%). This brine pool, which has since become known as the Orca Basin, is located within the vicinity of 26°55'N, 91°20'W, and contains brines at depths between ~2200 and 2400 m.

Salinity in the Orca Basin brine pool increases dramatically from ~37‰ to ~250‰ at depths between ~2000 and 2200 m, coincident with a slight increase in temperature (Shokes et al. 1977). While temperature and salinity both increase with depth in the hypersaline waters, oxygen concentrations decrease dramatically, becoming anoxic within the ~200 m thick brine layer. SRP and silicic acid concentrations both increase with depth in the basin, having profiles similar to that of salinity, while nitrate decreases to nearly 0 μM as a result of denitrification in the anoxic and hypoxic waters. More extensive considerations of brine pools and seeps, methane seeps, and anoxic and hypoxic environments are found in this book, focusing on nutrients (Chapter 3), trace metals (Chapter 4), and sediments (Chapter 7).

Orca Basin brines are characterized by a salinity-to-Cl ratio of 1.6823 (Millero et al. 1979), lower than the standard seawater ratio (1.80655, Equation 2.3) and higher than the ratio found in Red Sea brines (1.653425), which has similar salinity values of ~257‰ (Millero et al. 1982). The σ of Orca Basin brines is ~191.5 kg m^{-3} at 25 °C, an order of magnitude denser than standard seawater (σ = 26.300 kg m^{-3}; Millero et al. 1979) and slightly less dense than Red Sea brines (σ ≈ 197.7 kg m^{-3}; Millero et al. 1982). Besides the Orca Basin, there are a number of other brine pools located in the northern GOM (Joye et al. 2005), including Green Knoll to the east of the Orca Basin (salinity values of ~250‰; Aharon et al. 1992), and many others that remain unexplored to this day.

Conclusion

The physicochemical properties of GOM seawater are the same as in ocean waters. Many of the studies considered here focused on the upper water column, on the continental shelves, or in river plumes, with little consideration of the basin's abyssal waters. Considering programs like GO-SHIP, which reoccupies basin-wide hydrographic sections throughout the world's oceans, beginning with WOCE in the 1990s, it is surprising that there is not a single basin-wide hydrographic section study that has been conducted in the GOM. Basin-wide sections for S_P, θ, $σ_θ$, and oxygen were considered here based on WOD13 data, but these data are sampled sparsely in space and time, making it challenging to identify temporal and spatial trends. Furthermore, nutrient and carbonate system data from WOD13 are too sparse to compile basin-wide sections, which limits understanding of the state of the GOM's biogeochemistry on relatively large spatial scales. Further studies of the physicochemical properties of GOM seawater are needed, such as improved characterization of the varied contributions to S_A, particularly in regard to the contributions of dissolved organic carbon to density. Continued sampling of the GOM on small/regional scales is necessary to fully understand and characterize the subtleties of GOM seawater, while large basin-wide studies that include measurements of nutrients and carbonate system parameters are required to monitor long-term trends associated with anthropogenic climate change.

Acknowledgments

We thank all of those who participated in the GOMECC cruises, especially GOMECC-2, as well as Robert Byrne and Alfonso Mucci for their constructive reviews and insightful feedback. We also thank Thomas Bianchi for his encouragement with this chapter, and M. Henderson for supporting us through the writing and revision process. This work was supported by NSF grant DGE-1451511 to Margolin and OCE- 1436748 to Millero.

References

Aharon, P., H. H. Roberts, and R. Snelling. 1992. Submarine venting of brines in the deep Gulf of Mexico: Observations and geochemistry. *Geology* 20:483–486.

Anderson, L. A., and J. L. Sarmiento. 1994. Redfield ratios of remineralization determined by nutrient data analysis. *Global Biogeochemical Cycles* 8(1):65–80.

Barber, C. R. 1969. The International Practical Temperature Scale of 1968. *Meteorologia* 5(2):35–44.

Benway, H. M., and P. G. Coble (eds.). 2014. *Report of the US Gulf of Mexico Carbon Cycle Synthesis Workshop*. Woods Hole, MA: Ocean Carbon and Biogeochemistry Program and North American Carbon Program. 67 pp.

Bianchi, T. S., J. R. Pennock, and R. R. Twilley (eds.). 1999. *Biogeochemistry of Gulf of Mexico Estuaries*. New York: John Wiley and Sons. 427 pp.

Boyer, T. P., J. I. Antonov, O. K. Baranova, C. Coleman, H. E. Garcia, A. Grodsky, D. R. Johnson, et al. 2013. *World Ocean Database 2013*. S. Levitus (ed.), A. Mishonov (tech. ed.). Silver Spring, MD: NOAA Atlas NESDIS 72. 209 pp. http://doi.org/10.7289/V5NZ85MT.

Brewer, P. G., and A. Bradshaw. 1975. The effect of non-ideal composition of seawater on salinity and density. *Journal of Marine Research* 33:157–175.

Cai, Y., L. Guo, X. Wang, and G. Aiken. 2015. Abundance, stable isotopic composition, and export fluxes of DOC, POC, and DIC from the lower Mississippi River during 2006–2009. *Journal of Geophysical Research: Biogeosciences* 120:2273–2289.

Cai, W.-J., X. Hu, W.-J. Huang, L.-Q. Jiang, Y. Wang, T.-H. Peng, and X. Zhang. 2010. Alkalinity distribution in the western

North Atlantic Ocean margins. *Journal of Geophysical Research* 115(C8):1–15.

Carder, K. L., K. A. Fanning, P. R. Betzer, and V. Maynard. 1977. Dissolved silica and the deep circulation in the Yucatan Strait and deep eastern Gulf of Mexico. *Deep Sea Research* 24:1149–1160.

Cardona, Y., A. Bracco, T. A. Villareal, A. Subramaniam, S. C. Weber, and J. P. Montoya. 2016. Highly variable nutrient concentrations in the northern Gulf of Mexico. *Deep Sea Research II: Tropical Studies in Oceanography* 129:20–30.

Coble, P. G., L. L. Robbins, K. L. Daly, W.-J. Cai, K. Fennel, and S. E. Lohrenz. 2010. A preliminary carbon budget for the Gulf of Mexico. *Ocean Carbon and Biogeochemistry News* 3(3):1–4.

Cox, R. A., F. Culkin, and J. P. Riley. 1967. The electrical conductivity/chlorinity relationship in natural seawater. *Deep Sea Research* 14:203–220.

Dagg, M. J., and G. A. Breed. 2003. Biological effects of Mississippi River nitrogen on the northern Gulf of Mexico—a review and synthesis. *Journal of Marine Systems* 43:133–152.

DeHaan, C. J., and W. Sturges. 2008. Deep cyclonic circulation in the Gulf of Mexico. *Journal of Physical Oceanography* 35:1801–1812.

Dickson, A. G. 1981. An exact definition of total alkalinity and a procedure for the estimation of alkalinity and total inorganic carbon from titration data. *Deep Sea Research, Part A: Oceanographic Research Papers* 28A(6):609–623.

Dittmar, W. 1884. Report on the scientific results of the exploring voyage of HMS *Challenger*. Pp. 1–251 in *Physics and Chemistry* 1. London: Longmans.

Doney, S. C., V. J. Fabry, R. A. Feely, and J. A. Kleypas. 2009. Ocean acidification: The other CO_2 problem. *Annual Reviews of Marine Science* 1:169–192.

Ducklow, H. W., D. K. Steinberg, and K. O. Buesseler. 2001. Upper ocean carbon export and the biological pump. *Oceanography* 14(4):50–58.

Emerson, S. R., and J. I. Hedges. 2008. *Chemical Oceanography and the Marine Carbon Cycle*. Cambridge: Cambridge University Press. 453 pp.

Feely, R. A., S. C. Doney, and S. R. Cooley. 2009. Ocean acidification: Present conditions and future changes in a high-CO_2 world. *Oceanography* 22(4):36–47.

Feely, R. A., C. L. Sabine, K. Lee, F. J. Millero, M. F. Lamb, D. Greeley, J. L. Bullister, et al. 2002. In situ calcium carbonate dissolution in the Pacific Ocean. *Global Biogeochemical Cycles* 16(4). doi:10.1029/2002GB001866.

Fine, R. A., D. A. Willey, and F. J. Millero. 2017. Global variability and changes in ocean total alkalinity from Aquarius satellite data. *Geophysical Research Letters* 44(1):261–267.

Fofonoff, N. P., and H. Bryden. 1975. Specific gravity and density of seawater at atmospheric pressure. *Journal of Marine Research* 33(Supplement):69–82.

Forchhammer, G. 1865. On the composition of sea-water in the different parts of the ocean. *Philosophical Transactions* 155:203–262.

Garcia, H. E., R. A. Locarnini, T. P. Boyer, J. I. Antonov, O. K. Baranova, M. M. Zweng, J. R. Reagan, and D. R. Johnson. 2013. *World Ocean Atlas 2013*, vol. 3: *Dissolved Oxygen, Apparent Oxygen Utilization, and Oxygen Saturation*. S. Levitus (ed.), A. Mishonov (tech. ed.). Silver Spring, MD: NOAA Atlas NESDIS 75. 27 pp.

Georgian, S. E., D. DeLeo, A. Durkin, C. E. Gomez, M. Kurman, J. J. Lunden, and E. E. Cordes. 2016. Oceanographic patterns and carbonate chemistry in the vicinity of cold-water coral reefs in the Gulf of Mexico: Implications for resilience in a changing ocean. *Limnology and Oceanography* 61:648–665.

Gierach, M. M., J. Vazquez-Cuervo, T. Lee, and V. M. Tsontos. 2013. Aquarius and SMOS detect effects of an extreme Mississippi River flooding event in the Gulf of Mexico. *Geophysical Research Letters* 40:5188–5193.

Grodsky, S., N. Reul, G. Lagerloef, G. Reverdin, J. Carton, B. Chapron, Y. Quilfen, V. Kudryavstev, and H.-Y. Kao. 2012. Haline hurricane wake in the Amazon/Orinoco plume: Aquarius/SAC-D and SMOS observations. *Geophysical Research Letters* 39:L20603.

Hill, K., T. Dauphinee, and D. Woods. 1986. The extension of the practical salinity scale 1978 to low salinities. *IEEE Journal of Oceanic Engineering* 11(1):109–112.

Hitchcock, G. L., W. J. Wiseman Jr., W. C. Boicourt, A. J. Mariano, N. Walker, T. A. Nelsen, and E. Ryan. 1997. Property fields in an effluent plume of the Mississippi River. *Journal of Marine Systems* 12:109–126.

Huang, W.-J., W.-J. Cai, Y. Wang, S. E. Lohrenz, and M. C. Murrell. 2015. The carbon dioxide system on the Mississippi River–dominated continental shelf in the northern Gulf of Mexico. 1: Distribution and air-sea CO_2 flux. *Journal of Geophysical Research: Oceans* 120:1429–1445.

Johnson, G. C., and S. G. Purkey. 2009. Deep Caribbean Sea warming. *Deep Sea Research I: Oceanographic Research Papers* 56:827–834.

Joye, S. B., I. R. MacDonald, J. P. Montoya, and M. Peccini. 2005. Geophysical and geochemical signatures of Gulf of Mexico seafloor brines. *Biogeosciences* 2:295–309.

Key, R. M., A. Olsen, S. van Heuven, S. K. Lauvset, A. Velo, X. Lin, C. Schirnick, et al. 2015. *Global Ocean Data Analysis Project, Version 2 (GLODAPv2)*. Oak Ridge, TN: ORNL/

CDIAC-162, NDP-093, Carbon Dioxide Information Analysis Center, Oak Ridge National Laboratory, US Department of Energy. doi:10.3334/CDIAC/OTG.NDP093_GLODAPv2.

Kilpatrick, K. A., G. Podestá, S. Walsh, E. Williams, V. Halliwell, M. Szczodrak, O. B. Brown, P. J. Minnett, and R. Evans. 2015. A decade of sea surface temperature from MODIS. *Remote Sensing of the Environment* 165:27–41.

Knudsen, M. 1901. *Hydrographische Tabellen* [Hydrographic tables]. London: Williams Norgate. 63 pp.

Krishna Rao, P., W. L. Smith, and R. Koffler. 1972. Global sea-surface temperature distribution determined from an environmental satellite. *Monthly Weather Review* 100(1): 10–14.

Lagerloef, G., H.-Y. Kao, T. Meissner, and J. Vazquez. 2015. *Aquarius Salinity Validation Analysis; Data Version 4.0.* Aquarius Technical Report, NASA, August 2. ftp://podaac.jpl.nasa.gov/SalinityDensity/aquarius/docs/v4/AQ-014-PS-0016_AquariusSalinityDataValidationAnalysis_DatasetVersion4.0and3.0.pdf.

Lee, K., C. L. Sabine, T. Tanhua, T.-W. Kim, R. A. Feely, and H.-C. Kim. 2011. Role of marginal seas in absorbing and storing fossil fuel CO_2. *Energy and Environmental Science* 4:1133–1146.

Lee, K., L. T. Tong, F. J. Millero, C. L. Sabine, A. G. Dickson, C. Goyet, G.-H. Park, R. Wanninkhof, R. A. Feely, and R. M. Key. 2006. Global relationships of total alkalinity with salinity and temperature in surface waters of the world's oceans. *Geophysical Research Letters* 33(L19605):1–5.

Lewis, E. L., and R. G. Perkin. 1978. Salinity: Its definition and calculation. *Journal of Geophysical Research* 83(C1):466–478.

Lewis, E., and D. W. R. Wallace. 1998. *Program developed for CO_2 system calculations*. ORNL/CDIAC-105. Oak Ridge, TN: Carbon Dioxide Information Analysis Center, Oak Ridge National Laboratory. 38 pp.

Locarnini, R. A., A. V. Mishonov, J. I. Antonov, T. P. Boyer, H. E. Garcia, O. K. Baranova, M. M. Zweng, et al. 2013. *World Ocean Atlas 2013*, vol. 1: *Temperature*. S. Levitus (ed.), A. Mishonov (tech. ed.). Silver Spring, MD: NOAA Atlas NESDIS 73. 40 pp.

Lunden, J. J., S. E. Georgian, and E. E. Cordes. 2013. Aragonite saturation states at cold-water coral reefs structured by *Lophelia pertusa* in the northern Gulf of Mexico. *Limnology and Oceanography* 58(1):354–362.

Lyman, J., and R. H. Fleming. 1940. Composition of seawater. *Journal of Geophysical Science* 72:2665–2678.

Margolin, A. R. 2017. Environmental impacts on carbon biogeochemistry in marginal seas. PhD diss., University of Miami, Miami, FL. 137 pp.

McDougall, T. J., and P. M. Barker. 2011. *Getting Started with TEOS-10 and the Gibbs Seawater (GSW) Oceanographic Toolbox*. SCOR/IAPSO WG. 127. https://www.researchgate.net/publication/285000773_Getting_started_with_TEOS-10_and_the_Gibbs_Seawater_GSWOceanographic_Toolbox. 28 pp.

McDougall, T. J., D. R. Jackett, and F. J. Millero. 2009. An algorithm for estimating absolute salinity in the global ocean. *Ocean Science Discussions* 6:215–242.

McDougall, T. J., D. R. Jackett, F. J. Millero, R. Pawlowicz, and P. M. Barker. 2012. A global algorithm for estimating absolute salinity. *Ocean Science* 8:1117–1128.

Merino, M. 1997. Upwelling on the Yucatan shelf: Hydrographic evidence. *Journal of Marine Systems* 13:101–121.

Millero, F. J. 2001. *Physical Chemistry of Natural Waters*. New York: John Wiley and Sons. 654 pp.

———. 2010. History of the equation of state of seawater. *Oceanography* 23(3):18–33.

———. 2013. *Chemical Oceanography*. 4th ed. Boca Raton, FL: CRC Press. 571 pp.

Millero, F. J., R. Feistel, D. Wright, and T. J. McDougall. 2008. The composition of standard seawater and the definition of a reference-composition salinity scale. *Deep Sea Research II: Tropical Studies in Oceanography* 55:50–72.

Millero, F. J., A. Gonzalez, P. G. Brewer, and A. Bradshaw. 1976. The density of North Atlantic and North Pacific deep waters. *Earth and Planetary Science Letters* 32:468–472.

Millero, F. J., D. Lawson, and A. Gonzalez. 1976. The density of artificial river and estuarine waters. *Journal of Geophysical Research* 81(6):1177–1179.

Millero, F. J., A. Lo Surdo, P. Chetirkin, and N. L. Guinasso Jr. 1979. The density and speed of sound of Orca Basin waters. *Limnology and Oceanography* 24(2):218–225.

Millero, F. J., A. Mucci, J. Zullig, and P. Chetirkin. 1982. The density of Red Sea brines. *Marine Chemistry* 11:463–475.

Morrison, J. M., and W. J. Merrell Jr. 1983. Property distributions and deep chemical measurements within the western Gulf of Mexico. *Journal of Geophysical Research* 88(C4):2601–2608.

Orr, J. C., V. J. Fabry, O. Aumont, L. Bopp, S. C. Doney, R. A. Feely, A. Gnanadesikan, et al. 2005. Anthropogenic ocean acidification over the twenty-first century and its impact on calcifying organisms. *Nature* 437:681–686.

Pickard, G. L., and W. J. Emery. 1982. *Descriptive Physical Oceanography*. 4th ed. Oxford: Pergamon Press. 249 pp.

Pierrot, D., E. Lewis, and D. W. R. Wallace. 2006. *MS Excel Program Developed for CO_2 System Calculations*. ORNL/CDIAC-105a. Oak Ridge, TN: Carbon Dioxide Information Analysis Center, Oak Ridge National Laboratory, US Department of Energy. doi:10.3334/CDIAC/otg.CO2SYS_XLS_CDIAC105a.

Rabalais, N. N., R. E. Turner, and W. J. Wiseman Jr. 2002. Gulf of Mexico hypoxia, a.k.a. "the dead zone." *Annual Review of Ecology and Systematics* 33:235–263.

Ren, W., H. Tian, B. Tao, J. Yang, S. Pan, W.-J. Cai, S. E. Lohrenz, R. He, and C. S. Hopkinson. 2015. Large increase in dissolved inorganic carbon flux from the Mississippi River to Gulf of Mexico due to climatic and anthropogenic changes over the 21st century. *Journal of Geophysical Research: Biogeosciences* 120:724–736.

Reynolds, R. W., N. A. Rayner, T. M. Smith, D. C. Stokes, and W. Wang. 2002. An improved in situ and satellite SST analysis for climate. *Journal of Climate* 15:1609–1625.

Rivas, D., A. Badan, and J. Ochoa. 2005. The ventilation of the deep Gulf of Mexico. *Journal of Physical Oceanography* 35(10):1763–1781.

Saunders, P. 1990. The International Temperature Scale of 1990, ITS-90. *WOCE Newsletter* 10. Wormley, UK: WOCE International Project Office at Natural Environment Research Council, Institute of Oceanographic Sciences, Deacon Laboratory.

Schlitzer, R. 2017. Ocean Data View. https://odv.awi.de.

Schmitz, W. J., Jr., D. C. Biggs, A. Lugo-Fernandez, L.-Y. Oey, and W. Sturges. 2005. A synopsis of the circulation in the Gulf of Mexico and on its continental margins. *Circulation in the Gulf of Mexico: Observations and Models*, American Geophysical Union Monograph Series 161:11–29.

Schott, F. A., J. Fischer, and L. Stramma. 1998. Transports and pathways of the upper-layer circulation in the western tropical Atlantic. *Journal of Physical Oceanography* 28:1904–1928.

Shiller, A. M. 1999. An overview of the marine chemistry of the Gulf of Mexico. Pp. 132–148 in H. Kumpf, K. Steidinger, and K. Sherman (eds.), *The Gulf of Mexico Large Marine Ecosystem: Assessment, Sustainability, and Management*. Malden, MA: Blackwell Science.

Shokes, R. F., P. K. Trabant, B. J. Presley, and D. F. Reid. 1977. Anoxic, hypersaline basin in the northern Gulf of Mexico. *Science* 196(4297):1443–1446.

Smith, R. H. 2010. Atlantic-Caribbean exchange through the Windward Passage. Master's thesis, University of Miami, Miami, FL. 130 pp.

Talley, L. D., G. L., Pickard, W. J. Emery, and J. H. Swift. 2011. *Descriptive Oceanography: An Introduction*. 6th ed. Amsterdam: Elsevier. 555 pp.

Turner, R. E., N. N. Rabalais, R. B. Alexander, G. McIsaac, and R. W. Howarth. 2007. Characterization of nutrient, organic carbon, and sediment loads and concentrations from the Mississippi River into the northern Gulf of Mexico. *Estuaries and Coasts* 30(5):773–790.

van Heuven, S., D. Pierrot, J. W. B. Rae, E. Lewis, and D. W. R. Wallace. 2011. *MATLAB Program Developed for CO_2 System Calculations*. ORNL/CDIAC-105b. Oak Ridge, TN: Carbon Dioxide Information Analysis Center, Oak Ridge National Laboratory, US Department of Energy, https://doi.org/10.3334/CDIAC/otg.CO2SYS_MATLAB_v1.1.

Wang, Z. A., R. Wanninkhof, W.-J. Cai, R. H. Byrne, X. Hu, T.-H. Peng, and W.-J. Huang. 2013. The marine inorganic carbon system along the Gulf of Mexico and Atlantic coasts of the United States: Insights from a transregional coastal carbon study. *Limnology and Oceanography* 58(1):325–342.

Wanninkhof, R., L. Barbero, R. Byrne, W.-J. Cai, W.-J. Huang, J.-Z. Zhang, M. Baringer, and C. Langdon. 2015. Ocean acidification along the Gulf Coast and East Coast of the USA. *Continental Shelf Research* 98:54–71.

Woosley, R. J., F. Huang, and F. J. Millero. 2014. Estimating absolute salinity (S_A) in the world's oceans using density and composition. *Deep Sea Research I: Oceanographic Research Papers* 93:14–20.

Wooster, W. S., A. J. Lee, and G. Dietrich. 1969. Redefinition of salinity. *Limnology and Oceanography* 14(3):437–438.

Worthington, L. V. 1959. The 18° water in the Sargasso Sea. *Deep Sea Research* 5(2–4):297–305.

Yang, B., R. H. Byrne, and R. Wanninkhof. 2015. Subannual variability of total alkalinity distributions in the northeastern Gulf of Mexico. *Journal of Geophysical Research: Oceans* 120:3805–3816.

Zeebe, R. E., and D. Wolf-Gladrow. 2001. *CO_2 in Seawater: Equilibrium, Kinetics, Isotopes*. Elsevier Oceanography Series 65. Amsterdam: Elsevier. 346 pp.

Zhou, Z., B. Stolpe, L. Guo, and A. M. Shiller. 2016. Colloidal size spectra, composition and estuarine mixing behavior of DOM in river and estuarine waters of the northern Gulf of Mexico. *Geochimica et Cosmochimica Acta* 181:1–17.

Zweng, M. M., J. R. Reagan, J. I. Antonov, R. A. Locarnini, A. V. Mishonov, T. P. Boyer, H. E. Garcia, et al. 2013. *World Ocean Atlas 2013*, vol. 2: *Salinity*. S. Levitus (ed.), A. Mishonov (tech. ed.). Silver Spring, MD: NOAA Atlas NESDIS 74. 39 pp.

3

Nutrients in the Gulf of Mexico
Distributions, Cycles, Sources, Sinks, and Processes

Donald G. Redalje, James W. Ammerman, Jorge A. Herrera-Silveira, Angela N. Knapp, Jeffrey W. Krause, David S. Valdes, and Anna S. Hayward

Introduction

The Gulf of Mexico (GOM) is a semi-enclosed sea bounded on the north by 5 coastal states of the United States and on the south by 6 coastal states of Mexico and Cuba. It has been divided into waters of the US, Mexico, and Cuba Exclusive Economic Zones with 2 small sections of international waters (Fig. 3.1a). Multiple rivers deliver freshwater to the coastal areas of the Gulf, dominated by the transport from the Mississippi-Atchafalaya River system (Moretzsohn et al. 2016). There are 33 major rivers that enter the Gulf from the United States with many smaller rivers and creeks, only 4 being major rivers; Cuba has 13 larger rivers that enter the Gulf from the northern coast and another 21 from the southern coast, with many smaller creeks and streams (Moretzsohn et al. 2016). Most of these rivers enter into and pass through wetlands, marshes, estuaries, or lagoon systems before flowing into the GOM, often sequestering or transforming nutrients in these aquatic systems (Bianchi et al. 1999). The Mississippi River is the exception due to its extensive channelization that ensures river access to the Port of New Orleans and, thus, reduced opportunity for nutrients to be cycled through wetlands, marshes, or an estuary. In addition to these riverine freshwater inputs, there is generalized runoff from land and substantial submarine groundwater discharge into the Gulf from much of the surrounding coastal margins (Cable et al. 1996; Burnett et al. 2003; Burnett et al. 2006). Each of these freshwater inputs brings some level of inorganic and organic nutrients to the GOM.

In addition to inputs from freshwater discharges, both macronutrients and micronutrients can be introduced to GOM waters from atmospheric deposition (Goolsby et al. 1999). Rainwater often includes elevated concentrations of nitrogen (N) nutrients. It has been suggested that airborne deposition of dust from the Sahara brings iron (Fe) to Gulf waters during summer (Lenes et al. 2001; Walsh and Steidinger 2001; Trapp et al. 2010). This airborne-dust-derived Fe input has been suggested to help the initiation of harmful algal blooms in the eastern GOM and may provide the Fe necessary for di-nitrogen (N_2) fixation by bacteria and cyanobacteria (Walsh and Steidinger 2001; Heil, Dixon, et al. 2014; Mulholland et al. 2014). Biological and chemical processes active in the surface layers of seafloor sediments on both the continental shelf and deep GOM result in the remineralization of nutrients (e.g., Nunnally et al. 2014). These remineralized nutrients contribute to a measurable flux of inorganic nutrients from the seafloor to the overlying waters (Feist et al. 2016). The nutrients contained in these overlying waters can be mixed or upwelled into the mixed layer via a range of physical and circulation processes, thereby making them available for uptake by organisms and inclusion in biogeochemical cycles.

Tropical ocean waters enter the GOM from the Atlantic Ocean and Caribbean Sea in the Yucatan Current and the Caribbean Current through the Yucatan Channel, forming the Gulf of Mexico Loop Current that moves in an anticyclonic direction throughout the Gulf. Waters of the Loop Current exit the Gulf through the Florida Straits as the Florida Loop Current. The tropical waters that enter the Gulf from the Atlantic and Caribbean are oligotrophic, although there may be some dissolved silicon (Si) transported in these waters that originated in the plumes of the Amazon and Orinoco Rivers, especially during the Atlantic hurricane season (Field 2007; Coles et al. 2013). The Loop Current is the main circulation feature of the GOM. It flows deeper into the Gulf seasonally with its greatest northern extent in summer and fall when the winds are generally from the southeast. The Loop Current often forms large anticyclonic eddies that may break off from the current and flow both northward and westward in the northern GOM. Smaller cyclonic and anticyclonic eddies, sometimes with filaments and fronts, can also form and move into shelf waters. Each of these mesoscale and sub-mesoscale features will have upwelling and downwelling parts of the eddy: upwelling on the eddy margins for anticyclonic eddies and in the eddy center for cyclonic eddies that can bring nutrient-rich waters up to the surface mixed layer, promoting localized primary production (e.g., Salas-de-León and Monreal-Gómez 2004; Müller-Karger et al. 2015). Upwelling can also occur along the coastal shelves when the winds lead to offshore Ekman transport of shelf waters.

In this chapter, we explore how each of these circulation features and freshwater sources contributes to the input, cycling, fluxes, and sinks for inorganic and organic nutrients in the GOM. We discuss the nutrient sources, sinks, transformations, and cycling for dissolved N, phosphorus (P), and Si with a focus on the major macronutrients, using the latest major databases. Some of the material discussed here briefly is addressed in other chapters in more detail (e.g., Chapter 1 on biological oceanography, Chapter 4 on trace metals, and Chapter 6 on dissolved, colloidal, and particulate organic matter).

One of the most obvious conclusions that can be drawn from a search for available data on nutrient concentrations in the GOM is that we have a large pool of data for various locations in the coastal Gulf and a paucity of data from the open Gulf (Fig. 3.1A–D). There have been a number of major interdisciplinary studies in the US coastal Gulf that generated a time series of nutrient data for a series of stations covering large expanses of the northern GOM and the west Florida shelf. For example, in the mid- to late 1980s, the US Department of the Interior Mineral Management Service (MMS, now BOEM) supported the multifaceted LATEX program that explored coastal waters of the northern Gulf affected by discharge from the Mississippi and Atchafalaya Rivers covering the shelf areas of Louisiana and Texas (Nowlin et al. 1998). From 1997 through 2000, MMS supported the Northeastern Gulf of Mexico Physical Oceanography Program (NEGOM): Chemical Oceanography and Hydrography Study (NEGOM-COH) (Jochens and Nowlin 2000). The investigators in this project studied the shelf areas

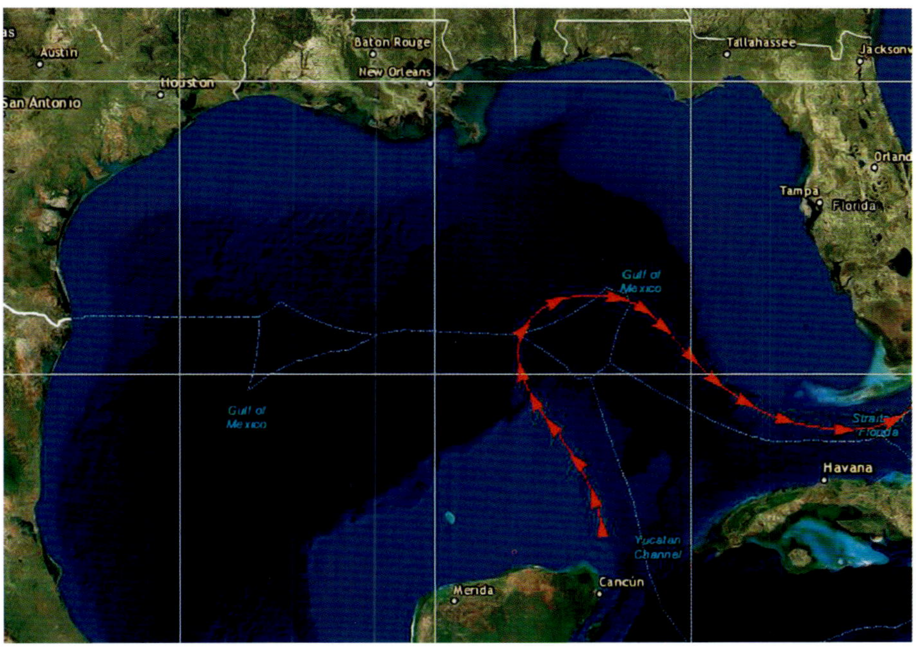

Figure 3.1A. The Gulf of Mexico includes waters that are part of the Exclusive Economic Zones (EEZ) for the United States, Mexico, and Cuba as well as 2 small zones of international jurisdiction (*light blue lines*). The climatological path of the Loop Current is indicated by red arrows (from the NOAA Gulf of Mexico Data Atlas, https://www.ncddc.noaa.gov/website/DataAtlas/atlas.htm).

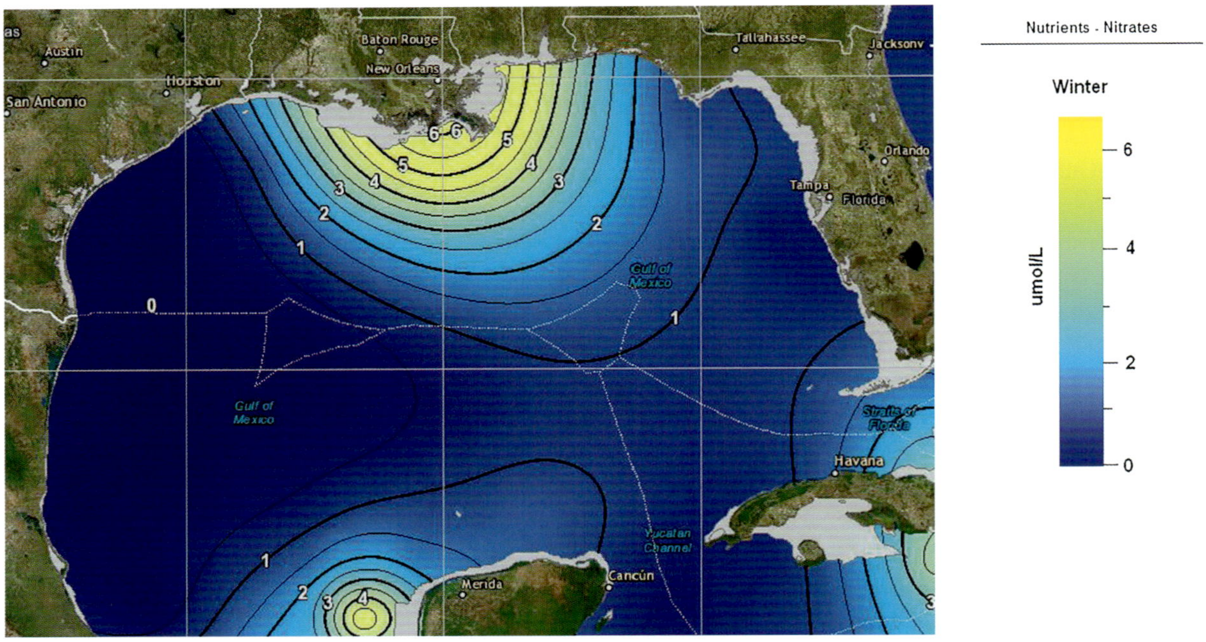

Figure 3.1B. The winter climatological average concentrations of nitrate at a depth of 10 m in the Gulf of Mexico are characterized by maxima from the Mississippi-Atchafalaya River system, the west coast of the Yucatan Peninsula, and the northern and southern coasts of Cuba (from the NOAA Gulf of Mexico Data Atlas, https://www.ncddc.noaa.gov/website/DataAtlas/atlas.htm).

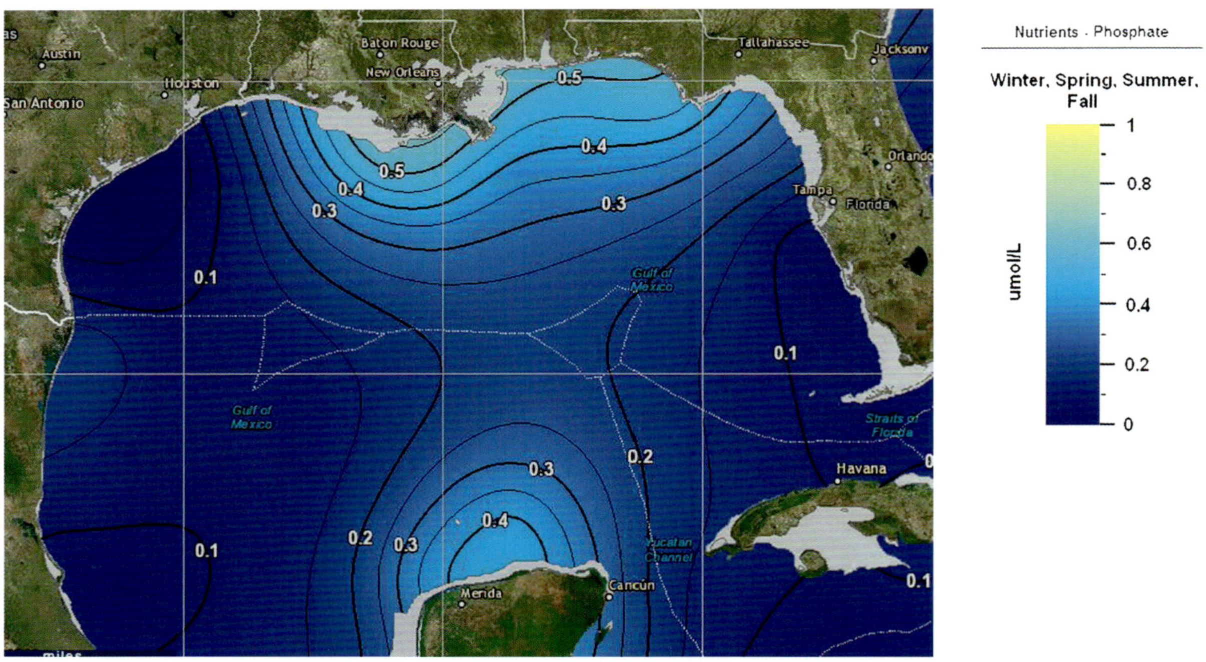

Figure 3.1C. The annual climatological average concentrations of phosphate at a depth of 10 m in the Gulf of Mexico are greatest near the mouths of the Mississippi-Atchafalaya River system, Mobile Bay, and the northern coast of the Yucatan Peninsula (from the NOAA Gulf of Mexico Data Atlas, https://www.ncddc.noaa.gov/website/DataAtlas/atlas.htm).

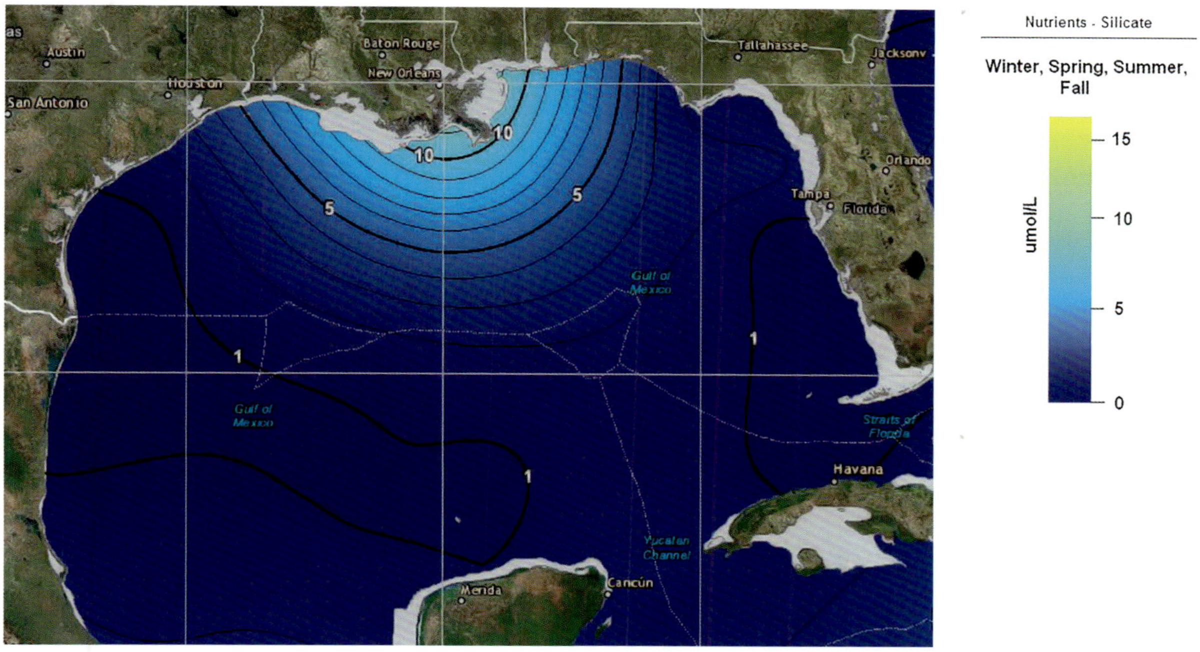

Figure 3.1D. The annual climatological average concentrations of silicate at a depth of 10 m in the Gulf of Mexico are greatest near the mouths of the Mississippi-Atchafalaya River system (from the NOAA Gulf of Mexico Data Atlas, https://www.ncddc.noaa.gov/website/DataAtlas/atlas.htm).

from the coastline out to the 1000 m isobath between the Mississippi River modern delta and Tampa, Florida (Jochens and Nowlin 2000; Jochens et al. 2002; Jochens and DiMarco 2008). The annual and final reports of these studies contain numerous nutrient measurements for the northern GOM coastal waters. From 1990 through 1996, the NOAA Coastal Ocean Program supported the Nutrient Enhanced Coastal Ocean Productivity (NECOP) program, which focused on the waters in the plume of the Mississippi River and the adjacent Louisiana shelf. The final report of that project also contains many nutrient concentration measurements (Wiseman et al. 1999). The study region covered in the NECOP program included the Louisiana shelf regions influenced by the Mississippi-Atchafalaya River system where seasonal hypoxia occurs (e.g., Rabalais et al. 2001; Turner et al. 2008). Morrison et al. (1983) conducted studies in the western Gulf of Mexico that included nutrient concentrations for stations both on the shelf and in the deep Gulf.

There have been a number of studies conducted as part of the Gulf of Mexico Research Initiative (GOMRI) that focused on the impacts of the Deepwater Horizon oil spill. Some of that data was sequestered for legal reasons, but much is becoming available from project websites or through publications (e.g., Cardona et al. 2016; Daly 2015). Most of the data for nutrients in the GOM can be accessed through the National Oceanic and Atmospheric Administration National Centers for Environmental Information (NOAA NCEI) website (https://www.ncei.noaa.gov/). The NOAA NCEI includes those data and metadata collections formerly available through the NOAA National Oceanographic Data Center (NODC) and the National Coastal Data Development Center (NCDDC). Of particular interest for GOM nutrient information is the NOAA Gulf of Mexico Data Atlas (www.ncddc.noaa.gov/website/DataAtlas/atlas.htm). Additional information can be obtained from the Gulf of Mexico Alliance (GOMA; www.gulfofmexicoalliance.org/). There are a number of very useful resources available on the websites for these organizations.

Nitrogen in the Gulf of Mexico

The GOM is a biogeochemically dynamic region where N cycling has been at the heart of several high-profile, societally relevant processes. The most well-known issue in the GOM is the hypoxic zone where oxygen concentrations on the shelf off the mouth of the Mississippi River drop to levels low enough to prohibit macrofauna such as fish from occupying the region (Turner and Rabalais 1991, 1994; Rabalais et al. 1996). This hypoxic zone primarily results from nutrient loading from the Mississippi-

Table 3.1. Nitrogen input fluxes to the Gulf of Mexico and its estuaries

N input flux	Location	Method	Type of N
Euphotic zone N_2 fixation			
	West FL shelf	$^{15}N_2$ incubations	
	Northwestern GOM	$^{15}N_2$ incubations	
	West FL shelf	$^{15}N_2$ incubations	
	Northern GOM	$^{15}N_2$ incubations	
Benthic N_2 fixation			
	TX estuaries	$^{15}N_2$ incubations	
	AL estuaries	N_2/Ar ratios	
	TX estuaries	$^{15}N_2$ incubations	
	AL estuaries	Acetylene reduction	
	AL estuaries	Acetylene reduction	
	Northern GOM	$^{15}N_2$ incubations	
Atmospheric deposition			
Aerosols	Sumatra, FL		NH_4^+
Aerosols	Sumatra, FL		NO_3^-+HNO_3
Wet + aerosols	Western GOM	Modeled	NOy
Wet + aerosols	Western GOM	Modeled	NH_4^+
Wet + aerosols	Eastern GOM	Modeled	NOy
Wet + aerosols	Eastern GOM	Modeled	NH_4^+
Wet	Forested N FL	Atmos dep. collectors	NH_4^+
Wet	Forested N FL	Atmos dep. collectors	NO_3^-
Wet	Agricultural N FL	Atmos dep. collectors	NH_4^+
Wet	Agricultural N FL	Atmos dep. collectors	NO_3^-
Wet	Sumatra, FL	Atmos dep. collectors	NO_3^-+NO_2^-
Wet	Sumatra, FL	Atmos dep. collectors	NH_4^+
Wet	Chassahowitzka Natl Wildlife Refuge, FL	Atmos dep. collectors	NO_3^-+NO_2^-
Wet	Chassahowitzka Natl Wildlife Refuge, FL	Atmos dep. collectors	NH_4^+
Wet	Corpus Christi, TX	Atmos dep. Collectors	NO_3^-+NO_2^-
Wet	Corpus Christi, TX	Atmos dep. collectors	NH_4^+

	Original units	mmol N m^{-2} yr^{-1}	Reference
	98.4 ± 79.9 pmol N L^{-1} hr^{-1}	3.6 ± 2.9 (a)	Mulholland et al. 2006
	84.5 ± 17.7 μmol N m^{-2} d^{-1}	30.8 ± 6.5 (a)	Holl et al. 2007
	2.3 ± 2.7 nmol N L^{-1} d^{-1}	8.4 ± 9.9 (a)	Mulholland et al. 2014
	0.07 to 0.37 nmol N L^{-1} d^{-1}	0.03 ± 0.14 (b)	Weber et al. 2016
	15.3 ± 53.0 μmol N m^{-2} hr^{-1}	134 ± 464 (c)	McCarthy et al. 2008
	12 μmol N m^{-2} hr^{-1}	105 (c)	Mortazavi et al. 2012
	749 ± 1740 μmol N m^{-2} hr^{-1}	6561 ± 15,242 (c)	Bruesewitz et al. 2013
	61.7 ± 4.5 μmol N m^{-2} hr^{-1}	540 ± 39 (c)	Bernard et al. 2014
	2.2 ± 0.5 μmol N m^{-2} hr^{-1}	19 ± 4 (c)	Bernard et al. 2015
	44.1 ± 58.8 μmol N m^{-2} hr^{-1}	386 ± 515 (c)	McCarthy et al. 2015
	kg N ha^{-1} yr^{-1}	1.4 ± 0.5 (d)	CASTNET
	kg N ha^{-1} yr^{-1}	18.8 ± 6.2 (d)	CASTNET
	364 kg N km^{-2} yr^{-1}	26 (e)	Howarth et al. 1996; Prospero et al. 1996
	224 kg N km^{-2} yr^{-1}	16 (e)	Howarth et al. 1996; Prospero et al. 1996
	812 kg N km^{-2} yr^{-1}	58 (e)	Howarth et al. 1996; Prospero et al. 1996
	280 kg N km^{-2} yr^{-1}	20 (e)	Howarth et al. 1996; Prospero et al. 1996
		4.4 (f)	Katz et al. 2009
		8 (f)	Katz et al. 2009
		5.2 (f)	Katz et al. 2009
		9.4 (f)	Katz et al. 2009
	mg L^{-1}	58 ± 12 (d)	NADP
	mg L^{-1}	10 ± 3 (d)	NADP
	mg L^{-1}	63 ± 13 (d)	NADP
	mg L^{-1}	9 ± 4 (d)	NADP
	mg L^{-1}	39 ± 6 (d)	NADP
	mg L^{-1}	15 ± 3 (d)	NADP

Continued

Table 3.1. Continued.

N input flux	Location	Method	Type of N
Riverine			
	Mississippi River basin		TN
	Mississippi and Atchafalaya Rivers		TN
	Mississippi and Atchafalaya Rivers		NO_3^-
	Mississippi and Atchafalaya Rivers		Organic N
	Mississippi and Atchafalaya Rivers		NH_4^+
	Franklin Lock		DIN
	Franklin Lock		TDN
	Atchafalaya River basin		TKN
	Atchafalaya River basin		$NO_3^-+NO_2^-$
Groundwater discharge			
	West FL shelf	Groundwater nutrients	$NO_3^-+NO_2^-$
	West FL shelf	Groundwater nutrients	NH_4^+
	West FL shelf	Groundwater nutrients	DON
	Key Largo, FL Keys	Groundwater nutrients	TIN
	Northern GOM	Groundwater nutrients	$NO_3^-+NO_2^-$
	Northern GOM	Groundwater nutrients	NH_4^+
	Northern GOM	Groundwater nutrients	DON
	Northern GOM	Groundwater nutrients	$NO_3^-+NO_2^-$
	Northern GOM	Groundwater nutrients	NH_4^+
	Northern GOM	Groundwater nutrients	DON
	West FL shelf	Groundwater nutrients	DIN
	West FL shelf	Groundwater nutrients	TDN
	Caloosahatchee River estuary	Submarine groundwater	DIN
	Caloosahatchee River estuary	Submarine groundwater	TDN
	Northern GOM	Shallow wells	DON
	Northern GOM	Shallow wells	$NO_3^-+NO_2^-$
	Northern GOM	Shallow wells	NH_4^+

Note: To convert between original flux units and the common units of mmol m^{-2} yr^{-1}, assumptions included the following:

(a) A 10 hr N_2 fixation period, and the mean of rates measured at multiple and seasons was representative of rates in the upper 10 m of water column over the year.

(b) A 24 hr N_2 fixation period, and the mean of rates measured at multiple and seasons was representative of rates over the year.

(c) Taking the mean of measurements over multiple sites and seasons.

(d) Taking the mean of 2000 to 2014 values.

(e) Modeled fluxes represent true fluxes.

(f) Summing biweekly samples from March 2006–February 2007.

(g) Mean from 1975 to 1987.

(h) Mean from 1980 to 1996.

(i) Mean from April 2009 to October 2010.

	Original units	mmol N m^{-2} yr^{-1}	Reference
	130 gmol N yr^{-1}	1.3 x 10^{14} mmol N yr^{-1}(g)	Howarth et al. 1996
	1,567,900 ± 58,470 metric tons yr^{-1}	1.1 x 10^{14} mmol N yr^{-1}(h)	Goolsby et al. 1999
	956480 tons yr^{-1}	6.8 x 10^{13} mmols N yr^{-1} (h)	Goolsby et al. 2001
	580,160 tons yr^{-1}	4.1 x 10^{13} mmols N yr^{-1} (h)	Goolsby et al. 2001
	31,360 tons yr^{-1}	2.2 x 10^{12} mmols N yr^{-1} (h)	Goolsby et al. 2001
	697 kg d^{-1}	328	Charette et al. 2013
	2819 kg d^{-1}	1330	Charette et al. 2013
	151 x 10^{3} tons yr^{-1}	1.1 x 10^{13} mmol N yr^{-1} (i)	Xu 2013
	179.3 x 10^{3} tons yr^{-1}	1.3 x 10^{13} mmol N yr^{-1} (i)	Xu 2013
	0.3 x 10^{-6} mol N m^{-2} d^{-1}	0.11	Hu et al. 2006
	695 x 10^{-6} mol N m^{-2} d^{-1}	254	Hu et al. 2006
	219 x 10^{-6} mol N m^{-2} d^{-1}	80	Hu et al. 2006
	0.46 g N m^{-2} d^{-1}	11,993	Paytan et al. 2006
	1.13 ± 0.87 mmol N m^{-2} d^{-1}	412 ± 318	Santos et al. 2008
	4.81 ± 0.18 mmol N m^{-2} d^{-1}	1756 ± 66	Santos et al. 2008
	2.28 ± 0.75 mmol N m^{-2} d^{-1}	832 ± 274	Santos et al. 2008
	1.7 ± 1.8 mmol N m^{-2} d^{-1}	621 ± 657	Santos et al. 2008
	3.1 ± 0.5 mmol N m^{-2} d^{-1}	1132 ± 183	Santos et al. 2008
	1.45 ± 0.8 mmol N m^{-2} d^{-1}	529 ± 292	Santos et al. 2008
	3160 ± 2080 μmol N m^{-2} d^{-1}	1153 ± 759(c)	Smith and Swarzenski 2012
	6990 ± 4853 μmol N m^{-2} d^{-1}	2551 ± 1771 (c)	Smith and Swarzenski 2012
	241 kg d^{-1}	113	Charette et al. 2013
	448 kg d^{-1}	210	Charette et al. 2013
	50 to 300 g m^{-2} yr^{-1}	3571 to 21,429 (c)	Bernard et al. 2014
	1 to 5 g m^{-2} yr^{-1}	71 to 357 (c)	Bernard et al. 2014
	2 to 10 g m^{-2} yr^{-1}	142 to 714 (c)	Bernard et al. 2014

Atchafalaya River system (MARS), although physical processes such as salinity stratification and circulation are also thought to contribute to the poor ventilation of these waters (Bianchi et al. 2010). Regardless, the nutrient load delivered to the northern GOM by the MARS from the upland agricultural areas of the US Midwest is significant (Table 3.1), and much of the research on N cycling in the GOM has consequently focused on the eutrophication and suboxia that has resulted from the respiration of phytoplankton biomass and riverine-derived organic matter (Dale et al. 2010). Additionally, on the west Florida shelf (WFS), regular occurrences of the harmful alga *Karenia brevis* have been linked to blooms of the N_2-fixing cyanobacteria *Trichodesmium* spp. that typically precede *K. Brevis* blooms, which affect the heath of humans as well as the health of other organisms in the ecosystem, prompting research into the processes that trigger and support *K. Brevis* blooms (Heil, Bronk, et al. 2014). Efforts to document the abundance, timing, and correlation of these two harmful algae blooms (HABs) has been supported by the Ecology of Harmful Algal Blooms (ECOHAB) project and has led to a better understanding of N cycling on the WFS (e.g., Lenes and Heil 2010; Bronk et al. 2014; Killberg-Thoreson et al. 2014; Mulholland et al. 2014).

After accounting for the focused interest in nutrient cycling in the northern GOM and on the WFS, however, N cycling in the other regions of the Gulf remains poorly characterized (Fig. 3.2). Most of the additional work that has been done in the GOM has been conducted in estuaries that transform and supply nutrients to the GOM, especially in Texas, Alabama, Louisiana, and northern Florida. Interest in N cycling in these nearshore regions is related both to the relevance of nutrient cycling to local fisheries and ecosystem health and to the relative accessibility of these locations compared to deepwater sites (Breitburg et al. 2009; de Mutsert et al. 2016). However, rate measurements made in these estuarine and nearshore regions are likely not representative of rates of these same processes in "open" Gulf settings; in many cases, rates of the same processes are significantly lower at locations slightly farther offshore. Moreover, in many cases, the sources of N to those estuaries are local, and N may be retained to differing degrees in each estuary and not enter the GOM proper. Together, these observations underscore the need for caution in extrapolating the rates reported here to broad regions of the GOM, as well as the need for more direct measurements in waters that are representative of the more spatially extensive open Gulf.

Sources of Nitrogen to the Gulf of Mexico

Riverine Fluxes

The majority of N in the GOM is imported by physical circulation from the western Atlantic, especially by the Loop Current. However, there are additional major fluxes of new N to the GOM, with the largest from the MARS (see Table 3.1). Other rivers also contribute N that may provide a locally significant source to the Gulf, although since these are relatively small contributions, we restrict our discussion here to the MARS. Most estimates of the total N (TN) flux from the MARS to the GOM are based on both nutrient concentration and water discharge measurements made by the US Geological Survey (USGS) at stations along the Mississippi and Atchafalaya Rivers and are estimated at ~1.1 to 1.3 $\times 10^{14}$ mmol N yr^{-1} (Howarth et al. 1996; Goolsby et al. 1999; Goolsby et al. 2001; Xu

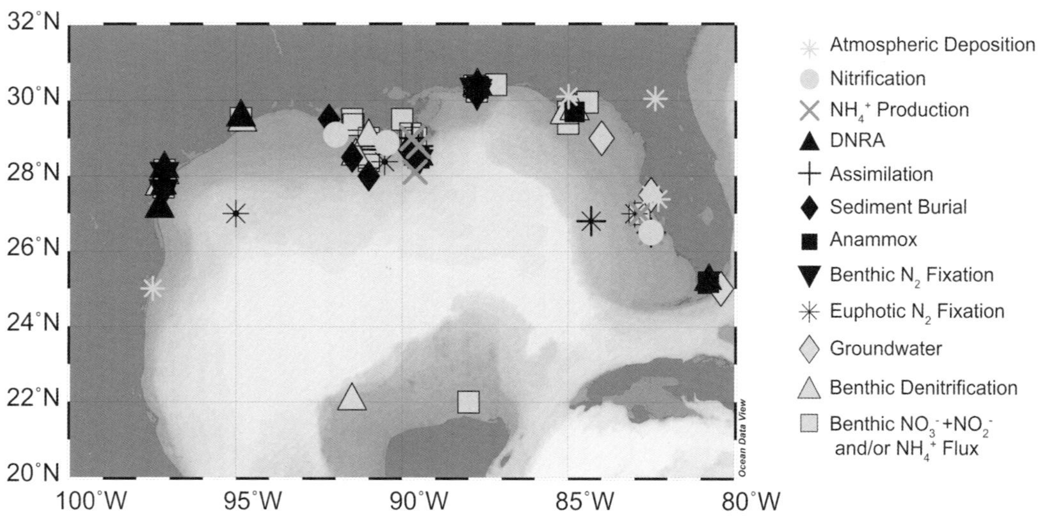

Figure 3.2. Locations of nitrogen flux measurements in the Gulf of Mexico and its estuaries (figure created using Ocean Data View [ODV]).

2013). Of that TN flux, nitrate+nitrite ($NO_3^-+NO_2^-$) contributes ~57% of the TN flux, or 6.8 $\times 10^{13}$ mmol N yr^{-1} (Goolsby et al. 2001); ammonium (NH_4^+) represents <1% of the TN flux (2.2 $\times$ 10^{12} mmol N yr^{-1}); and dissolved organic N (DON) represents ~43% of the TN flux, or 4.1 $\times$ 10^{13} mmol N yr^{-1} (Goolsby et al. 2001). These TN fluxes from the Mississippi to the North Atlantic are significant but are only ~50% to 60% of the TN inputs from the Amazon basin to the North Atlantic (Howarth et al. 1996).

While the Loop Current entrains nutrient-poor surface waters from the western Atlantic, the northern GOM receives enough inorganic and organic N from the MARS that surface waters are enriched in inorganic N at least seasonally, if not year-round. The concentration of $NO_3^-+NO_2^-$ in surface waters impacted by the river plume often exceeds 1 μM during the spring and summer, when peak river water and N discharge occur (Sylvan and Ammerman 2013), although nutrient concentrations in surface waters decrease with increasing salinity (e.g., Jochem et al. 2004; Sylvan et al. 2006; Quigg et al. 2011). Currently, the spatial extent over which these riverine-derived nutrients influence phytoplankton growth is unclear. However, the $\delta^{15}N$ ("$\delta^{15}N$," where ($\delta^{15}N = \{[(^{15}N/^{14}N)_{sample}/(^{15}N/^{14}N)_{reference}] - 1\}*1000$), and where the reference N_2 gas in air) of $NO_3^-+NO_2^-$ entering the GOM from the Mississippi River is between ~7‰ and 11‰ (Bryant-Mason et al. 2013; Fry and Allen 2003), which is significantly higher than that of mean deep-ocean NO_3^- $\delta^{15}N$, ~5‰ (Sigman et al. 2009). Together with salinity, measurements of the isotopic composition of NO_3^- in the GOM may prove useful in future efforts to trace the extent to which MARS-derived NO_3^- mixes into the GOM.

Of relevance to coastal nutrient loading in general and to the northern GOM hypoxic zone in particular is the fate of the riverine-derived N once it enters the Gulf. Given the problem of riverine-driven eutrophication, there is considerable interest in quantifying the extent to which N may be removed via denitrification and/or anaerobic ammonia oxidation (anammox) and burial of organic matter, compared with how much is retained in a bioavailable form such as NO_3^- and/or NH_4^+. While the oxygen concentration in the water column of the hypoxic zone does not drop low enough to support water-column denitrification and/or anammox (Childs et al. 2002), these N removal processes occur at significant rates in GOM sediments. This topic is addressed in more detail later, and while spatial and temporal heterogeneity precludes rigorous estimates, work in estuaries (An and Gardner 2002; Bernard et al. 2015; McCarthy et al. 2008) and in Florida Bay (Gardner and McCarthy 2009) indicates that under conditions of high organic matter loading a significant fraction of the N incorporated into the sediments may not be removed as N_2 gas but instead may be reduced to NH_4^+, the most biologically useful form of N (Burgin and Hamilton 2007). Consequently, the problem of eutrophication driven by riverine-derived N may not be mitigated by N loss processes commonly associated with low-oxygen environments, but instead a significant portion of the N delivered to the northern GOM may be retained and recycled as NH_4^+ and thus continue to aggravate nuisance algal blooms in the GOM.

Groundwater

After fluxes of N from the MARS, groundwater represents the next-largest source of N to the GOM. Fluxes of N in groundwater vary regionally, but estimates range from 0.1 to >10,000 mmol N m^{-2} yr^{-1} (see Table 3.1; Fig. 3.3). This magnitude and wide range in estimates of N flux from groundwater to the ocean are consistent with a comparison of prior estimates for groundwater N fluxes from pristine and anthropogenically impacted systems (Slomp and Van Cappellen 2004). In particular, significant groundwater fluxes have been identified on the WFS (up to 1153 ± 759 mmol dissolved inorganic N m^{-2} yr^{-1}) (Hu et al. 2006; Smith and Swarzenski 2012), in the Caloosahatchee River estuary (210 mmol TDN m^{-2} yr^{-1}) (Charette et al. 2013), and in the Florida Keys (11,993 mmol total inorganic N m^{-2} yr^{-1}) (Paytan et al. 2006). Even more modest groundwater N flux estimates, such as 1756 ± 66 mmol NH_4^+ m^{-2} yr^{-1} in the northern GOM (Santos et al. 2008) are 1 to 2 orders of magnitude larger than fluxes from other sources of new N to the GOM such as water-column N_2 fixation and atmospheric deposition. The role of groundwater as a significant source of N has been considered in the context of the nutrients required to support the *K. brevis* blooms on the WFS (Hu et al. 2006; Smith and Swarzenski 2012), although given the magnitude of the potential contribution of groundwater-derived N to the GOM, this flux appears poorly constrained with respect to its spatial and temporal variability, as well as to the fate of groundwater-derived N once it enters the GOM. Ongoing research in the Mississippi Bight has indicated that submarine groundwater fluxes may represent an episodic or seasonal input of N nutrients as well as trace metals, based on elevated levels of barium (Ba) and

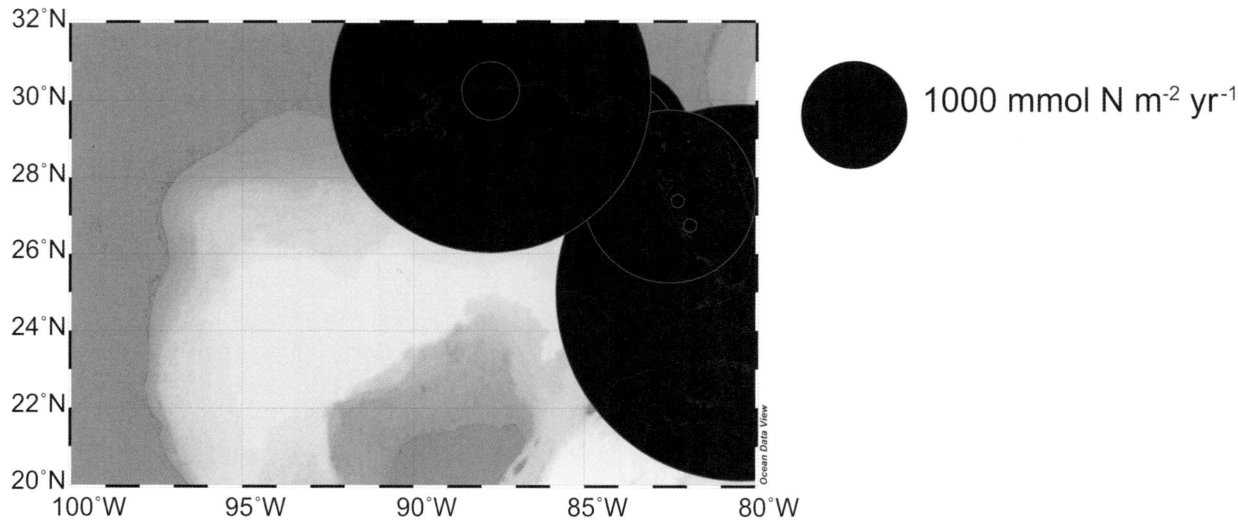

Figure 3.3. Groundwater nitrogen fluxes (figure created using ODV).

short-lived radium (Ra) isotopes that can be significant for the chemical mass balance in this part of the northern GOM (Joung et al. 2017).

N_2 Fixation

While biologically-mediated N_2 fixation is a process commonly associated with the euphotic zone of the water column, there is a growing appreciation of this process in benthic environments as well (Newell et al. 2016). Here we review measurements of N_2 fixation made in both the water column and benthic environments of the GOM.

WATER-COLUMN N_2 FIXATION

While euphotic-zone N_2 fixation is the primary source of new N to the ocean (Gruber 2004), significant challenges remain in constraining the spatial and temporal distribution of this process in the ocean, as well as in identifying the organisms responsible for quantitatively significant rates of marine N_2 fixation (Sohm et al. 2011). Unicellular and heterocystous diazotrophs have been observed in the GOM; however, their location, abundance, and N_2 fixation rates remain unconstrained (Luo et al. 2012). The globally significant, cyanobacterial diazotroph *Trichodesmium* spp. is perhaps the most conspicuous and well-studied marine N_2 fixer, and its distribution in the ocean is thought to be associated with warm, stratified waters where Fe is relatively abundant (Sohm et al. 2011). Perhaps unsurprisingly, then, *Trichodesmium* spp. are commonly observed in the GOM, especially on the WFS, where their occurrence has been linked to annual African dust deposition events (Lenes et al. 2001; Walsh et al. 2006; Lenes and Heil 2010), as well as in the northern GOM (Holl et al. 2007). Since *Trichodesmium* spp. colonies are macroscopic, they can be detected with the naked eye as well as from space (Walsh et al. 2006; Westberry and Siegel 2006), and documenting their abundance is more common than measuring rates of N_2 fixation. Estimating rates of water-column N_2 fixation has proven challenging; indeed, one of the reasons that so much of our understanding of marine N_2 fixation is based on *Trichodesmium* spp. is that this organism was identified before molecular techniques were available. Early measurements of N_2 fixation relied on hand-picking and concentrating *Trichodesmium* spp. colonies into densities sufficient to produce a signal in acetylene-reduction measurements (Capone 1993). However, N_2 fixation rates estimated using acetylene (C_2H_2) reduction to ethylene (C_2H_4) by the enzyme nitrogenase, the enzyme that reduces N_2 gas to NH_3, have proven complicated. Briefly, a 3:1 relationship is commonly assumed for the ratio of moles of acetylene reduced relative to the moles of N_2 fixed because during acetylene reduction to ethylene, only 1 of the triple bonds between the 2 carbon atoms is broken, whereas in measurements of $^{15}N_2$ fixation, all 3 bonds between the N atoms are broken (Montoya et al. 1996). In particular, studies in the GOM that have measured acetylene reduction and $^{15}N_2$ fixation in parallel have found that the ratio of acetylene reduced to ethylene compared to the quantity of $^{15}N_2$ gas incorporated into particulate N ranged from 3.3:1 to 15.8:1 (Mulholland et al. 2006). Additional methodological problems have been pointed out by studies showing that without allowing $^{15}N_2$ gas sufficient time

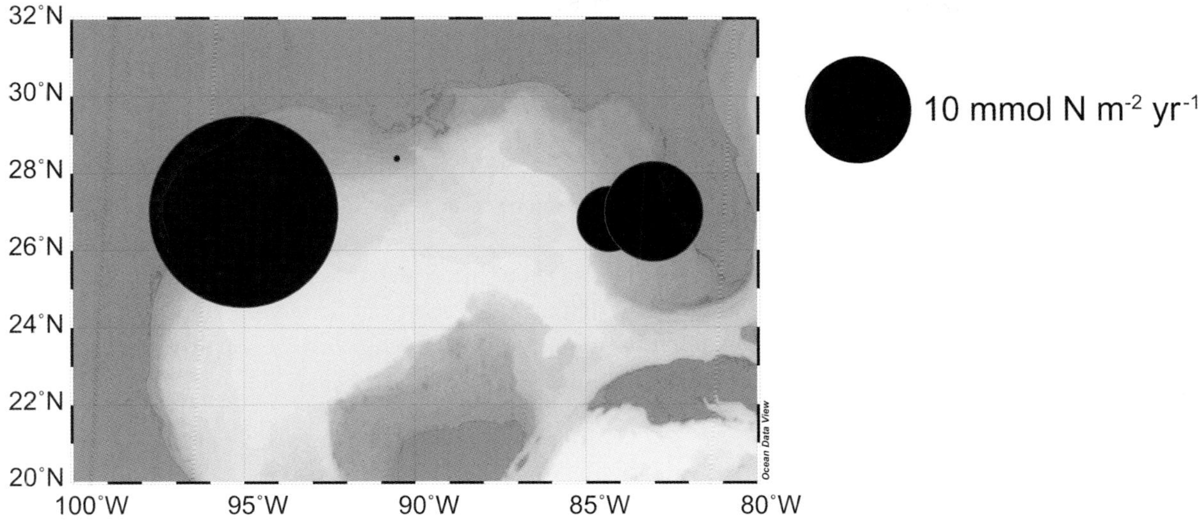

Figure 3.4. Water-column N_2 fixation (figure created using ODV).

to equilibrate in the water it is added to, N_2 fixation rates estimated using $^{15}N_2$ addition methods may systematically underestimate rates of N_2 fixation (Grosskopf et al. 2012; Wilson et al. 2012).

Given the methodological uncertainties in N_2 fixation-rate estimates, we restrict our discussion here to water-column-rate measurements made using $^{15}N_2$ fixation techniques, where bulk water-column N_2 fixation rates were measured (i.e., rates inclusive of all diazotrophic activity) and *Trichodesmium* spp. colonies, when observed, were not preferentially concentrated (see Table 3.1; Fig. 3.4). In the GOM, rates of N_2 fixation have been measured on the WFS and range from 3 to 8 mmol N m^{-2} yr^{-1} (Mulholland et al. 2006; Mulholland et al. 2014), although other reports document *Trichodesmium* spp. abundance on the WFS without reporting N_2 fixation rates (e.g., Lenes et al. 2001; Lenes and Heil 2010). N_2 fixation by *Trichodesmium* spp. has been invoked as an important N source for the *K. brevis* blooms that often follow *Trichodesmium* spp. blooms on the WFS (Heil, Dixon, et al. 2014; Mulholland et al. 2014). Additionally, significant N_2 fixation rates have been documented in the northern GOM, 30.8 ± 6.5 mmol N m^{-2} yr^{-1} (Holl et al. 2007) (see Fig. 3.3). Recent work has measured low rates of N_2 fixation, 0.03 ± 0.14 mmol N m^{-2} yr^{-1}, off the Mississippi River delta (Weber et al. 2016). We note that while the Holl et al. (2007) N_2 fixation rates are the highest reported from the GOM (see Fig. 3.4), that rate is likely an underestimate, as the value reported here converted their 250 m water-column integrated rate to surface-only rates. Since *Trichodesmium* spp. are known to focus their distribution and activity in surface waters, and our calculation assumes an equal distribution of N_2 fixation rates throughout the water column, we have likely underestimated the true surface-water N_2 fixation rates reported by Holl et al. (2007). Regardless, the rates of N_2 fixation thus far reported from the GOM are modest, and the high rates from the northern GOM are comparable to rates measured near Hawaii (Montoya et al. 2004), but even these highest rates from the GOM are several-fold lower than rates measured in the tropical North Atlantic (Capone et al. 2005).

In spite of the presumably significant role for diazotrophs as a N source for the *K. brevis* blooms on the WFS and the significant N_2 fixation rates measured in the northern GOM (Holl et al. 2007), few rates of N_2 fixation have been made in the open waters of the GOM. Quantifying N_2 fixation rates in these more expansive regions of the GOM could constrain the importance of the GOM as a source of new N to the ocean, in particular to the western North Atlantic, where geochemical signatures, including elevated thermocline NO_3^- to phosphate (PO_4^{3-}) concentration ratios and low NO_3^- $\delta^{15}N$ values, have been documented and associated with N_2 fixation in the Atlantic basin (Gruber and Sarmiento 1997; Hansell et al. 2004; Knapp et al. 2008). However, a portion of those geochemical signatures could originate in the GOM and be exported to the North Atlantic thermocline; quantifying this will require characterizing the spatial and temporal variability in N_2 fixation rates over a larger portion of the GOM.

BENTHIC N_2 FIXATION

While water-column N_2 fixation, especially by *Trichodesmium* spp. on the WFS, receives significant attention, prior work indicates that benthic N_2 fixation in the estuaries adjacent to the GOM is well documented and that these rates exceed those measured in the water column by 1 to 3 orders of magnitude (see Table 3.1; Fig. 3.5). In particular, benthic N_2 fixation-rate estimates in Texas estuaries range from 134 to 6561 mmol N m^{-2} yr^{-1} (Bruesewitz et al. 2013; McCarthy et al. 2008), and in Alabama estuaries, reported rates range from 19 to 540 mmol N m^{-2} yr^{-1}). In the northern GOM benthic N_2 fixation rates of 386 ± 515 mmol N m^{-2} yr^{-1} have been reported (McCarthy et al. 2015); rates of benthic N_2 fixation measured in other coastal locations are comparable (Knapp 2012). These considerable fluxes of new N to the GOM raise the question, Are benthic environments a net source or sink of fixed N? Similar to measuring N_2 fixation in the water column, quantifying N_2 fixation in benthic environments is challenging. Regardless, studies comparing the net flux of N_2 gas into and out of sediments in Alabama estuaries found that during the summer, when overlying water-column NO_3^- concentrations were <0.2 μM, the net flux of N_2 gas was into, not out of, the sediments (Bernard et al. 2014). While N_2 fixation carried out by cyanobacterial diazotrophs in both the water column and shallow sediments may be somewhat inhibited by water-column NO_3^- and especially by NH_4^+, it seems that N_2 fixation carried out in deep, dark sediments, often by sulfate-reducing bacteria and methanogens, is largely insensitive to dissolved inorganic N concentrations (Knapp 2012). Moreover, in the northern GOM, N_2 fixation and denitrification were observed simultaneously occurring in the same sediments (McCarthy et al. 2015). These observations of high rates of benthic N_2 fixation occurring in the same sediments that simultaneously support high rates of N loss and NO_3^- and NH_4^+ flux out of the same sediments indicate that we have much to learn, not only about the sensitivities of benthic N_2 fixation but also about the degree to which benthic N_2 fixation may diminish any "ecosystem services" provided by sediments to reduce N loading to coastal ecosystems (Newell et al. 2016).

Atmospheric Nitrogen Deposition

Measurements of atmospheric deposition of N to the Gulf are necessarily indirect—there are no atmospheric deposition sampling locations in the GOM itself. Consequently, estimates of both wet deposition and aerosol fluxes to the GOM are based on measurements made at terrestrial locations adjacent to the GOM, as well as from modeling studies that incorporate terrestrial sampling data. Interest in quantifying N deposition fluxes to marine surface waters has increased over the last decade as modeling studies have suggested that anthropogenic sources of atmospheric N, primarily from fossil-fuel consumption, have resulted in a 3- to 4-fold increase in total N deposition to the oceans since 1860 (Duce et al. 2008). However, detailed studies of the chemical and isotopic composition of organic N in atmospheric deposition conducted at Bermuda have suggested a potential marine source for a significant portion of atmospheric N deposition (Altieri et al. 2016). In this case, the cycling of N between the sur-

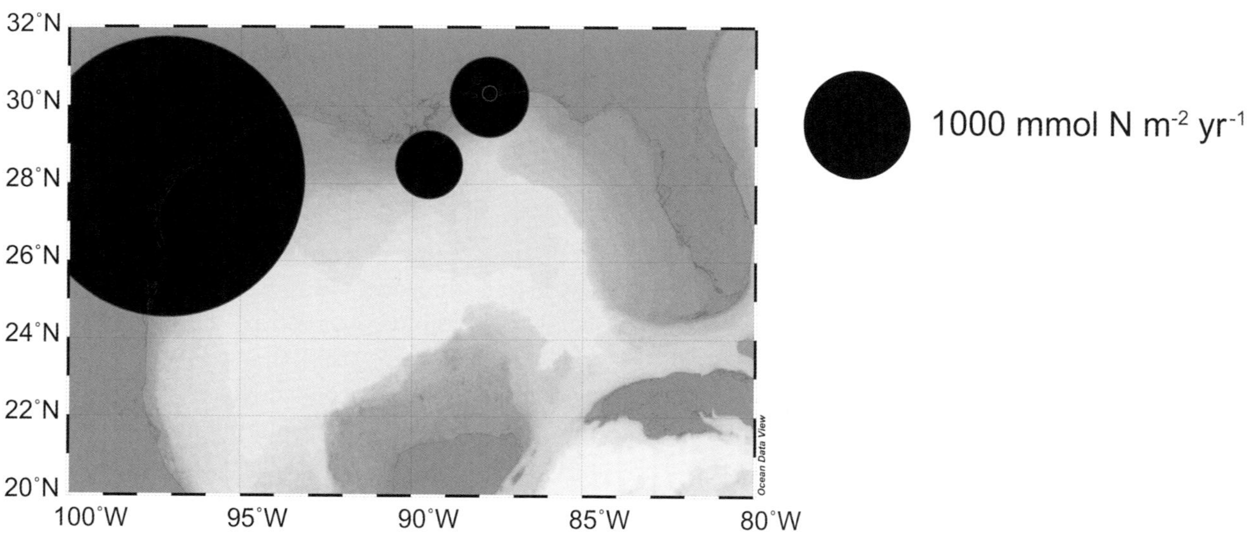

Figure 3.5. Benthic N_2 fixation (figure created using ODV).

face ocean and atmosphere would reduce the magnitude as well as ecological impact of the total N deposition flux. This may be relevant in the case of N deposition fluxes to the GOM, which are largely based on wet and dry deposition-flux measurements made on land at locations adjacent to the GOM. At these sites, N that had previously been attributed to anthropogenic and/or natural terrestrial sources may actually have a marine source, especially when air-mass back trajectories indicate that the depositing air mass was recently over the GOM, in particular during times of high rates of marine primary productivity (Altieri et al. 2016). To better constrain whether the ocean is a net source or sink for N in atmospheric deposition, future efforts will need to better constrain the spatial and temporal variation in atmospheric N deposition fluxes and continue to use tools such as the chemical and isotopic composition of N in atmospheric deposition, together with back-trajectory analysis, to identify the potential sources of N in atmospheric deposition (Hastings et al. 2003; Knapp et al. 2010; Altieri et al. 2012; Hastings et al. 2013).

Estimates of atmospheric N deposition fluxes to the GOM appear comparable to water-column N_2 fixation fluxes (see Table 3.1) and are based on measurements made through both the National Atmospheric Deposition Program (NADP; http://nadp.isws.illinois.edu/) and the US Environmental Protection Agency's Clean Air Status and Trends Network (CASTNET; https://www.epa.gov/castnet). Sampling efforts from the NADP in northern Florida at Sumatra and at the Chassahowitzka National Wildlife Refuge, as well as in Corpus Christi, Texas, indicate that wet deposition fluxes of N range from 10 to 63 mmol N m^{-2} yr^{-1}, with wet deposition fluxes of NO_3^- $+NO_2^-$ typically 3- to 6-fold higher than fluxes of NH_4^+ (National Atmospheric Deposition Program [NRSP-3] 2015) (see Fig. 3.6).

While fluxes of aerosols (dry deposition) to the ocean are more challenging to quantify, estimates from the CASTNET suggest that like wet deposition, $NO_3^-+NO_2^-$ in aerosols is more abundant than NH_4^+ and that the total contribution of N from aerosols is less than that from wet deposition. The average annual dry-deposition fluxes of NH_4^+ and $NO_3^-+NO_2^-$ measured at the CASTNET sampling location in Sumatra, Florida, are 1.4 ± 0.5 mmol N m^{-2} yr^{-1} and 18.8 ± 6.2 mmol N m^{-2} yr^{-1}, respectively (US Environmental Protection Agency Clean Air Markets Division 2016). These estimates are similar to those made in northern Florida at agricultural and forested locations, where the wet-deposition flux of NH_4^+ ranges from 4 to 5 mmol N m^{-2} yr^{-1} and the wet-deposition flux of NO_3^- $+NO_2^-$ were roughly twice that, ranging from 8 to 9 mmol N m^{-2} yr^{-1} (Katz et al. 2009; see Table 3.1). Additionally, modeling of data collected at the NADP and CASTNET locations has indicated that N deposition fluxes to the eastern GOM may exceed those to the western GOM. Modeling efforts from Prospero et al. (1996) as reported in Howarth et al. (1996) calculated atmospheric fluxes of NOy (NO_3^- + HNO_3) to the western GOM of 26 mmol N m^{-2} yr^{-1} and 58 mmol N m^{-2} yr^{-1} to the eastern GOM, while NH_4^+ deposition fluxes to the western and eastern GOM were calculated to be 20 and 16 mmol N m^{-2} yr^{-1}, respectively (see Fig. 3.6). Generally, these estimates

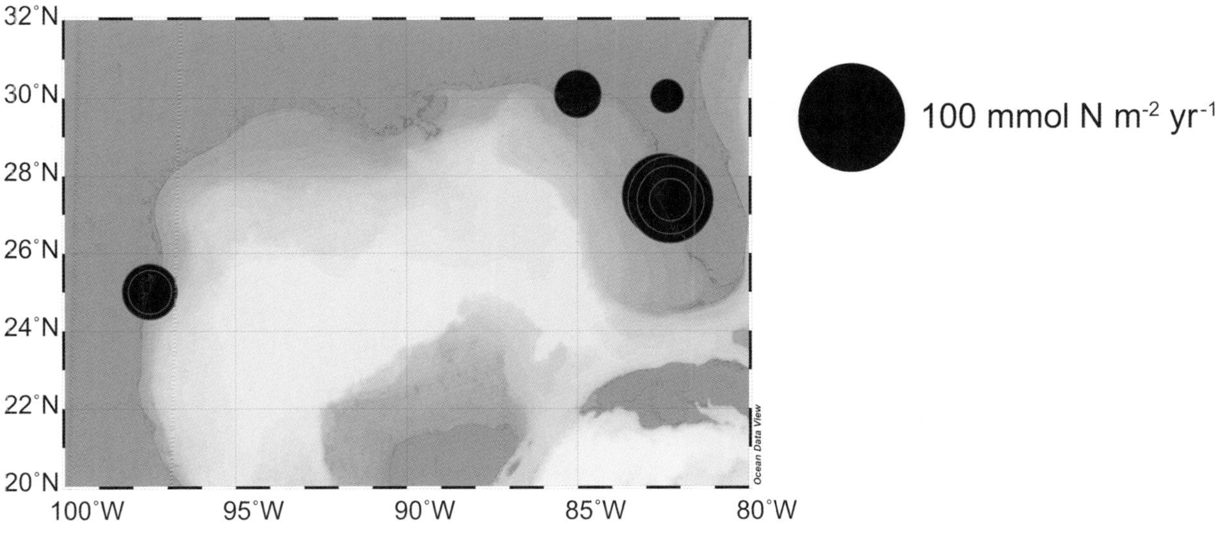

Figure 3.6. Atmospheric N deposition (figure created using ODV).

of N fluxes to the GOM are higher than for estimates of the NO_3^-+NO_2^- deposition to oligotrophic marine locations. For example, annual NO_3^-+NO_2^- flux estimates at Bermuda are 5.6 mmol N m^{-2} yr^{-1} (Hastings et al. 2003), and for the TN flux to Bermuda are 10.0 mmol N m^{-2} yr^{-1} (Knapp et al. 2010).

Sinks of Nitrogen in the Gulf of Mexico

Motivation for quantifying rates of N removal from the GOM are largely motivated by nutrient-driven eutrophication of the northern GOM and the impact that hypoxia has on the ecosystem and those deriving their livelihood from fisheries impacted by the nutrient loading (e.g., Turner and Rabalais 1991, 2003; Rabalais et al. 2001, 2002). Quantifying the degree to which natural biogeochemical processes may mitigate nutrient loading to the GOM by either reduction of inorganic N to N_2 gas or burial in sediments depends on quantifying the high degree of spatial and temporal variability in the strength of these N sinks. Additionally, an improved understanding is required of the sensitivities of biologically-mediated N sinks such as denitrification and anammox, and what triggers a switch between these processes and dissimilatory nitrate reduction to ammonia (DNRA), which will retain N in its most bioavailable form in these sensitive environments. Finally, we note that many studies of N loss and transformation processes use stable isotope labeling to track rates of processes such as denitrification, anammox, and DNRA. However, such studies often report "potential" rates of these processes because isotopically-labeled substrates may be added at concentrations that significantly exceed those typically observed in the environment, leading to experimentally derived potential rates that exceed rates occurring naturally in the sediments (Hou et al. 2012; McCarthy et al. 2015). For biologically-mediated N cycling and removal processes, Table 3.2 notes whether substrates were added to estimate a flux and, if so, which kind of substrate and how much.

Sedimentation

One of the primary pathways by which N is removed from the GOM is via burial of sinking particulate organic N (PON) into the sediments. Estimates for the N flux to the sediments span a range with highest fluxes resulting from measurements of sinking particulate N fluxes from the upper water column, followed by short-term sedimentary N accumulation rates, to short- and long-term burial rates, with significant decreases in the magnitude of the flux at each step (Table 3.2). Specifically, while the PON sinking flux in the northern GOM has been estimated to be 2581 ± 2164 mmol N m^{-2} yr^{-1} (Redalje et al. 1994), the mean of short-term benthic PON accumulation at multiple sites in the northern GOM was 3521 ± 3214 mmol N m^{-2} yr^{-1}, with short-term PON burial rates of 282 ± 229 mmol N m^{-2} yr^{-1} and long-term PON burial rates of 225 ± 179 mmol N m^{-2} yr^{-1} (Gordon et al. 2001; see Fig. 3.7). Similarly, studies in the northern GOM have found decreasing PON burial rates with increasing distance from the MARS system, with mean organic N burial rates across the inner shelf, outer shelf, and slope regions of 123 ± 134 mmol N m^{-2} yr^{-1} (Gordon and Goñi 2004). Carbon

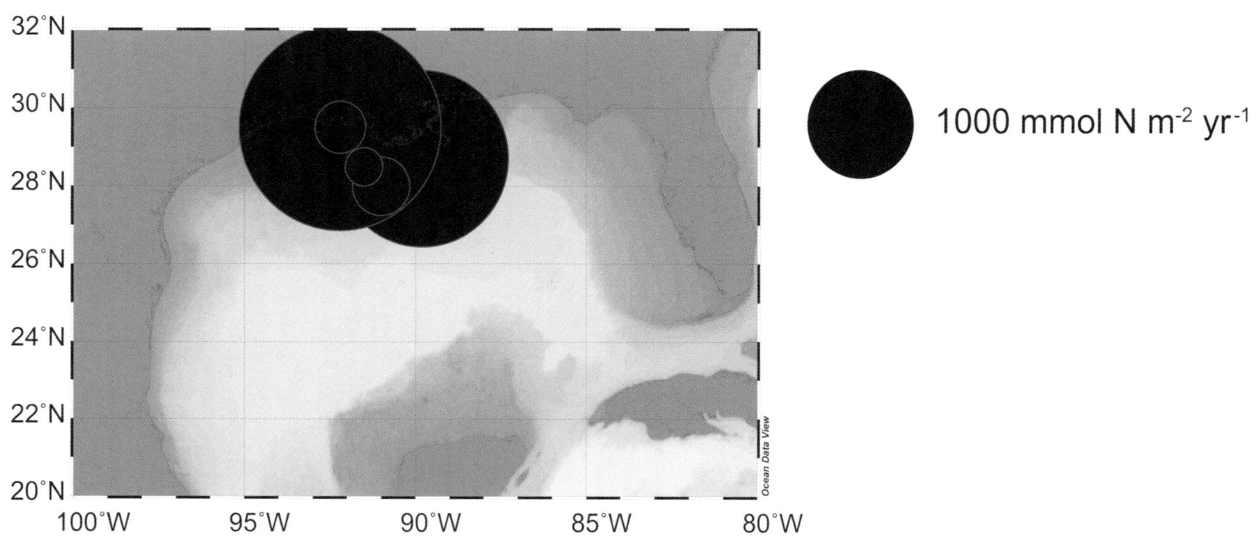

Figure 3.7. Sedimentary N burial (figure created using ODV).

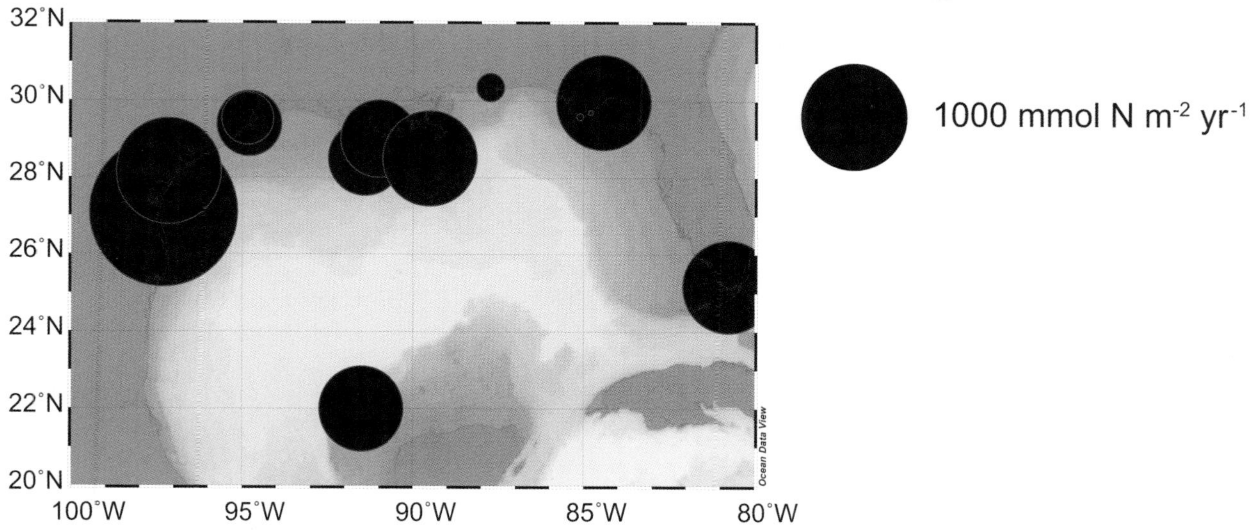

Figure 3.8. Benthic denitrification (figure created using ODV).

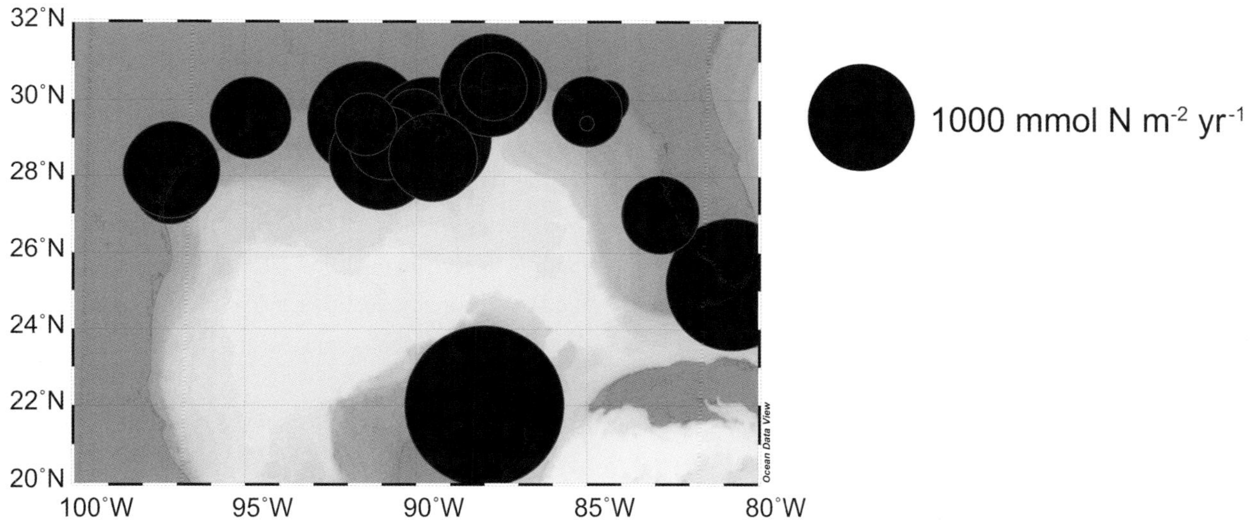

Figure 3.9. Benthic NH_4^+ flux (figure created using ODV).

isotopic measurements indicate that a large fraction of the organic matter buried in these sediments is terrestrial in origin (Goñi et al. 1998), and other work has estimated that burial of organic N accounts for ~5% to 20% of the organic N inputs from the MARS to the GOM (Gordon et al. 2001).

The magnitude of this N burial flux is reduced to the degree that benthic nutrients, especially NH_4^+, that were produced from the remineralization of organic N are released back to the water column. While benthic nutrient fluxes from the sediments to the water column exhibit large variation in space and time in the GOM (Figs. 3.8 and 3.9), it is worth noting that the flux of NH_4^+ from the sediments to the overlying water column is comparable to the magnitude of organic N burial in sediments (see Table 3.2). Benthic denitrification and anammox may also reduce the magnitude of organic N buried in the sediments through the consumption of the inorganic products of organic matter respiration, although the source of NO_3^- and/or NO_2^- for these processes may be the overlying water column, which would not diminish the magnitude of sedimentary N burial.

Benthic Denitrification

While the hypoxic zone in the shallow waters of the northern GOM is notorious for apparently having bottom

Table 3.2. Nitrogen loss fluxes from the Gulf of Mexico and its estuaries

N removal flux	Location	Method
Water column denitrification		
	Northern GOM	Acetylene reduction
Benthic denitrification		
	Northern GOM	Sediment core incubations, N_2 production
	TX estuaries	Sediment core incubations, N_2 production
	TX estuaries	Sediment core incubations, N_2 production
	TX estuaries	Benthic chambers, N_2 production
	TX estuaries	Sediment core incubations, ^{15}N labeling
	Northern GOM	Sediment core incubations; potential denitrification rate after 100 μM NO_3^-addition, 10% acetylene inhibition
	TX estuaries	Sediment core incubations; potential denitrification rate after 100 μM NO_3 addition, 10% acetylene inhibition
	TX estuaries	Sediment core incubations
	FL Bay	Sediment core incubations
	FL estuary	Sediment core incubations
	Northern GOM cold seeps	Sediment slurry incubation, 75 μM $^{15}NO_3^-$ addition
	Northern GOM	Sediment core incubations
	TX estuaries	Sediment core incubations, 10 μM $^{15}NO_3^-$ addition
	FL estuary	Sediment core incubations, 50 μM $^{15}NO_3^-$ addition
	AL estuaries	Sediment core incubations, 100 μM $^{15}NO_3^-$ addition
	Northern GOM	Sediment core incubations, 50 μM $^{15}NO_3^-$ addition
Benthic anammox		
	FL Bay	Sediment core incubations, 5 μM $^{15}NH_4^+$ addition
	Northern GOM	Sediment core incubations, 50 μM $^{15}NO_3^-$ addition
	AL estuaries	Sediment slurry with 100uM $Na^{15}NO_3^-$
	Northern GOM	Sediment core incubations, 5 μM $^{15}NH_4^+$ addition
Sinking N lux, burial		
	Northern GOM	Sinking PN flux
	Northern GOM	Sediment core
	Northern GOM	Sediment core
	Northern GOM	Sediment core
	Northern GOM	Sediment core

Note: To convert between original flux units and the common units of mmol m^{-2} yr^{-1}, assumptions included the following:

(a) Mean of multiple sites.

(b) Taking the mean of measurements over multiple sites and seasons.

(c) Mean of multiple seasons and diel cycles.

(d) Mean of multiple methods and sites.

(e) Mean from shallow and deep sediments.

(f) Includes N_2 produced by anammox.

(g) Mean of multiple seasons.

(h) Mean of short-term accumulation rates.

(i) Mean of long-term accumulation rates.

(j) Mean of long-term burial rates.

(k) Mean of inner shelf, outer shelf, and slope.

Type of N	Original units	mmol N m^{-2} yr^{-1}	Reference
N_2O	Below detection	0 (a)	Childs et al. 2002
N_2 gas	88 ± 54 μmol N m^{-2} hr^{-1}	771 ± 473 (b)	Seitzinger 1987
N_2 gas	27.4 ± 20.5 μmol N m^{-2} hr^{-1}	240 ± 180 (b)	Yoon and Benner 1992
N_2 gas	18.5 ± 15.8 μmol N m^{-2} hr^{-1}	162 ± 138 (b)	Zimmerman and Benner 1994
N_2 gas	1 ± 1 mmol N m^{-2} d^{-1}	365 ± 365 (c)	An and Joye 2001
28+29+30N_2 production	21.7 ± 40.0 μmol N m^{-2} hr^{-1}	1901 ± 3504 (b)	An and Gardner 2002
28+29+30N_2 production	70 ± 26.3 μmol N m^{-2} hr^{-1}	613 ± 230 (a)	Childs et al. 2002
	22.6 ± 10.5 μmol N m^{-2} hr^{-1}	198 ± 92 (d)	Bernot et al. 2003
Net N_2 fluxes	27.0 ± 109.2 μmol N m^{-2} hr^{-1}	237 ± 957 (b)	Gardner et al. 2006
N_2 gas production	82.6 ± 90.5 μmol N m^{-2} hr^{-1}	724 ± 793 (b)	Gardner and McCarthy 2009
N_2 gas production	9.6 ± 13.8 μmol N m^{-2} d^{-1}	4 ± 5 (b)	Gihring et al. 2010
28+29+30N_2 production	53.5 ± 37.5 μmol N m^{-2} hr^{-1}	469 ± 329 (e)	Bowles and Joye 2011
N_2 gas production	1.4 mmol N m^{-2} d^{-1}	511 (b)	Lehrter et al. 2012
28+29+30N_2 production	109 ± 120 μmol N m^{-2} hr^{-1}	955 ± 1051 (b)	Bruesewitz et al. 2013
N_2 gas	5.4 ± 4.9 μmol N m^{-2} d^{-1}	2 ± 2 (a)	Canion et al. 2014
29+30N_2	7.7 ± 15.2 μmol N m^{-2} hr^{-1}	67 ± 133 (b)	Bernard et al. 2015
28+29+30N_2production	88.1 μmol N m^{-2} hr^{-1}	772 (b), (f)	McCarthy et al. 2015
$^{29}N_2$	0	0 (b)	Gardner and McCarthy 2009
N_2 gas production	0 μmol N m^{-2} d^{-1}	0 (a)	Canion et al. 2014
	0.40 +/- 0.17 μmol N m^{-2} hr^{-1}	4 ± 2 (b)	Bernard et al. 2015
$^{29}N_2$ production	8.8 to 25. 2 μmol N m^{-2} hr^{-1}	77 to 221 (b)	McCarthy et al. 2015
	0.099 ± 0.083 g N m^{-2} d^{-1}	2581 ± 2164 (g)	Redalje et al. 1994
	4.93 ± 4.50 mg N cm^{-2} yr^{-1}	3521 ± 3214 (h)	Gordon et al. 2001
	0.395 ± 0.32 mg N cm^{-2} yr^{-1}	282 ± 229 (i)	Gordon et al. 2001
	0.315 ± 0.251 mg N cm^{-2} hr^{-1}	225 ± 179 (j)	Gordon et al. 2001
	1.71 ± 1.87 g N m^{-2} yr^{-1}	122 ± 134 (k)	Gordon and Goñi 2004

waters with insufficient oxygen to support macrofauna such as fish and shrimp, oxygen is never completely removed from the water column, so this topic remains controversial (see Bianchi et al. 2010). Consequently, processes such as denitrification and anammox that occur in water columns with <5 μM oxygen concentrations have not been documented in the northern GOM hypoxic zone (Childs et al. 2002). However, the sediments in the northern GOM become oxygen depleted quickly, and sedimentary N loss by denitrification has been well documented in this region. According to the other biologically-mediated processes reviewed in this chapter, a note on the methods used to quantify denitrification rates is warranted. In some cases, rates of denitrification have been quantified using isotopically-labeled NO_3^- (one of the primary substrates for denitrification) to better detect the end product of denitrification, N_2 gas. In some of these studies, concentrations of added $^{15}NO_3^-$ significantly exceed background levels, which may potentially overestimate rates of $^{15}N_2$ production. Table 3.2 notes if and how much $^{15}NO_3^-$ was added to calculate denitrification rates; these rates should be viewed as potential denitrification rates and likely represent an upper bound to actual benthic denitrification rates. For example, in Copano Bay, Texas, rates of benthic denitrification, anammox, and DNRA were near 0 until sediment cores received a 100 μM $^{15}NO_3^-$ spike (Hou et al. 2012). Still other methods for measuring denitrification include the "acetylene-block" technique, which inhibits the reduction of nitrous oxide (N_2O), an intermediate in denitrification, to N_2 gas (Yoshinari and Knowles 1976). N_2O is readily quantifiable on a gas chromatograph with much lower levels of atmospheric contamination than N_2, making the acetylene-block technique favorable for detecting denitrification. However, acetylene also inhibits nitrification (Hynes and Knowles 1978); to the extent that benthic nitrification provides the NO_3^- for denitrification (which is often the case in benthic environments), the acetylene-block method will underestimate benthic denitrification rates. Studies that use acetylene to estimate benthic denitrification rates have been noted in Table 3.2.

Regardless of the method used, numerous studies have found that benthic denitrification is a significant, if not dominant, pathway by which riverine-derived N is removed from the northern GOM. In a northern Florida bay, of the N leaving the sediments as NO_3^-, NO_2^-, NH_4^+, and N_2, 69% left as N_2, corresponding to denitrification rates of 771 mmol N m^{-2} yr^{-1} (Seitzinger 1987). In sediments in the northern GOM, prior studies have found that benthic denitrification removes 39% to 68% of the total N load from the MARS, corresponding to rates of 511 and 772 mmol N m^{-2} yr^{-1} (see Table 3.2; Fig. 3.10; Lehrter et al. 2012; McCarthy et al. 2015).

Denitrification has been measured in numerous estuaries encircling the northern GOM, with rates ranging from 4 ± 5 mmol N m^{-2} yr^{-1} in a northern Florida estuary (Gihring et al. 2010) to 1901 ± 3504 mmol N m^{-2} yr^{-1} in Texas estuaries (An and Gardner 2002), although typically denitrification rates in estuaries in the northern GOM and in Florida Bay range from 200 to 700 mmol N

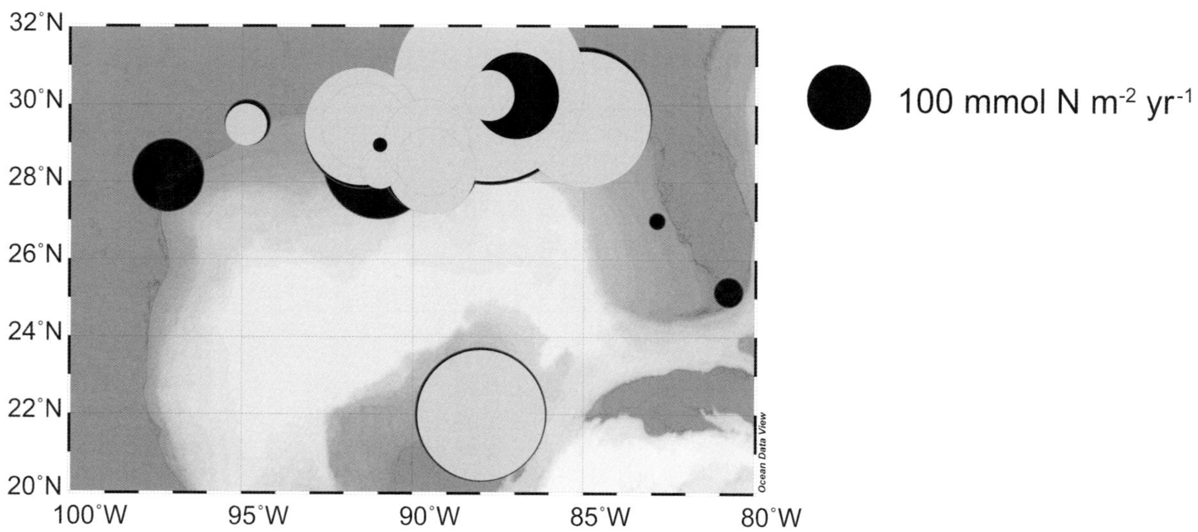

Figure 3.10. Benthic NO_3^-+NO_2^- flux. Fluxes to the water column are in black circles; fluxes to the sediments are in gray circles (figure created using ODV).

m^{-2} yr^{-1} (e.g., Bernot et al. 2003; Childs et al. 2002; Gardner and McCarthy 2009; Gardner et al. 2006; Yoon and Benner 1992; see Table 3.2; Fig. 3.10). Finally, substantial rates of benthic denitrification have been measured at cold seeps in the GOM in addition to that in estuaries and locations proximal to the MARS; mean rates of 469 ± 329 mmol N m^{-2} yr^{-1} have been found in sediments inhabited by *Beggiatoa* spp. (Bowles and Joye 2011). These rates of benthic denitrification largely fall within the range found in other studies (Devol 2015).

Benthic denitrification is a primary loss pathway for N in the GOM; however, as the process is biologically-mediated, the spatial and temporal distribution and rates of the process are all dependent on the sensitivities of the microbes carrying out denitrification. Benthic denitrification appears to be sensitive to temperature, both generally and in northern GOM estuaries (Canion et al. 2014). In Texas estuaries denitrification rates were nearly 10-fold higher in the summer than winter, and temperature could account for ~50% of the seasonal difference in rates (Zimmerman and Benner 1994). Perhaps unsurprisingly, benthic denitrification rates have also been positively correlated with sediment organic carbon content; in northern GOM sediments near an oil well head rates decreased with distance from well heads and as sediment organic content decreased (DeLaune et al. 2005).

Similar to rates in the water column, rates of denitrification in sediments are also sensitive to oxygen concentrations, although the relationship between denitrification rates and oxygen concentrations is not always straightforward. For example, while McCarthy et al. (2015) found that the highest N_2 fluxes out of northern GOM sediments occurred when bottom-water oxygen concentrations were low, An and Joye (2001) found that oxygen produced by benthic microalgae increased rates of denitrification because the increased oxygen supply enhanced rates of nitrification or the oxidation of organic N to NO_3^-, thus increasing the supply of one of the substrates for denitrification. Consequently, while oxygen may inhibit denitrification because oxygen is energetically more favorable as an electron acceptor than NO_3^-, insufficient oxygen may also limit the production of NO_3^- by nitrification. A similar relationship was observed in the northern GOM by McCarthy et al. (2015), who found evidence for a strong coupling between nitrification, denitrification, and benthic organic matter consumption.

Further complicating the relationship between oxygen and benthic denitrification is the degree to which anoxia has developed in the sediments and alternative electron acceptors are used to respire organic matter. In particular, when sediments support sulfate reduction, the sulfide that accumulates can inhibit denitrification to the extent that the NO_3^- being denitrified is supplied by the nitrification of benthic organic N. This is due to the ability of sulfide to inhibit nitrification, which has been shown to significantly limit the mitigation of N loading in benthic environments by denitrification (Joye and Hollibaugh 1995). Given that benthic denitrification represents a substantial loss term of N from the GOM, identifying the source of the NO_3^- being removed, that is, whether it is from the overlying water column or nitrified benthic organic matter, is relevant to estimates of how much nutrient loading is reduced by this process in the GOM. In a northern Florida bay, Gihring et al. (2010) found that at a relatively exposed site, the NO_3^- being consumed by denitrification originated from benthic nitrification of organic matter, while at a more protected site NO_3^- in the overlying water column supplied most of the NO_3^- for benthic denitrification. However, during the summer in an Alabama estuary when sulfide was abundant (i.e., mM concentrations), denitrification was supported by NO_3^- in the overlying water column (Bernard et al. 2014). In the same location, while sulfide especially inhibited denitrification during the summer, sulfide in the sediments reduced the N loss via denitrification year-round (Bernard et al. 2015). In the GOM, often when sediments become sulfidic, instead of being denitrified, NO_3^- is often reduced to NH_4^+ by DNRA, which is then released to the overlying water column where it will continue to support primary productivity (An and Gardner 2002). Quantifying this partitioning of NO_3^- reduction to N_2 gas (via denitrification and/or anammox) in comparison to NH_4^+ production (via DNRA) has increasingly been the focus of research in the GOM, especially at sites receiving considerable N input from the MARS, and has led some to question whether the mitigation of coastal N loading by denitrification has been overestimated (Burgin and Hamilton 2007).

Benthic Anaerobic Ammonia Oxidation

Anammox has only recently been measured in benthic environments. While there is evidence that anammox is an active N removal pathway in GOM sediments, rates of anammox are relatively low compared to benthic denitrification rates when the process is detected at all. For example, in sediment cores from the northern GOM spiked with 5 μM $^{15}NH_4^+$, potential anammox rates ranged from 77 to 221 mmol N m^{-2} yr^{-1}, and anammox was estimated to potentially remove up to 29% of the total N load from the MARS (McCarthy et al. 2015; see Table 3.2). However, significantly lower anammox rates were observed in Alabama estuary sediments spiked with 100 µM $^{15}NO_3^-$ (4 ± 2 mmol N m^{-2} yr^{-1}) (Bernard et al. 2015). In other studies in Florida Bay (Gardner and McCarthy 2009) and in the northern GOM (Canion et al. 2014) where denitrification was observed, anammox could not be detected. In general, rates of benthic anammox observed in the GOM are low relative to observations from other coastal locations (e.g., Thamdrup and Dalsgaard 2002). While anammox is a chemoautolithotrophic process and may thus be expected to occur at lower rates relative to denitrification, further investigation of anammox is warranted to better constrain the pathways by which inorganic N is removed from the GOM.

Nitrogen Transformations in the Gulf of Mexico

Benthic Nutrient Fluxes

Significant fluxes from the sediments to the overlying water column of N as NO_3^-+NO_2^-, dissolved organic N (DON), and especially as NH_4^+ have been documented in the GOM (Table 3.3). However, benthic nutrient fluxes do not represent a source of new N to the Gulf but a recycling of N that was previously deposited as organic N and/or NO_3^- that diffused into the sediments from the overlying water column. Consequently, the benthic nutrients emitted from the sediments to the overlying water column reduce the magnitude of N loss by sediment burial in the GOM. This is particularly relevant in the nearshore regions where high rates of N burial as well as large fluxes of NH_4^+ from the sediments to the water column have been observed (see Fig. 3.8). While there is significant spatial and temporal variability in these fluxes, on average there is a large net flux of NH_4^+ out of the sediments to the overlying water column; fluxes range from 18 mmol N m^{-2} yr^{-1} in a hypersaline Texas estuary (Cotner et al. 2004) to 2190 ± 1305 mmol N m^{-2} yr^{-1} on the Yucatan Peninsula (Valdes and Real 2004), although typically NH_4^+ fluxes from the sediments to the overlying water column range from 200 to 600 mmol N m^{-2} yr^{-1} in the GOM (Table 3.3; see Fig. 3.8). In contrast, the net flux of NO_3^- NO_2^- appears to be into the sediments, although at certain locations and times NO_3^-+NO_2^- can be emitted from the sediments to the water column, with these fluxes typically <150 mmol N m^{-2} yr^{-1} (Seitzinger 1987; Yoon and Benner 1992; Gardner et al. 2006; McCarthy et al. 2008; Lehrter et al. 2012; Bruesewitz et al. 2013; Nunnally et al. 2013). More commonly, however, net NO_3^-+NO_2^- fluxes are into the sediments, with rates ranging from -24 ± 75 mmol N m^{-2} yr^{-1} in Texas estuaries (Zimmerman and Benner 1994) to -815 mmol N m^{-2} yr^{-1} in Alabama estuaries (Bernard et al. 2015), although more common NO_3^-+NO_2^- fluxes into the sediments range from -100 to -400 mmol N m^{-2} yr^{-1} (see Fig. 3.9). This net flux of NO_3^-+NO_2^- into and NH_4^+ out of GOM sediments has been observed throughout the northern GOM (e.g., Mortazavi, Iverson, Huang, et al. 2000; Rowe et al. 2002; Lin et al. 2011; Lehrter et al. 2012; Nunnally et al. 2013; Nunnally et al. 2014). As described elsewhere in this chapter, this benthic sink for NO_3^- NO_2^- and source of NH_4^+ may be driven in part by DNRA in sediments once they become sulfidic but may also result from the respiration of organic N under anoxic sedimentary conditions, in which case NH_4^+, not NO_3^-+NO_2^-, will be produced during the remineralization of organic N.

These inorganic N fluxes into and out of the sediments exhibit temporal variability, including higher fluxes of NH_4^+ from the sediments to the water column during the summer in the northern GOM than during other seasons (Mortazavi, Iverson, Huang, et al. 2000; Nunnally et al. 2013), as well as switching the sign of the flux; Bruesewitz et al. (2013) found that the sediments in Texas estuaries switched from a net sink of N during floods to a net source of N during droughts. Regardless of how they may vary in space and time, any net flux of NO_3^-+NO_2^-, NH_4^+, and/or DON out of the sediments to the overlying water column can serve as a nutrient source for phytoplankton, which may further exacerbate the widespread eutrophication in the northern GOM.

Benthic Dissimilatory Nitrate Reduction to Ammonia

As described previously, DNRA is thought to be the NO_3^- reduction pathway favored by benthic microbes when sediments become sulfidic, as sulfide inhibits nitrification and thus limits any nitrification-driven denitrification (Joye and Hollibaugh 1995). While DNRA has not been measured as frequently as denitrification, when it has been measured in the GOM, the rates are comparable to those of benthic denitrification (see Table 3.3; Fig. 3.11) and range from 30 ± 40 mmol N m^{-2} yr^{-1} in Texas estuaries (Bruesewitz et al. 2013) up to 907 ± 519 mmol N m^{-2} yr^{-1} in Florida Bay (Gardner and McCarthy 2009). As occurs with benthic denitrification and anammox, many measurements of DNRA involve the addition of isotopically-labeled substrate (in this case, $^{15}NO_3^-$) above background levels, so many of the reported rates

Table 3.3. Rates of internal cycling of nitrogen in the Gulf of Mexico and its estuaries

Process	Location	Method	Type of N	Original units	mmol N m^{-2} yr^{-1}	Reference
Sediment nutrient flux						
	Northern GOM	Sediment core incubations	NH_4^+	18.7 ± 38.6 µg at m^{-2} hr^{-1}	164 ± 338 (a)	Seitzinger 1987
	Northern GOM	Benthic chambers	NH_4^+	129 µmol N m^{-2} hr^{-1}	1130 (a)	Teague et al. 1988
	TX estuaries	Sediment core incubations	NH_4^+	49.1 ± 34.5 µg atm m^{-2} hr^{-1}	430 ± 302 (a)	Yoon and Benner 1992
	Northern GOM	Benthic chambers	NH_4^+	28.6 ± 19.2 µmol N m^{-2} hr^{-1}	251 ± 168 (b)	Gardner et al. 1993
	Northern GOM	Sediment cores and benthic landers	NH_4^+	115 ± 30 µmol N m^{-2} hr^{-1}	1007 ± 263 (c)	Miller-Way et al. 1994
	TX estuaries	Sediment core incubations	NH_4^+	11.7 ± 16 µmol N m^{-2} hr^{-1}	102 ± 140 (a)	Zimmerman and Benner 1994
	Northern GOM	Benthic chambers	NH_4^+	3.5 ± 0.7 mmol N m^{-2} d^{-1}	1278 ± 256 (a)	Morse and Rowe 1999
	Northern GOM	Sediment core incubations	NH_4^+	16.1 ± 1.1 mg N m^{-2} d^{-1}	420 ± 29 (a)	Mortazavi et al. 2000
	LA shelf	Benthic chambers	NH_4^+	2.65 ± 1.3 mmol m^{-2} d^{-1}	967 ± 475 (a)	Rowe et al. 2002
	TX estuaries	Benthic chambers	NH_4^+	0.05 ± 0.42 mmol N m^{-2} d^{-1}	18 ± 153 (a)	Cotner et al. 2004
	Yucatan Peninsula coastline	Sediment core incubations	NH_4^+	250 ± 149 µmol N m^{-2} hr^{-1}	2190 ± 1305 (a)	Valdes and Real 2004
	TX estuaries	Sediment core incubations	NH_4^+	65 ± 65 µmol N m^{-2} hr^{-1}	569 ± 569 (a)	Gardner et al. 2006
	TX estuaries	Sediment core incubations	NH_4^+	52.1 ± 60.5 µmol N m^{-2} hr^{-1}	456 ± 530 (a)	McCarthy et al. 2008
	Florida Keys	Sediment core incubations	NH_4^+	171 ± 139 µmol N m^{-2} hr^{-1}	1498 ± 1218 (a)	Gardner and McCarthy 2009
	Northern GOM	Sediment core incubations	NH_4^+	1.11 ± 0.98 mmol N m^{-2} d^{-1}	405 ± 358 (a)	Murrell et al. 2009
	Northern GOM	Sediment core incubations	NH_4^+	35.5 ± 39.5 µmol N m^{-2} hr^{-1}	311 ± 346 (a)	Lin et al. 2011
	Northern GOM	Benthic chambers	NH_4^+	0.55 ± 0.89 mmol N m^{-2} d^{-1}	201 ± 325 (a)	Lehrter et al. 2012
	AL estuaries	Sediment core incubations	NH_4^+	102 µmol N m^{-2} hr^{-1}	894 (a)	Mortazavi et al. 2012
	TX estuaries	Sediment core incubations	NH_4^+	92 ± 116 µmol N m^{-2} hr^{-1}	806 ± 1016 (a)	Bruesewitz et al. 2013
	Northern GOM	Micro incubation chambers	NH_4^+	1.3 ± 0.9 mmol N m^{-2} d^{-1}	475 ± 329 (a)	Nunnally et al. 2013
	Northern GOM	Sediment core incubations	NH_4^+	3.7 ± 1.5 g m^{-2} yr^{-1}	264 ± 107 (a)	Bernard et al. 2014

Continued

Table 3.3. Continued

Process	Location	Method	Type of N	Original units	mmol N m^{-2} yr^{-1}	Reference
Sediment nutrient flux						
	West FL shelf	Benthic chambers	NH_4^+	1426 ± 959 μmol N m^{-2} d^{-1}	520 ± 350 (a)	Dixon et al. 2014
	AL estuaries	Sediment core incubations	NH_4^+	45 to 110 μmol N m^{-2} hr^{-1}	394 to 964 (d)	Bernard et al. 2015
	Northern GOM	Sediment core incubations	NH_4^+	77.4 ± 67.1 μmol N m^{-2} hr^{-1}	678 ± 588 (a)	McCarthy et al. 2015
	Northern GOM	Sediment core incubations	NH_4^+	0.91 ± 1.61 mmol N m^{-2} d^{-1}	332 ± 588 (a)	Roberts and Doty 2015
	Northern GOM	Sediment core incubations	NO_3^-+NO_2^-	16.0 ± 12.1 μg at m^{-2} hr^{-1}	140 ± 106 (a)	Seitzinger 1987
	Northern GOM	Benthic chambers	NO_3^-	-19 μmol N m^{-2} hr^{-1}	-166 (a)	Teague et al. 1988
	Northern GOM	Sediment core incubations	NO_3^-+NO_2^-	9.2 ± 7.7 μg atm m^{-2} hr^{-1}	81 ± 67 (a)	Yoon and Benner 1992
	Northern GOM	Benthic chambers	NO_3^-+NO_2^-	-3.9 ± 20.7 μmol N m^{-2} hr^{-1}	-34.1 ± 181.3 (b)	Gardner et al. 1993
	Northern GOM	Sediment cores and benthic landers	NO_3^-+NO_2^-	-16.3 ± 9.7 μmol N m^{-2} hr^{-1}	-142.8 ± 85 (c)	Miller-Way et al. 1994
	TX estuaries	Sediment core incubations	NO_3^-+NO_2^-	-2.7 ± 8.6 μmol N m^{-2} d^{-1}	-24 ± 75 (a)	Zimmerman and Benner 1994
	Northern GOM	Sediment core incubations	NO_3^-	-17 ± 3.6 mg N m^{-2} d^{-1}	-443 ± 94 (a)	Mortazavi et al. 2000
	LA shelf	Benthic chambers	NO_3^-+NO_2^-	-0.8 ± 0.7 mmol m^{-2} d^{-1}	-292 ± 256 (a)	Rowe et al. 2002
	Yucatan Peninsula coastline	Sediment core incubations	NO_3^-+NO_2^-	-47 ± 33 μmol N m^{-2} hr^{-1}	-412 ± 289 (a)	Valdes and Real 2004
	TX estuaries	Sediment core incubations	NO_3^-+NO_2^-	5.6 ± 10.1 μmol N m^{-2} hr^{-1}	49 ± 88 (a)	Gardner et al. 2006
	TX estuaries	Sediment core incubations	NO_3^-+NO_2^-	2.28 +/- 1.93 μmol N m^{-2} hr^{-1}	20.0 ± 16.9 (a)	McCarthy et al. 2008
	Florida Keys	Sediment core incubations	NO_3^-+NO_2^-	2.3 +/- 10.0 μmol N m^{-2} hr^{-1}	20.1 ± 87.6 (e)	Gardner and McCarthy 2009
	Northern GOM	Sediment core incubations	NO_3^-	0.58 ± 1.08 mmol N m^{-2} d^{-1}	212 ± 394 (a)	Murrell et al. 2009
	Northern GOM	Benthic chambers	NO_3^-+NO_2^-	0.017 ± 0.41 mmol N m^{-2} d^{-1}	6 ± 150 (a)	Lehrter et al. 2012
	AL estuaries	Sediment core incubations	NO_3^-+NO_2^-	-30 μmol N m^{-2} hr^{-1}	-263 (a)	Mortazavi et al. 2012
	TX estuaries	Sediment core incubations	NO_3^-+NO_2^-	14 ± 11 μmol N m^{-2} hr^{-1}	123 ± 96 (a)	Bruesewitz et al. 2013
	Northern GOM	Micro incubation chambers	NO_3^-+NO_2^-	-0.26 ± 0.05 mmol N m^{-2} d^{-1}	95 ± 18 (a)	Nunnally et al. 2013
	Northern GOM	Sediment core incubations	NO_3^-+NO_2^-	-1.8 ± 0.14 g m^{-2} yr^{-1}	-129 ± 10 (a)	Bernard et al. 2014

Continued

Table 3.3. Continued

Process	Location	Method	Type of N	Original units	mmol N m^{-2} yr^{-1}	Reference
Sediment nutrient flux						
	West FL shelf	Benthic chambers	NO_3^-+NO_2^-	17.5 ± 67.6 μmol N m^{-2} d^{-1}	6 ± 25 (a)	Dixon et al. 2014
	AL estuaries	Sediment core incubations	NO_3^-+NO_2^-	-93 to -121 μmol N m^{-2} hr^{-1}	-815 to -1060 (d)	Bernard et al. 2015
	Northern GOM	Sediment core incubations	NO_3^-+NO_2^-	-26.6 ± 45.5 μmol N m^{-2} hr^{-1}	-233 ± 399 (a)	McCarthy et al. 2015
	Northern GOM	Sediment core incubations	NO_3^-	-0.9 ± 1.3 mmol N m^{-2} d^{-1}	-329 ± 475 (a)	Roberts and Doty 2015
	Northern GOM	Sediment core incubations	DON	7 ± 2.9 g N m^{-2} yr^{-1}	500 ± 207 (a)	Bernard et al. 2014
	West FL shelf	Benthic chambers	DON	-301 ± 1390 μmol N m^{-2} d^{-1}	-110 ± 507 (a)	Dixon et al. 2014
	Northern GOM	Sediment core incubations	DON	0.44 ± 0.78 mmol N m^{-2} d^{-1}	161 ± 285 (a)	Roberts and Doty 2015
N assimilation						
	Northern GOM	$^{15}NH_4^+$ assimilation	^{15}N-Particulate N	0.34 ± 0.15 μM N hr^{-1}	2978 ± 1314 (f)	Gardner et al. 1997
	Northern GOM	$^{15}NH_4^+$ assimilation	^{15}N-Particulate N	0 to 0.03 μmol N L^{-1} hr^{-1}	0 to 263 (g)	Jochem et al. 2004
	West FL shelf	$^{15}NH_4^+$ assimilation	^{15}N-Particulate N	68.2 ± 63.1 nmol N L^{-1} hr^{-1}	597 ± 553 (a)	Mulholland et al. 2006
	West FL shelf	$^{15}NO_3^-$ assimilation	^{15}N-Particulate N	35.3 ± 75.6 nmol N L^{-1} hr^{-1}	309 ± 662 (a)	Mulholland et al. 2006
	West FL shelf	$^{15}NH_4^+$ assimilation	^{15}N-Particulate N	0.15 ± 0 μmol N L^{-1} hr^{-1}	1314 ± 0 (b)	Bronk et al. 2014
	West FL shelf	$^{15}NO_3^-$ assimilation	^{15}N-Particulate N	0.093 ± 0.045 μmol N L^{-1} hr^{-1}	815 ± 394 (b)	Bronk et al. 2014
	West FL shelf	^{15}N-urea assimilation	^{15}N-Particulate N	0.039 ± 0.001 μmol N L^{-1} hr^{-1}	342 ± 9 (b)	Bronk et al. 2014
	West FL shelf	^{15}N-AA assimilation	^{15}N-Particulate N	0.025 ± 0.013 μmol N L^{-1} hr^{-1}	219 ± 114 (b)	Bronk et al. 2014
DNRA						
	TX estuaries	Potential rate; 100 μM $^{15}NO_3^-$ addition	NH_4^+	58.4 ± 25.6 μmol N m^{-2} hr^{-1}	512 ± 224 (a)	An and Gardner 2002
	TX estuaries	Potential rate; 100 μM $^{15}NO_3^-$ addition	$^{15}NH_4^+$ production	14.3 ± 21.5 μmol N m^{-2} hr^{-1}	125 ± 188 (a)	Gardner et al. 2006
	TX estuaries	Sediment core incubation; potential rate; 100 μM $^{15}NO_3^-$ addition	$^{15}NH_4^+$ production	12.7 ± 12.0 μmol N m^{-2} hr^{-1}	111 ± 105 (a)	McCarthy et al. 2008
	FL Bay	Sediment core incubation	$^{15}NH_4^+$ production	103.5 ± 59.3 μmol N m^{-2} hr^{-1}	907 ± 519 (a)	Gardner and McCarthy 2009

Continued

Table 3.3. Continued

Process	Location	Method	Type of N	Original units	mmol N m^{-2} yr^{-1}	Reference
DNRA						
	TX estuaries	Sediment core incubation; potential rate; 20 to 40 µM $^{15}NO_3^-$ addition	$^{15}NH_4^+$ production	3.4 ± 4.6 µmol N m^{-2} hr^{-1}	30 ± 40 (a)	Bruesewitz et al. 2013
	AL estuaries	Sediment core incubations, 10 µM $^{15}NO_3^-$ addition	$^{15}NH_4^+$ production	52.1 ± 80.6 µmol N m^{-2} hr^{-1}	456 ± 706	Bernard et al. 2015
	Northern GOM	Sediment core incubations, 50 µM $^{15}NO_3^-$ addition	$^{15}NH_4^+$ production	9.6 ± 10.2 µmol N m^{-2} hr^{-1}	84 ± 89 (a)	McCarthy et al. 2015
NH_4^+ production						
	Northern GOM	$^{15}NH_4^+$ dilution		10 ± 14 nM hr^{-1}	88 ± 123 (f)	Cotner and Gardner 1993
	Northern GOM	$^{15}NH_4^+$ dilution		0.18 ± 0.09 µM N hr^{-1}	1577 ± 788 (f)	Gardner et al. 1997
	Northern GOM	$^{15}NH_4^+$ dilution		0 to 0.035 µmol N L^{-1} hr^{-1}	0 to 307 (g)	Jochem et al. 2004
	West FL shelf	Zooplankton excretion		0.71 ± 1.66 µmol N L^{-1} d^{-1}	259 ± 606 (h)	Vargo et al. 2008
	West FL shelf	$^{15}NH_4^+$ dilution		0.175 ± 0.03 µmol N L^{-1} hr^{-1}	1553 ± 263 (b)	Bronk et al. 2014
Nitrification						
	Northern GOM	$^{15}NH_4^+$ addition		0 to 150 nM L^{-1} hr^{-1}	0 to 1314 (h)	Carini et al. 2010
	West FL shelf	$^{15}NH_4^+$ addition		0.12 ± 0.05 µM N hr^{-1}	1051 ± 438 (h)	Bronk et al. 2014
	Northern GOM	$^{15}NH_4^+$ addition		9 to 494 nmol N L^{-1} d^{-1}	3 to 180 (i)	Bristow et al. 2015
	Northern GOM	$^{15}NO_2^-$ addition		1 to 411 nmol N L^{-1} d^{-1}	0.4 to 150 (i)	Bristow et al. 2015

Note: To convert between original flux units and the common units of mmol m^{-2} yr^{-1}, assumptions included the following:

(a) Taking the mean of measurements over multiple sites and seasons.

(b) Mean of multiple sites.

(c) Mean of multiple methods.

(d) Range of multiple sites and seasons.

(e) Mean from 2004 to 2006.

(f) Mean of seasons and sites with highest salinities.

(g) Range from sites with highest salinities.

(h) Mean rate applied over 24 hr, 365 days/yr, and 1 m water-column depth.

(i) Range of rates applied over 365 days/yr, and 1 m water column.

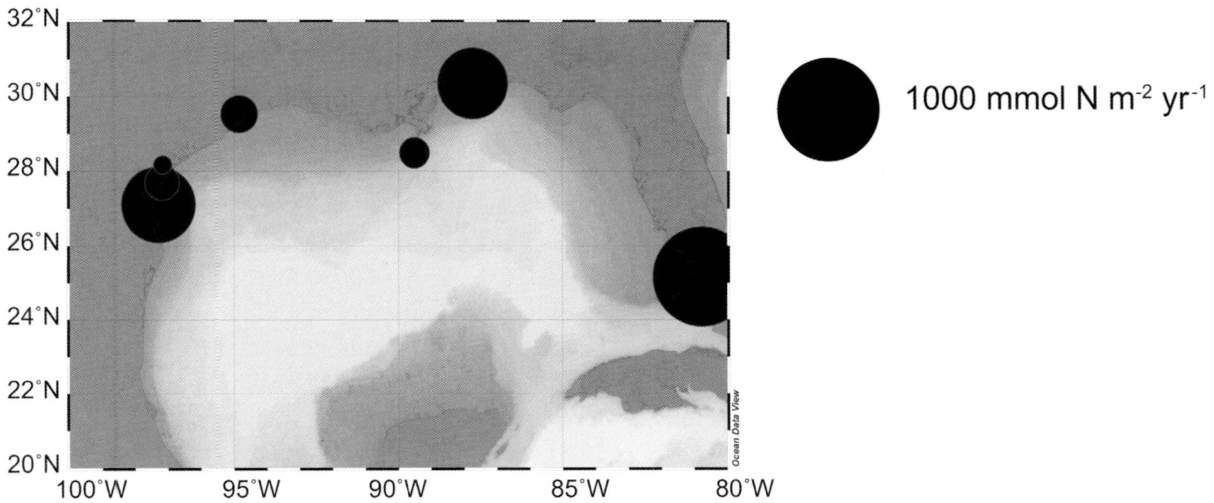

Figure 3.11. Benthic dissimilatory NO_3^- reduction to NH_4^+ (DNRA) (figure created using ODV).

are considered potential rates and are likely upper bounds for typical benthic DNRA rates. Clearly, better quantifying rates of this process in the GOM is critical to evaluating ecosystem services provided by coastal sediments for removing anthropogenic N loading.

Biological Nitrogen Assimilation

Compared to other N cycling processes in the GOM, few direct measurements of N assimilation by phytoplankton have been made, and many of the measurements made have focused on the WFS in conjunction with the ECOHAB project to identify the sources of N fueling blooms of the harmful algae *K. brevis* (Mulholland et al. 2006; Bronk et al. 2014). These measurements indicate that rates of NH_4^+ assimilation are typically highest and range from 597 to 1314 mmol N m^{-2} yr^{-1}, with rates of NO_3^- assimilation on the WFS ranging from 309 to 815 mmol N m^{-2} yr^{-1} (Mulholland et al. 2006; Bronk et al. 2014; see Table 3.3). Additionally, urea and amino acid uptake have been measured on the WFS, with average rates of 342 and 219 mmol N m^{-2} yr^{-1}, respectively (Bronk et al. 2014). In the Mississippi river plume in the northern GOM, reported rates of NH_4^+ assimilation range from 0 to 2978 mmol N m^{-2} yr^{-1} at high-salinity stations, with NH_4^+ assimilation rates significantly higher at stations with slightly lower surface-water salinities of ~28 (Gardner et al. 1997; Jochem et al. 2004).

While these direct measurements of N assimilation are useful for quantifying the demand of different N substrates in the GOM, other data inform the extent to which the availability of inorganic N relative to other nutrients may limit phytoplankton growth, especially in the waters affected by the MARS. In particular, due to the high N:P ratio in the MARS discharge that peaks from May through July (~50:1), seasonal shifts in N versus P limitation of phytoplankton in the northern GOM have been observed (Lohrenz et al. 1999; Lohrenz et al. 2008). For example, in phytoplankton incubation experiments at stations of intermediate salinity, additions of P were more effective at stimulating biomass than additions of N during the spring and summer, although N was found to limit phytoplankton in September through the winter (Sylvan et al. 2006, 2007). These observations are consistent with other metrics of phytoplankton nutrient limitation in the northern GOM (e.g., Smith and Hitchcock 1994; Lohrenz et al. 1999; Pakulski et al. 2000; Quigg et al. 2011; Sylvan and Ammerman 2013), although outside the Mississippi River plume, phytoplankton appear to be N limited (Quigg et al. 2011).

NH_4^+ Production

As in other studies of N cycling in the GOM, measurements of NH_4^+ production have been made in the northern GOM in association with the Mississippi River plume and on the WFS in the context of the *K. brevis* blooms. In the Mississippi River plume in the northern GOM, NH_4^+ production rates at the highest-salinity stations range from 0 to 1577 mmol N m^{-2} yr^{-1}, although rates typically peak at intermediate-salinity (15 to 29) stations (Cotner and Gardner 1993; Gardner et al. 1994, 1997; Jochem et al. 2004), so the rates reported in Table 3.3 can be considered a lower bound, especially for NH_4^+ production rates in MARS-influenced surface waters. Additionally, NH_4^+ regeneration rates were an order of magnitude higher in the summer than in the winter in the northern GOM (Cotner and Gardner 1993). On the WFS, rates of zooplankton excretion of NH_4^+ were estimated at 259 ± 606 mmol N m^{-2} yr^{-1} (Vargo et al. 2008), and average NH_4^+ regeneration rates on the WFS were 1553 ± 263 mmol N m^{-2} yr^{-1} (Bronk et al. 2014).

Nitrification

Nitrification rates in the GOM have been measured in only a handful of studies and include measurements of both NH_4^+ and NO_2^- oxidation rates. In the northern GOM rates of NH_4^+ oxidation range from 0 to 1314 mmol N m^{-2} yr^{-1} (Carini et al. 2010; Bristow et al. 2015), and NO_2^- oxidation rates range from 0.4 to 150 mmol N m^{-2} yr^{-1} (Bristow et al. 2015; see Table 3.3). NH_4^+ oxidation rates in the northern GOM were typically highest deeper in the water column and correlated with NH_4^+ and NO_2^- concentrations (Carini et al. 2010; Bristow et al. 2015). On the WFS, average NH_4^+ oxidation rates of 1261 ± 613 mmol N m^{-2} yr^{-1} have been reported and were roughly twice as high at coastal and estuarine sites as at offshore locations (Bronk et al. 2014).

Summary of Nitrogen in the Gulf of Mexico

The picture that emerges from our current understanding of N cycling in the GOM is somewhat obscure due to the dearth of measurements from nearly everywhere *except* the WFS and northern GOM sites immediately impacted by the MARS. Given the poor spatial coverage, it is perhaps premature to construct a quantitative N budget for the GOM. However, based on available measurements, the dominant inputs of N to the GOM are MARS-derived N, followed by groundwater fluxes, atmospheric deposition, benthic N_2 fixation, and water-column/euphotic-zone N_2 fixation (see Table 3.1). Losses of N are dominated by sediment burial occurring at very high rates near the mouth of the MARS, followed by benthic denitrification, with much smaller (although poorly documented) contributions from anammox (see Table 3.2). The magnitude of the sedimentary N burial term may be reduced by the extent to which organic N is respired in sediments and released as NH_4^+ and/or NO_3^- to the overlying water column, and importantly, this supply of inorganic N back to the overlying water column may exacerbate eutrophication, especially in the northern GOM (Burgin and Hamilton 2007).

What remains unknown about N cycling in the GOM is how most processes vary spatially in the Gulf, in particular in deep waters off the shelves. While there are reports of euphotic-zone N_2 fixation occurring at significant rates in the surface waters of some deepwater locations (Holl et al. 2007; Mulholland et al. 2006), it is premature to generate basin-wide N_2 fixation fluxes without better characterizing the rates of this process in other regions of the GOM. Additionally, given the magnitude of groundwater N fluxes to the GOM, this source of new N appears underappreciated given its potential significance for fueling phytoplankton blooms in otherwise nutrient-poor surface waters (Smith and Swarzenski 2012). Additionally, there are unique features in the GOM, including cold seeps (Bowles and Joye 2011) and salt domes such as in the Orca Basin (Meckler et al. 2011), that likely harbor unique microbial communities that may have distinct N cycling. In general, the Gulf represents a remarkably broad range of biogeochemical conditions in a relatively small space and thus is an ideal "model system" where the sensitivities of a number of biogeochemical processes, especially those involving the N cycle, can, and should, be tested.

Phosphorus in the Gulf of Mexico

Phosphorus (P), nitrogen, and silicon are the major macronutrients in seawater. Though nitrogen is usually considered the primary limiting nutrient in estuarine, coastal, and open-ocean waters, the role of P has become more important in marine nutrient cycling and as a potentially limiting nutrient in some marine environments. This is partly the result of much greater anthropogenic inputs of N relative to P, both globally (Peñuelas et al. 2013) and to coastal waters (Laurent and Fennel 2017). Recent general marine phosphorus-cycling reviews include Benitez-Nelson (2000) and Paytan and McLaughlin (2007). Dyhrman et al. (2007) and Hood et al. (2004) reviewed microbial aspects of marine P cycling. There are two different stories about P as a nutrient in the euphotic zone of the Gulf of Mexico, one concerning the open Gulf where P is a scarce nutrient in a very oligotrophic system (Cardona 2013), modified by Loop Current eddies that can bring nutrient-rich water to the surface (Cardona 2013); the second, about P in Gulf embayments, estuaries, and river-impacted coastal waters where P cycling can be very dynamic (Bianchi et al. 2009; Feist et al. 2016).

Phosphorus Sources

Approximately two-thirds of the freshwater flow into the GOM comes from the Mississippi River with a watershed that is the third largest in the world (see Fig. 3.1C). The Mississippi River is highly enriched in N and P, leading to high levels of primary production and carbon cycling in coastal waters and to the formation of an annual hypoxic zone off the coast of Louisiana (Rabalais et al. 2007; Bianchi et al. 2010). This hypoxic zone is second in size only to that of the Baltic Sea. Spatially referenced regression on watershed attributes (SPARROW) models have consistently shown that at least half of the N and P in the Mississippi is from nonpoint source agricultural activities, with animal waste contributing a higher fraction of P than N (Alexander et al. 2008; Robertson and Saad 2013). Benthic fluxes of phosphate have been measured in a number of Gulf estuaries (Twilley et al. 1999; DiDonato et al. 2006) and may provide enough phosphate to the phytoplankton to relieve P limitation in some locations as estuarine waters enter the coastal Gulf.

Processes and Cycling

Mississippi River N and P fluxes are extremely variable but are correlated generally with river flow (Cardona et al. 2016). They drive primary productivity on the Louisiana shelf, with the highest productivity in the spring leading to the development of bottom-water hypoxia (Lohrenz et al. 2008). Though N is clearly the major limiting nutrient on the Louisiana shelf (Turner and Rabalais 2013), several studies in the 1990s and since have suggested that spring P limitation of primary productivity occurs on the Louisiana shelf because of the high N loading (Sylvan et al. 2006; Sylvan et al. 2007; Sylvan and Ammerman 2013), particularly near the Mississippi and Atchafalaya River mouths. These studies used a variety of different approaches to demonstrate P limitation, both biogeochemical (nutrient concentrations and ratios) and physiological measurements (rapid phosphate uptake rate, induction of high alkaline phosphatase enzymatic activity, increased photosynthetic efficiency with added phosphate, and increased chlorophyll in bottle incubations with added phosphate) (Sylvan et al. 2006; Gaas and Ammerman 2007; Sylvan et al. 2007). When the complete surface chemical inventory of C, N, and P on the Louisiana shelf was measured during 3 different spring and summer months, it was demonstrated that total mean N/P ratios exceeded 50 on all 3 cruises (Sylvan and Ammerman 2013). These studies influenced the EPA Science Advisory Board 2008 report *Hypoxia in the Northern Gulf of Mexico: An Update by the EPA Science Advisory Board.* As a result of this report, the 2008 *Gulf Hypoxia Action Plan* (Mississippi River/Gulf of Mexico Watershed Nutrient Task Force 2008) for the first time developed a numerical target for the reduction of riverine total P as well as total N, calling for a 45% reduction of both to decrease the area of the hypoxic zone to 5000 km^2. Based on USGS data, the 5-year moving average for N loading from 2011 to 2014 has decreased by 18% over the 1980–96 baseline period. However, P loading during the same period has increased by 15% above the baseline (Mississippi River/Gulf of Mexico Watershed Nutrient Task Force 2015).

It is clear, though still disputed (Turner and Rabalais 2013), that on the Louisiana shelf between the Mississippi and Atchafalaya Deltas, primary productivity can be P limited during the spring and early summer, especially during high-flow years (Sylvan et al. 2006). Because this P limitation occurs during the period of high spring productivity, it may be important in hypoxia formation and should be further investigated. Sutula et al. (2004) demonstrated that P released from the suspended particles and sediments in the Mississippi River can be remobilized at times of high river flow and deposited onto the shelf sediments. Once deposited, the P in these particulates and sediments can be remineralized to provide substantial fluxes of reactive P to the overlying coastal water column, thus providing P to support primary production in the coastal GOM (Nunnally et al. 2014).

P limitation on the Louisiana shelf has also been simulated by models (Laurent et al. 2012; Laurent and Fennel 2014). The results suggest that P limitation reduces the size of the summer hypoxic area and shifts it toward the west. However, the degree of areal reduction was a function of how the model boundary conditions were estimated. More study of the benthic remineralization process in these coastal waters will be necessary to help formulate better boundary condition estimates and a greater understanding of benthic fluxes and their contribution to hypoxia.

Important questions remain, however, as highlighted in the recent 2013 *Reassessment* by the Mississippi River/Gulf of Mexico Watershed Nutrient Task Force. This reassessment lists 10 action items important to reducing the size of the Louisiana hypoxic zone in the future, including 2 that specifically target research on nutrient issues. Action Item 8 states, "Decrease scientific uncertainty on the source, fate, and transport of nitrogen and phosphorus," and Action Item 9 says, "Decrease scientific uncertainty of nitrogen and phosphorus effects on hypoxia" (Mississippi River/Gulf of Mexico Watershed Nutrient Task Force 2013, 42, 48). Both are very broad areas that will require significant research efforts to achieve. The deadline for reduction of the hypoxic area to 5000 km^2 was recently changed from 2015 to 2035, with an interim goal of a 20% reduction by 2025 (Mississippi River/Gulf of Mexico Watershed Nutrient Task Force 2015). For us to reach these reduction targets, there must be a concerted effort to generate political will to provide sufficient funding to support the necessary research.

Forms and Distributions of Phosphorus

Recent shelf surveys of surface N and P concentrations from 2010 to 2012 showed values 2 and 3 times smaller than 1985–2009 measurements (Cardona et al. 2016). Offshore (>200 m) measurements of surface phosphate in the Gulf are usually low and often within instrument detection limits (Cardona et al. 2016). Winter surface concentrations of dissolved inorganic P (see Fig. 3.1B) range from >0.5 µM or about 0.3–0.4 µM near sources' nutrient input, such as the Mississippi-Atchafalaya River system, Mobile Bay, and the Yucatan Peninsula. Open Gulf concentrations range from <0.1 to ~0.2 µM based on data in the NOAA Gulf of Mexico Data Atlas database for winter (see Fig. 3.1B). Concentrations determined for the surface waters near the site of the Deepwater Horizon oil spill range from <0.1 to ~0.2 µM, while deeper concentrations have been observed to range from low values of 0.6 µM up to >2.0 µM (Daly 2015). In the southern part of the GOM, surface inorganic P concentrations range from <0.1 µM up to ~0.4 µM in coastal waters (Valdes-Lozano et al. 2010; Herrera-Silveira et al. 2002). The deep southern Gulf has similar inorganic P concentrations as those found farther north, with concentrations ranging between 0.8 and 2.2 µM and greater (Valdes-Lozano et al. 2010).

There is limited information about dissolved organic and particulate P in the GOM, with most research focused on the Louisiana shelf region. Dissolved organic P is certainly an important source of P in the Mississippi plume, as shown by the high alkaline phosphatase activity (Gaas and Ammerman 2007). The available studies show gradients of dissolved organic P from near the mouths of the Mississippi and Atchafalaya Rivers with concentrations declining offshore (Pakulski et al. 2000; Sylvan and Ammerman 2013), much like the distribution of inorganic P. In general, DOM concentrations are somewhat higher near the mouth of the Atchafalaya River than the Mississippi River, presumably due to the great abundance of marshes in that river system. About two-thirds of the dissolved organic P in the Mississippi River plume region was colloidal organic P as measured by concentration with a ~1 kDalton ultrafiltration filter (Rinker and Powell 2006).

Silicon in the Gulf of Mexico

Si plays a key role in regulation of the global carbon cycle, and globally important fluxes occur in the small area of the land-sea interface (Tréguer and De La Rocha 2013, and references therein). Chemical weathering of Si bound in rock and the requirement of Si as a nutrient for autotrophs both result in the consumption of CO_2. But within sediments, Si plays an important role in the production of authigenic clays that, along with respiration of organic matter, reduce alkalinity from the production of CO_2. Therefore, there is a strong coupling between the Si and C cycles over a variety of timescales. The presence of major freshwater and sediment sources (e.g., Mississippi/ Atchafalaya Rivers, Mobile Bay collective rivers) and high biological productivity and sediment accumulation means that regional processes in the GOM have broader significance in the global Si cycle (Fig. 3.12).

In the past, it was often thought that the cycle for dissolved silicon (DSi) in the coastal and open ocean was simple. DSi concentrations in rivers and streams resulted from the weathering of rocks. This DSi would then enter coastal waters from river and stream discharge, most often after flowing through wetlands and estuaries where significant amounts of the DSi would be taken up by organisms and precipitated into biogenic silica (BSi, i.e., amorphous SiO_2-nH_2O). The remaining DSi would then flow into coastal and offshore waters, where it would be taken up by planktonic organisms, primarily diatoms with some smaller contribution from silicoflagellates and rhizarians (e.g., radiolarians, phaeodarians) or by benthic sponges. These organisms would cycle Si through a variety of food web pathways—Si typically is cycled ~40 times prior to burial (Tréguer and De La Rocha 2013)—and then be transported to the seafloor in aggregates or in zooplankton fecal pellets. In deepwater systems, at least 50% of the produced BSi dissolves as it sinks, increasing the concentrations of DSi in mid- and deeper waters. In the sediments, a variety of biological and chemical processes would result in the remineralization of the BSi, allowing for a flux of DSi into the overlying waters. These overlying waters would then be either mixed or upwelled into the surface mixed layer and reenter the food web once again. This older view of the DSi cycle in coastal

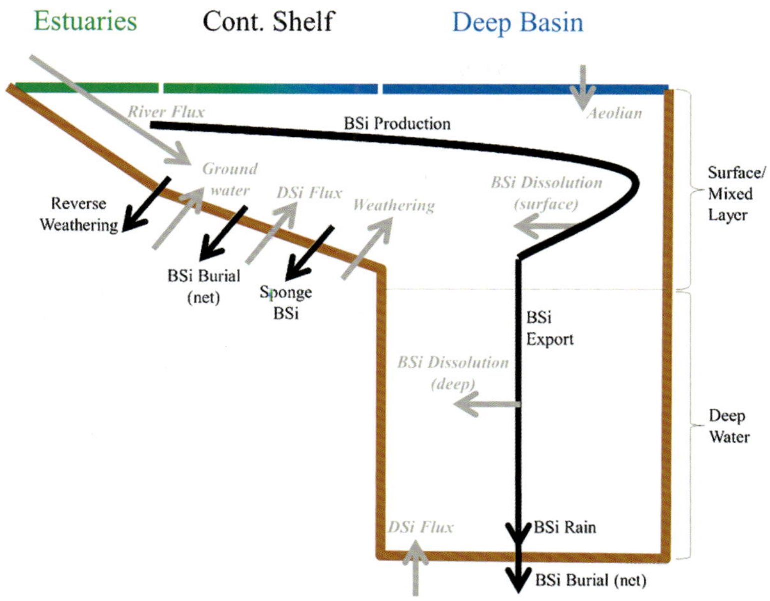

Figure 3.12. Conceptual figure of the Gulf of Mexico Si cycle (modified from global figure by Tréguer and De La Rocha 2013). DSi is delivered from the land to ocean via riverine, eolian, groundwater, ocean weathering, and benthic flux. Si is converted to BSi or amorphous clays (reverse weathering; Michalopoulos and Aller 1995) and buried in sediments. As occurs in the global Si cycle, BSi production mainly from diatoms is presumed to be the dominant source in the GOM Si budget.

and ocean waters is incomplete, and we are learning that the DSi cycle, especially in the river plume systems of the GOM, is much more complex (see Fig. 3.12).

The EPA Science Advisory Board suggested that DSi may have an important role in structuring phytoplankton communities and that exploration of Si limitation in the ocean should be a research priority (Dale et al. 2010). However, only 1 study to date (which preceded the EPA recommendation) has reported data bearing directly on biological transformation of Si in the GOM (Nelson and Dortch 1996). Despite numerous regional geochemical studies reporting some data related to the Si cycle (e.g., Turner et al. 2004; Presti and Michalopoulos 2008), there has been no concurrent examination of how both biotic and abiotic processes transform Si and the role Si plays within the GOM system. Of the studies published, most data are from the northern GOM with a more intense dearth of information in the southern domain. Overall, the understanding of Si lags far behind that of N and P in this system and represents a major knowledge gap. In the following discussion, we review the concentrations of Si, sources and sinks for Si, the complex cycling of DSi in the GOM, and the susceptibility to anthropogenic factors.

Forms and Distributions

Si has no dissolved organic phase, and the distribution is mainly dominated by orthosilicic acid (hereafter referred as silicic acid or DSi). Multiple forms of dissolved inorganic Si may occur in aquatic systems, including both ionic ($SiO(OH)_3^-$) and nonionic ($Si(OH)_4$) forms. However, at seawater pH (~8) silicic acid ($Si(OH)_4$), which is monomeric and nonionic, dominates (i.e., 97%) the total DSi pool. At this pH, DSi has a saturation of ~2000 µM (Stumm and Morgan 1996); thus, seawater, and even high DSi near river plumes, is grossly undersaturated for DSi. As pH increases, the proportion of ionic forms increases, but the opposite occurs when pH is reduced. Increases in pH (e.g., ~8.5–9.0) cause Si mineral phases to form, for example, sepiolite (Nelson et al. 1984). At higher pH (e.g., >9.0), which may occur in estuaries from an intense phytoplankton bloom, magnesium (Mg) precipitates in seawater, producing brucite, which may scavenge DSi—this same chemical reaction is used to concentrate phosphate (Karl and Tien 1992) or silicic acid (Brzezinski et al. 2003) in seawater. Lowering pH affects solubility, and DSi will polymerize. Such a condensation reaction resulting from lower pH is catalyzed by biological organisms to form BSi. It has been demonstrated that ($Si(OH)_4$) is the predominant chemical form of DSi taken up (Del Amo and Brzezinski 1999). Given this form is 97% of the DSi pool at marine pH, there is an ecological advantage to accessing the most concentrated form.

Concentrations of DSi in the GOM surface waters are variable. In the southern GOM values range from <1 µM in the open Gulf to 1–3 µM in the waters of the Yucatan Channel and Florida Straits (Carder et al. 1977; Daly 2015). Surface concentrations in the northern GOM are generally greater than in the southern portion and range from 2 to 8 µM in waters deeper than ~200 m, while in coastal waters of the northern Gulf that are <200 m in depth, DSi concentrations are generally in the 20–30 µM

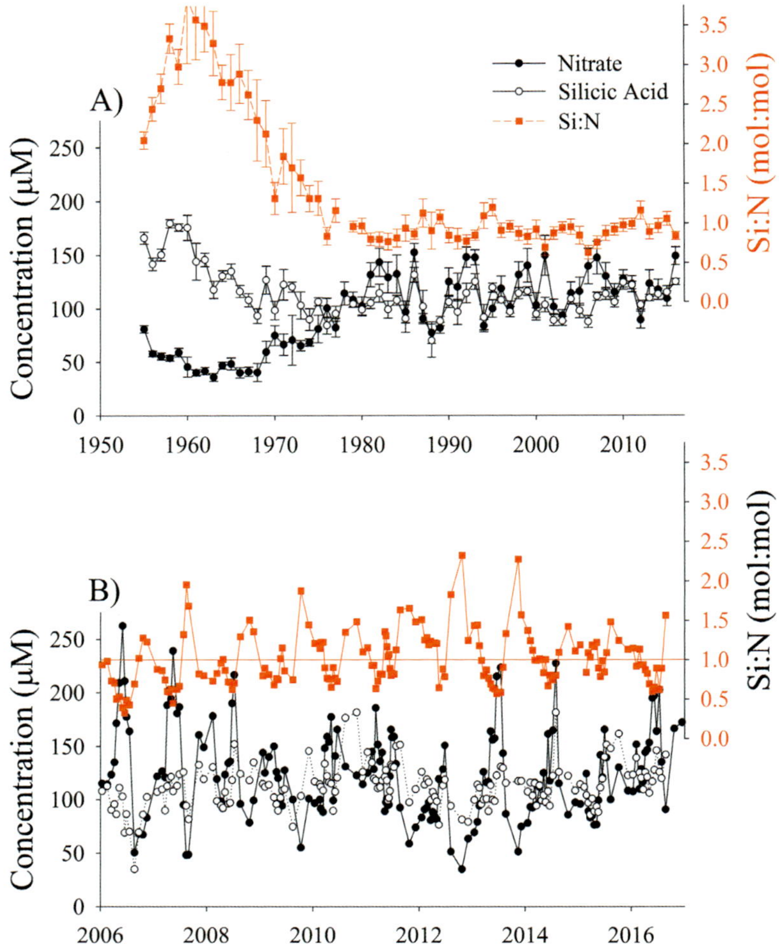

Figure 3.13A, B. Nitrate, silicic acid, Si:N ratio of MR water at St. Francisville, LA, USGS station 07373420. (*A*) Annual mean (±SE), 1955–2016. (*B*) Data on bimonthly resolution from 2006 to 2016. Red horizontal line is diatom Si:N ratio.

range but can reach as high as 50 μM (Nowlin et al. 1998; Lohrenz et al. 1999; Chen et al. 2000; Jochens and Nowlin 2000; Lohrenz et al. 2008; Daly 2015; Cardona et al. 2016). In the mesopelagic waters (200–500 m in depth) of the GOM, concentrations range from ~6 to 7 μM in the northern open Gulf and from ~3 μM up to 20 μM in the Yucatan Channel and Florida Straits (Carder et al. 1977; Daly 2015), with concentrations increasing with depth to the bottom of the permanent thermocline (Morrison et al. 1983). In contrast, DSi concentrations in the lower Mississippi River near north of New Orleans, although decreasing since the early to mid-1900s (Fig. 3.13A), are generally in the range of ~100 μM (Nelson and Dortch 1996; Rabalais et al. 1996; Dagg et al. 2005; Turner et al. 2007). In deeper waters of the northern Gulf (~500 m down to ~2000 m), concentrations of DSi range from ~18 μM up to ~30 μM, with relatively uniform concentrations from the upper to the lower parts of the deeper water column (Morrison et al. 1983; Daly 2015). In the waters of the Yucatan Channel, Florida Straits, and the Yucatan Basin, we see an interesting pattern in DSi concentrations. Deep-water concentrations in the eastern basin of the Gulf, the Yucatan Channel, and Florida Straits are greater, ~24–26 μM, while those of the Yucatan Basin are lower, in the range of 16–18 μM; this lower-DSi water is thought to be deep waters entering the Gulf from the deeper waters of the Caribbean Sea (Carder et al. 1977).

Sources and Sinks

Dissolved silicic acid enters the GOM from a variety of sources, including river and stream discharge, generalized runoff from land, airborne deposition of dust particularly in the eastern GOM, inflow of waters from the Caribbean Sea through the Yucatan Channel, and flux from the sediments to the overlying water column, including groundwater (Carder et al. 1977; Bernard et al. 2011; Tréguer and De La Rocha 2013). Land-sea interface regions are the dominant areas for input of Si to the oceans. Recent estimates indicate that 84% of the global net Si input to the ocean is from riverine and groundwater discharge (Tréguer and De La Rocha 2013). Freshwater containing DSi coming from rivers and streams may first pass through wetlands, estuaries, or lagoons where biogeochemical cycling may reduce the concentrations of DSi through a variety of processes, including uptake by organisms producing BSi. In the Mississippi River plume (Presti and Michalopoulos 2008) and coastal lagoons to the north (Krause et al. 2017), diagenetic reactions create authigenic precipitates around BSi (i.e., reverse weathering), which sequester Si rather than allow it to diffuse back into the water column.

Benthic flux of silicic acid may be an important Si source for BSi production over the continental shelf. Reported benthic fluxes of DSi on the Louisiana shelf near the Mississippi River plume range from a high of 6.6 mmol Si m^{-2} d^{-1} during April 1992 (Miller-Way et al. 1994) to 2.1 mmol Si m^{-2} d^{-1} (range 1.6–2.5) between April 2006 and September 2007 (Lehrter et al. 2012). Due to the highly stratified water column on the Louisiana shelf, the flux of DSi across the pycnocline may at times be a more important term than benthic flux for introducing DSi into the upper euphotic zone. But primary production also occurs below the pycnocline except in the most turbid waters near the Mississippi River plume (Lehrter et al. 2012). Lehrter et al. estimated that 43–66% of the required dissolved inorganic nitrogen (DIN) for primary production below the pycnocline could have been sustained from benthic flux, and if all the primary production was assumed to be from diatoms, the DSi from the benthic flux could have accounted for 47–100% of diatom Si demand.

Major Rate Processes

Silicifiers

Many biological groups aside from diatoms use Si in the GOM, including other phytoplankton such as silicoflagellates (Casey et al. 1979) and picocyanobacteria such as *Synechococcus* (Baines et al. 2012), rhizarians, ebridian flagellates (Jones 1974; Casey et al. 1979), sponges (Lessard-Pilon et al. 2010), copepods (i.e., have siliceous jaws; Miller et al. 1980; Miller et al. 1990), and coastal marsh grasses (Lanning and Eleuterius 1983; Eleuterius and Lanning 1987; Lu and Liu 2003, 2005). Later we discuss the most-studied groups—diatoms, sponges, and marsh grasses—and assess their potential roles in regional Si biogeochemistry.

Diatoms play important roles in coastal biogeochemistry and often dominate the phytoplankton assemblage biomass in coastal environments, including the northern GOM. Diatoms have an obligate requirement for Si, which is typically similar in magnitude to their requirement for N (Brzezinski 1985). Nelson and Dortch (1996) suggest that DSi concentrations may be low enough to limit diatom growth, a condition that has broad biogeochemical and ecological ramifications (see Cloern 2001). Nelson and Dortch observed that DSi concentration on the Louisiana shelf induced kinetic limitation of Si uptake (i.e., uptake rate was suboptimal) in 77% of samples during the spring and 20% of samples during the summer.

A prominent source of BSi in the GOM is sponge spicules. Sponges are found in both coastal and open Gulf habitats, and the production of spicules can be a major sink of DSi as well as a source of BSi (Storr 1964; Rützler and Macintyre 1978; Maldonado et al. 2011; Ugalde et al. 2015). Coastal sponges have been shown to take up DSi at rates of 0.01–0.9 mmol Si m^{-2} d^{-1}, representing a major sink for DSi on continental margins (Maldonado et al. 2011). In contrast, it has also been demonstrated that sponge spicules represent a substantial sink for BSi in sediments due to very slow rates of Si dissolution, depending on water temperatures. Depending on the sponge species, spicules lost no volume after a 6-month incubation in temperate or tropical waters, while others lost up to 47% of their volume after incubation (Bertolino et al. 2017). Although there is much research to be done regarding the regional role of sponges, if results from other coastal systems are representative, this group may be the major standing stock term for BSi in the GOM system.

Terrestrial grasses are also known to accumulate significant quantities of SI in the form of amorphous silica phytoliths and may be a significant portion of the riverine-delivered BSi flux and common in coastal sediments. In the GOM and southeastern United States, Lu and Liu (2003) identified 34 coastal grass species from 6 subfamilies that produce phytoliths. Coincidently, many grass phytoliths are similar in size to diatoms, and the radiations of diatoms and grasses appear to be nearly contemporaneous in the fossil record (Falkowski et al. 2004). The percentage of BSi in marsh plants is variable, but in Simmons Bayou (coastal Mississippi), Lanning and Eleuterius (1983) observed that leaf content ranged from 0.2% to 4.1% and stem content ranged from <0.1% to 6.1%. *Spartina cynosuroides* had BSi content >1% for both structures. Other common marsh grasses, such as *Juncus*, accumulate BSi, and the weight percentage increases with age (Eleuterius and Lanning 1987).

Given the role of marshes in buffering Si delivery to coastal waters (Carey and Fulweiler 2014), phytolith (rigid, decay-resistant, microscopic structures composed of silica, found in some higher plant tissues) dynamics may play an important role in the GOM Si cycle at the land-sea interface. Jendrzejewski and Hart (1978) observed that phytoliths are typically rare in the GOM abyssal plain but are more commonly observed in the sediments of the continental slope (they arise and are derived from the continents) in Mississippi-influenced regions. Additionally, these authors also observed a strong spatial difference, with higher phytolith frequency in the western GOM sediments and rarer in the eastern basin. The specific surface area of certain phytoliths ranges from 100 to 300 $m^2 g^{-1}$ (Fraysse et al. 2009), which may equal or exceed some diatoms by ~10–250 $m^2 g^{-1}$ (Kamatani et al. 1980; Dixit et al. 2001; Demarest et al. 2009). This suggests phytoliths may be highly reactive when in contact with seawater, which is significantly undersaturated relative to Si. Given the potential reactivity, phytoliths may be an important component of coastal sediment Si cycling in the GOM.

While terrestrial grasses are known to accumulate BSi, there is poor understanding of the broader biogeochemical or ecosystem consequences in the GOM. Significant study on this topic has been done in New England marshes. Based on Si flux measurements in Narragansett Bay, Carey and Fulweiler (2014) suggested that salt marshes act as a sponge for Si, in that Si is stored and released later for use by organisms (e.g., diatoms). This source of Si to the water column was especially important during low water-flow periods, when riverine nutrient delivery is minimized. Furthermore, another study linked the Si accumulation rate in New England marshes to the degree of N enrichment, where high-N marshes had higher Si accumulation (Carey and Fulweiler 2013). Considering the well-documented N loading in the region, future work in the GOM will need to address whether regional marshes act similarly given the major differences in physical forcing (e.g., less frequent and lower-amplitude tides, higher precipitation).

Abiotic Scavenging

The occurrence of adsorption and desorption of silicic acid from mineral surfaces has been known for decades. It was originally proposed that silicate minerals reacted with dissolved species (Sillen 1961), and later Mackenzie et al. (1967) demonstrated that DSi was bound (adsorption) in clay suspensions when DSi concentrations were high but released (desorption) when DSi was depleted. More recently, a similar effect was demonstrated in San Francisco Bay as the uptake of the radiotracer ^{32}Si from abiotic factors (e.g., authigenic formation) was substantial (i.e., more than biology) and required significant modifications to the Brzezinski and Philips (1997) ^{32}Si method to more accurately quantify ^{32}Si uptake by diatoms (Koch 2005). This finding is consistent with the observation that DSi, like phosphate, is readily adsorbed onto the surface of sediments and lost from the water column as sediments are deposited (Lane et al. 2004; Mayer and Gloss 1980).

It was suggested that abiotic uptake of silicic acid in the Mississippi River delta was prevalent during mixing between freshwater and salt water (Bien et al. 1958). Fanning and Pilson (1973) observed a maximum adsorption of ~7% for Mississippi River and Louisiana shelf conditions; however, most of the values were <3%, or no resolvable adsorption. Such results are consistent with the conclusions that adsorption and desorption reactions do not appear to be a dominant abiotic factor cycling Si in the water column (Chou and Wollast 2006), and this may be applicable to the GOM. However, within the sediments, much less is known about the rate of absorption/desorption reactions for Si, and given the importance of benthic flux to this environment, this requires future examination.

Physicochemical Dissolution

Silicic acid, which is remineralized from biogenic silica in the water column, supports 53% of euphotic zone BSi production globally (Nelson et al. 1995). In most systems >95% of the BSi produced in the euphotic zone will dissolve in the water column and/or sediments (Tréguer and De La Rocha 2013); but there is a scarcity of BSi dissolution data in coastal systems and no published rates in the GOM. Beucher et al. (2004) measured BSi production and dissolution rates in Bay of Brest (macrotidal region in northwestern France) surface waters and observed that dissolution could support 31% to 47% of BSi production in the spring and summer and up to 60% to 83% in the fall and winter when production was lower. Their analysis did not consider remineralized Si from benthic flux, which would increase these percentages. To date only 11 euphotic-zone multidepth profiles of BSi dissolution rates have been reported in non-upwelling coastal systems or the Southern Ocean, all of which were from the Amazon River plume (DeMaster et al. 1991; Nelson et al. 1995). Water-column BSi dissolution in the Amazon system ranged from 22% to 240% of BSi production (mean 65%). This indicates that at times remineralized silicic acid from BSi dissolution could sustain all BSi production.

Accumulation and Reverse Weathering in Sediments

Recently, Tréguer and De La Rocha (2013) report ~6 Tmol BSi is buried annually in the sediments; 52% of this burial occurs in the coasts and continental margins (DeMaster 2002; DeMaster 2003). This high BSi burial rate indicates a disproportionately important role for these regions relative to their size (~10% of the ocean area). Turner et al. (2008) report an increase in %BSi content of the northern GOM hypoxic-zone sediments of approximately a factor of 3 since the 1980s. Similarly, Parsons et al. (2002) observed increases in diatom preservation in the same sediments starting in the 1960s. Therefore, the accumulation rate in this system is clearly dynamic and may have been altered in the mid- to late 20th century. Using a more recent approach from Allison et al. (2007) for accumulation, Presti and Michalopoulos (2008) estimated that the BSi accumulation rate within the deltaic shelf area of the Mississippi River plume was 4.4×10^9 mol Si a^{-1}, about an order of magnitude lower than the accumulation of BSi in the Amazon deltaic shelf area (Michalopoulos and Aller 2004). While DeMaster (2002) suggested that continental margins were even more important for the accumulation of BSi globally, data in the Amazon and Mississippi River plume regions suggest that the total Si sequestration in these sediments has been systematically underestimated.

Reverse weathering, that is, the authigenic formation of clay minerals in marine sediments, was proposed in the 1960s but not verified in the laboratory until the 1990s (Michalopoulos and Aller 1995). The generalized reaction involves silica plus degraded clays, iron, organic matter, and cations to form new clay material along with CO_2 and water (Mackenzie and Kump 1995). Reverse weathering interacts with sediment BSi in 2 ways: (1) the formation of aluminosilicate minerals from silicic acid in pore water derived from amorphous silica dissolution (Michalopoulos and Aller 1995) and (2) the formation of authigenic clays around sediment-BSi fragments, that is, diatom frustules (Michalopoulos and Aller 2004). This latter process reduces BSi remineralization and may represent an important sink not only for Si but for potassium (K), Fe, and aluminum (Al) (Michalopoulos and Aller 2004). The formation of authigenic clays around BSi fragments renders the BSi less reactive to traditional alkaline leach procedures in the laboratory (DeMaster 1981; Mortlock and Froelich 1989). This transformation leads to an underestimation of the amount of BSi retained in land-sea interface sediments. Michalopoulos and Aller (2004), working in deltaic sediments of the Amazon, demonstrated that up to 90% of the BSi present in sediment deposits is authigenically altered. The inclusion of alteration products increases the Si storage estimates in the Amazon delta to 22% of riverine Si. Recently, an independent analysis of cosmogenic Si-32 ($t_{1/2}$ = 134 yr) in Amazon sediments found it occurred mainly in mineral Si pools (Rahman et al. 2016) rather than operational BSi pools quantified using typical methods (DeMaster 1981; Mortlock and Froelich 1989). This study provided the first independent validation of reverse weathering and the potential importance in Si cycling (Rahman et al. 2016). In the Mississippi River delta, similar measurements suggest that ~40% of the long-term reactive silica is stored as authigenic components (Presti and Michalopoulos 2008). Additional analysis in a northern GOM coastal lagoon (Mississippi Sound; Krause et al. 2017) has shown quantities of reactive Si similar to those observed by Presti and Michalopoulos (2008) near the Mississippi River delta. Future work should aim to understand the spatial extent to which this process is important to sediment Si sequestration and also the factors regulating the formation of these authigenic products.

Anthropogenic Perturbation

Hydrological Changes in the Mississippi River Watershed

Human activities upriver or upstream of the Mississippi River have led to major changes in the delivery of Si to the coastal GOM. Land-use changes have increased the amount of farmland at the expense of forests (Turner and Rabalais 1991; Turner et al. 2008), thereby increasing nutrient (e.g., N, P) loads in runoff. The building of dams and the resulting impoundments in the upper Mississippi River basin have contributed to the reduction in riverine DSi flux because of increased sedimentation of amorphous Si and slow dissolution rates at these installations (Triplett et al. 2008). Others have suggested that increased Si uptake and incorporation into biogenic Si in river waters also contribute to a decrease in DSi in the Mississippi River discharge to the Gulf (Houser et al. 2010). These types of anthropogenic perturbations have been observed in other regions where eutrophication in the watershed and damming of tributaries and rivers has led to reduced Si delivery to coastal waters (Conley et al. 1993; Humborg et al. 2000; Gilpin et al. 2004).

Eutrophication and Diatom Assemblage Feedbacks

In the Mississippi River, the annual average Si:N ratio has declined from >3 in 1960 to ~1 in 1980, driven by both increasing nitrate and decreasing silicic acid concentrations, and has remained around the latter value to the present day (Fig. 3.13A). In the last decade, the Mississippi River exhibited strong seasonal oscillations between favorable (>1.0) and unfavorable (<1.0) Si:N conditions for diatoms, with the lowest average monthly values observed in April and May (Fig. 3.13B). This period corresponds to the annual peak in Mississippi River discharge and elevated primary production rates on the Louisiana shelf from nitrate loading (Lohrenz et al. 1997). The Louisiana shelf system is not unique in this type of shift (Cloern 2001), as similar changes in N and P loadings relative to Si have been observed in various locations, including the Danube River (Humborg et al. 1997), Irish Sea (Allen et al. 1998), Bay of Brest (Lepape et al. 1996), East River/Long Island Sound (Gobler et al. 2006), and the German Bight (Radach et al. 1990), and some of the earliest observations were in lake systems (Conley et al. 1993, and references therein). In some systems, these changes have coincided with shifts to lower diatom abundance relative to other phytoplankton groups, such as in the Black Sea (Bodeanu 1993). A commonly cited DSi threshold for the net production of BSi is 2 μM (Egge and Aksnes 1992); however, more current research has shown that diatoms can be adapted to very low substrate and increase biomass when DSi is <1.5 μM (Krause et al. 2010, 2013). Low DSi concentrations in the northern GOM may result in an earlier onset of kinetic limitation for heavily silicified diatoms, which in turn can shift the diatom assemblage to favor large but lightly silicified groups (e.g., *Pseudo-nitzschia* spp.). This mechanism of lightly silicified diatom groups being more competitive in low DSi conditions is often evoked in the literature. But due to inadequate methods for isolating the effect of Si, which is complicated by the strong coupling of Si and N drawdown, this has yet to be shown directly in the marine environment.

The changes in Mississippi River–derived nutrient ratios since the 1950s coincide with ecosystem-wide population shifts on the Louisiana shelf. Such changes include declines in zooplankton biomass that prefer grazing on diatoms (i.e., copepods) and the fraction of C assimilated in zooplankton fecal pellets (Turner et al. 1998), increases in the abundance of phytoplankton species that can generate harmful algal blooms (Dortch et al. 1999), and the occurrence of more frequent and intense hypoxia events relative to those in the previous century (Rabalais et al. 2007). While the increase in sediment BSi has been shown to be across diatom groups, Parsons et al. (2002) observed the highest rate of increase in lightly silicified pennate diatoms from the genera *Pseudo-nitzschia* spp. This genus includes species that produce domoic acid, a neurotoxin responsible for amnesic shellfish poisoning (Liefer et al. 2009). The decline in DSi availability relative to nitrate and phosphate has been linked to shifts from diatom-dominated to flagellate-dominated ecosystems. This "DSi paradigm" (Ragueneau et al. 2006) was first proposed for coastal systems by Officer and Ryther (1980) and has been evoked as a potential mechanism for changes in many systems (Cloern 2001). However, the link between significant ecosystem perturbations in the Si cycle and additional biogeochemical shifts are largely built on observational rather than mechanistic evidence.

Comparisons with Nutrients in Other Atlantic Ocean Marginal Seas

It is useful to compare nutrient sources and observed concentrations in GOM waters with those found in other regional seas that border the Atlantic Ocean (e.g., the Baltic, Caribbean, Mediterranean, and North Seas) for similarities or differences. Nutrient concentrations and distributions in each of these seas are strongly influenced by anthropogenic activities that lead to increased loadings in water (e.g., Lucea et al. 2003; De Galan et al. 2004; Carruthers et al. 2005; Van Der Zee and Chou 2005; riverine, groundwater, treated effluents; Bernard et al. 2009; Ludwig et al. 2009; da Cunha and Buitenhuis 2013; Tanhua et al. 2013; Rodellas et al. 2015; Burson et al. 2016; Moon et al. 2016; Segura-Noguera et al. 2016) and atmospheric deposition (e.g., increased ammonium loading; United Nations Environment Program 2010, and references therein; Desboeufs et al. 2015; GESAMP 2012; Hansell and Follows 2008; Prospero et al. 1996; Ruoho-Airola et al. 2012). As we have demonstrated previously, eutrophication via similar mechanisms also is occurring in the GOM. Due to the lack of winter convection, as observed in higher latitudes, the open-sea region of the GOM is predominantly oligotrophic; thus, such eutrophication amplifies the magnitude of the nutrient gradient between the coastal waters and those off the continental shelf. The physical circulation (e.g., currents, tides, localized upwelling) within these systems plays a key role in distributing nutrients; however, the GOM is microtidal, which means that tidal mixing, which is dominant in many other systems, has a lower proportional effect and the nutrient enrichment may be confined to areas near coastal locations and surface plumes.

Each of the 3 European marginal seas has cities along coastlines that contribute to eutrophication in coastal areas as well as rivers that introduce nutrients to these marginal seas (Bouraoui and Grizzetti 2011). The European Union maintains a large database, the European Environment Information and Observation Network (EIONET—Water, formerly EUOWATERNET; https://www.eea.europa.eu/themes/water/europes-seas-and-coasts), of water-quality indicators among stations maintained by the European Environment Agency (EEA). Trends of data from 1985 were compared to those from 2012 for the major nutrients in the European Seas, including the Baltic, North, and Mediterranean Seas (https://www.eea.europa.eu/data-and-maps/indicators#c0=10&c12-operator=or&b_start=0). These data suggest that concentrations of DIN ($NO^{2+}NO^{3}$, NH^{4}) and PO_4 have not changed over the 1985–2012 time period at most sampling stations, while those for some stations decreased (14% for total DIN and 13% for PO^{4}). Concentrations at some coastal stations increased, while those for open waters generally decreased. This trend is similar to what is seen for the GOM; the open Gulf remains highly oligotrophic with some coastal stations showing higher concentrations (see Fig. 3.1 A–D). Nutrient delivery into the European regional seas from rivers has decreased between 1977 and 2006, while that delivered via groundwater discharge has increased substantially, a change attributed in part to changes in European Union regulations (Bouraoui and Grizzetti 2011). In the Mediterranean Sea, the nutrient concentrations found in the eastern and western basins are somewhat different, with greater N and P concentrations in the western Mediterranean (Lazzari et al. 2016; Moon et al. 2016; Macias et al. 2017). The Moon et al. (2016) study also determined that the influence of riverine inputs of nutrients was less than that of atmospheric deposition in both basins, likely due to the low baseline nutrient concentrations that make the system ultraoligotrophic. Concentrations of P in the Mediterranean appear to have declined more so than did those of N for both basins (Moon et al. 2016). In both the North Sea and the Baltic Sea, there are greater concentrations of both N and P near the coasts and much lower concentrations in the open seas (Andersen and Laamanen 2009; Paramor et al. 2009). For the Baltic Sea, nutrient inputs from rivers, the atmosphere, and the sediments are most important, but there can also be input of N from N fixation by cyanobacteria (Andersen and Laamanen 2009). In the North Sea, about 30% of the nutrient inputs come from atmospheric deposition, with the remainder from land-based sources (Paramor et al. 2009). Major nutrient inputs (about 75% of the N) enter the North Sea from the Rhine and Elbe Rivers (Paramor et al. 2009). Observations common to each of these European marginal seas is that nutrient concentrations have generally been declining over the past few decades; however, the inputs of dissolved N and dissolved P are out of balance in both coastal and open waters (Andersen and Laamanen 2009; Paramor et al. 2009). The dissolved N:P ratios in coastal waters can be very high (e.g., >200–300), greatly exceeding Redfield proportions, due to very high N inputs, while the ratio in open waters can be much lower than Redfield proportions, leading to strong P limitation of primary production in coastal waters and strong N limitation in the open seas (Andersen and Laamanen 2009; Paramor et al. 2009; Lazzari et al. 2016; Moon et al. 2016). These data, while not perfect analogs to the northern GOM, are consistent with regional trends and demonstrate how marginal seas in many parts of the Atlantic are prone to anthropogenic transformation.

Similar to the northern GOM, these increases in N and P have corresponded with declines in dissolved Si as a result of upstream modifications (e.g., dams, reservoirs), which retain diatom-based Si (e.g., Humborg et al. 2006). In the Rhine River system, Hartmann et al. (2011) observed that under some conditions, when P was reduced (eutrophied rivers for P), there was an increase in dissolved Si delivery, which has downstream implications for coastal diatom productivity. While not robust, these data suggest that in some of the affected European coastal systems, mitigation of eutrophication may lead to some amelioration of the downstream changes by shifting the system back to what is considered more "natural" dynamics among primary producers (e.g., diatoms vs. flagellates). Given the push from the GOM scientific community, the US government, and non-governmental organizations toward reduction of N loading to the region, it is possible an analogous situation may be resolvable in the region. However, resolving such a "rebound" signal will likely require a significant shift in dynamics and aggressive monitoring given the high degree of temporal/spatial variability in nutrient concentrations and primary production.

The Caribbean Sea is a more open environment than the GOM, with islands to the northeastern boundary and the coastlines of South America, Central America, and

Mexico as the western boundary. The open waters of the Caribbean Sea are, in general, characterized as being oligotrophic, while coastal areas are eutrophic (Gledhill et al. 2008; Scranton et al. 2014). The Cariaco Basin, in the southern portion of the Caribbean Sea off Venezuela, has been well studied in the past decade through the Cariaco Ocean Time-Series Program, with numerous publications from that program available there (Institute for Marine Remote Sensing, n.d.). A large body of research and nutrient data has been collected and synthesized on the website of the Caribbean Environmental Program (CEP) of the United Nations Environment Programme (UNEP 2010, and references therein). Nutrient inputs from land-based sources, including the two major rivers (the Orinoco and Amazon), are the greatest contributors to eutrophication in coastal Caribbean waters, delivering 13,000 tons of N and 5800 tons of P per year (UNEP 2000). Riverine inputs of nutrients increase during northern summer and fall, and during seasonal rains nutrient inputs from the Orinoco River lead to increases in coastal primary production (Corredor and Morell 2001). Atmospheric deposition of nutrients is growing and expected to increase 2–4-fold by the year 2020 over levels measured prior to 2000 (Desboeufs et al. 2015; GESAMP and Advisory Committee on Protection of the Sea 2001). Human activities in various bays in the Caribbean have had a major influence on nutrient levels in those coastal waters, with much of the nutrient inputs due to raw sewage discharges (UNEP et al. 2004). Little to no data are available for Si cycling in this region; however, we do know that the large rivers deliver dissolved Si to coastal waters that then flow into the Caribbean Sea and potentially into the Gulf of Mexico (Bernard et al. 2009; da Cunha and Buitenhuis 2013; Thunell et al. 2008).

Some routine observations can be made about the nutrient distribution patterns and sources for each of the marginal seas just discussed. With the exception of the Baltic Sea, perhaps, the open portions of the marginal seas tend to be oligotrophic with biological production most often limited by the availability of N. The coastal waters in these seas are eutrophic and often impacted by anthropogenic nutrient inputs, the primary sources coming from municipal and industrial effluents, rivers, and atmospheric deposition. Coastal waters in the seas are often thought to have greater inputs of N that can lead to limitation of biological productivity by P and at times Si (e.g., diatoms). Because of the large inputs of N near the coasts of these marginal seas and the lack of large inputs in the open portions of the seas, Redfield proportions of N and P are rarely, if ever, observed. The coastal areas generally exhibit very high N:P ratios, while the open seas have very low N:P ratios. This lack of balance between inputs and concentrations of essential nutrients can affect ecosystems and the abundances of living resources in these marginal seas.

The marginal seas that are most similar seem to be the GOM and the Mediterranean Sea. Both regions are microtidal and warm and have large river inputs of both freshwater and nutrients at coastal zones with oligotrophic deep waters. Atmospheric inputs of N are important in both the Gulf and the Mediterranean. Cities and industrial activities located on coasts or along the rivers provide high concentrations of nutrients, primarily N, to the coastal waters in both regions. In all of the regions that have coastlines in each of the regional seas discussed, there are efforts to reduce nutrient inputs to the coastal waters and ongoing monitoring programs to assess the effectiveness of nutrient reduction efforts. Management of nutrients and eutrophication remains the highest priority for environmental managers for all of the nations that border these regional seas. A greater understanding of the main processes of nutrient inputs, utilization, cycling, and sequestration, as well as enhanced monitoring of coastal eutrophication may help managers make more informed decisions on nutrient mitigation plans.

Conclusion

Nutrient cycles in the GOM are very complex and interactive. Inputs of nutrients include those from natural biological and oceanographic processes, such as regeneration both within the water column and in the sediments, and physical circulation of water masses, including currents and mesoscale features such as eddies and filaments. There are also atmospheric inputs of both macronutrients (N, P, and Si) and micronutrients (trace metals). Coastal regions receive inputs from rivers, streams, and estuaries through natural processes. These regions receive enhanced nutrient loading because of human activities within the river and stream watersheds that leads to enhanced levels of biological production. Losses include uptake and incorporation by organisms, sedimentation, and advective losses in currents, primarily the Loop Current. Within the water column and the sediments are numerous cycling, transformation, and exchange processes that result in various forms of N, P, and Si transitioning between inorganic and organic forms. All of these processes result in situations where nutrients appear to be found in balanced "Redfield stoichiometry" and many situations where there are deviations from that stoichiometry (Redfield et al. 1963). The deviations from balanced stoichiometry are usually associated with river plumes and other areas where nutrient input fluxes are affected by human activities on land and within the watersheds. As a result, the GOM can be characterized as highly eutrophic along the coastlines, particularly near river mouths, and highly oligotrophic in open waters, with conditions that from personal observation are more oligotrophic than those found in the North Pacific Central Gyre and the Sargasso Sea.

Acknowledgments

We thank Jaye E. Cable for her helpful suggestions and contributions to the effort. We also thank Rachel L. Coen, who helped in putting the references and citations into the appropriate format and proofreading drafts of the chapter. We also thank the two reviewers, James L. Pinckney and Jay A. Brandes, for their helpful and constructive comments allowing us to improve the quality of the chapter.

References

Alexander, R. B., R. A. Smith, G. E. Schwarz, E. W. Boyer, J. V. Nolan, and J. W. Brakebill. 2008. Differences in phosphorus and nitrogen delivery to the Gulf of Mexico from the Mississippi River basin. *Environmental Science & Technology* 42:822–830.

Allen, J. R., D. J. Slinn, T. M. Shammon, R. G. Hartnoll, and S. J. Hawkins. 1998. Evidence for eutrophication of the Irish Sea over four decades. *Limnology and Oceanography* 43:1970–1974.

Allison, M. A., T. S. Bianchi, B. A. McKee, and T. P. Sampere. 2007. Carbon burial on river-dominated continental shelves: Impact of historical changes in sediment loading adjacent to the Mississippi River. *Geophysical Research Letters* 34: L01606. doi:10.1029/2006GL028362.

Altieri, K. E., S. E. Fawcett, A. J. Peters, D. M. Sigman, and M. G. Hastings. 2016. Marine biogenic source of atmospheric organic nitrogen in the subtropical North Atlantic. *Proceedings of the National Academy of Sciences* 113:925–930.

Altieri, K. E., M. G. Hastings, A. J. Peters, and D. M. Sigman. 2012. Molecular characterization of water soluble organic nitrogen in marine rainwater by ultra-high resolution electrospray ionization mass spectrometry. *Atmospheric Chemistry and Physics* 12:3557–3571.

An, S. M., and W. S. Gardner. 2002. Dissimilatory nitrate reduction to ammonium (DNRA) as a nitrogen link, versus denitrification as a sink in a shallow estuary (Laguna Madre/Baffin Bay, Texas). *Marine Ecology Progress Series* 237:41–50.

An, S., and S. B. Joye. 2001. Enhancement of coupled nitrification-denitrification by benthic photosynthesis in shallow estuarine sediments. *Limnology and Oceanography* 46:62–74.

Andersen, J. H., and M. Laamanen. 2009. *Eutrophication in the Baltic Sea: An Integrated Thematic Assessment of the Effects of Nutrient Enrichment and Eutrophication in the Baltic Sea Region*. Baltic Sea Environment Proceedings 115B. Helsinki: Helsinki Commission. 149 pp.

Baines, S. B., B. S. Twining, M. A. Brzezinki, J. W. Krause, S. Vogt, D. A. Assael, and H. McDaniel. 2012. Significant silicon accumulation by marine picocyanobacteria. *Nature Geoscience* 5:886–891.

Benitez-Nelson, C. R. 2000. The biogeochemical cycling of phosphorus in marine systems. *Earth-Science Reviews* 51:109–135.

Bernard, C. Y., H. H. Dürr, C. Heinze, J. Segschneider, and E. Maier-Reimer. 2009. Contribution of riverine nutrients to the silicon biogeochemistry of the global ocean—a model study. *Biogeosciences Discussions* 6:1091–1119.

———. 2011. Contribution of riverine nutrients to the silicon biogeochemistry of the global ocean—a model study. *Biogeosciences* 8:551–564.

Bernard, R. J., B. Mortazavi, and A. A. Kleinhuizen. 2015. Dissimilatory nitrate reduction to ammonium (DNRA) seasonally dominates NO_3^- reduction pathways in an anthropogenically impacted sub-tropical coastal lagoon. *Biogeochemistry* 125:47–64.

Bernard, R. J., B. Mortazavi, L. Wang, A. C. Ortmann, H. MacIntyre, and W. C. Burnett. 2014. Benthic nutrient fluxes and limited denitrification in a sub-tropical groundwater-influenced coastal lagoon. *Marine Ecology Progress Series* 504:13–26.

Bernot, M. J., W. K. Dodds, W. S. Gardner, M. J. McCarthy, D. Sobolev, and J. L. Tank. 2003. Comparing denitrification estimates for a Texas estuary by using acetylene inhibition and membrane inlet mass spectrometry. *Applied and Environmental Microbiology* 69:5950–5956.

Bertolino, M., R. Cattaneo-Vietti, M. Pansini, C. Santini, and G. Bavestrello. 2017. Siliceous sponge spicule dissolution: In field experimental evidences from temperate and tropical waters. *Estuarine, Coastal and Shelf Science* 184:46–53.

Beucher, C., P. Tréguer, R. Corvaisier, A. M. Hapette, and M. Elskens. 2004. Production and dissolution of biosilica, and changing microphytoplankton dominance in the Bay of Brest (France). *Marine Ecology Progress Series* 267:57–69.

Bianchi, T. S., S. F. DiMarco, J. H. Cowan Jr., R. D. Hetland, P.

Chapman, J. W. Day, and M. A. Allison. 2010. The science of hypoxia in the northern Gulf of Mexico: A review. *Science of the Total Environment* 408:1471–1484.

Bianchi, T. S., S. F. DiMarco, R. W. Smith, and K. M. Schreiner. 2009. A gradient of dissolved organic carbon and lignin from Terrebonne-Timbalier Bay Estuary to the Louisiana shelf (USA). *Marine Chemistry* 117:32–41.

Bianchi, T. S., J. Pennock, and R. Twilley (eds.). 1999. *Biogeochemistry of Gulf of Mexico Estuaries*. Hoboken, NJ: John Wiley and Sons. 448 pp.

Bien, G., D. Contois, and W. Thomas. 1958. The removal of soluble silica from fresh water entering the sea. *Geochimica et Cosmochimica Acta* 14:35–54.

Bodeanu, N. 1993. Microalgal blooms in the Romanian area of the Black Sea and contemporary eutrophication conditions. Pp. 203–209 in T. Smayda and Y. Shimizu (eds.), *Toxic Phytoplankton Blooms in the Sea*. Amsterdam: Elsevier Science Publishers.

Bouraoui, F., and B. Grizzetti. 2011. Long term change of nutrient concentrations of rivers discharging in European seas. *Science of the Total Environment* 409:4899–4916.

Bowles, M., and S. Joye. 2011. High rates of denitrification and nitrate removal in cold seep sediments. *International Society for Microbial Ecology Journal* 5:565–567.

Breitburg, D. L., J. K. Craig, R. S. Fulford, K. A. Rose, W. R. Boynton, D. C. Brady, B. J. Ciotti, et al. 2009. Nutrient enrichment and fisheries exploitation: Interactive effects of estuarine living resources and their management. *Hydrobiologia* 629:31–47.

Bristow, L. A., N. Sarode, J. Cartee, A. Caro-Quintero, B. Thamdrup, and F. J. Stewart. 2015. Biogeochemical and metagenomic analysis of nitrite accumulation in the Gulf of Mexico hypoxic zone. *Limnology and Oceanography* 60:1733–1750.

Bronk, D. A., L. Killberg-Thoreson, R. E. Sipler, M. R. Mulholland, Q. N. Roberts, P. W. Bernhardt, M. Garrett, J. M. O'Neil, and C. A. Heil. 2014. Nitrogen uptake and regeneration (ammonium regeneration, nitrification and photoproduction) in waters of the west Florida shelf prone to blooms of *Karenia brevis*. *Harmful Algae* 38:50–62.

Bruesewitz, D. A., W. S. Gardner, R. F. Mooney, L. Pollard, and E. J. Buskey. 2013. Estuarine ecosystem function response to flood and drought in a shallow, semiarid estuary: Nitrogen cycling and ecosystem metabolism. *Limnology and Oceanography* 58:2293–2309.

Bryant-Mason, A., Y. J. Xu, and M. Altabet. 2013. Isotopic signature of nitrate in river waters of the lower Mississippi and its distributary, the Atchafalaya. *Hydrological Processes* 27:2840–2850.

Brzezinski, M. A. 1985. The Si:C:N ratio of marine diatoms: Interspecific variability and the effect of some environmental variables. *Journal of Phycology* 21:347–357.

Brzezinski, M. A., J. L. Jones, K. D. Bidle, and F. Azam. 2003. The balance between silica production and silica dissolution in the sea: Insights from Monterey Bay, California, applied to the global data set. *Limnology and Oceanography* 48:1846–1854.

Brzezinski, M. A., and D. R. Phillips. 1997. Evaluation of Si-32 as a tracer for measuring silica production rates in marine waters. *Limnology and Oceanography* 42:856–865.

Burgin, A. J., and S. K. Hamilton. 2007. Have we overemphasized the role of denitrification in aquatic ecosystems? A review of nitrate removal pathways. *Frontiers in Ecology and the Environment* 5:89–96.

Burnett, W. C., P. K. Aggarwal, H. Bokuniewicz, J. E. Cable, M. A. Charette, E. Kontar, S. Krupa, et al. 2006. Quantifying submarine groundwater discharge in the coastal zone via multiple methods. *Science of the Total Environment* 367:498–543.

Burnett, W. C., H. Bokuniewicz, M. Huettel, W. S. Moore, and M. Taniguchi. 2003. Groundwater and pore water inputs to the coastal zone. *Biogeochemistry* 66:3–33.

Burson, A., M. Stomp, L. Akil, C. P. D. Brussaard, and J. Huisman. 2016. Unbalanced reduction of nutrient loads has created an offshore gradient from phosphorus to nitrogen limitation in the North Sea. *Limnology and Oceanography* 61:869–888.

Cable, J. E., W. C. Burnett, J. P. Chanton, and G. L. Weatherly. 1996. Estimating groundwater discharge into the northeastern Gulf of Mexico using radon-222. *Earth and Planetary Science Letters* 144:591–604.

Canion, A., J. E. Kostka, T. M. Gihring, M. Huettel, J. E. E. van Beusekom, H. Gao, G. Lavik, and M. M. M. Kuypers. 2014. Temperature response of denitrification and anammox reveals the adaptation of microbial communities to in situ temperatures in permeable marine sediments that span 50 degrees in latitude. *Biogeosciences* 11:309–320.

Capone, D. G. 1993. Determination of nitrogenase activity in aquatic samples using the acetylene reduction procedure. Pp. 621–631 in P. F. Kemp, B. F. Sherr, E. B. Sherr, and J. J. Cole (eds.), *Handbook of Methods in Aquatic Microbial Ecology*. Boca Raton, FL: Lewis Publishers.

Capone, D. G., J. A. Burns, J. P. Montoya, A. Subramaniam, C. Mahaffey, T. Gunderson, A. F. Michaels, and E. J. Carpenter. 2005. Nitrogen fixation by *Trichodesmium* spp.: An important source of new nitrogen to the tropical and subtropical North Atlantic Ocean. *Global Biogeochemical Cycles* 19:GB2024. doi:10.1029/2004GB002331.

Carder, K. L., K. A. Fanning, P. R. Betzer, and V. Maynard. 1977. Dissolved silica and the circulation in the Yucatan Strait and deep eastern Gulf of Mexico. *Deep Sea Research* 24:1149–1160.

Cardona, Y. 2013. Mesoscale variability in the Gulf of Mexico, its impact and predictability. PhD diss., Georgia Institute of Technology. 113 pp.

Cardona, Y., A. Bracco, T. A. Villareal, A. Subramaniam, S. C. Weber, and J. P. Montoya. 2016. Highly variable nutrient concentrations in the northern Gulf of Mexico. *Deep Sea Research Part II: Tropical Studies in Oceanography* 129:20–30.

Carey, J., and R. Fulweiler. 2013. Nitrogen enrichment increases net silica accumulation in a temperate salt marsh. *Limnology and Oceanography* 58:99–111.

———. 2014. Salt marsh tidal exchange increases residence time of silica in estuaries. *Limnology and Oceanography* 59:1203–1212.

Carini, S. A., M. J. McCarthy, and W. S. Gardner. 2010. An isotope dilution method to measure nitrification rates in the northern Gulf of Mexico and other eutrophic waters. *Continental Shelf Research* 30:1795–1801.

Carruthers, T. J. B., B. I. van Tussenbroek, and W. C. Dennison. 2005. Influence of submarine springs and wastewater on nutrient dynamics of Caribbean seagrass meadows. *Estuarine, Coastal and Shelf Science* 64:191–199.

Casey, R., L. Gust, A. Leavesley, D. Williams, R. Reynolds, T. Duis, and J. M. Spaw. 1979. Ecological niches of radiolarians, planktonic foraminiferans and pteropods inferred from studies on living forms in the Gulf of Mexico and adjacent waters. *Transactions—Gulf Coast Association of Geological Societies* 29:216–223.

Charette, M. A., P. B. Henderson, C. F. Brier, and Q. Liu. 2013. Submarine groundwater discharge in a river-dominated Florida estuary. *Marine Chemistry* 156:3–17.

Chen, X., S. E. Lohrenz, and D. A. Wiesenburg. 2000. Distribution and controlling mechanisms of primary production on the Louisiana-Texas continental shelf. *Journal of Marine Systems* 25:179–207.

Childs, C. R., N. N. Rabalais, R. E. Turner, and L. M. Proctor. 2002. Sediment denitrification in the Gulf of Mexico zone of hypoxia. *Marine Ecology Progress Series* 240:285–290.

Chou, L., and R. Wollast. 2006. Estuarine silicon dynamics. Pp. 93–120 in V. Ittekkot, D. Unger, C. Humborg, and N. Tac An (eds.), *The Silicon Cycle: Human Perturbations and Impacts on Aquatic Systems*: Washington, DC: Island Press.

Cloern, J. E. 2001. Our evolving conceptual model of the coastal eutrophication problem. *Marine Ecology Progress Series* 210:223–253.

Coles, V. J , M. T. Brooks, J. Hopkins, M. R. Stukel, P. L. Yager, and R. R. Hood. 2013. The pathways and properties of the Amazon River plume in the tropical North Atlantic Ocean. *Journal of Geophysical Research: Oceans* 118:6894–6913.

Conley, D. J., C. L. Schelske, and E. F. Stoermer. 1993. Modification of the biogeochemical cycle of silica with eutrophication. *Marine Ecology Progress Series* 101:179–192.

Corredor, J. E., and J. M. Morell. 2001. Seasonal variation of physical and biogeochemical features in the eastern Caribbean surface water. *Journal of Geophysical Research* 106:4517–4525.

Cotner, J. B., and W. S. Gardner. 1993. Heterotrophic bacterial mediation of ammonium and dissolved free amino-acid fluxes in the Mississippi River plume. *Marine Ecology Progress Series* 93:75–87.

Cotner, J. B., M. W. Suplee, N. W. Chen, and D. E. Shormann. 2004. Nutrient, sulfur and carbon dynamics in a hypersaline lagoon. *Estuarine, Coastal and Shelf Science* 59:639–652.

da Cunha, L. C., and E. T. Buitenhuis. 2013. Riverine influence on the tropical Atlantic Ocean biogeochemistry. *Biogeosciences* 10:6357–6373.

Dagg, M. J., T. S. Bianchi, G. A. Breed, W.-J. Cai, S. Duan, B. A. McKee, R. T. Powell, and C. M. Stewart. 2005. Biogeochemical characteristics of the lower Mississippi River, USA, during June 2003. *Estuaries* 28:664–674.

Dale, V. H., C. L. Kling, J. L. Meyer, J. Sanders, H. Stallworth, T. Armitage, D. Wangsness, et al. (eds.). 2010. *Hypoxia in the Northern Gulf of Mexico*. New York: Springer. 284 pp.

Daly, K. L. 2015. Nutrient concentrations for the northeastern Gulf of Mexico and west Florida shelf: June 2012–August 2014. Scholar Commons, C-IMAGE data, Paper 14. http://scholarcommons.usf.edu/cimage_data/14.

De Galan, S., M. Elskens, L. Goeyens, A. Pollentier, N. Brion, and W. Baeyens. 2004. Spatial and temporal trends in nutrient concentrations in the Belgian continental area of the North Sea during the period 1993–2000. *Estuarine, Coastal and Shelf Science* 61:517–528.

de Mutsert, K., J. Steenbeck, K. Lewis, J. Buszowski, J. H. Cowan, and V. Christensen. 2016. Exploring effects of hypoxia on fish and fisheries in the northern Gulf of Mexico using a dynamic spatially explicit ecosystem model. *Ecological Modelling* 331:142–150.

Del Amo, Y., and M. A. Brzezinski. 1999. The chemical form of dissolved Si taken up by marine diatoms. *Journal of Phycology* 35:1162–1170.

Delaune, R. D., A. Jugsujinda, I. Devai, and A. X. Hou. 2005. Denitrification in bottom sediment near oil production facilities off the Louisiana Gulf Coast. *Chemistry and Ecology* 21:101–108.

Demarest, M. S., M. A. Brzezinski, and C. P. Beucher. 2009. Fractionation of silicon isotopes during biogenic silica dissolution. *Geochimica et Cosmochimica Acta* 73:5572–5583.

DeMaster, D. 1981. The supply and accumulation of silica in the marine environment. *Geochimica et Cosmochimica Acta* 45:1715–1732.

———. 2002. The accumulation and cycling of biogenic silica in the Southern Ocean: Revisiting the marine silica budget. *Deep Sea Research II: Tropical Studies in Oceanography* 49:3155–3167.

———. 2003. The diagenesis of biogenic silica: Chemical transformations occurring in the water column, seabed, and crust. *Treatise on Geochemistry* 7:87–98.

DeMaster, D. J., B. A. McKee, W. S. Moore, D. M. Nelson, W. J. Showers, and W. O. Smith Jr. 1991. Geochemical processes occurring in the waters at the Amazon River/ocean boundary. *Oceanography* 4:15–20.

Desboeufs, K., P. Formenti, S. Triquet, B. Laurent, C. Denjean, I. Gutteriez-Moreno, and O. L. Mayol-Bracero. 2015. Characterization of nutrients wet deposition under influence of Saharan dust at Puerto-Rico in Caribbean Sea. *Geophysical Research Abstracts* 17:EGU2015–11106.

Devol, A. H. 2015. Denitrification, anammox, and N_2 production in marine sediments. *Annual Review of Marine Science* 7:403–423.

DiDonato, G. T., E. M. Lores, M. C. Murrell, L. M. Smith, and J. M. Caffrey. 2006. Benthic nutrient flux in a small estuary in northwestern Florida (USA). *Gulf and Caribbean Research* 18:15–26.

Dixit, S., P. Van Cappellen, and A. J. Van Bennekom. 2001. Processes controlling solubility of biogenic silica and pore water build-up of silicic acid in marine sediments. *Marine Chemistry* 73:333–352.

Dixon, L. K., P. Murphy, and C. Charniga. 2014. The potential of benthic nutrient flux in support of *Karenia brevis* blooms off of west central Florida, USA. *Harmful Algae* 38:30–39.

Dortch, Q., R. E. Turner, M. L. Parsons, and N. N. Rabalais. 1999. What is the threat of harmful algal blooms in Louisiana coastal waters? Pp. 134–144 in L. P. Rozas, J. A. Nyman, C. E. Profitt, N. N. Rabalais, D. J. Reed, and R. E. Turner (eds.), *Recent Research in Coastal Louisiana: Natural System Function and Response to Human Influences*. Baton Rouge: Louisiana Sea Grant College Program.

Duce, R. A., J. LaRoche, K. Altieri, K. R. Arrigo, A. R. Baker, D. G. Capone, S. Cornell, et al. 2008. Impacts of atmospheric anthropogenic nitrogen on the open ocean. *Science* 320:893–897.

Dyhrman, S. T., J. W. Ammerman, and B. A. S. Van Mooy. 2007. Microbes and the marine phosphorus cycle. *Oceanography* 20:110–116.

Egge, J., and D. Aksnes. 1992. Silicate as regulating nutrient in phytoplankton competition. *Marine Ecology Progress Series* 83:281–289.

Eleuterius, L. N., and F. Lanning. 1987. Silica in relation to leaf decomposition of "*Juncus roemerianus*." *Journal of Coastal Research* 3:531–534.

EPA Science Advisory Board. 2008. *Hypoxia in the Northern Gulf of Mexico: An Update by the EPA Science Advisory Board*. Washington, DC: EPA Science Advisory Board. EPA-SAB-08-003. 333 pp. https://yosemite.epa.gov/sab/sabproduct.nsf/C3D2F27094E03F90852573B800601D93/$File/EPA-SAB-08-003complete.unsigned.pdf.

Falkowski, P. G., M. E. Katz, A. H. Knoll, A. Quigg, J. A. Raven, O. Schofield, and F. J. R. Taylor. 2004. The evolution of modern eukaryotic phytoplankton. *Science* 305:354–360.

Fanning, K. A., and M. E. Pilson. 1973. The lack of inorganic removal of dissolved silica during river-ocean mixing. *Geochimica et Cosmochimica Acta* 37:2405–2415.

Feist, T. J., J. J. Pauer, W. Melendez, J. C. Lehrter, P. A. DePetro, K. R. Rygwelski, D. S. Ko, and R. G. Kreis Jr. 2016. Modeling the relative importance of nutrient and carbon loads, boundary fluxes, and sediment fluxes on Gulf of Mexico hypoxia. *Environmental Science & Technology* 50:8713–8721.

Field, A. 2007. Amazon and Orinoco River plumes and NBC rings: Bystanders or participants in hurricane events? *Journal of Climate* 20:316–333.

Fraysse, F., O. S. Pokrovsky, J. Schott, and J. D. Meunier. 2009. Surface chemistry and reactivity of plant phytoliths in aqueous solutions. *Chemical Geology* 258:97–206.

Fry, B., and Y. C. Allen. 2003. Stable isotopes in zebra mussels as bioindicators of river-watershed linkages. *River Research and Applications* 19:683–696.

Gaas, B. M., and J. W. Ammerman. 2007. Automated high resolution ectoenzyme measurements: Instrument development and deployment in three trophic regimes. *Limnology and Oceanography Methods* 5:463–473.

Gardner, W. S., R. Benner, G. Chin-Leo, J. B. Cotner, B. J. Eadie, J. F. Cavaletto, and M. B. Lansing. 1994. Mineralization of organic material and bacterial dynamics in Mississippi River plume water. *Estuaries* 17:816–828.

Gardner, W. S., E. E. Briones, E. C. Kaegi, and G. T. Rowe. 1993. Ammonium excretion by benthic invertebrates and sediment-water nitrogen flux in the Gulf of Mexico near the Mississippi River outflow. *Estuaries* 16:799–808.

Gardner, W. S., J. F. Cavaletto, J. B. Cotner, and J. R. Johnson. 1997. Effects of natural light on nitrogen cycling rates in the

Mississippi River plume. *Limnology and Oceanography* 42:273–281.

Gardner, W. S., and M. J. McCarthy. 2009. Nitrogen dynamics at the sediment-water interface in shallow, sub-tropical Florida Bay: Why denitrification efficiency may decrease with increased eutrophication. *Biogeochemistry* 95:185–198.

Gardner, W. S., M. J. McCarthy, S. M. An, D. Sobolev, K. S. Sell, and D. Brock. 2006. Nitrogen fixation and dissimilatory nitrate reduction to ammonium (DNRA) support nitrogen dynamics in Texas estuaries. *Limnology and Oceanography* 51:558–568.

GESAMP (IMO, FAO, UNESCO-IOC, WMO, IAEA, UN, UNEP, UNDP Joint Group of Experts on the Scientific Aspects of Marine Environmental Protection). 2012. *The Atmospheric Input of Chemicals to the Ocean*. GESAMP Reports and Studies No. 84, GAW Report No. 203. 69 pp. https://library.wmo.int/pmb_ged/gaw_203.pdf.

GESAMP (IMO, FAO, UNESCO-IOC, WMO, WHO/IAEA, UN, UNEP Joint Group of Experts on the Scientific Aspects of Marine Environmental Protection) and Advisory Committee on Protection of the Sea. 2001. *Protecting the Oceans from Land-Based Activities: Land-Based Sources and Activities Affecting the Quality and Uses of the Marine, Coastal and Associated Freshwater Environment*. GESAMP Reports and Studies No. 71. 168 pp. http://www.gesamp.org/publications/protecting-the-oceans-from-land-based-activities.

Gihring, T. M., A. Canion, A. Riggs, M. Huettel, and J. E. Kostka. 2010. Denitrification in shallow, sublittoral Gulf of Mexico permeable sediments. *Limnology and Oceanography* 55:43–54.

Gilpin, L., K. Davidson, and E. Roberts. 2004. The influence of changes in nitrogen: Silicon ratios on diatom growth dynamics. *Journal of Sea Research* 51:21–35.

Gledhill, D. K., R. Wanninkhof, F. J. Millero, and M. Eakin. 2008. Ocean acidification of the greater Caribbean region 1996–2006. *Journal of Geophysical Research* 113:C10031.

Gobler, C. J., N. J. Buck, M. E. Sieracki, and S. A. Sanudo-Wilhelmy. 2006. Nitrogen and silicon limitation of phytoplankton communities across an urban estuary: The East River–Long Island Sound system. *Estuarine Coastal and Shelf Science* 68:127–138.

Goñi, M. A., K. C. Ruttenberg, and T. I. Eglinton. 1998. A reassessment of the sources and importance of land-derived organic matter in surface sediments from the Gulf of Mexico. *Geochimica et Cosmochimica Acta* 62:3055–3075.

Goolsby, D. A., W. A. Battaglin, B. T. Aulenbach, and R. P. Hooper. 2001. Nitrogen input to the Gulf of Mexico. *Journal of Environmental Quality* 30:329–336.

Goolsby, D. A., W. A. Battaglin, G. B. Lawrence, R. S. Artz, B. T. Aulenbach, R. P. Hooper, D. R. Keeney, and G. J. Stensland. 1999. *Flux and Sources of Nutrients in the Mississippi–Atchafalaya River Basin: Topic 3 Report for the Integrated Assessment on Hypoxia in the Gulf of Mexico*. NOAA Coastal Ocean Program Decision Analysis Series No. 17. Silver Spring, MD: NOAA Coastal Ocean Program. 329 pp.

Gordon, E. S., and M. A. Goñi. 2004. Controls on the distribution and accumulation of terrigenous organic matter in sediments from the Mississippi and Atchafalaya River margin. *Marine Chemistry* 92:331–352.

Gordon, E. S., M. A. Goñi, Q. N. Roberts, G. C. Kineke, and M. A. Allison. 2001. Organic matter distribution and accumulation on the inner Louisiana shelf west of the Atchafalaya River. *Continental Shelf Research* 21:1691–1721.

Grosskopf, T., W. Mohr, T. Baustian, H. Schunck, D. Gill, M. M. M. Kuypers, G. Lavik, R. A. Schmitz, D. W. R. Wallace, and J. LaRoche. 2012. Doubling of marine dinitrogen-fixation rates based on direct measurements. *Nature* 488:361–364.

Gruber, N. 2004. The dynamics of the marine nitrogen cycle and its influence on atmospheric CO_2 variations. Pp. 97–148 in M. Follows and T. Oguz (eds.), *The Ocean Carbon Cycle and Climate*. Dordrecht, Netherlands: Kluwer Academic.

Gruber, N., and J. L. Sarmiento. 1997. Global patterns of marine nitrogen fixation and denitrification. *Global Biogeochemical Cycles* 11:235–266.

Hansell, D. A., N. R. Bates, and D. B. Olson. 2004. Excess nitrate and nitrogen fixation in the North Atlantic Ocean. *Marine Chemistry* 84:243–265.

Hansell, D. A., and M. J. Follows. 2008. Nitrogen in the Atlantic Ocean. Pp. 597–630 in D. G. Capone, D. A. Bronk, M. R. Mulholland, and E. J. Carpenter, *Nitrogen in the Marine Environment*, 2nd ed. Oxford: Elsevier. doi:10.1016/B978-0-12-372522-6-6.00013-X.

Hartmann, J., J. Levy, and S. Kempe. 2011. Increasing dissolved silica trends in the Rhine River: An effect of recovery from high P loads. *Limnology* 12:63–73.

Hastings, M. G., K. L. Casciotti, and E. M. Elliott. 2013. Stable isotopes as tracers of anthropogenic nitrogen sources, deposition, and impacts. *Elements* 9:339–344.

Hastings, M. G., D. M. Sigman, and F. Lipschultz. 2003. Isotopic evidence for source changes of nitrate in rain at Bermuda. *Journal of Geophysical Research: Atmospheres* 108:D24.

Heil, C. A., D. A. Bronk, L. K. Dixon, G. L. Hitchcock, G. J. Kirkpatrick, M. R. Mulholland, J. M. O'Neil, J. J. Walsh, R. Weisberg, and M. Garrett. 2014. The Gulf of Mexico

ECOHAB: Karenia Program 2006–2012. *Harmful Algae* 38:3–7.

Heil, C. A., L. K. Dixon, E. Hall, M. Garrett, J. M. Lenes, J. M. O'Neil, B. M. Walsh, et al. 2014. Blooms of *Karenia brevis* (Davis) G. Hansen & O. Moestrup on the west Florida shelf: Nutrient sources and potential management strategies based on a multi-year regional study. *Harmful Algae* 38:127–140.

Herrera-Silveira, J. A., I. Medina-Gomez, N. Aranda-Cirerol, A. Zaldivar, J. Ramirez, and J. Trejo. 2002. Trophic status in coastal waters of the Yucatan Peninsula (SE, Mexico) using water quality indicators. Pp. 351–359 in C. A. Brebbia (ed.), *Environment Problems in Coastal Regions IV*. Southampton, UK: Wit-Press.

Holl, C. M., T. A. Villareal, C. D. Payne, T. D. Clayton, C. Hart, and J. P. Montoya. 2007. *Trichodesmium* in the western Gulf of Mexico: $^{15}N_2$-fixation and natural abundance stable isotope evidence. *Limnology and Oceanography* 52:2249–2259.

Hood, R. R., V. J. Coles, and D. G. Capone. 2004. Modeling the distribution of *Trichodesmium* and nitrogen fixation in the Atlantic Ocean. *Journal of Geophysical Research* 109(C6):C06007. doi:10.1029/2002JC001754.

Hou, L., M. Liu, S. A. Carini, and W. S. Gardner. 2012. Transformation and fate of nitrate near the sediment-water interface of Copano Bay. *Continental Shelf Research* 35:86–94.

Houser, J. N., D. W. Bierman, R. M. Burdis, and L. A. Soeken-Gittinger. 2010. Longitudinal trends and discontinuities in nutrients, chlorophyll, and suspended solids in the upper Mississippi River: Implications for transport, processing, and export by large rivers. *Hydrobiologia* 651:127–144.

Howarth, R. W., G. Billen, D. Swaney, A. Townsend, N. Jaworski, K. Lajtha, J. A. Downing, et al. 1996. Regional nitrogen budgets and riverine N & P fluxes for the drainages to the North Atlantic Ocean: Natural and human influences. *Biogeochemistry* 35:75–139.

Hu, C., F. E. Muller-Karger, and P. W. Swarzenski. 2006. Hurricanes, submarine groundwater discharge, and Florida's red tides. *Geophysical Research Letters* 33:L11601. doi:10.1029/2005GL025449.

Humborg, C., D. J. Conley, L. Rahm, F. Wulff, A. Cociasu, and V. Ittekkot. 2000. Silicon retention in river basins: Far-reaching effects on biogeochemistry and aquatic food webs in coastal marine environments. *Ambio* 29:45–50.

Humborg, C., V. Ittekkot, A. Cociasu, and B. Vonbodungen. 1997. Effect of Danube River dam on Black Sea biogeochemistry and ecosystem structure. *Nature* 386:385–388.

Humborg, C., M. Pastuszak, J. Aigars, H. Siegmund, C-M. Mörth, and V. Ittekkot. 2006. Decreased silica land-sea fluxes through damming in the Baltic Sea catchment—significance of particle trapping and hydrological alterations. *Biogeochemistry* 77: 265–281.

Hynes, R. K., and R. Knowles. 1978. Inhibition by acetylene of ammonia oxidation in *Nitrosomonas europaea*. *FEMS Microbiology Letters* 4:319–321.

Institute for Marine Remote Sensing. n.d. Cariaco Ocean Time-Series Program. Accessed October 4, 2018. www.imars.usf.edu/cariaco.

Jendrzejewski, J. P., and G. F. Hart. 1978. Distribution of siliceous microfossils in surficial bottom sediments of the Gulf of Mexico. *Palynology* 2:159–166.

Jochem, F. J., M. J. McCarthy, and W. S. Gardner. 2004. Microbial ammonium cycling in the Mississippi River plume during the drought spring of 2000. *Journal of Plankton Research* 26:1265–1275.

Jochens, A. E., and S. F. DiMarco. 2008. Physical oceanographic conditions in the deepwater Gulf of Mexico in summer 2000–2002. *Deep Sea Research Part II: Tropical Studies in Oceanography* 55:2541–2554.

Jochens, A. E., S. F. DiMarco, W. D. Nowlin Jr., R. O. Reid, and M. C. Kennicutt II. 2002. *Northeastern Gulf of Mexico Chemical Oceanography and Hydrography Study: Synthesis Report*. OCS Study MMS 2002–055. New Orleans: US Department of the Interior, Minerals Management Service, Gulf of Mexico OCS Region. 586 pp.

Jochens, A. E., and W. D. Nowlin Jr. (eds.). 2000. *Northeastern Gulf of Mexico Chemical Oceanography and Hydrography Study, Annual Report: Year 3*. OCS Study MMS 2000-078. New Orleans: US Department of the Interior, Minerals Management Service, Gulf of Mexico OCS Region. 89 pp.

Jones, E. E. 1974. *The Protozoa of Mobile Bay, Alabama*. University of South Alabama Monograph 1. Mobile: University of South Alabama. 113 pp.

Joung, D., H. Box, P. Ho, L. Whitmore, W. Moore, and A. M. Shiller. 2017. Seasonal dynamics of trace elements and Ra isotopes in Mississippi coastal waters (abstract). 2017 Gulf of Mexico Ecosystem Science Conference, New Orleans, Louisiana, February 6–9.

Joye, S. B., and J. T. Hollibaugh. 1995. Influence of sulfide inhibition of nitrification on nitrogen regeneration in sediments. *Science* 270:623–625.

Kamatani, A., J. P. Riley, and G. Skirrow. 1980. The dissolution of opaline silica of diatom tests in sea water. *Journal of the Oceanographical Society of Japan* 36:201–208.

Karl, D. M., and G. Tien. 1992. MAGIC: A sensitive and precise method for measuring dissolved phosphorus in aquatic environments. *Limnology and Oceanography* 37:105–116.

Katz, B. G., A. A. Sepulveda, and R. J. Verdi. 2009. Estimating nitrogen loading to ground water and assessing vulnerability to nitrate contamination in a large karstic springs basin, Florida. *Journal of the American Water Resources Association* 45:607–627.

Killberg-Thoreson, L., M. R. Mulholland, C. A. Heil, M. P. Sanderson, J. M. O'Neil, and D. A. Bronk. 2014. Nitrogen uptake kinetics in field populations and cultured strains of *Karenia brevis*. *Harmful Algae* 38:73–85.

Knapp, A. N. 2012. The sensitivity of marine N_2 fixation to dissolved inorganic nitrogen. *Frontiers in Microbiology* 3:374.

Knapp, A. N., P. J. DiFiore, C. Deutsch, D. M. Sigman, and F. Lipschultz. 2008. Nitrate isotopic composition between Bermuda and Puerto Rico: Implications for N_2 fixation in the Atlantic Ocean. *Global Biogeochemical Cycles* 22:GB3014. doi:10.1029/2007GB003107.

Knapp, A. N., M. G. Hastings, D. M. Sigman, F. Lipschultz, and J. N. Galloway. 2010. The flux and isotopic composition of reduced and total nitrogen in Bermuda rain. *Marine Chemistry* 120:83–89.

Koch, F. 2005. Silicate uptake and utilization by diatoms in Central San Francisco Bay. Master's thesis, San Francisco State University. 127 pp.

Krause, J. W., E. S. Darrow, R. A. Pickering-Turner, R. H. Carmichael, A. M. Larson, and J. Basaldua. 2017. Reactive silica fractions in coastal lagoon sediments from the northern Gulf of Mexico. *Continental Shelf Research* 151:8–14. doi:10.1016/j.csr.2017.09.014.

Krause, J. W., D. M. Nelson, and M. W. Lomas. 2010. Production, dissolution, accumulation and potential export of biogenic silica in a Sargasso Sea mode-water eddy. *Limnology and Oceanography* 55:569–579.

———. 2013. Biogenic silica cycling during summer phytoplankton blooms in the North Pacific subtropical gyre. *Deep Sea Research Part I: Oceanographic Research Papers* 71:49–60.

Lane, R. R., J. W. Day, D. Justic, E. Reyes, J. N. Day, and E. Hyfield. 2004. Changes in stoichiometric Si, N and P ratios of Mississippi River water diverted through coastal wetlands to the Gulf of Mexico. *Estuarine, Coastal and Shelf Science* 60:1–10.

Lanning, F., and L. Eleuterius. 1983. Silica and ash in tissues of some coastal plants. *Annals of Botany* 51:835–850.

Laurent, A., and K. Fennel. 2014. Simulated reduction of hypoxia in the northern Gulf of Mexico due to phosphorus limitation. *Elementa: Science of the Anthropocene* 2:22. doi:10.12952/journal.elementa.000022.

———. 2017. Modeling river-induced phosphorus limitation in the context of coastal hypoxia. Pp. 149–171 in D. Justic, K. A. Rose, R. D. Hetland, and K. Fennel (eds.), *Modeling Coastal Hypoxia*. Basel, Switzerland: Springer.

Laurent, A., K. Fennel, J. Hu, and R. Hetland. 2012. Simulating the effects of phosphorus limitation in the Mississippi and Atchafalaya River plumes. *Biogeosciences* 9:4707–4723.

Lazzari, P., C. Solidoro, S. Salon, and G. Bolzon. 2016. Spatial variability of phosphate and nitrate in the Mediterranean Sea: A modeling approach. *Deep Sea Research Part I: Oceanographic Research Papers* 108:39–52.

Lehrter, J. C., D. L. Beddick Jr., R. Devereux, D. F. Yates, and M. C. Murrell. 2012. Sediment-water fluxes of dissolved inorganic carbon, O_2, nutrients, and N_2 from the hypoxic region of the Louisiana continental shelf. *Biogeochemistry* 109:233–252.

Lenes, J. M., B. P. Darrow, C. Cattrall, C. A. Heil, M. Callahan, G. A. Vargo, and R. H. Byrne. 2001. Iron fertilization and the *Trichodesmium* response on the west Florida shelf. *Limnology and Oceanography* 46:1261–1277.

Lenes, J. M., and C. A. Heil. 2010. A historical analysis of the potential nutrient supply from the N_2 fixing marine cyanobacterium *Trichodesmium* spp. to *Karenia brevis* blooms in the eastern Gulf of Mexico. *Journal of Plankton Research* 32:1421–1431.

Lepape, O., Y. Delamo, A. Menesguen, A. Aminot, B. Quequiner, and P. Tréguer. 1996. Resistance of a coastal ecosystem to increasing eutrophic conditions: The Bay of Brest (France), a semi-enclosed zone of Western Europe. *Continental Shelf Research* 16:1885–1907.

Lessard-Pilon, S. A., E. L. Podowski, E. E. Cordes, and C. R. Fisher. 2010. Megafauna community composition associated with *Lophelia pertusa* colonies in the Gulf of Mexico. *Deep Sea Research Part II: Tropical Studies in Oceanography* 57:1882–1890.

Liefer, J. D., H. L. MacIntyre, L. Novoveska, W. L. Smith, and C. P. Dorsey. 2009. Temporal and spatial variability in *Pseudo-nitzschia* spp. in Alabama coastal waters: A "hot spot" linked to submarine groundwater discharge? *Harmful Algae* 8:706–714.

Lin, X. A., M. J. McCarthy, S. A. Carini, and W. S. Gardner. 2011. Net, actual, and potential sediment-water interface NH_4^+ fluxes in the northern Gulf of Mexico (NGOMEX): Evidence for NH_4^+ limitation of microbial dynamics. *Continental Shelf Research* 31:120–128.

Lohrenz, S. E., G. L. Fahnenstiel, D. G. Redalje, G. A. Lang, X. Chen, and M. J. Dagg. 1997. Variations in primary production of northern Gulf of Mexico continental shelf waters linked to nutrient inputs from the Mississippi River. *Marine Ecology Progress Series* 155:45–54.

Lohrenz, S. E., G. L. Fahnenstiel, D. G. Redalje, G. A. Lang, M. J. Dagg, T. E. Whitledge, and Q. Dortch. 1999. Nutrients, irradiance, and mixing as factors regulating primary production in coastal waters impacted by the Mississippi River plume. *Continental Shelf Research* 19:1113–1141.

Lohrenz, S. E., D. G. Redalje, W.-J. Cai, J. Acker, and M. Dagg. 2008. A retrospective analysis of nutrients and phytoplankton productivity in the Mississippi River plume. *Continental Shelf Research* 28:1466–1475.

Lu, H., and K.-B. Liu. 2003. Phytoliths of common grasses in the coastal environments of southeastern USA. *Estuarine, Coastal and Shelf Science* 58:587–600.

———. 2005. Phytolith assemblages as indicators of coastal environmental changes and hurricane overwash deposition. *The Holocene* 15:965–972.

Lucea, A., C. M. Duarte, S. Agustí, and M. Søndergaard. 2003. Nutrient (N, P and Si) and carbon partitioning in the stratified NW Mediterranean. *Journal of Sea Research* 49:157–170.

Ludwig, W., E. Dumont, M. Meybeck, and S. Heussner. 2009. River discharges of water and nutrients to the Mediterranean and Black Sea: Major rivers for ecosystem changes during past and future decades? *Progress in Oceanography* 80:199–217.

Luo, Y. W., S. C. Doney, L. A. Anderson, M. Benavides, I. Berman-Frank, A. Bode, S. Bonnet, et al. 2012. Database of diazotrophs in global ocean: Abundance, biomass and nitrogen fixation rates. *Earth System Science Data* 4:47–73.

Macias, D., E. Garcia-Gorriz, and A. Stips. 2017. Major fertilization sources and mechanisms for Mediterranean Sea coastal ecosystems. *Limnology and Oceanography* 63:897–914.

Mackenzie, F. T., R. M. Garrels, O. P. Bricker, and F. Bickley. 1967. Silica in sea water: Control by silica minerals. *Science* 155:1404–1405.

Mackenzie, F. T., and L. R. Kump. 1995. Reverse weathering, clay mineral formation, and oceanic element cycles. *Science* 270:586.

Maldonado, M., L. Navarro, A. Grasa, A. Gonzalez, and I. Vaquerizo. 2011. Silicon uptake by sponges: A twist to understanding nutrient cycling on continental margins. *Scientific Reports* 1:30. doi:10.1038/srep00030.

Mayer, L. M., and S. P. Gloss. 1980. Buffering of silica and phosphate in a turbid river. *Limnology and Oceanography* 25:12–22.

McCarthy, M. J., K. S. McNeal, J. W. Morse, and W. S. Gardner. 2008. Bottom-water hypoxia effects on sediment-water interface nitrogen transformations in a seasonally hypoxic, shallow bay (Corpus Christi Bay, TX, USA). *Estuaries and Coasts* 31:521–531.

McCarthy, M. J., S. E. Newell, S. A. Carini, and W. S. Gardner. 2015. Denitrification dominates sediment nitrogen removal and is enhanced by bottom-water hypoxia in the northern Gulf of Mexico. *Estuaries and Coasts* 38:2279–2294.

Meckler, A. N., H. J. Ren, D. M. Sigman, N. Gruber, B. Plessen, C. J. Schubert, and G. H. Haug. 2011. Deglacial nitrogen isotope changes in the Gulf of Mexico: Evidence from bulk sedimentary and foraminifera-bound nitrogen in Orca Basin sediments. *Paleoceanography* 26:PA4216.

Michalopoulos, P., and R. C. Aller. 1995. Rapid clay mineral formation in Amazon delta sediments: Reverse weathering and oceanic elemental cycles. *Nature* 270:614–617.

———. 2004. Early diagenesis of biogenic silica in the Amazon delta: Alteration, authigenic clay formation, and storage. *Geochimica et Cosmochimica Acta* 68:1061–1085.

Miller, C. B., D. M. Nelson, R. R. Guillard, and B. L. Woodward. 1980. Effects of media with low silicic acid concentrations on tooth formation in *Acartia tonsa* Dana (Copepoda, Calanoida). *Biological Bulletin* 159:349–363.

Miller, C., D. Nelson, C. Weiss, and A. Soeldner. 1990. Morphogenesis of opal teeth in calanoid copepods. *Marine Biology* 106:91–101.

Miller-Way, T., G. S. Boland, G. T. Rowe, and R. R. Twilley. 1994. Sediment oxygen consumption and benthic nutrient fluxes on the Louisiana continental shelf: A methodological comparison. *Estuaries* 17:809–815.

Mississippi River/Gulf of Mexico Watershed Nutrient Task Force. 2008. *Gulf Hypoxia Action Plan 2008 for Reducing, Mitigating, and Controlling Hypoxia in the Northern Gulf of Mexico and Improving Water Quality in the Mississippi River Basin*. Washington, DC: United States Environmental Protection Agency. 64 pp.

———. 2013. *Reassessment 2013: Assessing Progress Made since 2008*. Washington, DC: United States Environmental Protection Agency. 86 pp.

———. 2015. *2015 Report to Congress*. Washington, DC: United States Environmental Protection Agency. 93 pp.

Montoya, J. P., C. M. Holl, J. P. Zehr, A. Hansen, T. A. Villareal, And D. G. Capone. 2004. High rates of N_2 fixation by unicellular diazotrophs in the oligotrophic Pacific Ocean. *Nature* 430:1027–1031.

Montoya, J. P., M. Voss, P. Kahler, and D. G. Capone. 1996. A simple, high-precision, high-sensitivity tracer assay for N_2 fixation. *Applied and Environmental Microbiology* 62:986–993.

Moon, J.-Y., K. Lee, T. Tanhua, N. Kress, and I.-N. Kim. 2016. Temporal nutrient dynamics in the Mediterranean Sea in response to anthropogenic inputs. *Geophysical Research Letters* 43(10):5423–5251. doi:10.1002/2016GL068788.

Moretzsohn, F., J. A. Sanchez Chavez, and J. W. Tunnell Jr. (eds.). 2016. Gulfbase: Resource database of Gulf of Mexico research. http://www.gulfbase.org.

Morrison, J. M., W. J. Merrell Jr., R. M. Key, and T. C. Key. 1983. Property distributions and deep chemical measurements within the western Gulf of Mexico. *Journal of Geophysical Research* 8 (C4):2601–2608.

Morse, J. W., and G. T. Rowe. 1999. Benthic biogeochemistry beneath the Mississippi River plume. *Estuaries* 22:206–214.

Mortazavi, B., R. L. Iverson, W. R. Huang, F. G. Lewis, and J. M. Caffrey. 2000. Nitrogen budget of Apalachicola Bay, a bar-built estuary in the northeastern Gulf of Mexico. *Marine Ecology Progress Series* 195:1–14.

Mortazavi, B., R. L. Iverson, W. M. Landing, and W. Huang. 2000. Phosphorus budget of Apalachicola Bay: A river-dominated estuary in the northeastern Gulf of Mexico. *Marine Ecology Progress Series* 198:33–42.

Mortazavi, B., A. A. Riggs, J. M. Caffrey, H. Genet, and S. W. Phipps. 2012. The contribution of benthic nutrient regeneration to primary production in a shallow eutrophic estuary, Weeks Bay, Alabama. *Estuaries and Coasts* 35:862–877.

Mortlock, R. A., and P. N. Froelich. 1989. A simple method for the rapid determination of biogenic opal in pelagic marine sediments. *Deep Sea Research, Part A: Oceanographic Research Papers* 36:1415–1426.

Mulholland, M. R., P. W. Bernhardt, C. A. Heil, D. A. Bronk, and J. M. O'Neil. 2006. Nitrogen fixation and release of fixed nitrogen by *Trichodesmium* spp. in the Gulf of Mexico. *Limnology and Oceanography* 51:1762–1776.

Mulholland, M. R., P. W. Bernhardt, I. Ozmon, L. A. Procise, M. Garrett, J. M. O'Neil, C. A. Heil, and D. A. Bronk. 2014. Contribution of diazotrophy to nitrogen inputs supporting *Karenia brevis* blooms in the Gulf of Mexico. *Harmful Algae* 38:20–29.

Müller-Karger, F. E., J. P. Smith, S. Werner, R. Chen, M. Roffer, Y. Liu, N. Muhling, et al. 2015. Natural variability of surface oceanographic conditions in the offshore Gulf of Mexico. *Progress in Oceanography* 134:54–76.

Murrell, M. C., J. G. Campbell, J. D. Hagy III, and J. M. Caffrey. 2009. Effects of irradiance on benthic and water column processes in a Gulf of Mexico estuary: Pensacola Bay, Florida, USA. *Estuarine, Coastal and Shelf Science* 81:501–512.

National Atmospheric Deposition Program (NRSP-3). 2015. *State Water Survey*. Champaign: NADP Program Office, Illinois State Water Survey, University of Illinois at Urbana-Champaign. 28 pp.

Nelson, D. M., and Q. Dortch. 1996. Silicic acid depletion and silicon limitation in the plume of the Mississippi River: Evidence from kinetic studies in spring and summer. *Marine Ecology Progress Series* 136:163–178.

Nelson, D. M., G. F. Riedel, R. Millan-Nunez, and J. R. Lara-Lara. 1984. Silicon uptake by algae with no known Si requirement. I. True cellular uptake and pH-induced precipitation by *Phaeodactylum tricornutum* (Bacillariophyceae) and *Platymonas* sp. (Prasinophyceae). *Journal of Phycology* 20:141–147.

Nelson, D. M., P. Tréguer, M. A. Brzezinski, A. Leynaert, and B. Queguiner. 1995. Production and dissolution of biogenic silica in the ocean: Revised global estimates, comparison with regional data and relationship to biogenic sedimentation. *Global Biogeochemical Cycles* 9:359–372.

Newell, S. E., M. J. McCarthy, W. S. Gardner, and R. W. Fulweiler. 2016. Sediment nitrogen fixation: A call for re-evaluating coastal N budgets. *Estuaries and Coasts* 39:1626–1638.

Nowlin, W. D., Jr., A. E. Jochens, R. O. Reid, and S. F. DiMarco. 1998. *Texas-Louisiana Shelf Circulation and Transport Processes Study: Synthesis Report*. Vol. 1, *Technical Report*. OCS Study MMS 98–0035. New Orleans: US Department of the Interior, Minerals Management Service, Gulf of Mexico OCS Region. 502 pp.

Nunnally, C. C., A. Quigg, S. DiMarco, P. Chapman, and G. T. Rowe. 2014. Benthic-pelagic coupling in the Gulf of Mexico hypoxic area: Sedimentary enhancement of hypoxic conditions and near bottom primary production. *Continental Shelf Research* 85:143–152.

Nunnally, C. C., G. T. Rowe, D. C. O. Thornton, and A. Quigg. 2013. Sedimentary oxygen consumption and nutrient regeneration in the northern Gulf of Mexico hypoxic zone. In *Understanding and Predicting Change in the Coastal Ecosystems of the Northern Gulf of Mexico, Journal of Coastal Research*, special issue, *Journal of Coastal Research* 63:84–96.

Officer, C. B., and J. H. Ryther. 1980. The possible importance of silicon in marine eutrophication. *Marine Ecology Progress Series* 3:83–91.

Pakulski, J. D., R. Benner, T. Whitledge, R. Amon, B. Eadie, L. Cifuentes, J. Ammerman, and D. Stockwell. 2000. Microbial metabolism and nutrient cycling in the Mississippi and Atchafalaya River plumes. *Estuarine Coastal and Shelf Science* 50:173–184.

Paramor, O. A. L., K. A. Allen, M. Aanesen, C. Armstrong, T. Hegland, W. Le Quesne, G. J. Piet, et al. 2009. *MEFPRO North Sea Atlas*. Liverpool: University of Liverpool. 80 pp. www.liverpool.ac.uk/media/livacuk/mefepo/documents/wp1/atlases/NS_Atlas_English.pdf.

Parsons, M. L., Q. Dortch, and R. E. Turner. 2002. Sedimentological evidence of an increase in *Pseudo-nitzschia* (Bacillariophyceae) abundance in response to coastal eutrophication. *Limnology and Oceanography* 47:551–558.

Paytan, A., and K. McLaughlin. 2007. The oceanic phosphorus cycle. *Chemical Reviews* 107:563–576.

Paytan, A., G. G. Shellenbarger, J. H. Street, M. E. Gonneea, K. Davis, M. B. Young, and W. S. Moore. 2006. Submarine groundwater discharge: An important source of new inorganic nitrogen to coral reef ecosystems. *Limnology and Oceanography* 51:343–348.

Peñuelas, J., B. Poulter, J. Sardans, P. Ciais, M. van der Velde, L. Bopp, O. Boucher, et al. 2013. Human-induced nitrogen-phosphorus imbalances alter natural and managed ecosystems across the globe. *Nature Communications* 4:2934. doi:10.1038/ncomms3934.

Presti, M., and P. Michalopoulos. 2008. Estimating the contribution of the authigenic mineral component to the long-term reactive silica accumulation on the western shelf of the Mississippi River delta. *Continental Shelf Research* 28:823–838.

Prospero, J. M., K. Barrett, T. Church, F. Dentener, R. A. Duce, J. N. Galloway, H. Levy, J. Moody, and P. Quinn. 1996. Atmospheric deposition of nutrients to the North Atlantic basin. *Biogeochemistry* 35:27–73.

Quigg, A., J. B. Sylvan, A. B. Gustafson, T. R. Fisher, R. L. Oliver, S. Tozzi, and J. W. Ammerman. 2011. Going west: Nutrient limitation of primary production in the northern Gulf of Mexico and the importance of the Atchafalaya River. *Aquatic Geochemistry* 17:519–544.

Rabalais, N. N., R. E. Turner, D. Justic, Q. Dortch, W. J. Wiseman, and B. K. Sen Gupta. 1996. Nutrient changes in the Mississippi River and system responses on the adjacent continental shelf. *Estuaries* 19:386–407.

Rabalais, N. N., R. E. Turner, B. K. Sen Gupta, E. Platon, and M. L. Parsons. 2007. Sediments tell the history of eutrophication and hypoxia in the northern Gulf of Mexico. *Ecological Applications* 17:S129–S143.

Rabalais, N. N., R. E. Turner, and W. J. Wiseman. 2001. Hypoxia in the Gulf of Mexico. *Journal of Environmental Quality* 30:320–329.

———. 2002. Gulf of Mexico hypoxia, aka "the dead zone." *Annual Review of Ecology and Systematics* 33:235–263.

Radach, G., J. Berg, and E. Hagmeier. 1990. Long-term changes of the annual cycles of meteorological, hydrographic, nutrient and phytoplankton time-series at Helgoland and at LE ELBE 1 in the German Bight. *Continental Shelf Research* 10:305–328.

Ragueneau, O., D. J. Conley, A. Leynaert, S. N. Longphuirt, and C. P. Slomp. 2006. Responses of coastal ecosystems to anthropogenic perturbations of silicon cycling. *SCOPE* 66:197–213.

Rahman, S., R. C. Aller, and J. K. Cochran. 2016. Cosmogenic 32Si as a tracer of biogenic silica burial and diagenesis: Major deltaic sinks in the silica cycle. *Geophysical Research Letters* 43:7124–7132.

Redalje, D. G., S. E. Lohrenz, and G. L. Fahnenstiel. 1994. The relationship between primary production and the vertical export of particulate organic matter in a river-impacted coastal ecosystem. *Estuaries* 17:829–838.

Redfield, A. C., B. H. Ketchum, and F. A. Richards. 1963. The influence of organisms on the composition of seawater. Pp. 26–77 in M. N. Hill (ed.), *The Sea*. London: Interscience.

Rinker, K. R., and R. T. Powell. 2006. Dissolved organic phosphorus in the Mississippi River plume during spring and fall 2002. *Marine Chemistry* 102:170–179.

Roberts, B. J., and S. M. Doty. 2015. Spatial and temporal patterns of benthic respiration and net nutrient fluxes in the Atchafalaya River delta estuary. *Estuaries and Coasts* 38:1918–1936.

Robertson, D. M., and D. A. Saad. 2013. SPARROW models used to understand nutrient sources in the Mississippi/Atchafalaya River basin. *Journal of Environmental Quality* 42:1422–1440.

Rodellas, V., J. Garcia-Otellana, P. Masqué, M. Feldman, and Y. Weinstein. 2015. Submarine groundwater discharge as a major source of nutrients to the Mediterranean Sea. *Proceedings of the National Academy of Sciences* 112:3926–3930.

Rowe, G. T., M. E. C. Kaegi, J. W. Morse, G. S. Boland, and E. G. E. Briones. 2002. Sediment community metabolism associated with continental shelf hypoxia, northern Gulf of Mexico. *Estuaries* 25:1097–1106.

Ruoho-Airola, T., K. Eilola, O. P. Savchuk, M. Parviainen, and V. Tarvainen. 2012. Atmospheric nutrient input to the Baltic Sea from 1850 to 2006: A reconstruction from modeling results and historical data. *AMBIO* 41:549–557.

Rützler, K., and I. G. Macintyre. 1978. Siliceous sponge spicules in coral reef sediments. *Marine Biology* 49:147–159.

Salas-de-León, D. A., and M. A. Monreal-Gómez. 2004. Anticyclonic-cyclonic eddies and their impact on near-surface chlorophyll stocks and oxygen supersaturation over the Campeche Canyon, Gulf of Mexico. *Journal of Geophysical Research* 109:C05012. doi:10.1029/2002JC001614.

Santos, I. R., W. C. Burnett, J. Chanton, B. Mwashote, I. Suryaputra, and T. Dittmar. 2008. Nutrient biogeochemistry in a Gulf of Mexico subterranean estuary and groundwater-derived fluxes to the coastal ocean. *Limnology and Oceanography* 53:705–718.

Scranton, M. I., G. T. Taylor, R. Thunell, C. R. Benitez-Nelson, F. Muller-Karger, K. Fanning, L. Lorenzoni, E. Montes, R. Varela, and Y. Astor. 2014. Interannual and subdecadal variability in the nutrient geochemistry of the Cariaco Basin. *Oceanography* 27:148–159.

Segura-Noguera, M., A. Cruzado, and D. Blasco. 2016. The biogeochemistry of nutrients, dissolved oxygen and chlorophyll-*a* in the Catalan Sea (NW Mediterranean Sea). *Scientia Marina* 80(S1):39–56.

Seitzinger, S. P. 1987. Nitrogen biogeochemistry in an unpolluted estuary—the importance of benthic denitrification. *Marine Ecology Progress Series* 41:177–186.

Sigman, D. M., P. J. DiFiore, M. P. Hain, C. Deutsch, Y. Wang, D. M. Karl, A. N. Knapp, M. F. Lehmann, and S. Pantoja. 2009. The dual isotopes of deep nitrate as a constraint on the cycle and budget of oceanic fixed nitrogen. *Deep Sea Research Part I: Oceanographic Research Papers* 56:1419–1439.

Sillen, L. G. 1961. The physical chemistry of seawater. Pp. 549–581 in M. Sears (ed.), *Oceanography*. Washington, DC: Association for the Advancement of Publication.

Slomp, C. P., and P. Van Cappellen. 2004. Nutrient inputs to the coastal ocean through submarine groundwater discharge: Controls and potential impact. *Journal of Hydrology* 295:64–86.

Smith, C. G., and P. W. Swarzenski. 2012. An investigation of submarine groundwater-borne nutrient fluxes to the west Florida shelf and recurrent harmful algal blooms. *Limnology and Oceanography* 57:471–485.

Smith, S. M., and G. L. Hitchcock. 1994. Nutrient enrichments and phytoplankton growth in the surface waters of the Louisiana Bight. *Estuaries* 17:740–753.

Sohm, J. A., E. A. Webb, and D. G. Capone. 2011. Emerging patterns of marine nitrogen fixation. *Nature Reviews Microbiology* 9:499–508.

Storr, J. F. 1964. *Ecology of the Gulf of Mexico Commercial Sponges and Its Relation to the Fishery*. Special Scientific Report 446. Washington, DC: US Department of Interior, Fish and Wildlife Service. 73 pp.

Stumm, W., and J. J. Morgan. 1996. *Aquatic Chemistry: Chemical Equilibria and Rates in Natural Waters*. 3rd ed. Hoboken, NJ: Wiley-Interscience. 1022 pp.

Sutula, M., T. S. Bianchi, and B. A. McKee. 2004. Effect of seasonal sediment storage in the lower Mississippi River on the flux of reactive particulate phosphorus to the Gulf of Mexico. *Limnology and Oceanography* 49:2223–2235.

Sylvan, J. B., and J. W. Ammerman. 2013. Seasonal distributions of organic nutrients on the Louisiana continental shelf and their implications for nutrient limitation and hypoxia formation. *Marine Chemistry* 154:113–123.

Sylvan, J. B., Q. Dortch, D. M. Nelson, M. Brown, W. Morrison, and J. W. Ammerman. 2006. Phosphorus limits phytoplankton growth on the Louisiana shelf during the period of hypoxia formation. *Environmental Science & Technology* 40:7548–7553.

Sylvan, J. B., A. Quigg, S. Tozzi, and J. W. Ammerman. 2007. Eutrophication-induced phosphorus limitation in the Mississippi River plume: Evidence from fast repetition rate fluorometry. *Limnology and Oceanography* 52:2679–2685.

Tanhua, T., D. Hainbucher, K. Schroeder, V. Cardin, M. Álvarez, and G. Civitarese. 2013. The Mediterranean Sea system: A review and an introduction to the special issue. *Ocean Science* 9:789–803.

Teague, K. G., C. J. Madden, and J. W. Day. 1988. Sediment-water oxygen and nutrient fluxes in a river-dominated estuary. *Estuaries* 11:1–9.

Thamdrup, B., and T. Dalsgaard. 2002. Production of N^2 through anaerobic ammonium oxidation coupled to nitrate reduction in marine sediments. *Applied and Environmental Microbiology* 68:1312–1318.

Thunell, R., C. Benitez-Nelson, F. Muller-Karger, L. Lorenzoni, K. Fanning, M. Scranton, R. Varela, and Y. Astor. 2008. Si cycle in the Cariaco Basin, Venezuela: Seasonal variability in silicate availability and the Si:C:N composition of sinking particles. *Global Biogeochemical Cycles* 22:GB4001. doi:10.1029/2007GB003096.

Trapp, J. M., F. J. Millero, and J. M. Prospero. 2010. Temporal variability of the elemental composition of African dust measured in trade wind aerosols at Barbados and Miami. *Marine Chemistry* 120:71–82.

Tréguer, P. J., and C. L. De La Rocha. 2013. The world ocean silica cycle. *Annual Review of Marine Science* 5:477–501.

Triplett, L., D. Engstrom, D. Conley, and S. Schellhaass. 2008. Silica fluxes and trapping in two contrasting natural impoundments of the upper Mississippi River. *Biogeochemistry* 87:217–230.

Turner, R. E., C. S. Milan, and N. N. Rabalais. 2004. A retrospective analysis of trace metals, C, N and diatom remnants in sediments from the Mississippi River delta shelf. *Marine Pollution Bulletin* 49:548–556.

Turner, R. E., N. Qureshi, N. N. Rabalais, Q. Dortch, D. Justic, R. F. Shaw, and J. Cope. 1998. Fluctuating silicate:nitrate ratios and coastal plankton food webs. *Proceedings of the National Academy of Science U.S.A.* 95:13048–13051.

Turner, R. E., and N. N. Rabalais. 1991. Changes in Mississippi River water quality this century: Implications for coastal food webs. *BioScience* 41:140–147.

———. 1994. Coastal eutrophication near the Mississippi River delta. *Nature* 368:619–621.

———. 2003. Linking landscape and water quality in the Mississippi River basin for 200 years. *Bioscience* 53:563–572.

———. 2013. Nitrogen and phosphorus phytoplankton growth limitation in the northern Gulf of Mexico. *Aquatic Microbial Ecology* 68:159–169.

Turner, R. E., N. N. Rabalais, R. B. Alexander, G. McIsaac, and R. W. Howarth. 2007. Characterization of nutrient, organic carbon, and sediment loads and concentrations from the Mississippi River into the northern Gulf of Mexico. *Estuaries and Coasts* 30:773–790.

Turner, R. E., N. N. Rabalais, and D. Justic. 2008. Gulf of Mexico hypoxia: Alternate states and a legacy. *Environmental Science & Technology* 42:2323–2327.

Twilley, R., J. Cowan, T. Miller-Way, P. Montagna, and B. Mortazavi. 1999. Benthic nutrient fluxes in selected estuaries in the Gulf of Mexico. Pp. 163–209 in T. S. Bianchi, J. R. Pennock, and R. R. Twilley (eds.), *Biogeochemistry of Gulf of Mexico*: New York: Wiley and Sons.

Ugalde, D., P. Gomez, and N. Simoes. 2015. Marine sponges (Porifera: Demospongiae) from the Gulf of México, new records and redescription of *Erylus trisphaerus* (de Laubenfels, 1953). *Zootaxa* 3911:151–183.

United Nations Environment Programme. 2000. Global Environment Outlook: Latin America and Caribbean-Environment Outlook 2000. Mexico City: UNEP Regional Office for Latin America and the Caribbean. http://web.unep.org/geo/assessments/global-assessments/geo-2000.

———. 2014. The Caribbean Environment Programme. http://www.cep.unep.org/publications-and-resources/marine-and-coastal-issues-links/nutrients.

United Nations Environment Programme, A. Villasol, and J. Beltrán. 2004. *GIWA Regional Assessment 4*. M. Fortnam and P. Blime (eds.). Kalmar, Sweden: University of Kalmar.

US Environmental Protection Agency. 2016. Clean air status and trends network (CASTNET). www.epa.gov/castnet.

Valdes, D. S., and E. Real. 2004. Nitrogen and phosphorus in water and sediments at Ria Lagartos coastal lagoon, Yucatan, Gulf of Mexico. *Indian Journal of Marine Sciences* 33:338–345.

Valdés-Lozano, D. S., E. Real, G. García-Uribe, and G. Olguín. 2010. Monitoreo de la Calidad del Agua en el Sur del Golfo de México [Monitoring water quality in the southern Gulf of Mexico]. In L. M. López Acuña, H. Bustos Serrano, and J. S. Palleiro Nayar (eds.), *ASOCEAN: XV Congreso Nacional de Oceanografía*.

Van Der Zee, C., and L. Chou. 2005. Seasonal cycling of phosphorus in the Southern Bight of the North Sea. *Biogeosciences, European Geosciences Union* 2:27–42.

Vargo, G. A., C. Heil, K. Fanning, L. K. Dixon, M. B. Neely, K. Lester, D. Ault, S. Murasko, J. Havens, J. Walsh, and S. Bell. 2008. Nutrient availability in support of *Karenia brevis* blooms on the central west Florida shelf: What keeps *Karenia* blooming? *Continental Shelf Research* 28:73–98.

Walsh, J. J., J. K. Jolliff, B. P. Darrow, and J. M. Lenes. 2006. Red tides in the Gulf of Mexico: Where, when, and why? *Journal of Geophysical Research: Oceans* 111:C11.

Walsh, J. J. and K. A. Steidinger. 2001. Saharan dust and Florida red tides: The cyanophyte connection. *Journal of Geophysical Research* 106 (C6):11597–11611.

Weber, S. C., L. Peterson, J. J. Battles, B. J. Roberts, R. N. Peterson, D. J. Hollander, J. P. Chanton, S. B. Joye, and J. P. Montoya. 2016. Hercules 265 rapid response: Immediate ecosystem impacts of a natural gas blowout incident. *Deep Sea Research Part II: Tropical Studies in Oceanography* 129:66–76.

Westberry, T. K., and D. A. Siegel. 2006. Spatial and temporal distribution of *Trichodesmium* blooms in the world's oceans. *Global Biogeochemical Cycles* 20:GB4016. doi:10.1029/2005GB002673.

Wilson, S. T., D. Bottjer, M. J. Church, and D. M. Karl. 2012. Comparative assessment of nitrogen fixation methodologies, conducted in the oligotrophic North Pacific Ocean. *Applied and Environmental Microbiology* 78:6516–6523.

Wiseman, W. J., Jr., N. N. Rabalais, M. J. Dagg, and T. E. Whitledge (eds.). 1999. *Nutrient Enhanced Coastal Ocean Productivity in the Northern Gulf of Mexico: Understanding the Effects of Nutrients on a Coastal Ecosystem*. NOAA Coastal Ocean Program Decision Analysis Series No. 14. Silver Spring, MD: National Oceanic and Atmospheric Association Coastal Ocean Program. 156 pp.

Xu, Y. J. 2013. Transport and retention of nitrogen, phosphorus and carbon in North America's largest river swamp basin, the Atchafalaya River basin. *Water* 5:379–393.

Yoon, W. B., and R. Benner. 1992. Denitrification and oxygen-consumption in sediments of two South Texas estuaries. *Marine Ecology Progress Series* 90:157–167.

Yoshinari, T., and R. Knowles. 1976. Acetylene inhibition of nitrous oxide reduction by denitrifying bacteria. *Biochemical and Biophysical Research Communications* 69:705–710.

Zimmerman, A. R., and R. Benner. 1994. Denitrification, nutrient regeneration and carbon mineralization in sediments of Galveston Bay, Texas, USA. *Marine Ecology Progress Series* 114:275–288.

4

Trace Metals in the Gulf of Mexico
Synthesis and Future Directions

Christopher T. Hayes, Liang-Saw Wen, Chih-Ping Lee, Peter H. Santschi, and Karen H. Johannesson

Introduction

The dynamic material fluxes to the Gulf of Mexico (GOM), such as the time-variable freshwater inputs from the Mississippi River or the shedding of Loop Current eddies within the Gulf, clearly have the potential to impact the cycling of dissolved trace metals in a unique way compared to typical oceanic conditions. In the oceanographic sense, the physical framework of the Gulf challenges the community to determine trace-metal distributions in a complex system but also affords a pathway to learn more about the processes regulating trace-metal behavior under a wider set of environmental conditions. Building on previous reviews (Shiller 1999; Wen et al. 1999), in this chapter we provide an updated synthesis of trace-metal biogeochemistry in the GOM and highlight areas that are still in need of future research.

We discuss 3 categories of trace-metal studies: bioactive metals, contaminants, and geochemical tracers. While most work on bioactive metals to date has focused

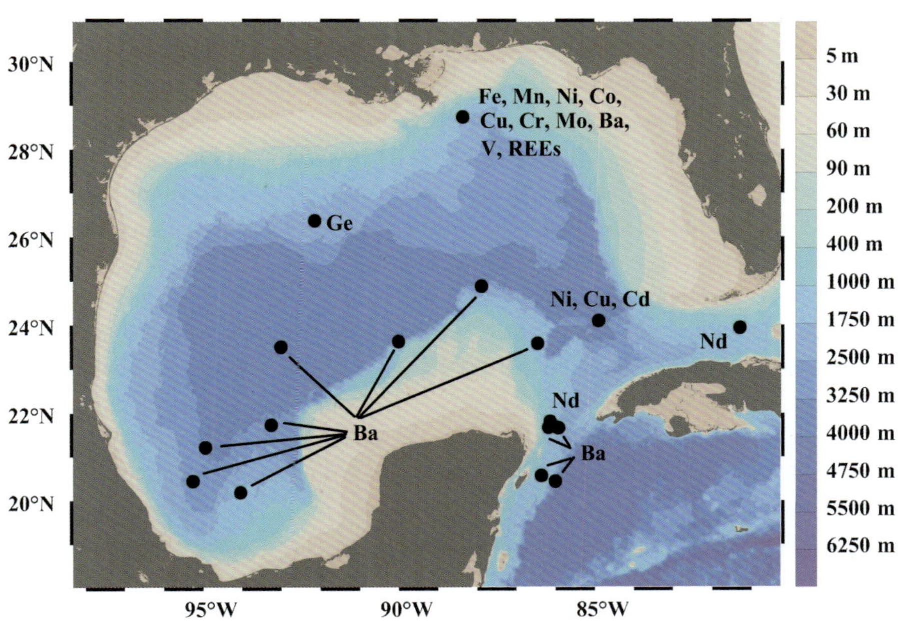

Figure 4.1. All sites in the Gulf of Mexico with trace-metal measurements in the water column with data deeper than 550 m (Baronas et al. 2016; Boyle et al. 1984; Buerkert 1999; Joung and Shiller 2013; Osborne et al. 2014; Osborne et al. 2015; Shiller et al. 2017). With the exception of the Ba data, the sites generally hug the continental shelf break, emphasizing the lack of knowledge of the metal distribution in the deep Gulf (figure made with GeoMapApp, http://www.geomapapp.org, using GMRT bathymetry; Ryan et al. 2009).

on estuarine and coastal systems, there have been significant advances, such as in the field of metal speciation, including colloidal size classes and organic ligands, that shed light onto a range of relevant oceanographic processes. For contaminants, the past 2 decades have seen a large research program on the distribution and cycling of mercury in addition to other industrial pollutants in and around the Gulf. In geochemical tracers, the GOM has been the site of work furthering the understanding of modern processes that may be useful in paleoceanographic reconstructions. In all cases, our knowledge of the distributions of trace elements is sparse for the deep Gulf (Fig. 4.1), and this proves to be a fruitful area for future research.

Bioactive Metals

Background

Micronutrient trace metals such as manganese (Mn), iron (Fe), cobalt (Co), nickel (Ni), copper (Cu), zinc (Zn), and cadmium (Cd) are taken up by phytoplankton in surface waters and released back into the deep ocean through the oxidation of organic matter. Because of their critical role on the growth of phytoplankton (e.g., Morel and Price 2003), a field was spawned, beginning in the late 1970s, to measure trace-metal distributions at nanomolar (10^{-9} M) to picomolar (10^{-12} M) concentrations and to determine the processes regulating these bioactive metals in seawater across the globe (Sohrin and Bruland 2011).

In addition to characterizing bulk concentrations, more recently researchers have interrogated the speciation of these bioactive trace metals in seawater. Metal speciation, defined as the distribution of the element between different chemical forms (Tessier and Turner 1995), is key to understanding metal availability to biota. Whereas only a small portion of the dissolved trace metals is present as truly free hydrated cationic species or complexes with inorganic ligands, a large fraction of many trace metals is complexed to organic ligands. Another important factor in dissolved metal speciation is size, ranging from truly dissolved species small enough to pass through cell membranes (<1 nm) to larger colloids that may act more like particles in some reactions (see also Chapters 5 and 6 for more details colloids as they relate to organic matter and radionuclides).

Colloids (from Greek *Kolla*, meaning "glue") are nanoparticles or macromolecules, operationally defined as having at least 1 size dimension between 1 nm and 1 μm (Everett 1972). Conceptually, colloids have intermediate properties between truly dissolved species and truly particulate species. The exact size range used in referring to colloids is almost always defined by the specifics of the filters used to investigate them. For instance, the upper limit is usually set by the filter pore size defining the filter-passing, "dissolved" species, typically 0.2 or 0.45 μm. The lower limit, defining the largest of the truly dissolved or "soluble" species, is set by the pore size or molecular weight size cutoff used in ultrafiltration, typically 0.02 μm or 10 kDa. For instance, a protein with a molecular weight of 1–10 kDa is roughly equivalent to a particle with a diameter of 1–3 nm (Erickson 2009). Doucet et al. (2006) reviewed the wide-ranging study of colloidal associations of trace metals in aquatic systems.

As a result of the advances in both chemical and physical metal speciation, these tools are currently used in conjunction with spatial concentration fields to discover how bioactive metals participate in biogeochemical cycles. We now turn more specifically to how GOM research has been featured in the development of the chemical oceanography of trace metals.

The Deep Oceanic Gulf of Mexico

The pioneering work on the bioactive metals in the deep Gulf can be traced back to Boyle et al. (1984), in which Cu, Ni, and Cd concentrations were accurately determined for the first time in this basin (Fig. 4.2). The bioactive metals in surface water of the coastal, northern GOM were enriched, especially Cu, likely due to the source of river input. A regional flux balance model was made to illustrate the contribution of the Cu-enriched water to the Florida Loop Current (Boyle et al. 1984). In contrast, the metal concentrations in the deep GOM, observed at one location between the west Florida shelf and the Campeche Bank, were similar to those observed in the North Atlantic (e.g., Middag et al. 2015). As in most places in the open ocean, nutrient-like profiles were observed for Cu, Ni, and Cd (Fig. 4.2), indicating biological uptake in surface water, export to depth as organic matter, and regeneration of the dissolved metals after organic matter respiration at intermediate depths.

In addition to biological control of the dynamics of bioactive metals in surface water, the impact from physical events should be taken into consideration, as shown by Springer-Young et al. (1992). When an atmospheric high-pressure system passed through the northern part of the open GOM, the surface seawater became less saline and warmer. This change in hydrography resulted in a change

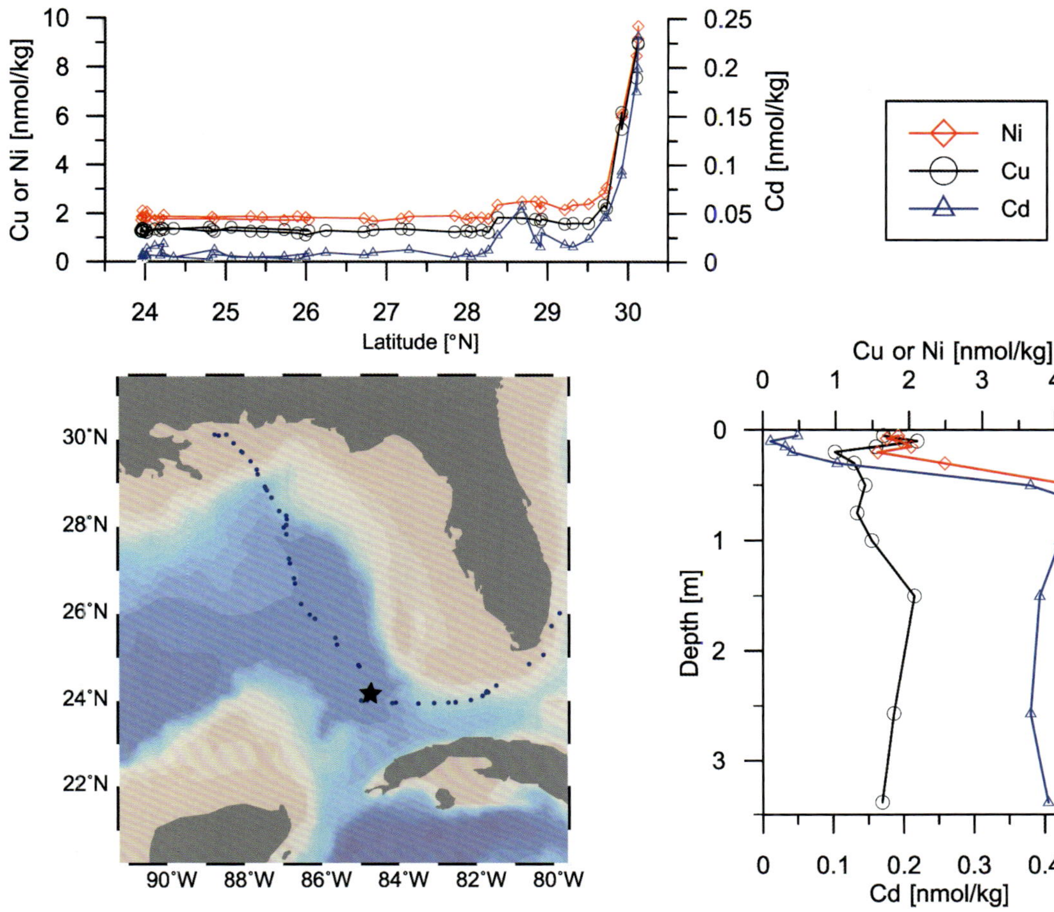

Figure 4.2. Trace-metal (Cu, Ni, Cd) surface transect, showing coastal enrichment near the Mississippi River outflow (*top*) and depth profiles, showing nutrient-like distributions (*bottom right*), published by Boyle et al. (1984) from a research cruise in April 1981. These data were produced from unfiltered seawater. Note that all data are in nmol/kg, but Cd concentrations are considerably lower and are plotted on their own axis. The cruise track is depicted by the water-sampling stations in blue dots in the inset map, and the profiles are from the station indicated by a black star (also mapped in Figure 4.1) (inset map created using Ocean Data View [ODV]).

in carbon uptake and Cu and Zn speciation, identified by electrochemistry. Accordingly, these authors concluded that reoccurring atmospheric and/or hydrographic events must be considered for any study of metal species in this dynamic setting.

Lenes et al. (2001) also found coastal enrichment in the surface-water distribution of dissolved Fe on the west Florida shelf near Tampa. Interestingly, these authors observed a midshelf minimum in Fe, with offshore waters also elevated in dissolved Fe by up to 3–4 nmol/kg. This was likely due to a summertime Saharan dust deposition event. Saharan dust may be an important source of Fe to Gulf waters, and its effect has been modeled based on atmospheric time-series measurements in Miami (Lenes et al. 2012). Nevertheless, this prediction is in dire need of more seawater measurements for model verification. Additionally, the relative importance of Saharan dust in relation to river/groundwater sources for the Fe in the Gulf is still unconstrained.

As an example of the insight provided by advances in chemical speciation in the GOM, Wen et al. (2011) undertook a study of surface water on a pair of warm- and cold-core eddies spanning from the open Gulf to the shelf zone of the Mississippi River (Fig. 4.3). On the whole, these authors observed the same coastal enrichment found by Boyle et al. (1984) and that the distributions of Cd, Ni, and Cu were controlled mainly by continuous mixing processes. In other words, the trace metals were roughly inversely correlated with salinity, indicating mixing between fresher coastal water with high trace-metal concentrations and saltier offshore water with low trace-metal concentrations. Providing further detail, Wen et al. (2011) performed physical size separation of the metals based on filtration at 0.4 μm, defining the dissolved phase, and at 10 kDa, defining the soluble phase. The difference in concentration between the dissolved phase (<0.4 μm) and the soluble phase (<10 kDa) is then interpreted as the concentration of the colloidal phase. In addition, characterization of the molecular reactivity of the metals was performed based on their chemical affinity toward a chelating resin (Chelex-100) and an anionic resin (AGMP-1). These techniques produced, among other parameters, an estimate of the percentage of the metal that was chemically unreactive (inert).

Aside from the obvious coastal enrichment, Wen et al. (2011) found subtle differences in the concentration of the metals in a warm-core eddy, recently shed from the Loop Current, and the paired cold-core eddy to its north (Fig. 4.3). More interestingly, very little Cd was found in either the colloidal (10 kDa–0.4 μm) size class or in the inert chemical reactivity class, whereas the dissolved phase of both Cu and Ni was roughly 30% colloidal and a similar percentage was inert. This trend applied to both the more coastal cold-core eddy and to the oligotrophic warm-core eddy. The coupled physical and chemical speciation suggests the possibility that humic substances that bind and lower the reactivity of metals have a source in rivers and coastal sediments. These humic substances must then be transported off the shelf, accounting for the persistence of inert, colloidal metals in the central GOM (Wen et al. 2011). This behavior is possibly mediated through complexation with reduced sulfur species, especially for Cu (Tang et al. 2002).

The only other study, aside from that by Boyle et al. (1984), to determine dissolved bioactive metals (Fe, Cu, Ni, Mn, and Co) in the deep waters (<550 m) of the GOM was carried out by Joung and Shiller (2013) for assessing the impact of the Deepwater Horizon well blowout of 2010. Their results demonstrated, as in the Wen et al. (2011) study, that surface-water concentrations correlated with salinity, implying the importance of mixing of Mississippi River water with shelf waters. No significant input

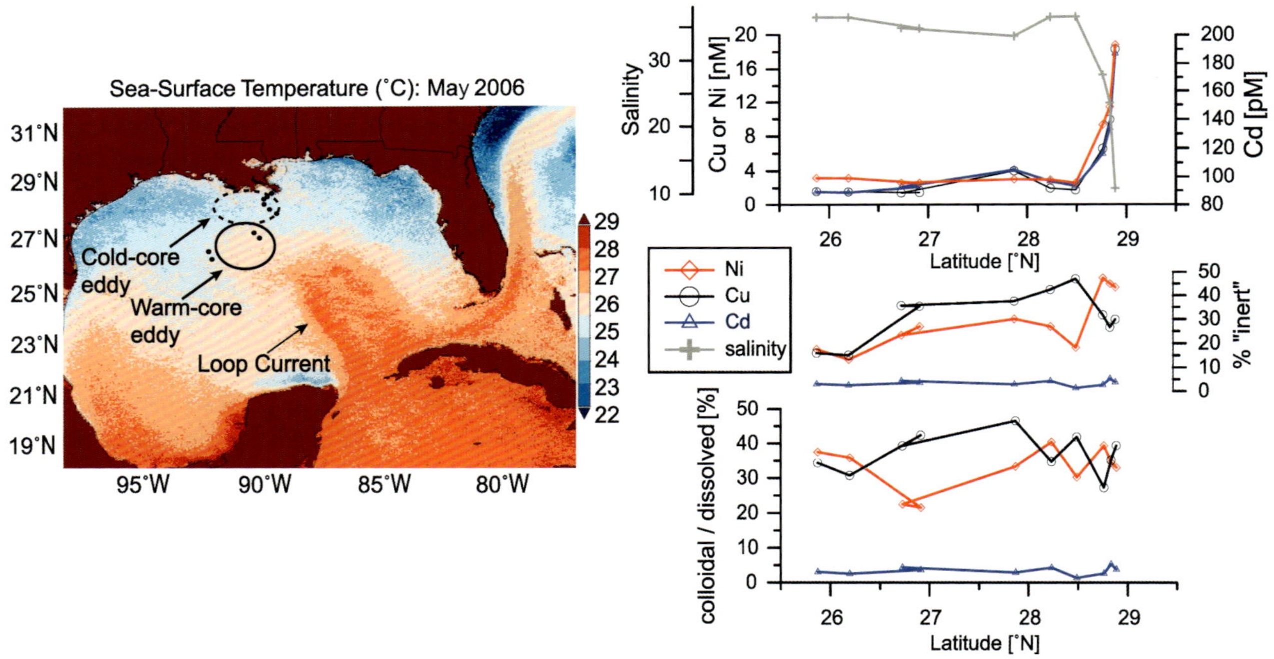

Figure 4.3. Surface trace-metal (Cd, Cu, Ni) transect from a cruise in May 2006 reported by Wen et al. (2011). The inset map is sea-surface temperature from Aqua-MODIS, level 3, 4 km resolution data extracted from Giovanni version 4, using a monthly average for the time of the cruise. The temperature field shows the warm-core eddy, recently shed from the Loop Current, and a paired cold-core eddy to the north of it that were sampled (sampling stations indicated by black dots). Plotting the dissolved trace metals (<0.4μm) as a function of the latitude (*right 2 plots*) makes the enrichment in the coastally influenced cold-core eddy prominent. The percentage of the dissolved phase that was colloidal (10 kDa–0.4 μm) was distinct between Cd and Cu/Ni for all samples and may have been slightly depleted in the warm-core eddy for Ni.

of the trace metals to the surface waters from the oil contamination was detected. In a deep oil/gas plume (1000–1300 m) resulting from the blowout, there was also no significant elevation for most metals (aside from Co and barium [Ba], discussed in the section on contaminants). No apparent contamination of Fe and Mn was observed, but only benthic, likely natural, input of these elements (Fig. 4.4). Joung and Shiller (2013) also considered the possibility of enhanced biological removal of Fe and Cu by microbial activity in the submerged oil plume; while finding it plausible, these authors did not find definitive evidence for this process. Overall, the impact of trace-metal input from the Deepwater Horizon well blowout appeared minor, though there is little information on the background reference concentrations one should expect for the deep Gulf.

Finally, a review of bioactive trace metals in the deep Gulf would be remiss to not describe one of the most exotic environments existing there, which are the benthic, anoxic brine pools (see also Chapter 2). These are created when tectonic forces expose ancient evaporite deposits to ambient seawater in a newly formed basin. The most

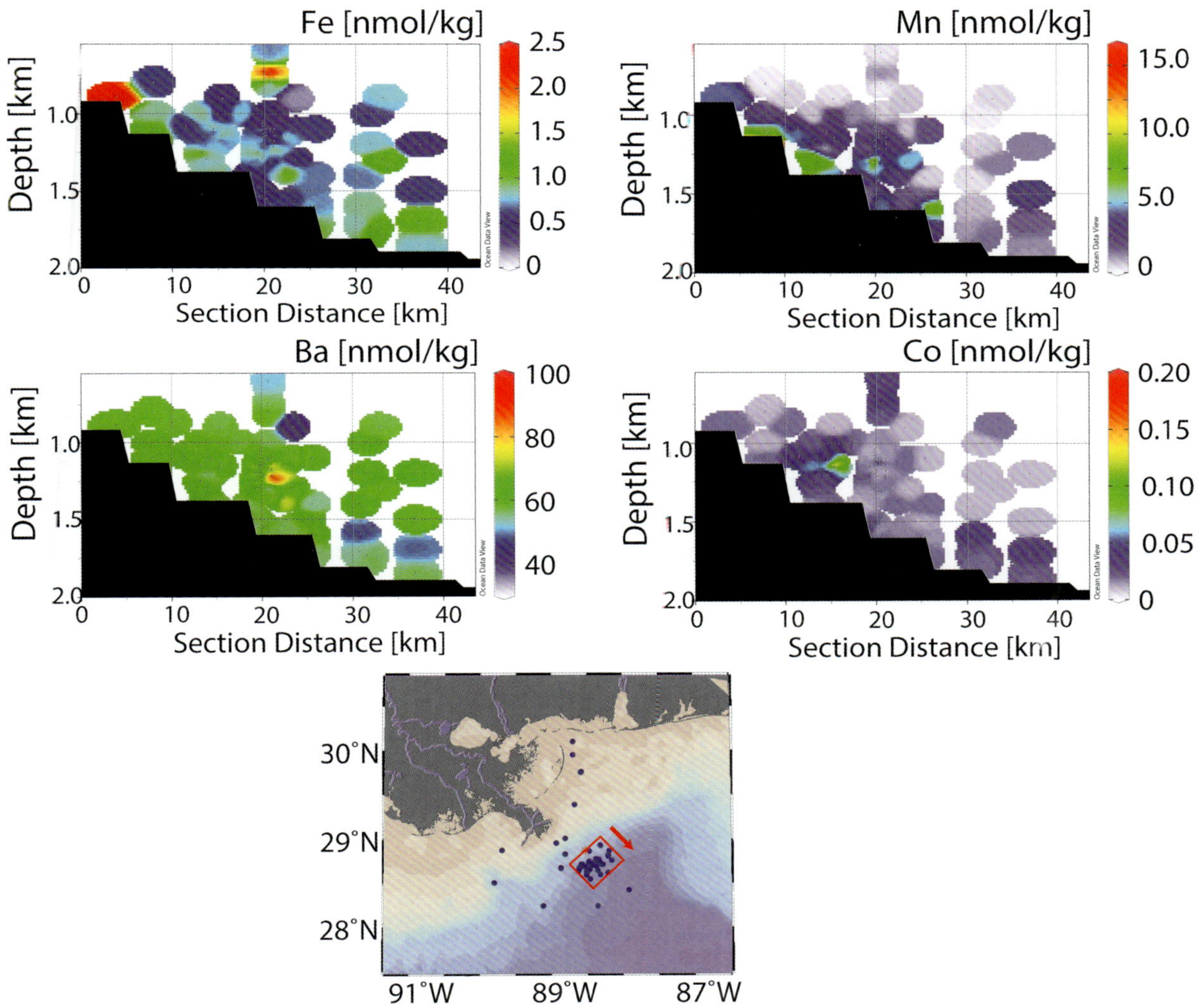

Figure 4.4. Dissolved trace-metal data collected and reported by Joung and Shiller (2013) from surveys of the north-central Gulf of Mexico following the Deepwater Horizon oil-well blowout in April–July 2010. The inset map shows the location of all samples reported from that study as blue dots. In the red box, the deepwater data have been condensed onto one downslope section, running northwest to southeast, as shown by the red arrow. This compiled data from 4 surveys, in early and late May 2010, October 2010, and October 2011. For Fe and Mn, no apparent contamination from the oil well could be discerned, and natural increases in concentration near the seafloor were suspected. For Co and Ba, clear contamination was detected at the level of the deep oil plume (1–1.3 km depth), particularly in late May 2010 (figure created using ODV).

well studied of these in the GOM is the Orca Basin in the northern Gulf (~27°N, 91°W) which has salinity of up to 250 g/kg in a 200 m thick layer of water in a 400 km^2 depression at 2400 m water depth (Shokes et al. 1977). The extreme stratification of the brine pool leads to complete anoxia, high accumulation of methane (Wiesenburg et al. 1985), and reduced forms of many important trace metals, Fe and Mn in particular (Trefry et al. 1984). Microbial reduction of solid manganese oxides leads to dissolved Mn(II) levels of up to 400 μM just below the level of anoxia (2250 m water depth), and there is some Mn removal by carbonate precipitation below this, leading to ~180 μM in the bulk brine pool (Van Cappellen et al. 1998). Solid iron oxyhydroxides are reduced by reaction with dissolved sulfide present, leading to >30 μM levels of dissolved Fe(II) (Van Cappellen et al. 1998). These concentrations are staggering, considering the pools are overlain by water with dissolved Fe and Mn at the nanomolar level (Joung and Shiller 2013). The Orca Basin is thus a potential source of these trace metals to bottom water through diffusive loss. The Orca Basin brine also provides a classic example for the complex cycling of selenium, technically a nonmetal in the chalcogen group, which undergoes a rapid change of redox state (from VI to IV) and an increased association with organic ligands going into the anoxic brine (Takayanagi and Wong 1985). Similar brine pools are being documented elsewhere on the northern GOM continental slope (Joye et al. 2005) and could also be important for many other metals, including barium (Schijf 2007).

In summary, the data on bioactive metals in the deep Gulf are still sparse. The geographic distribution of trace metals across the Gulf and the Caribbean is still largely unknown. Therefore, it is not yet possible to determine mass balances for sources and sinks of these trace metals in terms of the Gulf's connection to its adjacent seas. Because of the essential requirement of trace metals for phytoplankton, the base of the marine food web, greater knowledge of trace-metal supplies could potentially aid prediction of species at higher trophic levels, including commercial fish stocks. Microbial production, including "red-tide"-forming dinoflagellates, may also be modulated by trace-metal (Fe) supply (Walsh et al. 2006) as well as nutrients delivered by submarine groundwater discharge (SGD; Hu et al. 2006; see also Chapter 3 for more detail on nutrients associated with SGD). Moreover, the available data are also inadequate to assess the input of the metals from future episodic events, such as the Deepwater Horizon spill, as there is no comprehensive understanding yet of the distributions, sources, sinks, and variability of bioactive metals. Further metal speciation information in the GOM will also be necessary to properly assess their reactivity.

Estuarine and Coastal Environments

Trace-metal chemistry has been much more intensely studied in the estuarine environments of the Gulf than in the deep, oceanic Gulf. Here we update a previous and comprehensive review on this topic (Wen et al. 1999) with research completed since its publication. In particular, research has been focusing on chemical and size speciation to elucidate the reactivity of the bioactive metals. Also included is a discussion on the relatively unique bioactive metals tungsten (W) and molybdenum (Mo). Finally, we close with a note on efforts to monitor bioactive metals in the Gulf over time.

Colloidal Associations

Estuarine studies of colloidal associations of selected trace metals demonstrate a spectrum of behaviors across the GOM. Colloidal associations of trace elements seem to be highest in shallow estuaries such as Galveston Bay, Texas (Benoit et al. 1994; Wen et al. 1999; Tang et al. 2001, 2002; Wen et al. 2011) or Ochlockonee Estuary, Florida (Powell et al. 1996), whereas in the Mississippi River mixing zone on the continental shelf, colloidal fractions seem to be lower and more restricted to the low-salinity region (Joung and Shiller 2016; Shim et al. 2012; Stolpe et al. 2014; Wen et al. 2011). In contrast to this general observation, a large-scale chemical oceanographic survey, called LATEX, of the Louisiana-Texas shelf in 1992–94, out to the 200 m isobath, reported measurements of dissolved, colloidal, particulate, and sedimentary trace elements, including bioactive and contaminant metals (Murray 1997). This author found coastal enrichment of dissolved and colloidal Cu, Cd, Ba, and lead (Pb), with interannual variations, but no clear difference in the observations between Texas and Louisiana coasts. In making these studies, Murray used slightly different colloidal size definitions that may limit the comparison. We now turn further into the great deal of trace-metal research over the past 2 decades that was done specifically in the estuaries of Texas and Louisiana.

In the Trinity River/Galveston Bay system, most of the dissolved metals (Fe, Mn, Cd, Cu, Ni, Pb, Zn, and silver [Ag]; size ≤0.45 μm) showed nonconservative estuarine mixing behavior, with midsalinity maxima com-

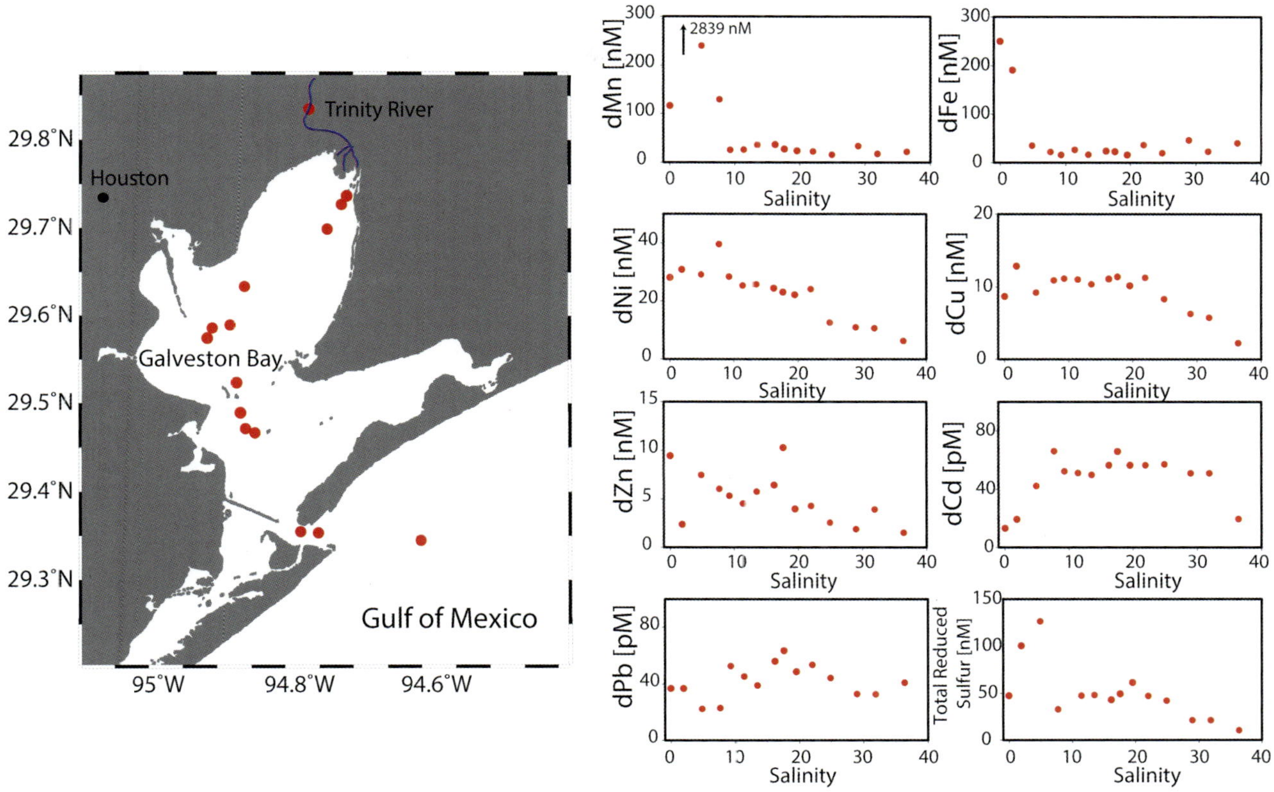

Figure 4.5. Galveston Bay and the Trinity River estuary have been sites of numerous studies on bioactive metal processes. These figures are modified from Tang et al. (2002) and demonstrate the nonconservative behavior of dissolved (d) metals (Mn, Fe, Ni, Cu, Zn, Cd, and Pb, from the top, left to right, followed by total reduced sulfur). This is evident in the midsalinity maxima, observed for most elements (Nu, Cu, Zn, Cd, and Pb). Because of a similar pattern of reduced sulfur species, Tang et al. (2002) hypothesized that these metals may be controlled by sulfur complexation. In contrast, Mn and Fe appear to be strongly removed in the Trinity River outlet (where the freshest water was sampled). Galveston Bay is on average 1.8 m deep, and its maximum depth is about 3 m (figure created using ODV).

monly observed (Wen et al. 1997; Wen et al. 1999; Tang et al. 2002; Fig. 4.5). Midsalinity maxima in a trace-metal-to-salinity cross-plot, in contrast to the linear associations seen in the open Gulf studies, imply an input of the metal during mixing, such as bottom sediments or upwelling (e.g., Morris 1985; Shiller 1996). The proportion of dissolved metals to the total metals (dissolved plus particulate) in the Trinity River/Galveston Bay system decreased in the following trend: Ni > Cu > Cd > Zn > Pb > Mn > Fe, with the particle-water partition coefficient (K_d) increasing in a similar order (Tang et al. 2002). K_d is usually defined as the metal concentration per mass of particulate material over the dissolved metal concentration per mass, or sometimes volume, of seawater.

Molecular weight distribution of the dissolved phase of metals in Galveston Bay exhibited the presence of colloidal (1 kDa–0.45 µm) metals but with various extents: Zn (91%) > Fe (79%) > Pb (64%) > Cu (55%) > Cd (45%) > Ni (36%) > Co (19%) (Wen et al. 1999). Colloidal Ag was also observed in the range of 15–70% of the dissolved Ag, and the proportion decreased with increasing salinity (Wen et al. 1997). The colloidal-to-soluble (<1 kDa) partition coefficients (K_c) were higher than the particle-to-soluble partitioning (K_p), implying the high complexing capacity for colloids. Fe was mostly associated with high-molecular-weight colloids (10 kDa–0.45 µm), whereas Cu, Ni, and Pb preferentially resided in the low-molecular-weight phase (1 kDa–10 kDa). Of course, these high-molecular-weight colloids could be either organic in nature or inorganic, such as oxyhydroxide nanoparticles (e.g., Stolpe et al. 2010). Tang et al. (2001) showed that the strongest Cu-binding ligands occur in the colloidal fraction, leading also to a close relationship between the colloidal fraction of Cu to that of organic carbon. In this case, however, organic carbon was likely only a proxy for reduced sulfur complexes with B-type, soft metals such as Ag and Cu, as Tang et al. (2002) were able to show a significant relationship between the sum (Cu + Ni + Zn) and the total reduced sulfur concentration in the filter-passing fraction.

Guo et al. (2000) determined colloidal metal concentrations from Galveston Bay, the GOM, and the Middle Atlantic Bight that were determined after extensive diafiltration of the retentate. Metal/aluminum ratios, Me/Al, in isolated marine colloids were significantly higher than those for average soils and continental crust. In contrast, colloidal metal–to–organic carbon (Me/OC) ratios were similar to those of average humic substances and marine plankton, suggesting that these colloids are composed of predominantly organic components. The Me/OC ratios of Galveston Bay colloids followed the sequence of Cu > Ni, chromium (Cr), Zn > Mn > Co > Pb, Cd, which is similar to the Irving-Williams order except for Mn. This suggests that the interaction of metals with marine colloids, at least in the environments studied, is determined by the relative affinity of metals for specific organic ligands as predicted by thermodynamics (Irving and Williams 1953).

In addition to the bioactive trace metals, the colloidal content of the long-lived isotopes of uranium (U) and thorium (Th) have been characterized in Galveston Bay and the Louisiana-Texas shelf. ^{238}U, ^{232}Th, and ^{230}Th have radioactive half-lives much longer than their residence time in the water column (half-lives of 4.5×10^9, 14.1×10^9, and 76×10^3 years versus ocean-residence times of 4×10^5, ~20, and ~20 years, respectively) (Rutgers van der Loeff and Geibert 2008). Thus, the radioactive decay within the water column of these isotopes can be neglected in their ocean cycling, though Th is still orders of magnitude more particle reactive or scavenging prone than U. Guo et al. (2007) found that colloidal U was negligible in the higher-salinity coastal waters off Galveston Bay and accounted for only 15% of the total dissolved U in Trinity River waters. The association of dissolved U with nanoparticles and macromolecular organic matter in higher-salinity seawater thus seems minimal, and most dissolved U in seawater should be in the form of anionic U carbonato complexes with a molecular weight ≤1 kDa. In contrast, Baskaran et al. (1992) found that dissolved ^{232}Th was roughly 60% colloidal in the water column of the Texas-Louisiana shelf, with potentially similar results for ^{230}Th (Guo et al. 1995). In comparison, in the open Atlantic near Bermuda, dissolved ^{232}Th was found to be <6% colloidal (Hayes et al. 2017); and in another oceanic setting near Hawaii, both dissolved ^{232}Th and ^{230}Th were <20% colloidal (Hayes et al. 2015). These results perhaps suggest that shelf processes in the GOM may lead to enriched colloidal contents for Th and for particle-reactive metals more generally. See Chapter 5 for the similar, extensive work on the shorter-lived ^{234}Th (half-life 24 days) and its colloidal contents in the GOM.

Intense organic-matter cycling in Galveston Bay was also found to significantly change the well-established relationship between Fe and Al in suspended particulate matter. In this relationship in coastal sediments, most trace metals tend to have a positive linear correlation with the major metals making up natural detrital minerals, for example, Fe or Al. Therefore, normalization of other metals to Fe or Al can be used in determining the balance of natural and anthropogenic sources (Summers et al. 1996; Windom et al. 1988; Windom et al. 1989). Correlations of metal to Fe ratios with the Fe content of particles from Galveston Bay were found to be significant and positive for particle-reactive elements such as Pb, but actually negative for nutrient-type elements such as Cu, Ni, and Cd (Wen et al. 2008). In Galveston Bay waters, in fact, Cu, Ni, and Cd showed departures from their source soil composition, that is, enrichments with respect to Al or Fe in samples with lower Fe (or Al) or higher OC content. This study by Wen et al. (2008) demonstrated, therefore, that autochthonous organic matter formation in Galveston Bay can cause alterations in the particulate matter elemental composition, even with no evidence for anthropogenic sources. Finally, whereas benthic fluxes of bioactive metals (Mn, Fe, Ni, and Zn) have been measured in Galveston Bay (Warnken et al. 2001; Warnken et al. 2003), it was found that only a small percentage of the Trinity River watershed metal load was being transmitted out into the GOM (Warnken and Santschi 2009). The percentage to which this flux contributes to the total inventory of metals in the Gulf still awaits a proper survey of the metal distribution throughout the open Gulf.

In regard to the lower Mississippi and Atchafalaya Rivers in Louisiana, using flow field-flow fractionation coupled to online detection by ICP-MS, Stolpe et al. (2010) examined the continuous size spectra of colloidal metals in the size range of 0.5–40 nm, including Fe, Mn, Cu, and Zn. This technique is a significant advance in size speciation work because the colloidal phase can be subdivided into smaller size ranges, which are commonly associated with specific molecular compositions. These authors found that dissolved organic matter (DOM) composition and abundance in the river significantly regulated the partitioning of trace metals between dissolved and colloidal phases and their colloidal size spectra. In general, colloidal size spectra of Cu in the Mississippi and Atchafalaya Rivers resembled those detected for chromophoric dissolved organic matter (CDOM), with a major peak in the 0.5–4 nm size range, indicating a high association of Cu with this type of organic matter. In contrast, the colloidal size spectra of Mn also had an additional peak in the 5–25 nm size range, similar to the size spectra of colloidal Fe, which was likely associated with a contribution from an inorganic (oxy)hydroxide phase. In rivers with higher dissolved organic carbon (DOC) concentrations and more terrestrial input, colloids were mostly small (0.5–4 nm) and likely composed of terrestrial fulvic acids. These small colloids accounted for most of the colloidal size class for most metals investigated (Fe, Mn, Cu, Zn, Pb, and U). The midsize colloids (5–40 nm) were Fe rich, likely reflecting an association with the (oxy) hydroxide component, and were more closely associated with binding P. In addition, the larger-sized colloids (>40 nm) were found to contain protein-rich organic matter, derived from in situ biological production (Stolpe et al. 2010). It was suggested that colloidal stability depends on background DOC concentration, pH, and calcium ion concentration and, further, that seasonal variation in colloidal size-class abundance could be expected. Determining the relative importance of these processes is an interesting target for future research.

In more recent work in the Mississippi River estuary and the Louisiana shelf (Joung and Shiller 2014, 2016), dissolved Fe was predominantly in the colloidal phase (0.02–0.45 μm), followed by Co, Cu, Mn, and Ni, whose colloidal concentrations were <30% of the dissolved concentrations. Nonconservative behaviors of these trace metals, as in Galveston Bay, were found and may be associated with various physical and/or biological processes: colloidal flocculation, biological activity, photochemical reactions, or benthic mobilization. Interestingly, the hypoxia in Louisiana shelf bottom waters detected in June–July 2009 led to elevation (Co, Cu, Ni, Mn, Fe, Ba) and depletion (Cr, vanadium [V]) of metals, which may be accomplished by benthic diffusion from sediments and/or adsorptive removal onto particles due to reducing conditions. Another recent investigation presented evidence that interdistributary bays within the Mississippi River delta, such as Barataria and Terrebonne Bays, act as sinks for V because of diffusion into the sediments, whereas these same environments may be sources of arsenic (As), again by diffusive fluxes, to the surface water and, hence, possibly the GOM (Telfeyan et al. 2017). Others have noted that sediments from both of these bays can exhibit elevated As contents (Rabalais et al. 1995). Again, the full impact of these fluxes on the open Gulf has yet to be adequately determined.

Estuarine Cycling of Molybdenum and Tungsten

Two bioactive elements that are not typically studied along with the first-row transition metals (Fe, Mn, Cu, etc.), Mo and W, were recently investigated in Terrebonne Bay, a shallow, well-mixed estuary along the coast of Louisiana located west of the Mississippi River delta (Mohajerin et al. 2016). Although Mo, but not W, is an essential trace metal for humans, both elements are found in microbial enzymes that catalyze redox reactions (Kletzin and Adams 1996; Pilato and Stiefel 1999; Andreesen and Makdessi 2008; Gonzalez et al. 2013). Mo and W occur in oxic, oceanic waters as oxyanions, $Mo^{VI}O_4^{2-}$ and $W^{VI}O_4^{2-}$, both of which are conservative in the ocean (Firdaus et al. 2008; Sohrin et al. 1987; Sohrin et al. 1999). In sulfidic waters, MoO_4^{2-} and WO_4^{2-} undergo sulfidization reactions, producing thioanions of the form $MO_xS_{4-x}^{2-}$ (where M = Mo or W and x = 0–4; Erickson and Helz 2000; Mohajerin et al. 2014; Mohajerin et al. 2016).

Mohajerin et al. (2016) measured Mo and W concentrations along the salinity gradient in surface waters and pore waters of the Terrebonne Bay estuary. Similar to what occurred in other estuaries, Mo exhibited nearly conservative behavior in the surface waters plotted along the same, linear binary mixing line across the salinity gradient independent of season. Extrapolation of the surface-water Mo data to a seawater salinity results in Mo concentrations of about 95 nM in the seawater endmember for April data and about 116 nM for August data. We note that both estimates are similar to Mo concentrations in global seawater (100–110 nM; Collier 1985; Firdaus et al. 2008; Sohrin et al. 1987; Sohrin et al. 1999), including GOM seawater (Joung and Shiller 2016; Yao and Byrne 1999). In contrast, W appeared to exhibit strong seasonal differences in the surface waters. More specifically, W concentrations varied linearly with salinity during the spring, whereas W concentrations in 3 of the estuarine samples were greatly in excess of the linear trend, which likely reflects a sediment source (Mohajerin et al. 2016). Nevertheless, extrapolation of the linear portion of the mixing line to seawater salinity leads to projected W concentration in GOM seawater of 40 ± 30 pM, which is similar to currently accepted values of W in seawater (50–60 pM; Sohrin et al. 1987). The surface-water data, in conjunction with other studies of these trace metals in river waters, strongly suggest that a substantial portion of the dissolved (i.e., 0.45 μm filter passing) W is actually associated with Fe^{III} oxyhydroxide colloids, whereas Mo is chiefly transported through the surface estuary as truly dissolved MoO_4^{2-} (Pokrovsky and Schott 2002; Stolpe et al. 2005; Taylor et al. 2012; Mohajerin et al. 2016).

The geochemical behaviors of Mo and W in the highly sulfidic pore water from the Terrebonne Bay estuary suggest limited communication with the surface waters. More specifically, Mo concentrations in the pore water were depleted relative to the surface waters, whereas W concentrations in pore water were enormously enriched compared to the surface waters (Mohajerin et al. 2016). Both trace metals varied systematically with dissolved sulfide concentrations in the pore water. Mo concentrations exhibited an inverse relationship with dissolved sulfide concentrations, and W concentrations increased with increasing dissolved sulfide concentrations above 60 μM hydrogen sulfide ($H_2S_{(aq)}$) (Mohajerin et al. 2016). Removal of Mo from solution in the pore water appears to be controlled by a precipitation of a putative Fe-Mo-S mineral (Helz et al. 2011; Mohajerin et al. 2016). The formation of this postulated mineral may involve reduction of Mo^{VI} to Mo^{IV}, perhaps via ligand-induced reduction by S^0 (Vorlicek et al. 2004; Dahl et al. 2013). The exact mechanism(s), however, remains to be determined conclusively. In contrast, the orders-of-magnitude enhancement of dissolved W concentrations in Terrebonne Bay pore water compared to the surface waters likely reflects reductive dissolution of carrier mineral phases, such as ferrihydrite, followed by thiolation of WO_4^{2-} to thiotungstate anions. Thiotungstate appears to be less particle reactive than WO_4^{2-} and readily desorbs from mineral surfaces (Cui and Johannesson 2017).

Temporal Monitoring of Bioactive Metals

As in the deep Gulf, the distributions and speciation of bioactive metals in the estuary and coastal system of the GOM are sensitive to episodic events and seasonal fluctuations. A study of several trace metals in the Mississippi River outflow after hurricanes clearly demonstrated the decrease of some metal concentrations, which could be attributed to the temporary loss of stratification and the diminishment of hypoxia (Shim et al. 2012). On slightly longer timescales, seasonal variations of trace metals were observed to correlate with the varying mixing ratios of the tributaries in addition to effects of stratification and seasonal hypoxia (Joung and Shiller 2016). Both these examples highlight the importance of meteorological conditions affecting the dynamics of these bioactive metals in estuary and coastal environments.

Of course, temporal variability in bioactive metals is impacted by both natural (fluvial, groundwater discharge) and anthropogenic (industrial) forces. After elevation above "natural" or "background" levels, some bioactive metals may be considered contaminants. The variability in contaminant metals is covered more fully in the next section, but 2 recent time-series studies from southern GOM estuaries illustrate this interplay. In the Veracruz Reef system, from June 2004 to February 2005, dissolved Cu and Ni were found to be mostly controlled by input from the Jamapa River, which naturally drains volcanic rocks, but dissolved V and Pb were derived from urban discharges, mainly in the rainy season (July–October) (Rosales-Hoz et al. 2009). Similarly, in the Veracruz Reefs, benthic macroalgae were analyzed in the rainy season of 2008 as a proxy for contamination in the dissolved phase (Horta-Puga et al. 2013). These authors found that macroalgae Pb and Cd also varied with Jamapa River flow. In contrast, Cd and Cu appeared lowest late in the rainy season (September), likely due to adsorption to a higher suspended matter load in the river.

Contaminant Metals

The GOM produces a substantial proportion of the commercial and recreational fish catch in the United States and Mexico. Additionally, tourism and recreational use of the Gulf are large parts of the regional economies. Concern over the health of seafood consumers and the economic viability of marine resources more generally has motivated targeted monitoring of contaminant metals in the Gulf. We first summarize a selection of work on one of the most heavily studied contaminant metals, mercury (Hg), and its chemical forms. Second, we recount a series of work on a range of contaminant metals in sediments with sources largely from the petroleum industry (see also Chapter 8). Third, we include the few studies that have investigated contaminant metals in the water column.

Mercury Cycling

Hg is a naturally occurring trace element in many geological reservoirs. Humans burning coal has largely mobilized Hg quantities of environmental concern into the surface environment (Selin 2009). Hg is emitted to the atmosphere as gaseous elemental Hg (Hg^0) or its oxidation product, Hg^{II}, which is subject to wet and dry deposition to the land and ocean surface (De Simone et al. 2014). In the aquatic environment, these inorganic forms can be converted, likely through microbial mediation, into monomethyl Hg (CH_3Hg^+), a lipophilic dissolved ion, or dimethyl Hg [$(CH_3)_2Hg$], a dissolved gas. Monomethyl Hg presents the largest health hazard, as this species bioaccumulates (Fitzgerald and Lamborg 2003; Harris et al. 2012; Mason et al. 2012) and is known to interfere with neurological development in children (Karagas et al. 2012).

While the ultimate source of Hg species to the GOM is atmospheric deposition, the Gulf receives inputs of Hg deposited in "upstream" areas via continental runoff through rivers and estuaries and seawater inflowing from the Atlantic Ocean via the Caribbean Sea. For instance, in areas influenced by the Mississippi River outflow in the northern Gulf, it was estimated that 75% of the inorganic Hg is riverine in source, the remainder being direct atmospheric deposition (Rice et al. 2009). A similar fraction of Hg sourced from runoff was found in Florida Bay (Kang et al. 2000), although it is generally believed that the preponderance of riverine Hg is carried as sediments that are deposited before traveling very far from the coast.

In the Gulf as a whole, one study (Harris et al. 2012) estimated that the Atlantic inflow dominates (85–90%) the sources of Hg, based on the volumetric transport in the Loop Current and average surface-water Hg concentrations from the North, equatorial, and South Atlantic Oceans (Sunderland and Mason 2007). Much of the Hg brought in with the Loop Current, however, likely leaves again as the current exits through the Florida Straits. The importance of the Loop Current as a source of Hg to the Gulf is therefore dependent on the degree of mixing that occurs between the Loop Current and the surrounding water. Interestingly, as anthropogenic Hg is now invading the deep ocean through North Atlantic Deep Water circulation (Lamborg et al. 2014), transport in the deepwater overflow from the Caribbean to the Gulf over the Yucatan Channel may not be a negligible source of Hg. There are no reported measurements of Hg in the open GOM or Caribbean Sea to better constrain the inflows and outflows of Hg due to ocean circulation.

Hg in GOM coastal waters is better constrained by measurements. Atmospheric deposition is well-monitored in US cities surrounding the Gulf (Prestbo and Gay 2009). There are more limited atmospheric measurements available in Mexico (Ruelas-Inzunza et al. 2013). Overall it is likely that Gulf-wide atmospheric Hg deposition is in the range of 50–150 nmol Hg m^{-2} yr^{-1} (Harris et al. 2012). This deposition is sourced from local and nonlocal emissions.

A coordinated monitoring program in Florida, the Florida Atmospheric Mercury Study, determined that about 40% of the atmospheric deposition to the Everglades was locally emitted, the balance being a global/regional background (Landing et al. 1995; Guentzel et al. 2001). In Pensacola, Florida, on the northern Gulf Coast, a slightly smaller fraction (~25%) of total atmospheric deposition was attributed to local sources (Landing et al. 2010). Local emissions of Hg, at least in South Florida, appear to have accelerated in the 1980s (Frederick et al. 2004) and declined to minimal values by 2000, after emissions from municipal waste combustors and medical waste incinerators were reduced (Atkeson et al. 2006). Today, <25% of total Hg deposition in the GOM is attributed to local emission (Harris et al. 2012).

Surveys of Hg in estuarine waters and sediments has been generally localized by state, such as Texas (Stordal et al. 1996; Bloom et al. 1999; Gill et al. 1999; Santschi et al. 1999); Louisiana (Delaune et al. 2008; Delaune et al. 2009; Kongchum et al. 2006; White et al. 2009; Yu et al. 2008); Florida (Guentzel et al. 1996; Kang et al. 2000; Rumbold et al. 2011; Bergamaschi et al. 2012); and the Veracruz, Tabasco, and Campeche states of Mexico (Ruelas-Inzunza et al. 2013, and references therein). Typical water Hg concentrations are in the low pM range, and sediment Hg concentrations range between ~25 and 3000 pmol/g, from relatively pristine sites such as Lake Borgne, Louisiana (Delaune et al. 2008), to known contaminated sites, such as Lavaca Bay, Texas (Santschi et al. 1999), respectively.

Methyl mercury concentrations typically contribute less than a few percent of total mercury in the water, but of course this level can still present a toxicity concern. Thus, efforts are made to understand the processes regulating the conversion of Hg to methylated species. For instance, in Lavaca Bay, which necessitated extensive Hg pollution remediation measures in the 1990s, a relatively large aqueous flux of inorganic Hg and monomethyl Hg from sediments to the water column was produced. This was determined at several sites in the bay, using pore-water gradients that could be compared to direct flux measurements (Gill et al. 1999). Strong diurnal fluxes of monomethyl Hg were documented, with most of the flux occurring during dark periods. Gill et al. (1999) suggested that the Fe redox cycling related to benthic microbial activity may control methylated Hg fluxes in Lavaca Bay.

As in the bioactive metal research, there have been advances in Hg cycling related to increased knowledge of its complexation and size speciation, particularly again in Galveston Bay. Both dissolved Hg (Stordal et al. 1996) and dissolved methyl Hg (Choe and Gill 2001) were found to be substantially colloidal in size (averaging 40–50%), likely bound to macromolecular organic matter. Thus, for both species, there are likely large macromolecular organic ligands that play a significant role in Hg transport in the environment. Han et al. (2007) provided further evidence for the presence of colloidal Hg in Offats Bayou of Galveston Bay. Han et al. (2006) determined the organic complexation of Hg using a competitive ligand technique and found binding constants similar to those of thiolic organic compounds. As further knowledge of the molecular identities of these large ligands develops, this will likely be an important topic in future Hg research.

In a more recent concerted effort encompassing all 5 US Gulf States, using Hg concentration gradients in the estuaries of the 10 largest American rivers draining into the Gulf, the flux of total Hg from estuaries to the northern Gulf was estimated to be about 1 order of magnitude less than that of direct atmospheric deposition (Buck et al. 2015). These authors also found that river discharges of Hg into estuaries were slightly greater than the Hg flux from estuaries to the Gulf (Fig. 4.6). This indicates that rivers can play a significant role in setting the Hg concentrations within estuaries and that some Hg must be removed and stored in estuarine sediments. The latter presents an additional environmental concern, as estuaries may be "hot spots" for the production of methyl Hg (Marvin-DiPasquale and Oremland 1998; Delaune et al. 2004; Hall et al. 2008; Rumbold et al. 2011; Bergamaschi et al. 2012;), and coastal disturbances, such as hurricanes, may remobilize sedimentary Hg as an additional source of Hg (and methyl Hg) to the water column (Liu et al. 2009). Bioaccumulation of Hg in pelagic fish of the northern GOM was documented by Cai et al. (2007), who found that Hg was greatly enriched in fish that are feeding on other fish or sediments.

Contaminant Metals in Sediments

The GOM has been a site of major oil exploration, drilling, and extraction since the 1940s, largely on the Texas-Louisiana continental shelf and on the Campeche Bank. This industry presents 2 pathways for trace metal contamination: (1) release of metals enriched in petroleum (e.g., V, Ni, Co) and (2) contamination resulting from the use of drilling mud, which is composed mostly of barite ($BaSO_4$) but also is known to contain other metals such as Cu, Hg, and Pb at enriched levels compared to uncontaminated marine sediments (Neff 2008). In addition to petroleum and its extraction process, other sources of contaminant metals (as do Hg sources) include point and nonpoint industrial sources via atmospheric deposition, rivers, estuaries, or ocean currents. Finally, contaminated sediments themselves can be a source of contaminant metals to the water column, such as in the Mississippi River delta where sediments are highly reducing, indicated by significant releases of dissolved Mn (Trefry and Presley 1982).

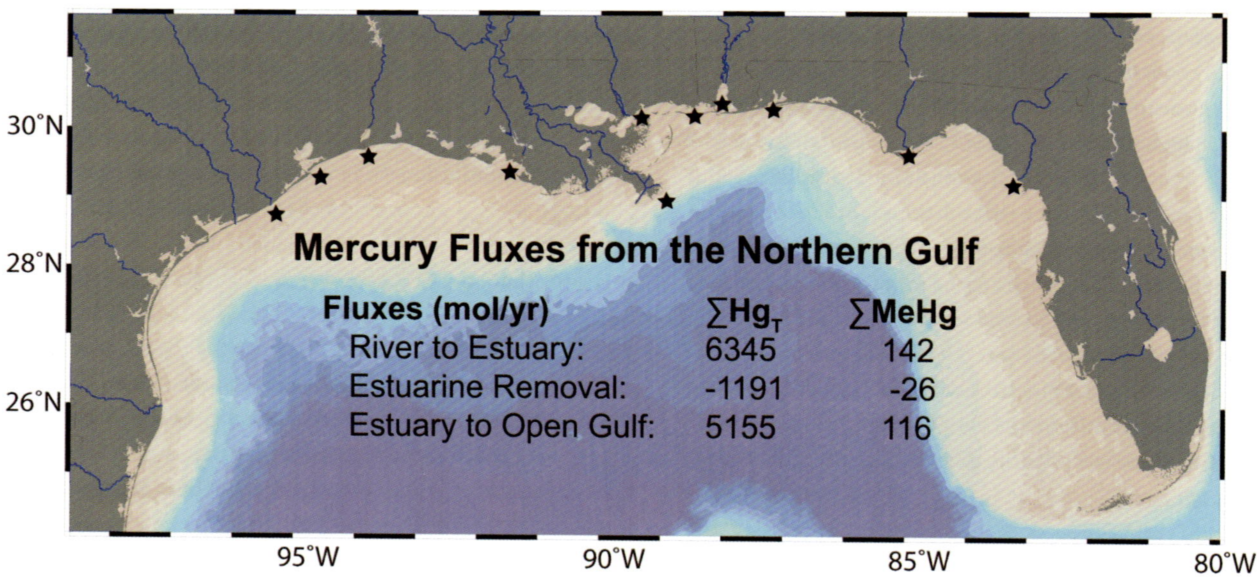

Figure 4.6. Source fluxes of total Hg (ΣHgT, including all Hg species in the dissolved, <0.4 μm phase) and total methylmercury (ΣMeHg) based on measurements of Hg across salinity gradients in 11 rivers/estuaries (*starred on the map*), including the 10 largest, draining into the northern Gulf of Mexico. The lower flux between estuaries into the open Gulf rather than between river water into estuaries is interpreted as Hg removal in estuaries, or "estuarine removal" (figure modified from Buck et al. 2015).

This reducing sediment can remobilize many contaminant metals.

Most work on trace metals in GOM sediments has been done in estuaries (Wen et al. 1999). As mentioned in the bioactive element section, an element with a predominantly natural source, such as Al, is used to normalize other metal concentrations, and elevations above crustal metal-to-aluminum ratios are interpreted as signs of anthropogenic input. The comprehensive survey of American estuaries around the GOM by Summers et al. (1996) is a classic example. Since publication of the Wen et al. (1999) review, there have been a number of similar studies in other Gulf areas as well as some historical reconstructions of anthropogenic input from sediment cores.

In the southwestern GOM, Ni enrichment detected in sediments was due to natural oil seeps on the Veracruz coast (Vazquez et al. 1991). From a sediment record on the Veracruz shelf, it was shown that Pb contamination has increased over the past 50 years, potentially sourced from Veracruz City (Celis-Hernandez et al. 2017). In the southeast, on the Campeche shelf, one study also found apparent Ni contamination in sediments (Vazquez and Sharma 2004) potentially related to oil extraction there, while another broader surface-sediment survey of the area found both Ni and V to be within expected natural levels (Macías-Zamora et al. 1999).

Moving from Campeche counterclockwise around the Gulf, a study in Florida Bay found some localized Ni enrichment near a boat marina, but otherwise overall trace-metal concentrations were low, and the heavy metals were most strongly correlated with OC content (Caccia et al. 2003). After completion of the circumference of the Gulf, enrichment of V on the order of 70% above pre-anthropogenic levels was found in recent sediments from the Louisiana shelf (Swarzenski et al. 2006). V contamination is suspected from oil sources, but this element's redox chemistry is such that it tends to precipitate or be adsorbed under low oxygen conditions (Shiller and Mao 1999). Therefore, the hypoxia of the Louisiana shelf may also contribute to sedimentary V enrichment there. Invoking yet another source, Turner et al. (2004) concluded that the V enrichment on the Louisiana shelf may be due in part (10–20% of total V source) to the metal industry (as V is used in steel alloys) rather than the oil industry. In their sediment records, V enrichment begins much earlier (~1920s) than the acceleration of Gulf oil extraction (~1960) and coincides with a large increase in industrial V production. Subsequently, it was confirmed that Louisiana sweet crude oil is relatively low in V (Joung and Shiller 2013).

Historically, in industrialized nations, contamination of coastal sediments with almost all trace metals has increased since the 1800s and accelerated between 1940 and 1970 (Valette-Silver 1993). Since the advent of environmental regulations such as the US Clean Water Act in the 1970s, trace-metal contamination has decreased. This pattern is clearly seen for Ag, Cd, and Zn contamination in well-dated sediment records from the Texas, Louisiana, and Florida coasts (Santschi et al. 2001). A similar temporal pattern is also seen for sedimentary Pb, which nearly mirrors the use and abandonment of leaded gasoline in the United States (Trefry et al. 1985; Santschi et al. 2001). Furthermore, a high-resolution coral archive revealed that high concentrations of Pb correlated well with the industrial production of Pb (1900–1940) and the use of alkyl-lead-containing gasoline in Mexico (Horta-Puga and Carriquiry 2014; Fig. 4.7), which had a distinct time history compared with the inputs to the western North Atlantic (Lazareth et al. 2000).

Trends in sedimentary Ba in the GOM are somewhat unique due to the extensive use of barite drilling fluids in oil/gas operations. On the Louisiana shelf, roughly twice the pre-anthropogenic levels of Ba are found in recent sediments (Swarzenski et al. 2006). In fact, corals from the northern GOM on the Flower Garden Banks (Deslarzes et al. 1995), the southern Gulf from the Veracruz Reef (Carriquiry and Horta-Puga 2010), and Cayo Arcas on the Campeche Bank (Cram et al. 2009) display higher Ba/calcium (Ca) ratios beginning in the 1970s and 1980s, coincident with the rise in the use of barite drilling fluid (Fig. 4.7). Whereas the percentage increase in coral Ba/Ca is not overwhelming (~10%), these data suggest that drilling activities may have increased the dissolved Ba concentrations across the entire Gulf, as corals likely take up Ba as a dissolved ion rather than as particulate barite.

Because some component of this drilling mud is obviously being dispersed from platforms, several studies have investigated other metals in adjacent sediments to detect contamination. Trocine and Trefry (1983) found that, with the exception of Ba, there were no significant anomalies in Cu, Cr, Mn, Ni, Pb, V, or Zn in sediments at a drill site in 90 m of water near the Flower Garden Banks. Similarly, whereas excess Hg was associated with spent drilling mud (Trefry et al. 2007), there was no apparent total Hg contamination in sediments in the vicinity (up to dis-

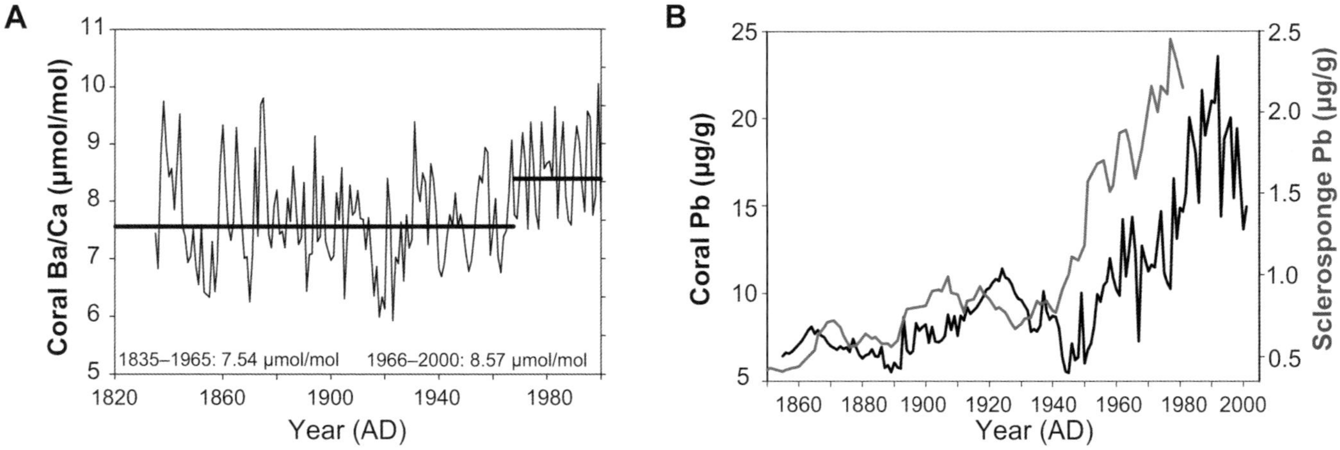

Figure 4.7. Coral records from the Veracruz Reef system in the southern Gulf of Mexico. Plot A is modified from Carriquiry and Horta-Puga (2010), and plot B from Horta-Puga and Carriquiry (2014). (*A*) The Veracruz Reef Ba/Ca record of the coral *Montastraea faveolata* is shown in black, with decadal averages noted. (*B*) The Pb concentration of the coral *Orbicella faveolata* (*black line*) from the Veracruz Reef is compared with that in the sclerosponge *Ceratoporella nicholsoni* (*gray line*) from the Bahamas (Lazareth et al. 2000).

tances of hundreds of meters) of several platforms on the Louisiana shelf (Trefry et al. 2007; Delaune et al. 2008). Delaune et al. (2008), however, did find elevated methyl Hg in sediments within 50 m of a platform and suggested that platforms may create sedimentary conditions conducive to methylation, such as increased organic matter availability.

The Deepwater Horizon oil spill (April–July 2010) also influenced the trace-metal content of the surrounding sediments. After the spill, the degradation of oil deposited on the seafloor changed the sediment redox conditions to cause authigenic enrichments in Mn, rhenium (Re), and Cd near the sediment-water interface (Hastings et al. 2016). These authors also found enrichment of Ba in the upper few centimeters of sediment cores taken proximal to the wellhead, likely brought in by the drilling muds applied to the leaking well. Furthermore, contaminant-level Ba was detected in deep (1.1 km) sediment trap samples, 7.4 km away from the wellhead, as late as January 2011, 4 months after drilling muds were last applied (Yan et al. 2016). Since barite ($BaSO_4$) is relatively dense, it should have settled very quickly after being added to the water column. The extended deposition of contaminant Ba was attributed to a combination of factors, including very fine (<5 µm) barite mud particles that persisted in a suspended phase; sorption of Ba onto components of less dense, marine snow particles; and incorporation of Ba into biogenic calcium carbonate tests and shells (Yan et al. 2016).

Contaminant Metals in the Water Column

Echoing other discussions in this chapter, there have been very few analyses of dissolved contaminant metals, particularly in the deep Gulf. As done for sediments, most work to date has been in estuaries. This includes a recent report of stable cesium (Cs) contamination in coastal Mississippi, likely sourced from a titanium refinery on St. Louis Bay (Bera et al. 2015), and Cu/Ni/Pb contamination in a lagoon adjacent to a power plant in coastal Veracruz, Mexico (Garduno-Ruiz et al. 2016). Where data in the deep Gulf do exist, they usually reflect a highly visible environmental concern, such as an oil spill or drilling rigs.

After the Deepwater Horizon oil spill, an attempt called "top kill" was made to plug the leaking well with massive amounts (on the order of 300,000 barrels) of barite drilling fluid (Graham et al. 2011). A few days after top kill, water sampling near the well site confirmed elevated dissolved Ba concentrations at the depth of the well (~1100 m) of up to 100 nmol/kg (Joung and Shiller 2013) over a natural background of about 60 nmol/kg (Buerkert 1999). Joung and Shiller (2013) also reported elevated dissolved Co in the same samples with enriched Ba and determined this Co was likely sourced from the crude oil itself (see Fig. 4.4). Other metals (Fe, Mn, Ni, Cu, V, Mo, Cr), as mentioned previously, did not appear significantly influenced by the spill because their concentrations were low in the leaking crude oil.

Another concern for water contamination is the annual discharging of 50–100 million cubic meters of produced water from drilling platforms within the Gulf (Neff et al. 2011). This "produced water" is extracted as a by-product in oil or gas mining, which can be highly elevated in heavy metals (Ag, Ba, Cd, Cu, Hg, Pb, V, Zn) by factors of 10 to 10,000 above levels in the ambient seawater (Trefry et al. 1996). At least at the Texas-Louisiana shelf sites studied by Trefry et al. (1996), it was determined that these metals were diluted by a factor of 1000 within 20–30 m of the well. This likely rendered the additions of the most toxic metals, Hg, Cd, and Pb, insignificant to the surrounding waters. Ba, Fe, and Mn, however, were enriched in the produced water at factors greater than 10,000; therefore, produced water may perturb the distribution of these metals in the Gulf as well. This is another area where future surveys of the large-scale distribution of contaminant metals in the Gulf are highly justifiable.

Geochemical Tracers

Barium: Coastal Inputs

As mentioned previously, Ba levels in the Gulf have been locally, and potentially more broadly, contaminated by industrial activities. However, natural marine Ba distributions are informative for a number of oceanographic processes. In wide-ranging geochemical applications in the modern ocean and its geologic past, as outlined by Joung and Shiller (2014), Ba has been used as a tracer for paleoproductivity when deposited in the sediments as biogenic barite (e.g., Dymond et al. 1992); a tracer of deepwater masses as recorded in the Ba/Ca ratio of benthic foraminifera (e.g., Lea and Boyle 1990; Hall and Chan 2004); or as a tracer of coastal, freshwater input, both in seawater itself or as recorded by planktonic foraminifera (e.g., Guay and Falkner 1997; Schmidt and Lynch-Stieglitz 2011). Surprisingly, data published by Beurkert (1999) indicates that Ba may have the best-resolved distribution of all elements in the deep Gulf (see location of profiles in Fig. 4.1).

Buerkert (1999) found that in general, the depth profile of dissolved Ba in the deep Gulf is nutrient-like, as found in the open ocean, having correlations with both dissolved silica and alkalinity (Chan et al. 1977). Below 1500 m, all across the southern GOM, dissolved Ba is nearly invariant around 62 nmol/kg (Fig. 4.8). This was attributed to the relatively rapid (~200 yr) flushing of the deep water in the Gulf by North Atlantic Deep Water, as modified on its passage through the Caribbean, similar to what was observed in the relatively homogeneous deepwater concentrations of Cd, Cu, and Ni at the single deep site studied by Boyle et al. (1984). We stress, however, this apparent homogeneity due to the flushing of deep water is likely to apply only to metals with long residence times (>500 yr) with respect to removal processes such as scavenging. Shorter-lived metals, like Fe or Co, may likely exhibit much more intrabasin variability.

In the upper 500 m of the southern Gulf sites, Buerkert (1999) reported surface enrichments of Ba, unlike a typical nutrient, that could be traced using salinity indications to fluvial input in addition to Ba-rich groundwater discharge from the Yucatan Platform (Fig. 4.8). Shaw et al. (1998) demonstrated that Ba is enriched in the coastal ocean by submarine groundwater discharge off the southeastern United States. The association of barium with submarine groundwater discharge in the Yucatan area was later confirmed by a detailed process study on the Yucatan Peninsula and coastal waters (Gonneea et al. 2014). Submarine groundwater discharge (Burnett et al. 2006) is also a well-known source of radium to the ocean (Moore 1996; Moore et al. 2008) and is being increasingly seen as an important source of other metals to the coastal ocean (Charette et al. 2016; see also Chapter 5 as this relates to radionuclides).

In general, it is thought that the enrichment of Ba in freshwater sources is due to the desorption of Ba from suspended sediments during estuarine mixing, as demonstrated in the lower Mississippi River (Hanor and Chan 1977). A comprehensive study from the Louisiana shelf demonstrates a more complex picture (Joung and Shiller 2014), however, finding changes in desorption and sediment input during different seasons or different river flow stages. Furthermore, in the deep GOM in particular, the Ba cycle may be significantly impacted by the so-called mud volcano barite mounds found near methane seeps (Castellini et al. 2006), as well as by benthic anoxic brines (Schijf 2007). Thus, while it is tempting to interpret paleorecords of increased Ba/Ca in planktic foraminifera in GOM surface waters at the end of the last ice age as evidence of massive postglacial flooding from the Mississippi River (Schmidt and Lynch-Stieglitz 2011), it has been rightly suggested that Ba is more reliably used in conjunction with other proxies of freshwater input (Joung and Shiller 2014).

Rare Earth Elements and Neodymium Isotopes: Circulation and Metal Sources

Rare earth element (REE) concentrations of GOM waters have recently been reported in the peer-reviewed litera-

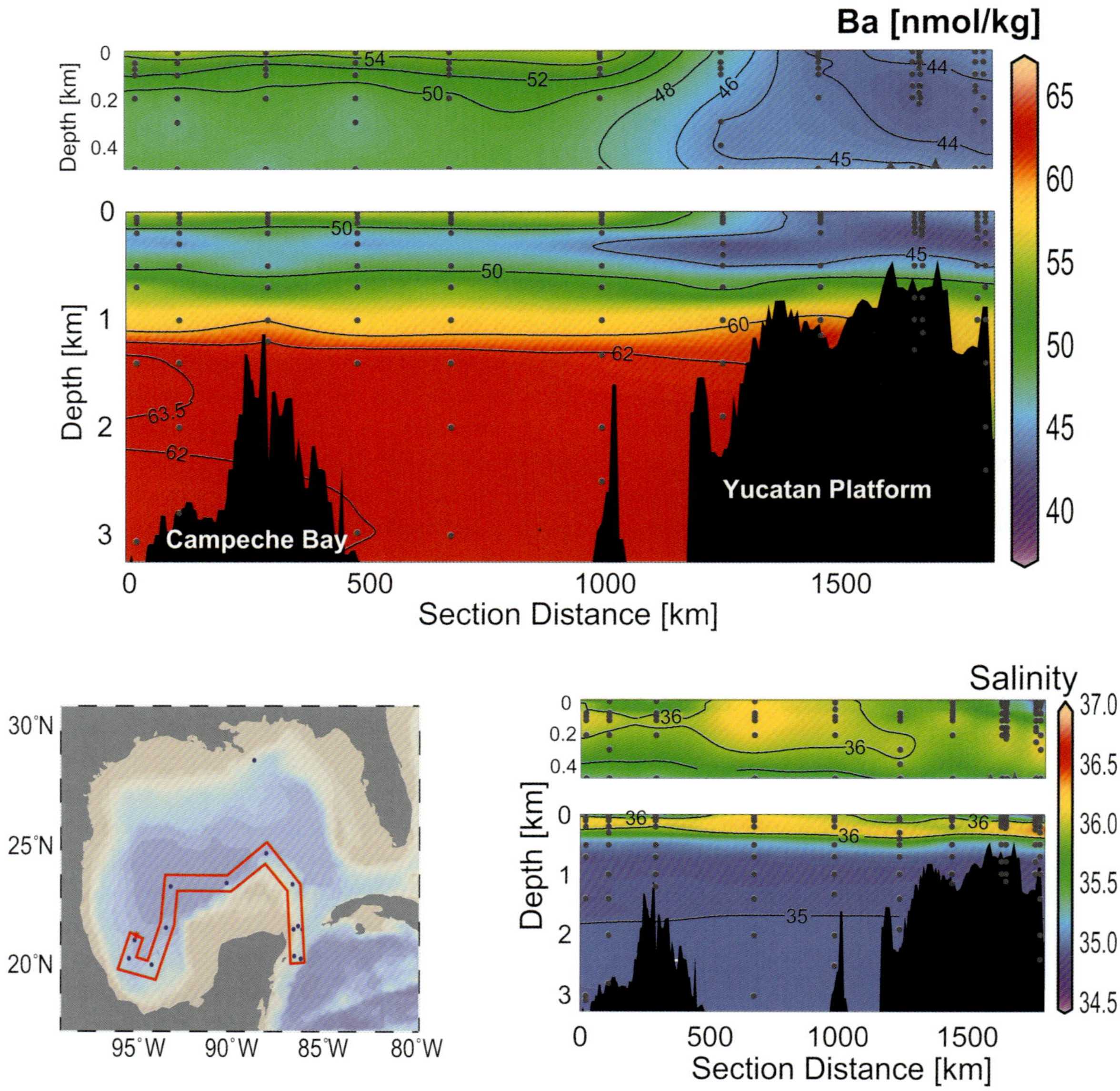

Figure 4.8. Dissolved (<0.45 µm) barium concentrations (*top*) and salinity distribution (*bottom right*) across the southern Gulf of Mexico (Buerkert 1999); transect indicated in the inset map. Note each section plots the full water column below and an enlarged section of the upper 500 m above (figure was created using ODV).

ture (Osborne et al. 2014; Osborne et al. 2015; Shiller et al. 2017). Shiller et al. (2017) found that the light REEs (lanthium [La], cerium [Ce], praseodymium [Pr], and neodymium [Nd]) were depleted from the dissolved phase, unlike the heavy REEs (e.g., holmium [Ho], erbium [Er], thulium [Tm], and ytterbium [Yb]), in a deep plume of methane-rich water during the Deepwater Horizon blowout, indicating biological uptake by methanotrophic bacteria. Significantly, this was one of the first reported cases of biological activity directly influencing REE distributions in the ocean.

Osborne et al. (2014) analyzed the Nd isotopic composition (ε_{Nd}) at a few GOM sites: the Yucatan Channel (Station 166-1), the Mississippi River outflow (Station 182-1), the slope of the west Florida shelf (Station 194-13), and the Florida Straits (Station 200-2). See Figure 4.9 for the location of these stations (also see Nd locations in Fig. 4.1). The epsilon notation represents parts per 10,000 deviation from a standard $^{143}Nd/^{144}Nd$ ratio. ^{143}Nd is produced by radioactive samarium (Sm) decay, whereas ^{144}Nd is primordial. Thus, higher ε_{Nd} values are referred to as more radiogenic; and lower values, unradiogenic.

Note that Osborne et al. (2014) used a more recent standard Nd ratio (JNdi-1), but here, in order to compare to literature data, we used the Chondritic Uniform Reservoir ($^{143}Nd/^{144}Nd = 0.512638$; Jacobsen and Wasserburg 1980).

The Osborne et al. (2014) and Osborne et al. (2015) stations were well placed because they sampled the downstream progression of the Caribbean Current entering the Gulf as it flows through the Yucatan Channel to the Loop Current, carrying water clockwise around the eastern Gulf, to the exit of water through the Florida Straits into the Atlantic, forming the Florida Loop Current (Fig. 4.9). There was a progression of seawater ε_{Nd} values (given as the mean and standard deviation of samples taken throughout the water column) from the Yucatan Channel (-9.55 ± 1.30), to the more radiogenic west Florida shelf (-8.92 ± 0.89) and Florida Straits (-8.85 ± 1.81) stations (Osborne et al. 2014). At these sites and Station 182-1 proximal to the Mississippi River outflow, the REE concentrations were generally highest in the surface waters (0–11 m), had subsurface minima between 100 and 200 m depth, and increased with depth toward the seafloor below 200 m (Osborne et al. 2015). The increase of REE concentration with depth is also reported for the open ocean (e.g., Elderfield and Greaves 1982; Elderfield 1988; Zhang and Nozaki 1996). The chief exception to this trend is Ce, which decreased with depth throughout the water column (Osborne et al. 2015), presumably due to scavenging of Ce(IV).

Osborne et al. (2014) focus on the ε_{Nd} data of Caribbean seawater in terms of the potential water masses and

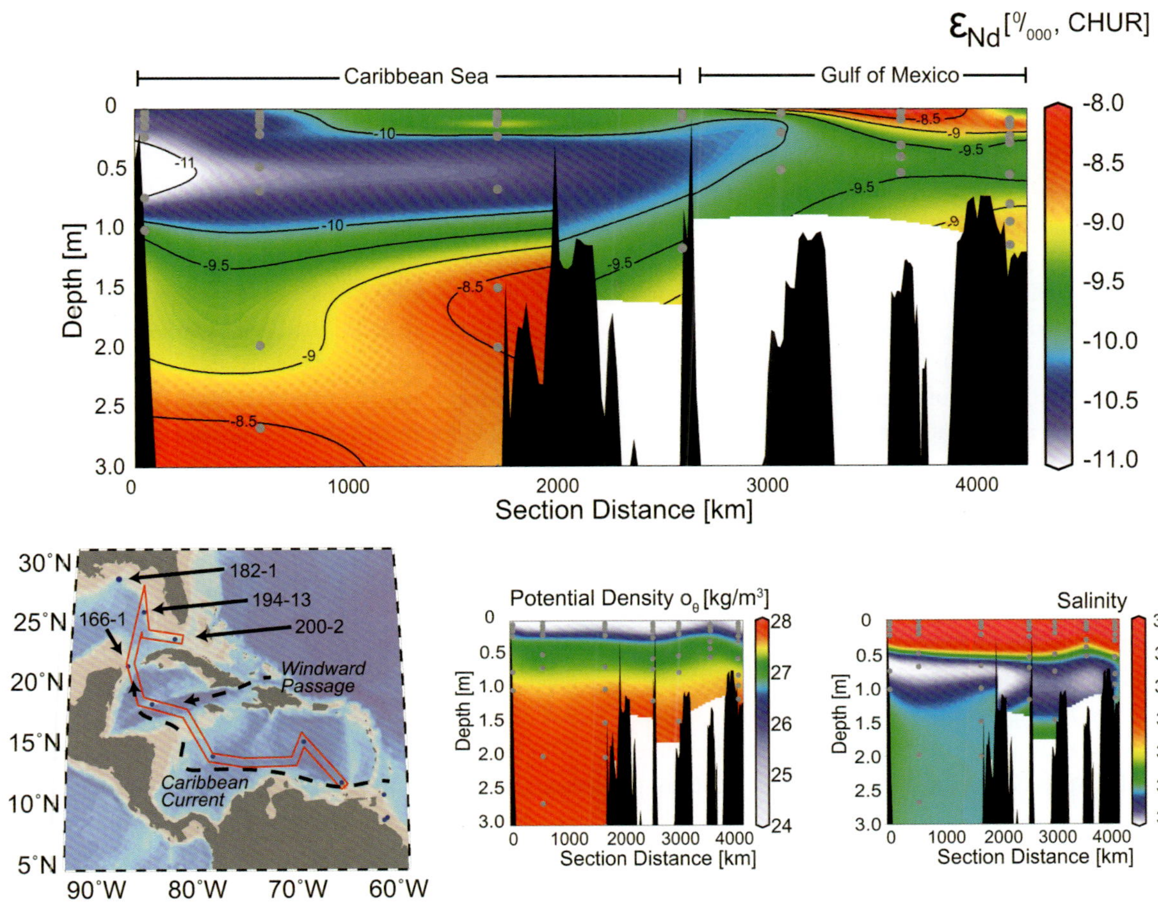

Figure 4.9. The isotopic distribution of dissolved Nd in a section spanning the Caribbean Sea and the Gulf of Mexico from the work of Osborne et al. (2014) and Osborne et al. (2015). Station numbers mentioned in the text are indicated on the inset map. Nd isotopic ratios are expressed in epsilon notation, parts per 10,000 deviation from the reference 14^{4}Nd/14^{4}Nd ratio in the Chondritic Uniform Reservoir (CHUR; Jacobsen and Wasserburg, 1980). Inset sections show the potential density and salinity distributions, roughly showing inflow of Atlantic water masses over sills through the Caribbean Current and Windward Passage (*shown in dashed arrows*). These flows combine to form the Loop Current, roughly in the shape of the section outlined in red through the Gulf, which after passing through the Florida Straits becomes the Gulf Stream (figure created using ODV).

general circulation within the Caribbean basin, demonstrating that the Amazon and Orinoco Rivers influence the ε_{Nd} signature of surface waters carried by the Caribbean Current and that waters below the thermocline likely formed in the North Atlantic (more unradiogenic ε_{Nd}). These interpretations are further supported by the elevated REE concentrations in the surface waters of both the Caribbean Sea and the GOM, although additional sources of REEs to the Gulf include the Mississippi River and Saharan dust (Osborne et al. 2015). For example, the Caribbean Sea and GOM water samples exhibit broad enrichments in the medium-weight REEs (e.g., Sm, europium [Eu], gadolinium [Gd], and terbium [Tb]) when normalized to open-ocean water from the Atlantic Ocean (i.e., 2000 m sample from the Bermuda Atlantic Time-series station; van der Fliert et al. 2012), which are greatest near the Orinoco River. Moreover, waters from the GOM station closest to the Mississippi River outlet are similarly enriched in the heavy REEs compared to waters from the Yucatan Channel and Caribbean Sea. Specifically, shale-normalized Yb/Nd ratios of the Mississippi River (Shiller 2002) are similar to those from the Osborne et al. stations at the Mississippi River outflow, the west Florida shelf, and the Florida Straits. This suggests that the Mississippi River outflow modifies the REE concentrations and fractionation patterns of GOM waters, particularly in surface waters (Osborne et al. 2015).

In contrast, the ε_{Nd} data of intermediate-depth waters, along with density distributions, at the west Florida shelf and Florida Straits sites indicated that this water formed within the subtropical gyre as North Atlantic subtropical water or "18-degree water" (σ_Θ = 26.5 kg/m^3) and flowed in though the Windward Passage (see Fig. 4.9). In deep waters, below 1500 m, sampled only in the Caribbean, ε_{Nd} was significantly more radiogenic (-8.3 to -9.2) than inflowing Atlantic water (-13). Osborne et al. (2014) suggest this could be explained by a release of a very radiogenic Nd (+4.2) from Caribbean sediments that does not significantly enrich Nd concentrations, a process referred to as boundary exchange (Lacan and Jeandel 2005; Siddall et al. 2008). This process could also be active in the GOM through documented near-bottom nepheloid layers (e.g., Baskaran et al. 1996).

Overall, the Osborne et al. (2014) and Osborne et al. (2015) studies only sparsely sampled the GOM, and therefore tracing of circulation and metal supply within the Gulf could be better constrained with further Nd isotope characterization. Nevertheless, it is tempting to return to the mean ε_{Nd} values for the Yucatan Channel (-9.6), the west Florida shelf (-8.9), and Florida Straits (-8.9) as a first estimate of the controlling processes. Starting from the Yucatan Channel waters, it is likely there is a radiogenic source of Nd to the Gulf that raises the epsilon values. The single measurement of ε_{Nd} of the dissolved phase of the Mississippi River water (-8.55; Goldstein and Jacobsen 1988) indicates that the Mississippi River likely exerts important controls on the Nd isotope composition of GOM waters. The more unradiogenic ε_{Nd} composition of bulk suspended matter (-10.02; Goldstein and Jacobsen 1988) and clay and silt (-10.8 and -12.3, respectively; Bayon et al. 2015) from the Mississippi River suggests that dissolution of river-derived suspended particulates may contribute less to the Nd isotope ratios of GOM waters than the dissolved phases (see, however, Jeandel and Oelkers (2015), for instances in many other locations where the opposite case holds).

The chemistry of Mississippi River waters differs from that of most large rivers and global average river water in terms of its high pH (7.8) and alkalinity (2 meq/L; Shiller 2002). Consequently, a greater fraction of each REE is strongly complexed with carbonate ions and/or natural organic ligands in Mississippi River water than in global average river water. This could conceivably facilitate the transfer of dissolved REEs from the river to the Gulf and hence explain, in part, the similarity between the ε_{Nd} values of GOM waters and the dissolved phase of the Mississippi River (Goldstein and Jacobsen 1987; Osborne et al. 2014; Shiller 2002; Tang and Johannesson 2003). Sedimentary records from the deep Gulf could be investigated to test the idea that the extent to which the Mississippi River supplies REEs (and metal supply more generally) could vary over climatic or geologic timescales.

Germanium: Reverse Weathering Tracer

Germanium (Ge) is a metalloid that has a geochemical cycle intimately linked with that of Si (see Chapter 3 for GOM Si distribution). These 2 elements are added to the ocean from similar sources, including rivers, hydrothermal vents, and aerosol dust, and they are both removed by incorporation into biogenic opal (SiO_2), Ge substituting for Si as a trace constituent in this amorphous mineral, made by diatoms, radiolaria, and siliceous sponges (Froelich et al. 1985). Since glacial-interglacial variations in sedimentary Ge/Si ratios were first observed (Froelich et al. 1992; Mortlock et al. 1991), there has been controversy on how to interpret this variation: that is, what processes control the cycling of Ge relative to Si in modern seawater? Some have argued that seawater Ge/Si ratios respond to changes in the weathering supply of Ge over Si from the continents, while it is also possible there were changes in the removal of Ge relative to Si into the sediments (e.g., Lin and Chen 2002; Hammond et al. 2004). Motivated by this problem, Baronas et al. (2016) chose the northern GOM to perform a critical Ge/Si study, since this area is influenced by a large flux of continental weathering products from the Mississippi River and is broadly representative of continental margins with high OC accumulation rates (Allison et al. 2007), where much of the Ge removal likely occurs. Baronas et al. (2016) analyzed sediments and water-column profiles for Ge (and Si) concentrations at several sites across the Louisiana shelf and 1 proximal deep site (2000 m), indicated in Figure 4.1.

Because of the similarity in sources and sinks, dissolved Ge and Si depth profiles generally resemble each other quite closely. Baronas et al. (2016) confirmed this in their northern GOM stations, but interestingly they found that the dissolved Ge/Si ratio in the Gulf (1.9–25 μmol/mol) was significantly elevated over the average for global seawater (0.7 μmol/mol). Providing another example of a contaminant metal, the authors attribute this enrichment to Ge contamination from industrial activities along the Mississippi River, as Ge is easily leached from coal and fly ash waste ponds along the river (Froelich et al. 1985; Froelich and Lesley 2001). Additionally, it was found there was some discrimination of Ge relative to Si in biogenic opal formation. That is, radiolarians and sponges (and to a lesser extent, diatoms) incorporate Ge and Si at a lower ratio than found in the ambient seawater, enriching the seawater with Ge (Baronas et al. 2016).

After sinking and being deposited on the seafloor, much of the biogenic opal dissolves within the upper centimeters of the sediments, especially in the shallow warm waters, corrosive to biogenic opal, of the Louisiana shelf (Tréguer and De la Rocha 2013). In terms of removal processes, Baronas et al. (2016) found that the majority of dissolved Ge released from opal dissolution was removed by incorporation into authigenic clay minerals, a process called reverse weathering. This process was relatively insensitive to the redox state of the sediments and is likely controlled by the available supply of dissolved Si (whether delivered by biogenic opal or other lithogenic minerals). In sum, by putting the northern GOM example into context with the global ocean, Baronas et al. (2016) concluded that it is variability in the reverse weathering removal of Ge that most likely controls the variability in seawater Ge/Si ratios. Thus, sedimentary records of Ge/Si over the Late Pleistocene more likely reflect changes in Ge removal than in weathering supply.

Conclusion

In this chapter, we discuss how the range of chemical properties exhibited by trace metals can be utilized to learn about important oceanographic processes in the GOM, including currents and eddy transport, sediment sources and burial, and biological productivity. While research over the past 15 years has greatly increased our knowledge of their chemical and physical speciation of trace metals and their cycling in coastal settings, there is still a great deal to learn about how bioactive metals are distributed in, added to, and removed from the oceanic GOM. The high ratio of margin to open ocean in this basin means that most metals will be influenced by margin sources, the Mississippi River, nepheloid layer transport, and submarine groundwater discharge from the Yucatan Platform being prime candidates. Before the full influence of these sources can be assessed, however, a proper determination of the deepwater distribution of metals such as Fe, Zn, and Cu needs to be pursued. For contaminant metals, we are reminded that the Gulf is one of the most peopled and heavily trafficked bodies of water in the world, leading to clear contamination of Ba, Hg, Pb, and Ge throughout the Gulf and many other metals close to point sources. The petroleum industry, in particular, will likely be a continuing source of contaminant metals to the Gulf, and monitoring to assess its ecological impacts should be maintained. Finally, for geochemical tracers, the chemical properties of Ba, REEs, and Ge will prove to be useful in future large-scale surveys of the bioactive metals in the Gulf, especially in their utility in distinguishing provenance of metals sources (Ba, ε_{Nd}) or metals sinks (Ge). The combined study of a spectrum of metals with different chemical properties is always more fruitful than isolated pursuits, and this is especially true in the chemically dynamic GOM.

Acknowledgments

Christopher T. Hayes acknowledges funding from the University of Southern Mississippi; thanks J. Rosen for editorial assistance; and thanks A. M. Shiller, W. M. Landing, W. S. Moore, and T. S. Bianchi for helping the lead author delve into the Gulf of Mexico literature. Peter H. Santschi acknowledges funding from the National Science Foundation (OCE#0851191). Karen H. Johannesson thanks

Matthilda and Michael Cochran for endowing the Cochran Family Professorship, which helped support her research.

References

Allison, M. A., T. S. Bianchi, B. A. McKee, and T. P. Sampere. 2007. Carbon burial on river-dominated continental shelves: Impact of historical changes in sediment loading adjacent to the Mississippi River. *Geophysical Research Letters* 34:L01606.

Andreesen, J. R., and K. Makdessi. 2008. Tungsten, the surprisingly positively acting heavymetal element for prokaryotes. *Annals of the New York Academy of Sciences* 1125:215–229.

Atkeson, T. D., C. D. Pollman, and D. M. Axelrad. 2006. Recent trends in mercury emissions, deposition, and biota in the Florida Everglades: A monitoring and modelling analysis. Chap. 26 in N. Pirrone and K. R. Mahaffey (eds.), *Dynamics of Mercury Pollution on Regional and Global Scales*. New York: Springer.

Baronas, J. J., D. E. Hammond, W. M. Berelson, J. McManus, and S. Severmann. 2016. Germanium-silicon fractionation in a river-influenced continental margin: The northern Gulf of Mexico. *Geochimica et Cosmochimica Acta* 178:124–142.

Baskaran, M., P. H. Santschi, G. Benoit, and B. D. Honeyman. 1992. Scavenging of thorium isotopes by colloids in seawater of the Gulf of Mexico. *Geochimica et Cosmochimica Acta* 56:3375–3388.

Baskaran, M., P. H. Santschi, L. Guo, T. S. Bianchi, and C. Lambert. 1996. ^{234}Th:^{238}U disequilibria in the Gulf of Mexico: Importance of organic matter and particle concentration. *Continental Shelf Research* 16:353–380.

Bayon, G., S. Toucanne, C. Skonieczny, L. Andre, S. Bermell, S. Cheron, B. Dennielou, et al. 2015. Rare earth elements and neodymium isotopes in world river sediments revisited. *Geochimicia et Cosmochimica Acta* 170:17–38.

Benoit, G., S. D. Oktay-Marshall, A. Cantu, E. M. Hood, C. H. Coleman, M. O. Corapcioglu, and P. H. Santschi. 1994. Partitioning of Cu, Pb, Ag, Zn, Fe, Al, and Mn between filter-retained particles, colloids, and solution in six Texas estuaries. *Marine Chemistry* 45:307–336.

Bera, G., K. M. Yeager, M. Shim, and A. M. Shiller. 2015. Anthropogenic stable cesium in water and sediment of a shallow estuary, St. Louis Bay, Mississippi. *Estuarine, Coastal and Shelf Science* 157:32–41.

Bergamaschi, B. A., D. P. Krabbenhoft, G. R. Aiken, E. Patino, D. G. Rumbold, and W. H. Orem. 2012. Tidally driven export of dissolved organic carbon, total mercury, and methylmercury from a mangrove-dominated estuary. *Environmental Science & Technology* 46:1371–1378.

Bloom, N. S., G. A. Gill, S. Cappellino, C. Dobbs, L. McShea, C. Driscoll, R. Mason, and J. Rudd. 1999. Speciation and cycling of mercury in Lavaca Bay, Texas, sediments. *Environmental Science & Technology* 33:7–13.

Boyle, E. A., D. F. Reid, S. S. Huested, and J. Hering. 1984. Trace metals and radium in the Gulf of Mexico: An evaluation of river and continental shelf sources. *Earth and Planetary Science Letters* 69:69–87.

Buck, C. S., C. R. Hammerschmidt, K. L. Bowman, G. A. Gill, and W. M. Landing. 2015. Flux of total mercury and methylmercury to the northern Gulf of Mexico from U.S. estuaries. *Environmental Science & Technology* 49:13992–13999.

Buerkert, T. P. 1999. Barium in water and foraminiferal shells: Indicator of oceanographic conditions in the Gulf of Mexico since the Late Pleistocene. PhD diss., Louisiana State University. 200 pp.

Burnett, W. C., H. Dulaiova, C. Stringer, and R. Peterson. 2006. Submarine groundwater discharge: Its measurement and influence on the coastal zone. *Journal of Coastal Research* 39:35–38.

Caccia, V. G., F. J. Millero, and A. Palanques. 2003. The distribution of trace metals in Florida Bay sediments. *Marine Pollution Bulletin* 46:1420–1433.

Cai, Y. J., J. R. Rooker, G. A. Gill, and J. P. Turner. 2007. Bioaccumulation of mercury in pelagic fishes from the northern Gulf of Mexico. *Canadian Journal of Fisheries and Aquatic Sciences* 64:458–469.

Carriquiry, J. D., and G. Horta-Puga. 2010. The Ba/Ca record of corals from the southern Gulf of Mexico: Contributions from land-use changes, fluvial discharge and oil-drilling muds. *Marine Pollution Bulletin* 60:1625–1630.

Castellini, D. G., G. R. Dickens, G. T. Snyder, and C. D. Ruppel. 2006. Barium cycling in shallow sediment above active mud volcanoes in the Gulf of Mexico. *Chemical Geology* 226:1–30.

Celis-Hernandez, O., L. Rosalez-Hoz, A. B. Cundy, and A. Carranza-Edwards. 2017. Sedimentary heavy metal(loid) contamination in the Veracruz shelf, Gulf of Mexico: A baseline survey from a rapidly developing tropical coast. *Marine Pollution Bulletin* 119:204–213.

Chan, L. H., D. Drummond, J. M. Edmond, and B. Grant. 1977. On the barium data from the Atlantic GEOSECS Expedition. *Deep Sea Research* 24:613–619.

Charette, M. A., P. J. Lam, M. C. Lohan, E. Y. Kwon, V. Hatje, C. Jeandel, A. M. Shiller, et al. 2016. Coastal ocean and shelf-sea biogeochemical cycling of trace elements and isotopes: Lessons learned from GEOTRACES. *Philosophical Transactions of the Royal Society A* 374:20160076.

Choe, K., and G. A. Gill. 2001. Isolation of colloidal monomethyl mercury in natural waters using cross-flow ultrafiltration techniques. *Marine Chemistry* 76:305–318.

Collier, R. W. 1985. Molybdenum in the northeast Pacific Ocean. *Limnology and Oceanography* 30:1351–1354.

Cram, S., C. A. Ponce de León, I. Sommer, S. Miceli, P. Fernandez, H. Rivas, and L. Galicia. 2009. Metal distribution in coral reef complex Cayo Arcas in the Gulf of Mexico. *Environmental Monitoring and Assessment* 151:413–424.

Cui, M., and K. H. Johannesson. 2017. Comparison of tungstate and tetrathiotungstate adsorption onto pyrite. *Chemical Geology* 464:57–68.

Dahl, T. W., A. Chappaz, J. P. Fitts, and T. W. Lyons. 2013. Molybdenum reduction in a sulfidic lake: Evidence from X-ray absorption fine-structure spectroscopy and implications for the Mo paleoproxy. *Geochimica et Cosmochimica Acta* 103:213–231.

De Simone, F., C. N. Gencarelli, L. M. Hedgecock, and N. Pirrone. 2014. Global atmospheric cycle of mercury: A model study on the impact of oxidation mechanisms. *Environmental Science and Pollution Research* 21:4110–4123.

Delaune, R. D., I. Devai, A. Hou, and A. Jugsujinda. 2008. Total and methyl Hg in sediment adjacent to offshore platforms of the Gulf of Mexico. *Soil and Sediment Contamination* 17:98–106.

Delaune, R. D., R. P. Gambrell, I. Devai, A. Jugsujinda, and M. Kongchum. 2009. Total Hg and methyl Hg distribution in sediments of selected Louisiana water bodies. *Journal of Environmental Science and Health, Part A* 44:557–567.

Delaune, R. D., R. P. Gambrell, A. Jugsujinda, I. Devai, and A. Hou. 2008. Total Hg, methyl Hg and other toxic heavy metals in a northern Gulf of Mexico estuary: Louisiana Pontchartrain basin. *Journal of Environmental Science and Health, Part A* 43:1006–1015.

Delaune, R. D., A. Judsujinda, I. Devai, and W. H. Patrick Jr. 2004. Relationship of sediment redox conditions to methyl mercury in surface sediment of Louisiana lakes. *Journal of Environmental Science and Health, Part A* 39:1925–1933.

Deslarzes, K. J. P., P. N. Boothe, B. J. Presley, and G. L. Steinmetz. 1995. Historical incorporation of barium in *Montastrea annularis* at the Flower Garden Banks, north-west Gulf of Mexico. *Marine Pollution Bulletin* 30:1–123.

Doucet, F., J. R. Lead, and P. H. Santschi. 2006. Colloid-trace element interactions in aquatic systems. Chap. 3 in K. J. Wilkinson and J. R. Lead (eds.), *Environmental Colloids and Particles*, vol. 10. Chichester, UK: John Wiley and Sons.

Dymond, J., E. Suess, and M. Lyle. 1992. Barium in deep-sea sediment: A geochemical proxy for paleoproductivity. *Paleoceanography* 7:163–181.

Elderfield, H. 1988. The oceanic chemistry of the rare-earth elements. *Philosophical Transactions of the Royal Society A* 325:105–126.

Elderfield, H., and M. J. Greaves. 1982. The rare earth elements in seawater. *Nature* 296:214–219.

Erickson, B. E., and G. R. Helz. 2000. Molybdenum (VI) speciation in sulfidic waters: Stability and lability of thiomolybdates. *Geochimica et Cosmochimica Acta* 64:1149–1158.

Erickson, H. P. 2009. Size and shape of protein molecules at the nanometer level determined by sedimentation, gel filtration, and electron microscopy. *Biological Procedures Online* 11:32–51.

Everett, D. H. 1972. Manual of symbols and terminology for physicochemical quantities and units. Appendix II: Definitions, terminology and symbols in colloid and surface chemistry. *Pure and Applied Chemistry* 31:577–638.

Firdaus, M. L., K. Norisuye, Y. Nakagawa, S. Nakatuska, and Y. Sohrin. 2008. Dissolved and labile particulate Zr, Hf, Nb, Ta, Mo and W in the western North Pacific Ocean. *Journal of Oceanography* 64:247–257.

Fitzgerald, W. F., and C. H. Lamborg. 2003. Geochemistry of mercury in the environment. Chap. 9.04 in H. D. Holland and K. K. Turekian (eds.), *Treatise on Geochemistry*. Oxford: Elsevier-Pergamon.

Frederick, P. C., B. Hylton, J. A. Heath, and M. G. Spalding. 2004. A historical record of mercury contamination in southern Florida (USA) as inferred from avian feather tissue. *Environmental Toxicology and Chemistry* 32:1474–1478.

Froelich, P. N., V. Blanc, R. A. Mortlock, S. N. Chillirud, W. Dunstan, A. Udomkit, and T.-H. Peng. 1992. River fluxes of dissolved silica to the ocean were higher during glacials: Ge/Si in diatoms, rivers and oceans. *Paleoceanography* 7:739–767.

Froelich, P. N., G. A. Hambrick, M. O. Andreae, R. A. Mortlock, and J. M. Edmond. 1985. The geochemistry of inorganic germanium in natural waters. *Journal of Geophysical Research* 90:1133–1141.

Froelich, P. N., and P. Lesley. 2001. Tracing germanium contamination from coal-fired powerplants down the Chattahoochee-Apalachicola River: Implications for the toxic metalloids arsenic and selenium. Pp. 488–491 in K. J. Hatcher (ed.), *Proceedings of the 2001 Georgia Water Resources Conference*. Athens: University of Georgia.

Garduno-Ruiz, E. P., L. Rosales-Hoz, and A. Carranze-Edwards. 2016. Dissolved copper, nickel and lead in Tampamachoco Lagoon and Tuxpan River Estuary in the SW Gulf of Mexico. *Bulletin of Environmental Contamination and Toxicology* 97:490–496.

Gill, G. A., N. S. Bloom, S. Cappellino, C. T. Driscoll, C. Dobbs, L. McShea, R. Mason, and J. W. M. Rudd. 1999. Sediment-water fluxes of mercury in Lavaca Bay, Texas. *Environmental Science & Technology* 33:663–669.

Goldstein, S. J., and S. B. Jacobsen. 1987. The Nd and Sr isotopic systematics of river-water dissolved material: Implications for the sources of Nd and Sr in seawater. *Chemical Geology* 66:245–272.

———. 1988. Nd and Sr isotopic systematics of river water suspended material: Implications for crustal evolution. *Earth and Planetary Science Letters* 87:249–265.

Gonneea, M. E., M. A. Charette, Q. Liu, J. A. Herrera-Silveira, and S. M. Morales-Ojeda. 2014. Trace element geochemistry of groundwater in a karst subterranean estuary (Yucatan Peninsula, Mexico). *Geochimica et Cosmochimica Acta* 132:31–49.

Gonzalez, P. J., M. G. Rivas, C. S. Mota, C. D. Brondin, I. Moura, and J. J. G. Moura. 2013. Periplasmic nitrate reductases and formate dehydrogenases: Biological control of the chemical properties of Mo and W for fine tuning of reactivity, substrate specificity and metabolic role. *Coordination Chemistry Reviews* 257:315–331.

Graham, B., W. K. Reilly, F. Beinecke, D. F. Boesch, T. D. Garcia, and C. A. Murray. 2011. *Deep Water: The Gulf Oil Disaster and the Future of Offshore Drilling*. National Commission on the BP Deepwater Horizon Oil Spill and Offshore Drilling, Report to the President. Washington, DC: US Government Printing Office. 381 pp.

Guay, C. K., and K. K. Falkner. 1997. Barium as a tracer of Arctic halocline and river waters. *Deep Sea Research II: Tropical Studies in Oceanography* 44:1543–1569.

Guentzel, J. L., W. M. Landing, G. A. Gill, and C. D. Pollman. 2001. Processes influencing rainfall deposition of mercury in Florida. *Environmental Science & Technology* 35:863–873.

Guentzel, J. L., R. T. Powell, W. M. Landing, and R. P. Mason. 1996. Mercury associated with colloidal material in an estuarine and an open-ocean environment. *Marine Chemistry* 55:177–188.

Guo, L., P. H. Santschi, M. Baskaran, and A. Zindler. 1995. Distribution of dissolved and particulate ^{230}Th and ^{232}Th in seawater from the Gulf of Mexico and off Cape Hatteras as measured by SIMS. *Earth and Planetary Science Letters* 133:117–128.

Guo, L., P. H. Santschi, and K. W. Warnken. 2000. Trace metal composition of colloidal material in estuarine and marine environments. *Marine Chemistry* 70:257–275.

Guo, L., K. W. Warnken, and P. H. Santschi. 2007. Retention behavior of dissolved uranium during ultrafiltration: Implications for colloidal uranium in surface waters. *Marine Chemistry* 107:156–166.

Hall, B. D., G. R. Aiken, D. P. Krabbenhoft, M. Marvin-DiPasquale, and C. M. Swarzenski. 2008. Wetlands as principal zones of methyl mercury production in southern Louisiana and the Gulf of Mexico region. *Environmental Pollution* 154:124–134.

Hall, J. M., and L. Chan. 2004. Ba/Ca in benthic foraminifera: Thermocline and middepth circulation in the North Atlantic during the last glaciation. *Paleoceanography* 19:PA4018.

Hammond, D. E., J. McManus, and W. M. Berelson. 2004. Oceanic germanium/silicon ratios: Evaluation of the potential overprint of temperature on weathering signals. *Paleoceanography* 19:PA2016.

Han, S., G. A. Gill, R. D. Lehman, and K. Y. Choe. 2006. Complexation of mercury by dissolved organic matter in surface waters of Galveston Bay, Texas. *Marine Chemistry* 98:156–166.

Han, S., R. D. Lehman, K. Y. Choe, and G. A. Gill. 2007. Chemical and physical speciation of mercury in Offatts Bayou: A seasonally anoxic bayou in Galveston Bay. *Limnology and Oceanography* 52:1380–1392.

Hanor, J. S., and L. H. Chan. 1977. Non-conservative behavior of barium during mixing of Mississippi River and Gulf of Mexico waters. *Earth and Planetary Science Letters* 37:242–250.

Harris, R., C. Pollman, D. Hutchinson, W. Landing, D. Axelrad, S. L. Morey, D. Dukhovskoy, and K. Vijayaraghavan. 2012. A screening model analysis of mercury sources, fate and bioaccumulation in the Gulf of Mexico. *Environmental Research* 119:53–63.

Harris, R., C. Pollman, W. Landing, D. Evans, D. Axelrad, D. Hutchinson, S. L. Morey, et al. 2012. Mercury in the Gulf of Mexico: Sources to receptors. *Environmental Research* 119:42–52.

Hastings, D. W., P. T. Schwing, G. R. Brooks, R. A. Larson, J. L. Morford, T. Roeder, K. A. Quinn, T. Bartlett, I. C. Romero, and D. J. Hollander. 2016. Changes in sediment redox conditions following the BP DWH blowout event. *Deep Sea Research II: Tropical Studies in Oceanography* 129:167–178.

Hayes, C. T., J. N. Fitzsimmons, E. A. Boyle, D. McGee, R. F. Anderson, R. Weisend, and P. L. Morton. 2015. Thorium isotopes tracing the iron cycle at the Hawaii Ocean Time-series Station ALOHA. *Geochimica et Cosmochimica Acta* 169:1–16.

Hayes, C. T., J. Rosen, D. McGee, and E. A. Boyle. 2017. Thorium distributions in high- and low-dust regions and the significance for iron supply. *Global Biogeochemical Cycles* 31:328–347.

Helz, G. R., E. Bura-naki, N. Mikac, and I. Ciglene. 2011. New model for molybdenum behavior in euxinic waters. *Chemical Geology* 284:323–332.

Horta-Puga, G., and J. D. Carriquiry. 2014. The last two centuries of lead pollution in the southern Gulf of Mexico recorded in the annual bands of the scleractinian coral *Orbicella faveolata*. *Bulletin of Environmental Contamination and Toxicology* 92:567–573.

Horta-Puga, G., S. Cházaro-Olvera, I. Winfield, M. Avila-Romero, and M. Moreno-Ramírez. 2013. Cadmium, copper and lead in macroalgae from the Veracruz Reef system, Gulf of Mexico: Spatial distribution and rainy season variability. *Marine Pollution Bulletin* 68:127–133.

Hu, C., F. E. Muller-Karger, and P. W. Swarzenski. 2006. Hurricanes, submarine groundwater discharge, and Florida's red tides. *Geophysical Research Letters* 33:L11601.

Irving, H., and R. J. P. Williams. 1953. The stability of transition-metal complexes. *Journal of the Chemical Society* 637:3192–3210.

Jacobsen, S. B., and G. J. Wasserburg. 1980. Sm-Nd isotopic evolution of chondrites. *Earth and Planetary Letters* 50:139–155.

Jeandel, C., and E. H. Oelkers. 2015. The influence of terrigenous particulate material dissolution on ocean chemistry and global element cycles. *Chemical Geology* 395:50–66.

Joung, D., and A. M. Shiller. 2013. Trace element distributions in the water column near the Deepwater Horizon well blowout. *Environmental Science & Technology* 47:2161–2168.

———. 2014. Dissolved barium behavior in Louisiana shelf waters affected by the Mississippi/Atchafalaya River mixing zone. *Geochimica et Cosmochima Acta* 141:303–313.

———. 2016. Temporal and spatial variations of dissolved and colloidal trace elements in Louisiana shelf waters. *Marine Chemistry* 181:25–43.

Joye, S. B., I. R. Macdonald, J. P. Montoya, and M. Peccini. 2005. Geophysical and geochemical signatures of Gulf of Mexico seafloor brines. *Biogeosciences* 2:295–309.

Kang, W.-J., J. H. Trefry, T. A. Nelsen, and H. R. Wanless. 2000. Direct atmospheric inputs versus runoff fluxes of mercury to the Lower Everglades and Florida Bay. *Environmental Science & Technology* 34:4058–4063.

Karagas, M. R., A. L. Choi, E. Oken, M. Horvat, R. Schoeny, E. Kamai, W. Cowell, P. Grandjean, and S. Korrick. 2012. Evidence on the human health effects of low-level methyl mercury exposure. *Environmental Health Perspectives* 120:799–806.

Kletzin, A., and M. W. W. Adams. 1996. Tungsten in biological systems. *Federation of European Microbiological Societies Microbiology Reviews* 18:5–63.

Kongchum, M., I. Devai, R. D. Delaune, and A. Jugsujinda. 2006. Total mercury and methyl mercury in freshwater and salt marsh soils of the Mississippi River deltaic plain. *Chemosphere* 63:1300–1303.

Lacan, F., and C. Jeandel. 2005. Neodymium isotopes as a new tool for quantifying exchange fluxes at the continent-ocean interface. *Earth and Planetary Science Letters* 232:245–257.

Lamborg, C. H., C. R. Hammerschmidt, K. L. Bowman, G. J. Swarr, K. M. Munson, D. C. Ohnemus, P. J. Lam, L.-E. Heimbürger, M. J. A. Rijkenberg, and M. A. Saito. 2014. A global ocean inventory of anthropogenic mercury based on water column measurements. *Nature* 512:65–68.

Landing, W. M., J. M. Caffrey, S. D. Nolek, K. J. Gosnell, and W. C. Parker. 2010. Atmospheric wet deposition of mercury and other trace elements in Pensacola, Florida. *Atmospheric Chemistry and Physics* 10:4867–4877.

Landing, W. M., J. J. Perry, J. L. Guentzel, G. A. Gill, and C. D. Pollman. 1995. Relationships between the atmospheric deposition of trace elements, major ions and mercury in Florida: The FAMS Project (1992–1993). *Water, Air, and Soil Pollution* 80:343–352.

Lazareth, C. E., P. Willenz, J. Navez, E. Keppens, F. Dehairs, and L. André. 2000. Sclerosponges as a new potential recorder of environmental changes: Lead in *Ceratoporella nicholsoni*. *Geology* 28:515–518.

Lea, D. W., and E. A. Boyle. 1990. Foraminiferal reconstruction of barium distributions in water masses of the glacial oceans. *Paleoceanography* 5:719–742.

Lenes, J. M., B. P. Darrow, C. Cattrall, C. A. Heil, M. Callahan, G. A. Vargo, and R. H. Byrne. 2001. Iron fertilization and the *Trichodesmium* response on the west Florida shelf. *Limnology and Oceanography* 46:1261–1277.

Lenes, J. M., J. M. Prospero, W. M. Landing, J. I. Virmani, and J. J. Walsh. 2012. A model of Saharan dust deposition to the eastern Gulf of Mexico. *Marine Chemistry* 134–135:1–9.

Lin, H., and C. Chen. 2002. A late Pliocene diatom Ge/Si record from the Southeast Atlantic. *Marine Geology* 180:151–161.

Liu, B., L. A. Schaider, R. P. Mason, M. S. Bank, N. N. Rabalais, P. W. Swarzenski, J. P. Shine, T. Holiweg, and D. B. Senn. 2009. Disturbance impacts on mercury dynamics in northern Gulf of Mexico sediments. *Journal of Geophysical Research: Biogeosciences* 114:1–12.

Macías-Zamora, J. V., J. A. Villaescusa-Celaya, A. Muñoz-Barbosa, and G. Gold-Bouchot. 1999. Trace metals in sediment cores from the Campeche shelf, Gulf of Mexico. *Environmental Pollution* 104:69–77.

Marvin-DiPasquale, M. C., and R. S. Oremland. 1998. Bacterial methyl mercury degradation in Florida Everglades peat sediment. *Environmental Science & Technology* 32:2556–2563.

Mason, R. P., A. L. Choi, W. F. Fitzgerald, C. R. Hammerschmidt, C. H. Lamborg, A. L. Soerensen, and E. M. Sunderland, 2012. Mercury biogeochemical cycling in the ocean and policy implications. *Environmental Research* 119:101–117.

Middag, R., R. Séférian, T. M. Conway, S. G. John, K. W. Bruland, and H. J. W de Baar. 2015. Intercomparison of dissolved trace elements at the Bermuda Atlantic Time series station. *Marine Chemistry* 177:476–489.

Mohajerin, T. J., G. R. Helz, and K. H. Johannesson. 2016. Tungsten-molybdenum fractionation in estuarine environments. *Geochimica et Cosmochimica Acta* 177:105–119.

Mohajerin, T. J., G. R. Helz, C. D. White, and K. H. Johannesson. 2014. Tungsten speciation in sulfidic waters: Determination of thiotungstate formation constants and modeling their distribution in natural waters. *Geochimica et Cosmochimica Acta* 144:157–172.

Moore, W. S. 1996. Large groundwater inputs to coastal waters revealed by ^{226}Ra enrichments. *Nature* 380:612–615.

Moore, W. S., J. L. Sarmiento, and R. M. Key. 2008. Submarine groundwater discharge revealed by ^{228}Ra distribution in the upper Atlantic Ocean. *Nature Geoscience* 1:309–311.

Morel, F. M. M., and N. M. Price. 2003. The biogeochemical cycles of trace metals. *Science* 300:944–947.

Morris, A. W. 1985. Estuarine chemistry and general survey strategy. Pp. 1–60 in P. C. Head (ed.), *Practical Estuarine Chemistry: A Handbook*: Cambridge: Cambridge University Press.

Mortlock, R. A., C. D. Charles, P. N. Froelich, M. A. Zibello, J. Saltzman, J. D. Hays, and L. H. Burckle. 1991. Evidence for lower productivity in the Antarctic Ocean during the last glaciation. *Nature* 351:220–223.

Murray, S. P. 1997. *An Observational Study of the Mississippi-Atchafalaya Coastal Plume: Final Report*. OCS Study MS-98-0040. Prepared for the Minerals Management Service, Gulf of Mexico OCS Region, under MMS contract 14-35-0001-30632. New Orleans: US Department of Interior. 513 pp.

Neff, J. M. 2008. Estimation of bioavailability of metals from drilling mud barite. *Integrated Environmental Assessment and Management* 4:184–193.

Neff, J., K. Lee, and E. DeBlois. 2011. Produced water: Overview of composition, fates and effects. Chap. 1 in K. J. Lee and J. Neff (eds.), *Produced Water*. New York: Springer.

Osborne, A. H., B. A. Haley, E. C. Hathorne, S. Flögel, and M. Frank. 2014. Neodymium isotopes and concentrations in Caribbean seawater: Tracing water mass mixing and continental input in a semi-enclosed ocean basin. *Earth and Planetary Science Letters* 406:174–186.

Osborne, A. H., B. A. Haley, E. C. Hathorne, Y. Plancherel, and M. Frank. 2015. Rare earth element distribution in Caribbean seawater: Continental inputs versus lateral transport of distinct REE compositions in subsurface water masses. *Marine Chemistry* 177:172–183.

Pilato, R. S., and E. L. Stiefel. 1999. Molybdenum and tungsten enzymes. Chap. 6 in J. Reedijk and E. Bouwman (eds.), *Bioinorganic Catalysis*. New York: Marcel Dekker.

Pokrovsky, O. S., and J. Schott. 2002. Iron colloids/organic matter associated transport of major and trace elements in small boreal rivers and their estuaries (NW Russia). *Chemical Geology* 190:141–179.

Powell, R. T., W. M. Landing, and J. E. Bauer. 1996. Colloidal trace metals, organic carbon and nitrogen in a southeastern U.S. estuary. *Marine Chemistry* 55:165–176.

Prestbo, E. M., and D. A. Gay. 2009. Wet deposition of mercury in the U.S. and Canada, 1996–2005: Results and analysis of the NADP mercury deposition network (MDN). *Atmospheric Environment* 43:4223–4233.

Rabalais, N. N., Q. Dortch, D. Justic, M. B. Kilgen, P. L. Klerks, P. H. Templet, and R. E. Turner. 1995. *Status and Trends of Eutrophication, Pathogen Contamination, and Toxic Substances in the Barataria-Terrebonne Estuarine System*. Barataria-Terrebonne National Estuary Program, Publication No. 22. Thibodaux, LA: Barataria-Terrebonne National Estuary Program. 265 pp.

Rice, G. E., D. B. Senn, and J. P. Shine. 2009. Relative importance of atmospheric and riverine mercury sources to the northern Gulf of Mexico. *Environmental Science & Technology* 43:415–422.

Rosales-Hoz, L., A. Carranza-Edwards, L. Sanvicente-Anorve, M. A. Alatorre-Mendieta, and F. Rivera-Ramirez. 2009. Distribution of dissolved trace metals around the Sacrificos Coral Reef Island, in the southwestern Gulf of Mexico. *Bulletin of Environmental Contamination and Toxicology* 83:713–719.

Ruelas-Inzunza, J., C. Delgado-Alvarez, M. Frias-Espericueta, and F. Paez-Osuna. 2013. Mercury in the atmospheric and coastal environments of Mexico. *Reviews of Environmental Contamination and Toxicology* 226:65–99.

Rumbold, D. G., D. W. Evans, S. Niemczyk, L. E. Fink, K. A. Laine, N. Howard, D. P. Krabbenhoft, and M. Zucker. 2011. Source identification of Florida Bay's methyl mercury problem: Mainland runoff versus atmospheric deposition and in situ production. *Estuaries and Coasts* 34:494–513.

Rutgers van der Loeff, M. M., and W. Geibert. 2008. U- and Th-series nuclides as tracers of particle dynamics, scavenging and biogeochemical cycles in the oceans. Chap. 7 in S. Krishnaswami and J. K. Cochran (eds.), *Radioactivity in the Environment*, vol. 13. Oxford: Elsevier.

Ryan, W. B. F., S. M. Carbotte, J. O. Coplan, S. O. Hara, A. Melkonian, R. Arko, R. A. Weissel, et al. 2009. Global Multi-resolution Topography synthesis. *Geochemistry, Geophysics, Geosystems* 10:Q03014.

Santschi, P. H., M. A. Allison, S. Asbill, A. B. Perlet, S. Cappellino, C. Dobbs, and L. McShea. 1999. Sediment transport and Hg recovery in Lavaca Bay, as evaluated from radionuclide and Hg distributions. *Environmental Science & Technology* 33:378–391.

Santschi, P. H., B. J. Presley, T. L. Wade, B. Garcia-Romero, and M. Baskaran. 2001. Historical contamination of PAHs, PCBs, DDTs, and heavy metals in Mississippi River delta, Galveston Bay and Tampa Bay sediment cores. *Marine Environmental Research* 52:51–79.

Schijf, J. 2007. Alkali elements (Na, K, Rb) and alkaline earth elements (Mg, Ca, Sr, Ba) in the anoxic brine of Orca Basin, northern Gulf of Mexico. *Chemical Geology* 243:255–274.

Schmidt, M. W., and J. Lynch-Stieglitz. 2011. Florida Straits deglacial temperature and salinity change: Implications for tropical hydrologic cycle variability during the Younger Dryas. *Paleoceanography* 26:PA4205.

Selin, N. E. 2009. Global biogeochemical cycling of mercury: A review. *Annual Review of Environment and Resources* 34:43–63.

Shaw, T. J., W. S. Moore, J. Kloepfer, and M. A. Sochaski. 1998. The flux of barium to the coastal waters of the southeastern USA: The importance of submarine groundwater discharge. *Geochimica et Cosmochimica Acta* 62:3047–3054.

Shiller, A. M. 1996. The effect of recycling traps and upwelling on estuarine chemical flux estimates. *Geochimica et Cosmochimica Acta* 60:3177–3185.

———. 1999. An overview of the marine chemistry of the Gulf of Mexico. Chap. 8 in H. Kumpf, K. Steidinger, and K. Sherman (eds.), *The Gulf of Mexico Large Marine Ecosystem: Assessment, Sustainability, and Management*. Malden, MA: Blackwell Science.

———. 2002. Seasonality of dissolved rare earth elements in the lower Mississippi River. *Geochemistry, Geophysics, Geosystems* 3:1068.

Shiller, A. M., E. W. Chan, D. J. Joung, M. C. Redmond, and J. D. Kessler. 2017. Light rare earth element depletion during Deepwater Horizon blowout methanotrophy. *Scientific Reports* 7:10389.

Shiller, A. M., and L. Mao. 1999. Dissolved vanadium on the Louisiana shelf: Effect of oxygen depletion. *Continental Shelf Research* 19:1007–1020.

Shim, M., P. W. Swarzenski, and A. M. Shiller. 2012. Dissolved and colloidal trace elements in the Mississippi River delta outflow after Hurricanes Katrina and Rita. *Continental Shelf Research* 42:1–9.

Shokes, R. F., P. K. Trabant, B. J. Presley, and D. F. Reid. 1977. Anoxic, hypersaline basin in the northern Gulf of Mexico. *Science* 196:1443–1446.

Siddall, M., S. Khatiwala, T. van de Flierdt, K. Jones, S. L. Goldstein, S. Hemming, and R. F. Anderson. 2008. Towards explaining the Nd paradox using reversible scavenging in an ocean general circulation model. *Earth and Planetary Science Letters* 274:448–461.

Sohrin, Y., and K. W. Bruland. 2011. Global status of trace elements in the ocean. *Trends in Analytical Chemistry* 30:1291–1307.

Sohrin, Y., K. Isshiki, T. Kuwamoto, and E. Nakayama. 1987. Tungsten in North Pacific waters. *Marine Chemistry* 22:95–103.

Sohrin, Y., M. Matsui, and E. Nakayama. 1999. Contrasting behavior of tungsten and molybdenum in the Okinawa Trough, the East China Sea and the Yellow Sea. *Geochimicia et Cosmochimica Acta* 63:3457–3466.

Springer-Young, M., S. R. Piotrowicz, and G. R. Harvey. 1992. Impact of an episodic event on copper and zinc concentrations and speciation in surface waters of the Gulf of Mexico. *Deep Sea Research* 39:953–964.

Stolpe, B., L. Guo, A. M. Shiller, and M. Hassellöv. 2010. Size and composition of colloidal organic matter and trace elements in the Mississippi River, Pearl River and the northern Gulf of Mexico, as characterized by flow field-flow fractionation. *Marine Chemistry* 118:119–128.

Stolpe, B., M. Hassellöv, K. Andersson, and D. R. Turner. 2005. High resolution ICPMS as a non-line detector for flow field-flow fractionation: Multi-element determination of colloidal size distributions in a natural water sample. *Analytica Chimica Acta* 535:109–121.

Stolpe, B., Z. Zhou, L. Guo, and A. M. Shiller. 2014. Colloidal size distribution of humic- and protein-like fluorescent organic matter in the northern Gulf of Mexico. *Marine Chemistry* 164:25–37.

Stordal, M. C., G. A. Gill, L.-S. Wen, and P. H. Santschi. 1996. Mercury phase speciation in the surface waters of three Texas estuaries: Importance of colloidal forms. *Limnology and Oceanography* 41:52–61.

Summers, J. K., T. L. Wade, V. D. Engle, and Z. A. Malaeb. 1996. Normalization of metal concentrations in estuarine sediments from the Gulf of Mexico. *Estuaries* 19:581–594.

Sunderland, E. M., and R. P. Mason. 2007. Human impacts on open ocean mercury concentrations. *Global Biogeochemical Cycles* 21:GB4022.

Swarzenski, P. W., M. Baskaran, R. J. Rosenbauer, and W. H. Orem. 2006. Historical trace element distribution in sediments from the Mississippi River delta. *Estuaries and Coasts* 29:1094–1107.

Takayanagi, K., and G. T. F Wong. 1985. Dissolved inorganic and organic selenium in the Orca Basin. *Geochimicia et Cosmochimica Acta* 49:539–546.

Tang, J., and K. H. Johannesson. 2003. Speciation of rare earth elements in natural terrestrial waters: Assessing the role of dissolved organic matter from the modeling approach. *Geochimica et Cosmochimica Acta* 67:2321–2339.

Tang, D., K. W. Warnken, and P. H. Santschi. 2001. Organic complexation of copper in surface waters of Galveston Bay. *Limnology and Oceanography* 46:321–330.

———. 2002. Distribution and partitioning of trace metals (Cd, Cu, Ni, Pb, Zn) in Galveston Bay waters. *Marine Chemistry* 78:29–45.

Taylor, H. E., R. C. Antweiler, D. A. Roth, C. N. Alpers, and P. Dileanis. 2012. Selected trace elements in the Sacramento River, California: Occurrence and distribution. *Archives of Environmental Contamination and Toxicology* 62:557–569.

Telfeyan, K., A. Breaux, J. Kim, J. E. Cable, A. S. Kolker, D. A. Grimm, and K. H. Johannesson. 2017. Arsenic, vanadium, iron and manganese biogeochemistry in a deltaic wetland, southern Louisiana, USA. *Marine Chemistry* 192:32–48.

Tessier, A., and D. R. Turner. 1995. *Metal Speciation and Bioavailability in Aquatic Systems*. New York: John Wiley and Sons. 696 pp.

Trefry, J. H., S. Metz, and R. P. Trocine. 1985. A decline in lead transport by the Mississippi River. *Science* 20:439–441.

Trefry, J. H., K. L. Naito, R. P. Trocine, and S. Metz. 1996. Assessing the potential for enhanced bioaccumulation of heavy metals from produced water discharges to the Gulf of Mexico. Pp. 339–354 in M. Reed and S. Johnsen (eds.), *Produced Water 2: Environmental Issues and Mitigation Technologies*. New York: Plenum Press.

Trefry, J. H., and B. J. Presley. 1982. Manganese fluxes from Mississippi Delta sediments. *Geochimica et Cosmochimica Acta* 46:1715–1726.

Trefry, J. H., B. J. Presley, W. L. Keeney-Kennicutt, and R. P. Trocine. 1984. Distribution and chemistry of manganese, iron, and suspended particulates in Orca Basin. *Geo-Marine Letters* 4:125–130.

Trefry, J. H., R. P. Trocine, M. L. McElvaine, R. D. Rember, and L. T. Hawkins. 2007. Total mercury and methyl mercury in sediments near offshore drilling sites in the Gulf of Mexico. *Environmental Geology* 53:375–385.

Tréguer, P. J., and C. L. De la Rocha. 2013. The world ocean silica cycle. *Annual Review of Marine Science* 5:477–401.

Trocine, R. P., and J. H. Trefry. 1983. Particulate metal tracers of petroleum drilling mud dispersion in the marine environment. *Environmental Science & Technology* 17:507–512.

Turner, R. E., C. S. Milan, and N. N. Rabalais. 2004. A retrospective analysis of trace metals, C, N and diatom remnants in sediments from the Mississippi River delta shelf. *Marine Pollution Bulletin* 49:548–556.

Valette-Silver, N. J. 1993. The use of sediment cores to reconstruct historical trends in contamination of estuarine and coastal sediments. *Estuaries* 16:577–588.

Van Cappellen, P., E. Viollier, A. Roychoudhury, L. Clark, E. Ingall, K. Lowe, and T. Dichristina. 1998. Biogeochemical cycles of manganese and iron at the oxic-anoxic transition of a stratified marine basin (Orca Basin, Gulf of Mexico). *Environmental Science & Technology* 32:2931–2939.

van der Fliert, T., K. Pahnke, H. Amakawa, P. Andersson, C. Basak, B. Coles, C. Colin, et al. 2012. GEOTRACES intercalibration of neodymium isotopes and rare earth element concentrations in seawater and suspended particles. Part 1: Reproducibility of results for the international intercomparison. *Limnology and Oceanography: Methods* 10:234–251.

Vazquez, F., M. Sanchez, H. Alexander, and D. Delgado. 1991. Distribution of Ni, V, and petroleum hydrocarbons in recent sediments from the Veracruz coast, Mexico. *Bulletin of Environmental Contamination and Toxicology* 46:774–781.

Vazquez, F. G., and V. K. Sharma. 2004. Major and trace elements in sediments of the Campeche Sound, southeast Gulf of Mexico. *Marine Pollution Bulletin* 48:87–90.

Vorlicek, T. P., M. D. Kahn, Y. Kasuya, and G. R. Helz. 2004. Capture of molybdenum in pyrite-forming sediments: Role of ligand-induced reduction by polysulfides. *Geochimica et Cosmochimica Acta* 68: 547–556.

Walsh, J. J., J. K. Jolliff, B. P. Darrow, J. M. Lenes, S. P. Milroy, A. Remsen, D. A. Dieterle, et al. 2006. Red tides in the Gulf of Mexico: Where, when, and why? *Journal of Geophysical Research: Oceans* 111:1–46.

Warnken, K. W., G. A. Gill, L. L. Griffin, and P. H. Santschi. 2001. Sediment-water exchange of Mn, Fe, Ni and Zn in Galveston Bay, Texas. *Marine Chemistry* 73:215–231.

Warnken, K. W., G. A. Gill, R. D. Lehman, T. M. Dellapenna, and M. A. Allison. 2003. The effects of shrimp trawling on sediment oxygen demand and the release of trace metals and nutrients from estuarine sediments. *Estuarine, Coastal and Shelf Science* 57:25–42.

Warnken, K. W., and P. H. Santschi. 2009. Delivery of trace metals (Al, Fe, Mn, V, Co, Ni, Cu, Cd, Ag, Pb) from the Trinity River watershed towards the ocean. *Estuaries and Coasts* 32:158–172.

Wen, L.-S., P. H. Santschi, G. A. Gill, and C. L. Paternostro. 1999. Estuarine trace metal distributions in Galveston Bay: Importance of colloidal forms in the speciation of the dissolved phase. *Marine Chemistry* 63:185–212.

Wen, L.-S., P. H. Santschi, G. A. Gill, C. L. Paternostro, and R. D. Lehman. 1997. Colloidal and particulate silver in river and estuarine waters of Texas. *Environmental Science & Technology* 31:727–731.

Wen, L.-S., P. H. Santschi, K. W. Warnken, W. Davison, H. Zhang, H. P. Li, and K. T. Jiann. 2011. Molecular weight and chemical reactivity of dissolved trace metals (Cd, Cu, Ni) in surface waters from the Mississippi River to Gulf of Mexico. *Estuarine, Coastal and Shelf Science* 92:649–658.

Wen, L.-S., A. M. Shiller, P. H. Santschi, and G. Gill. 1999. Trace element behavior in Gulf of Mexico estuaries. Pp. 303–346 in T. S. Bianchi, J. R. Pennock, and R. R. Twilley (eds.), *Biogeochemistry of Gulf of Mexico Estuaries*. New York: John Wiley and Sons.

Wen, L.-S., K. W. Warnken, and P. H. Santschi. 2008. The role of organic carbon, iron, and aluminum oxyhydroxides as trace metal carriers: Comparison between the Trinity River and the Trinity River Estuary (Galveston Bay, Texas). *Marine Chemistry* 112:20–37.

White, J. R., R. D. Delaune, C. Y. Li, and S. J. Bentley. 2009. Sediment methyl and total mercury concentrations along the Georgia and Louisiana inner shelf, USA. *Analytical Letters* 42:1219–1231.

Wiesenburg, D. A., J. M. Brooks, and B. B. Bernard. 1985. Biogenic hydrocarbon gases and sulfate reduction in the Orca Basin brine. *Geochimica et Cosmochimica Acta* 49:2069–2080.

Windom, H. L., R. G. Smith, and C. Rawlinson. 1989. Particulate trace metal composition and flux across the southeastern U.S. continental shelf. *Marine Chemistry* 27:283–297.

Windom, H., R. Smith, C. Rawlinson, M. Hungspreugs, S. Dharmvanij, and G. Wattayhorn. 1988. Trace metal transport in a tropical estuary. *Marine Chemistry* 24:293–305.

Yan, B., U. Passow, J. P. Chanton, E. Nöthig, V. Asper, and J. Sweet. 2016. Sustained deposition of contaminants from the Deepwater Horizon spill. *Proceedings of the National Academy of Sciences of the United States of America* 113:E3332–E3340.

Yao, W., and R. H. Byrne. 1999. Determination of trace chromium (VI) and molybdenum (VI) in natural and bottled mineral waters using long path length absorbance spectroscopy (LPAS). *Talanta* 48:277–282.

Yu, K., R. D. Delaune, I. Devai, R. Tao, and A. Jugsujinda. 2008. Total and methyl mercury in wetland soils and sediments of Louisiana's Pontchartrain Basin (USA). *Journal of Environmental Science and Health, Part A* 43:1657–1662.

Zhang, J., and Y. Nozaki. 1996. Rare earth elements and yttrium in seawater: ICP-MS determinations in the East Caroline, Coral Sea, and South Fiji basins of the western South Pacific Ocean. *Geochimica et Cosmochimica Acta* 60:4631–4644.

5

Radionuclides as Tracers for Geochemical and Sedimentary Processes in the Gulf of Mexico

D. Reide Corbett, John P. Walsh, and Peter H. Santschi

Introduction

The world population currently stands at more than 7 billion people, roughly 40% of them living within 100 km of a coastline (Kummu et al. 2016). Coastal population growth in conjunction with climate change puts increasing pressure on coastal and marine ecosystems. Natural (e.g., floods, hurricanes, and their aftermath) and anthropogenic drivers (e.g., Deepwater Horizon oil spill, coastal overdevelopment, nutrient loading) create additional change to coastal systems across varying spatial and temporal scales (e.g., Day et al. 2007; Rabalais et al. 2014). Understanding the dynamic interplay between the ocean, climate, and anthropogenic influences on the natural cycles is critical to planning and managing for the future. Advances in our knowledge of these influences have benefited through the application of natural and artificial radionuclides in studies throughout the Gulf of Mexico (GOM) and the global ocean. The vast range in chemistry and half-life allows these radioactive elements to be used to study processes across a wide spectrum of spatial and temporal scales.

Radionuclides can be broadly classified based on their origin: (1) primordial radionuclides, those (and their progeny) with long half-lives formed at the time of the formation of the solar system and Earth; (2) cosmogenic radionuclides, produced in the atmosphere through interaction with cosmic radiation; and (3) anthropogenic radionuclides, produced and introduced to the environment through various modern processes (Ivanovich and Harmon 1992; Swarzenski et al. 2000).

Primordial radionuclides are defined by decay chains initiated by parent isotopes with very long half-lives (>68 million years) and the progeny of which have a wide distribution throughout the ocean and Earth. The origin of these radionuclides is associated with Earth's nucleosynthesis, leading to the ubiquitous nature of these isotopes. The most relevant here are the 3 natural decay chains defined by the long-lived parents ^{238}U, ^{232}Th, and ^{235}U (Fig. 5.1). These long-lived parents decay to produce a series of shorter-lived daughters, including isotopes of thorium, radium, radon, polonium, bismuth, and lead. Chemically, these elements range from an inert gas (i.e., radon [Rn]) to a readily sorbed, tetravalent cation (i.e., thorium [Th]) (Rutherford 1904; Swarzenski et al. 2000; Bourdon et al. 2003; Turekian 2003). In addition, the half-lives of radionuclides within the 3 decay chains range from milliseconds to millions of years. The chemical properties and half-lives determine the fate of these radionuclides in the ocean, and the differences in these defining properties allow these radioactive elements to provide insight into how the ocean operates across a broad range of spatial and temporal scales.

Cosmic irradiation of the atmosphere produces neutrons and protons that react with N_2 and O_2, argon (Ar), and so on, resulting in the production of radionuclides, including those often used to study ocean processes (e.g., tritium [^{3}H], beryllium [^{7}Be], radiocarbon [^{14}C]). These

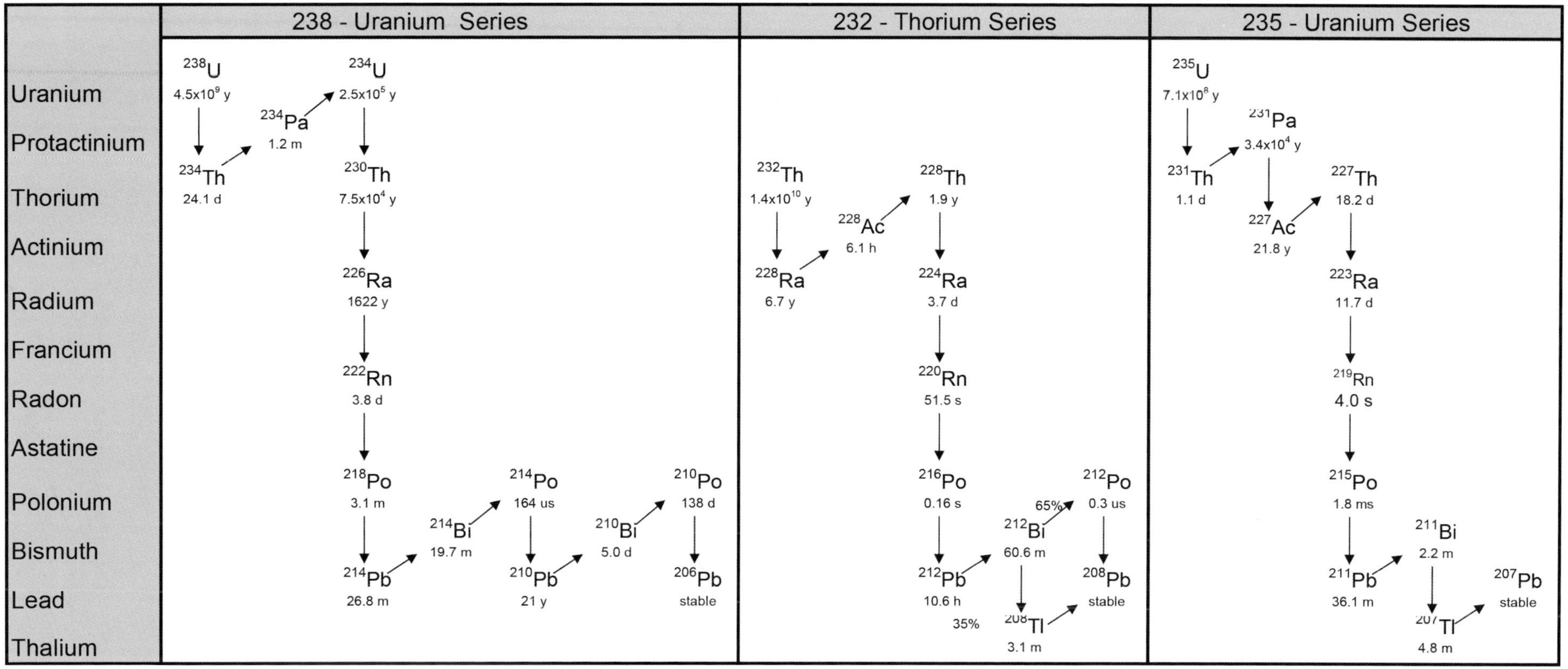

Figure 5.1. U-Th decay series isotopes and associated half-lives. Vertical arrows indicate alpha (α) decay, while beta (β) decay is defined by diagonal arrows. ^{212}Bi has a split decay chain with 65% via β- to ^{212}Po and 36% by α to ^{208}Tl.

radionuclides are naturally produced at relatively constant rates and delivered to Earth's surface through wet and dry fallout (Anderson and Libby 1951; Craig and Lal 1961; Suess 1965; Kaste and Baskaran 2011). However, the nuclear age (e.g., bomb testing, nuclear power) and some industrial applications have complicated tritium and radiocarbon sources/contributions since the 1950s. Regardless, these radionuclides allow the tracing and dating of water circulation in the oceans and continents (Begemann and Libby 1957; Doney et al. 1997; Macdonald et al. 1993). Sediments, groundwater, organic matter, and continental ice can all be dated using different cosmogenic radionuclides (Wang et al. 1999; Adolphi and Muscheler 2016; Subt et al. 2016). In addition, radioisotopes of hydrogen and carbon are important biological elements and can therefore be used to study both chemical and biological processes. This chapter focuses on geochemical and sedimentary applications.

Finally, anthropogenic radionuclides are produced either through nuclear reactions or industrial activity. Human-made radionuclides are usually introduced into the environment by the fission of ^{235}U and/or the detonation of fission and fusion devices belowground or in the atmosphere. These radionuclides (e.g., iodine [^{129}I], cesium [^{137}Cs], plutonium [$^{239,240}Pu$]) mark a specific time event that can show up in a particular horizon within a water column or a geologic record. Although, it should be noted that some anthropogenic radionuclides (e.g., ^{129}I, ^{14}C) have a natural source as well, so both processes must be considered when applying these as process-based tracers.

The radioactive elements described are delivered to the ocean through several primary pathways (Fig. 5.2; summarized by Swarzenski et al. 2000):

- Rivers are a source for both dissolved and particulate radionuclides. Relatively soluble species, including uranium, are largely dissolved and/or colloidal-sized (defined loosely here as <0.4 μm) in freshwater. U-Th decay chains are represented in river water, albeit to varying degrees of fractionation or disequilibrium.

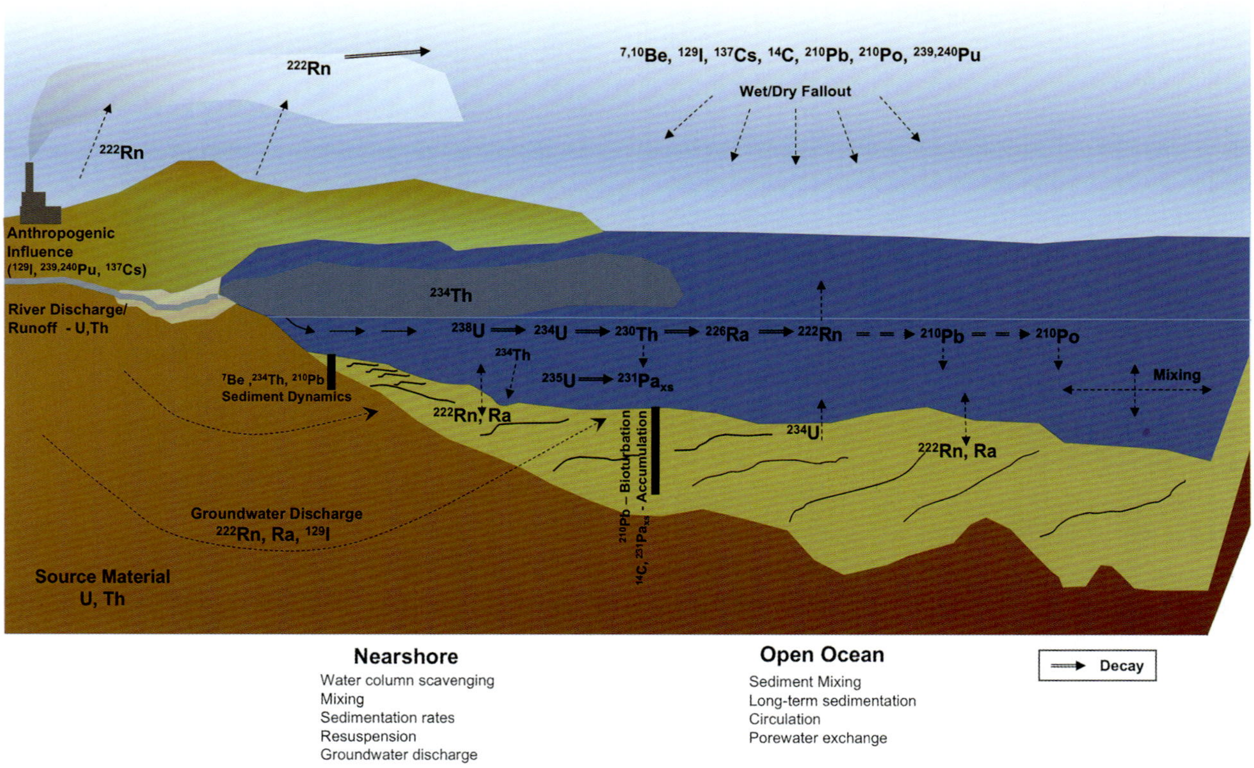

Figure 5.2. Schematic of oceanic inputs and cycles of select radionuclides often used in coastal and open-ocean setting. Vertical bold black lines in sea sediments represent sedimentary sections/cores. The "Ra" symbols reflect the radium quartet commonly used to track groundwater and pore-water exchange (i.e., ^{223}Ra, ^{224}Ra, ^{226}Ra, ^{228}Ra).

- Highly particle-reactive thorium isotopes settling out of the water column to the seafloor continually produce radium-radon (Ra-Rn) isotopes. Coastal and offshore marine sediments continuously release Ra and Rn into bottom waters, largely by diffusive/advective processes.
- Groundwater (either fresh or recirculated marine water) may discharge into coastal waters where hydrologic gradients are favorable. Wave pumping and tidal flushing also may deliver radionuclides into overlying waters.
- Autochthonous production of daughter isotopes by the radioactive decay of a long-lived parent can be major oceanic sources, especially in basins where water residence times are long.
- Wet/dry deposition is an important source for cosmogenic and anthropogenic radionuclides to the surface ocean.

This chapter summarizes the geochemistry of several of the commonly used radionuclides in understanding ocean processes and provides examples of their application in the GOM.

Radionuclides and Sedimentary Processes

Rivers annually transport approximately 15 billion tons of sediment to the ocean (Syvitski et al. 2005). Much of this material is supplied by only a few dozen rivers to river-dominated ocean margins (Walsh et al. 2013), which are considered to be major repositories of organic carbon in marine sediments and an important part of global biogeochemical cycles (McKee et al. 2004; Corbett et al. 2006). The dispersal of this sediment also impacts water-column oxygen levels (Rabalais et al. 2014) and the fate of particle-reactive pollutants to coastal environments as pesticides, polycyclic aromatic hydrocarbons (PAHs), and other organic pollutants that exhibit an affinity to adsorb onto particulate matter (Meade 1996; Arzayus et al. 2002; also see Chapter 8 for more detail on organic contaminants). Sediment dynamics in coastal systems are extremely complex, as sediment can undertake numerous sequences of transport, deposition, and resuspension before long-term net accumulation (Wright and Nittrouer 1995; McKee et al. 2004; Walsh and Nittrouer 2009).

The nature of sediment cycling on ocean margins is poorly understood because of spatially and temporally variable processes—physical (winds, waves, tides, currents, and river discharge), biological (bioturbation), and sedimentary (aggregation, settling, trapping, and resuspension)—that remobilize and redistribute alluvial and marine sediments (Coleman 1988; Meade 1996; Corbett et al. 2004; Corbett et al. 2007). Consequently, the study of observed events (storms, floods, etc.) and their effect on sedimentary processes, as well as the measurement and modeling of the progression of sedimentary transport, is very difficult (McKee et al. 2004). Nevertheless, it is critical to understand how sediments and associated pollutants are transported and distributed on continental margins because profound differences in sedimentation are observed (Nittrouer et al. 2007; Walsh and Nittrouer 2009). An understanding of the environmental fate of pollutants is necessary to monitor and manage the health of ecosystems. Additionally, understanding sediment transport processes advances our knowledge of the evolution of continental margins and connections with the deep ocean. Indeed, radionuclides are critical as chronometers for sedimentary processes, and the correct chronometer is needed to address the research question (e.g., flood sedimentation vs. postglacial lake drainage). It must be remembered that the temporal and spatial nature of a sedimentary source signals (e.g., a watershed perturbation such as a flooding event), and the factors affecting its propagation, will determine if and how it is recorded in the marine system (i.e., the sink) (Romans et al. 2016).

The particle-laden waters of most coastal systems provide a unique environment for using radionuclides as tools for understanding processes acting to disturb and redistribute sediments. Many elements are adsorbed onto settling particles or make up a fraction of the sediment (e.g., carbon) and, therefore, follow the pathways of the particles. Carbon-14, for example, is incorporated into inorganic or organic material during formation (e.g., the shell or plant structure). Several other naturally occurring and anthropogenically introduced radionuclides (e.g., ^{210}Pb, ^{234}Th, ^{7}Be, ^{210}Pb, and ^{137}Cs) associate strongly with particles (and colloids) and are therefore also useful tracers of particle transport and fate. The half-life of the tracer, integrated over 6 to 7 half-lives, determines the approximate temporal scale for which a tracer can be used. Many investigators have used these tracers in both coastal and open-ocean environments of the Gulf (Baskaran and Santschi 1993; Baskaran et al. 1996; Baskaran et al. 1997; Corbett et al. 2004; Corbett et al. 2006; Corbett et al. 2007; Dail et al. 2007; Guo et al. 1995; Guo et al. 1997; Guo et al. 2002; Hung et al. 2004; Hung et al. 2010; McKee et al. 1984; Santschi et al. 1999; Santschi et al. 2001; Santschi and Rowe 2008; Yeager et al. 2004).

^{210}Pb ($t_{1/2}$ = 22.3 yr) is a naturally occurring radionuclide that is part of the ^{238}U decay series and is often used for age-dating sediments (Krishnaswamy et al. 1971; Oldfield and Appleby 1984). In broad terms, the marine cycle of ^{210}Pb is characterized by supply and/or production of dissolved ^{210}Pb that is then scavenged by solid surfaces and thus removed from the water column. ^{210}Pb is delivered to the water column by 3 sources: (1) atmospherically derived from the decay of ^{222}Rn; (2) the in situ decay of ^{226}Ra and from ^{222}Rn in the water column; and (3) direct leaching of rocks and soils during weathering. ^{210}Pb has been used extensively as a tracer for particle-reactive elements and for geochronological studies of lake, estuarine, and coastal sediments.

^{234}Th is a naturally occurring radionuclide continuously produced from the decay of its direct parent ^{238}U in the water column and in seabed sediment (Aller and Cochran 1976; Aller et al. 1980; McKee et al. 1984; Waples et al. 2006). Uranium concentrations are typically small in river water and increase with salinity to a maximum value in open-ocean waters. Unlike Th, U is not readily scavenged from the water column by particles in marine waters, producing "excess" ^{234}Th on the particles in the seabed (DeMaster et al. 1986). The rapid removal of ^{234}Th onto particles relative to its half-life (24.1 days) makes it useful for examining rates of geochemical and sedimentological processes in the upper water column and in shallow seafloor sediments. Researchers have utilized ^{234}Th in estuaries, the coastal ocean, and the deep sea as an indicator of particle export from surface waters (e.g., Coale and Bruland 1985; Moran and Smith 2000; Savoye et al. 2004), particle mixing in recently deposited sediments (Aller and Cochran 1976; McKee et al. 1986; Santschi et al. 2001; Green et al. 2002), short-term deposition in high-deposition environments (Kuehl et al. 1986; Kuehl et al. 1996; Day et al. 1995; Schmidt et al. 2001; Corbett et al. 2007), and seasonal variations in the processes controlling the fate of particle-reactive species (Smoak et al. 1996; Corbett et al. 2004; Dail et al. 2007).

^{7}Be ($t_{1/2}$ = 53.3 days) is a naturally occurring, particle-reactive radionuclide produced by cosmic ray spallation processes of nitrogen and oxygen within Earth's atmosphere (Olsen et al. 1986). It is produced primarily in the stratosphere; however, due to the long residence times of stratospheric air masses, most of the ^{7}Be delivered to Earth's surface comes from the troposphere (Dutkiewicz and Husain 1985; Canuel et al. 1990). Deposition on Earth's surface occurs primarily through precipitation and dry deposition of ^{7}Be associated with aerosols. Residence times of ^{7}Be in the water column can vary from a few days to several weeks depending on the amount of fine particles in suspension, sediment accumulation rates, and frequency of sediment resuspension (Olsen et al. 1986; Baskaran and Santschi 1993). The short half-life of ^{7}Be makes it a useful particle tracer for sediment dynamic activities on timescales of days to months, and it has been applied extensively as a proxy for fluvial material in estuaries and continental shelf systems (Olsen et al. 1986; Canuel et al. 1990; Baskaran and Santschi 1993; Baskaran et al. 1997; Sommerfield et al. 1999; Allison et al. 2000; Corbett et al. 2004; Corbett et al. 2007; Palinkas et al. 2005; Kolker et al. 2012). Riverine input of ^{7}Be, compared to direct atmospheric deposition, has been shown to be a more important source in environments with a high ratio of drainage basin to estuarine surface area (Baskaran and Santschi 1993; Baskaran et al. 1997; Sommerfield et al. 1999). Corbett et al. (2004) showed a very high ratio (~700) between drainage basin size and the depositional area on the Louisiana shelf, suggesting the Mississippi River provides a strong signal, relative to local atmospheric input.

^{137}Cs ($t_{1/2}$ = 30.1 yr) and ^{239,240}Pu ($t_{1/2}$ = 24,110, and 6500 yr, respectively) are by-products of fission reactions. These radionuclides derived from nuclear weapons testing were first detectible in sediments in approximately 1953, with the maximum fallout occurring in 1963 in the Northern Hemisphere (UNSCEAR 2000). A similar pattern, although with significantly reduced activities, is present in the Southern Hemisphere, making this technique much less useful south of the equator. This historic annual flux provides the possibility of 2 dated horizons, one reflecting the initial onset of deposition in the early 1950s and the 1963 peak in fallout. Some areas (also in the Northern Hemisphere) had additional input of ^{137}Cs in 1986 following the Chernobyl incident in the former Soviet Union (Aarkrog 1988). The annual variation allows ^{137}Cs to be a viable stratigraphic marker in a wide variety of depositional environments, including reservoirs, lakes, wetlands, coastal areas, and floodplains (Pennington et al. 1973; Chmura and Kosters 1994; Neubauer et al. 2002). Three assumptions must be made when using these radionuclides to measure sediment accumulation in aquatic systems: (1) the radionuclides entering the water body are quickly adsorbed to suspended material in the water column; (2) in cases where the radionuclide is deposited in the catchment area of a water body, the radionuclide is firmly adsorbed onto soil particles, and any redistribution results chiefly from erosion and deposition; and (3) once deposited in the sediment column, the radionuclide is immobile and not affected by reworking or diffusional movement up or down in the substrate (Hutchinson 1995).

This suite of radionuclides can be used to evaluate particle flux in the water column, short-term sediment deposition and resuspension, or decadal-scale accumulation. These tracers have been used extensively throughout the GOM over the last several decades. The following examples illustrate the application of these tools in the system and, by no means, is this an exhaustive review of their usage.

^{14}C, as noted previously, is produced naturally and has also formed as a result of nuclear bomb testing. With a half-life of about 5730 years, it is a powerful tool for dating material over the last ~50,000 years. When applying it for dating, the year 1950 is considered to be the "present" because after that time nuclear testing significantly altered the amount of ^{14}C in the atmosphere. The conversion of ^{14}C age to a calendar age is accomplished using a calibration curve (e.g., Suess 1967; Stuiver et al. 1998), and these curves as well as the software programs to use them have evolved over time, including CALIB (Stuiver and Reimer 1993; Stuiver et al. 2013, 2017) and OXCAL (Bronk Ramsey 1995; Bronk Ramsey and Lee 2013). The latest software employs statistical techniques (e.g., Bayesian modeling) to more quantitatively determine calendar ages. A correction for a "reservoir effect" is often necessary when the material measured has a carbon origin other than the atmosphere, such as in situations where old carbon may be incorporated. For example, a marine reservoir correction of 400 years is commonly used for the modern ocean (e.g., Richey et al. 2007), or a localized value for marine reservoir may be applied (e.g., Reimer and Reimer 2001; Siani et al. 2001; Kovanen and Easterbrook 2002; Pearce et al. 2017). Also, the fractionation of ^{14}C may need to be accounted for depending on the material being dated.

Sediment Dynamics on a River-Dominated Margin: From Days to Decades

The Mississippi-Atchafalaya River system (MARS) is a major world river that dominates coastal processes in the northern GOM. It has the third-largest drainage basin area and the seventh-largest water discharge and suspended load among the world's rivers (Milliman and Meade 1983; Meade 1996). Approximately 60% of the total suspended matter and 66% of the total dissolved load transported from the conterminous United States to the ocean is carried by the MARS (Presley et al. 1980). Draining approximately 41% of the conterminous United States, the MARS delivers approximately 210×10^6 metric tons (t) of suspended matter to the shelf of the northern GOM each year (measured at Tarbert Landing, 495 km upriver from the river mouth) (Milliman and Meade 1983; Meade 1996; Meade and Parker 1985; Mossa 1996). Due to the buildup of dams along the river, sediment delivery and delivery of organic matter at the Mississippi River delta have changed over time (Santschi et al. 2007).

Forcing agents (e.g., waves, fluvial discharge, and currents) vary along the northern Gulf Coast and exhibit large seasonal fluctuations. To help distinguish the importance of these varying processes affecting coastal systems, ^{7}Be has been used independently and in concert with ^{234}Th (Corbett et al. 2004; Corbett et al. 2007). ^{7}Be is a key indicator of fluvial-borne sediments in this system, whereas ^{234}Th is purely a marine source derived from ^{238}U in seawater, and together they describe the fate of fluvial sediment as it enters the coastal ocean. Corbett et al. (2004) used radionuclide inventories, obtained by integrating the tracer activity with depth in the seabed and accounting for bulk density and porosity, derived from box cores collected during 2 cruises in April and October 2000 (i.e., high- vs. low-discharge conditions) to provide insights on the transport and fate of terrestrial and marine material. This initial study demonstrated that the transport of even fine sediment by surface plumes is generally limited to kilometers (<30 km) from the river mouth before initial deposition. Further, seasonal variations in ^{7}Be and ^{137}Cs indicated significant remobilization of sediment and potential export off- and along-shelf during the high-energy (e.g., wind, wave) winter months. As Wright and Nittrouer (1995) pointed out a decade earlier, shelf sediments can undergo numerous episodes of resuspension, transport, and deposition before long-term net accumulation.

Additional radionuclide data collected along the shelf further substantiated the importance of seasonal variations of waves and currents in redistributing sediments. High-frequency sampling (~monthly) by Corbett et al. (2007) of short-lived radiotracers at 3 sites of increasing distance from the Mississippi Delta (i.e., proximal, intermediate, distal) was conducted to isolate the relative influence of the dominant drivers in seasonal sediment

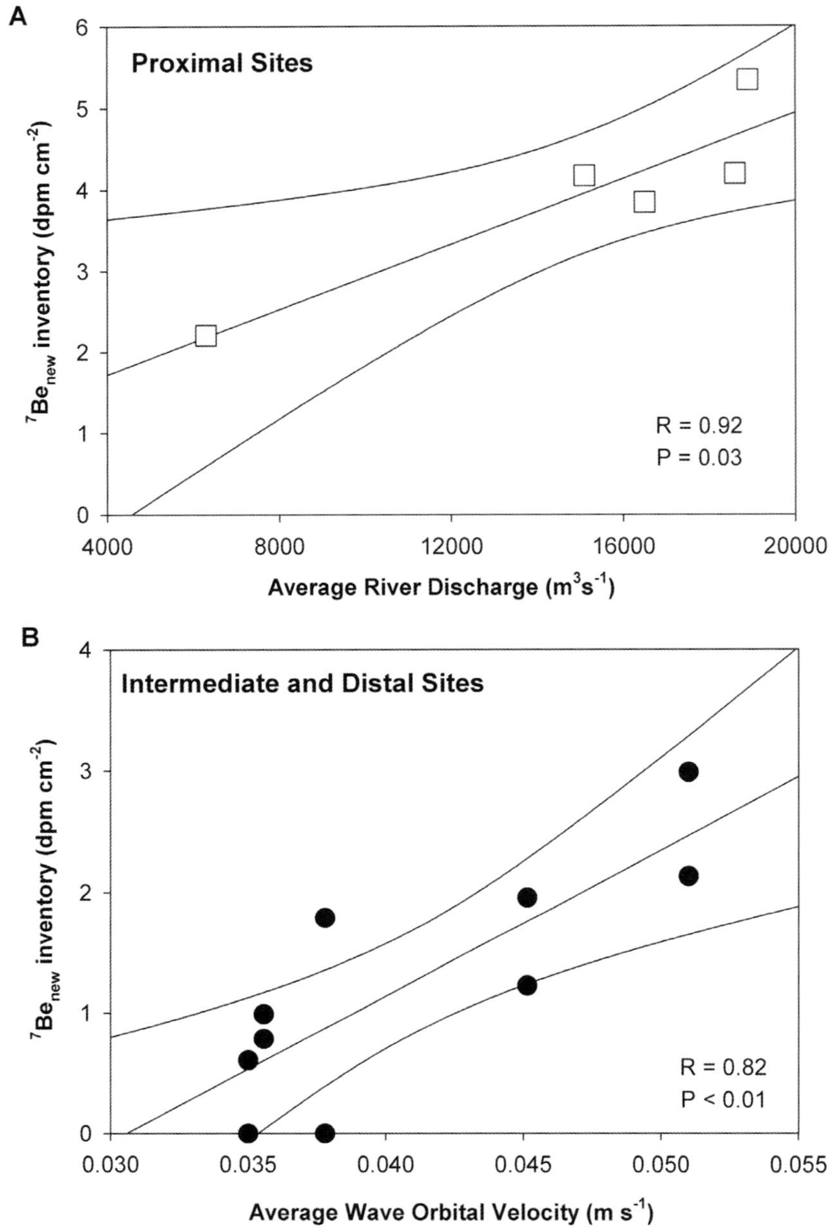

Figure 5.3. $^7Be_{new}$ inventories measured at the proximal (*A*) and intermediate/distal (*B*) sites west of Southwest Pass off the Mississippi Delta show a significant correlation with average river discharge and wave orbital velocity, respectively (modified from Corbett et al. 2007).

dynamics. ^7Be was the primary tracer used to evaluate the importance of fluvial discharge in relation to wave orbital velocity on transporting sediment seaward of Southwest Pass. $^7Be_{new}$ inventories (e.g., accounting for decay between cruises) showed a significant positive relationship with river discharge at the proximal site (Fig. 3A). In contrast, $^7Be_{new}$ inventories and wave-orbital velocity showed no clear statistical relationship at the proximal site but a significant positive correlation at the intermediate and distal sites (Fig. 3B). Thus, sediment supply (e.g., river flux) does not appear to be as important as wave-orbital velocities in controlling the transport and distribution of fluvial sediment across the shelf. The influence of these 2 drivers (e.g., river discharge and waves) on sediment processes can be further differentiated using the ratio between ^7Be and ^{234}Th through time (Fig. 5.4). Generally, the $^7Be_{new}/^{234}Th_{new}$ activity ratio increased with greater fluvial influence and decreased when sediment resuspension played a dominant role. This ratio approach reiterates that the transport of fluvial sediment across the shelf is dependent on sea conditions and not river discharge, with increased across-/along-shelf sediment transport during periods of high wave energy.

Although waves and currents associated with winter storms have been shown to seasonally move sediments along and across the Louisiana continental shelf (Allison

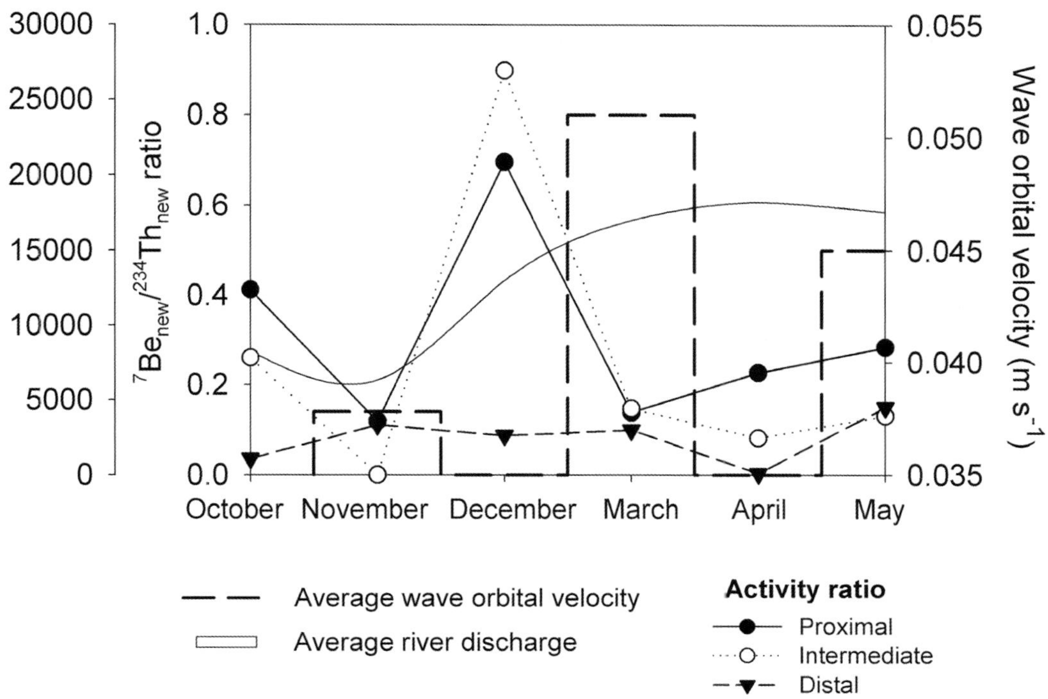

Figure 5.4. The variation of $^{7}Be_{new}/^{234}Th_{new}$ activity ratio at each sampling site (i.e., Proximal, Intermediate, Distal) west of Southwest Pass off the Mississippi Delta during each sampling period demonstrates the relationship in sediment-forcing mechanisms. Ratios are plotted with the average river discharge and average wave orbital velocities for each sampling period. $^{7}Be_{new}/^{234}Th_{new}$ ratios are used to evaluate fluctuating sediment supply as a result of different sources for ^{7}Be (fluvial) and $^{234}Th_{xs}$ (marine). Thus, a comparatively high $^{7}Be_{new}/^{234}Th_{new}$ ratio indicates a greater influence of fluvial-sourced material relative to marine (modified from Corbett et al. 2007).

et al. 2000; Corbett et al. 2004; Corbett et al. 2007), the passage of several tropical cyclones (e.g., Hurricanes Ivan, Katrina, and Rita) provided a rare opportunity to investigate the seafloor impacts associated with major events occurring in quick succession. During major storms, large waves and strong currents can resuspend sediment previously deposited at shallower depths during non-storm periods and redeposit farther seaward, affecting regions of the seabed that rarely experience such energetic conditions. As a result, large-magnitude storms can play a key role in the cross-shelf transport of materials and thus the sediment storage budget of a margin over geologic time. Insights on the effects of great hurricanes on the seabed were obtained by evaluating sediment cores for naturally occurring radionuclides (^{7}Be, ^{234}Th, ^{210}Pb), x-radiography, and organic content and stable isotopic composition (Dail et al. 2007; Goñi et al. 2007; Walsh et al. 2006).

Rapid-response sampling efforts following these tropical cyclones provided the opportunity to collect box cores throughout the Mississippi-Atchafalaya continental shelf to determine the nature and areal extent of the impacted region (Dail et al. 2007; Goñi et al. 2007; Walsh et al. 2006). Naturally-occurring radionuclide tracers were key to interpreting the influence of these weather systems on the seabed. Storm layers within the sediment column were identified by hiatal surfaces on x-radiographs, measurable excess (activity above that supported by its direct parent) $^{234}Th_{xs}$, and ^{7}Be, and a decrease in the $^{210}Pb_{xs}$ activity (Fig. 5.5). Observations on the shelf indicated up to 5 distinguishable layers totaling as much as 40 cm of storm-layer deposition (Goñi et al. 2007). When $^{234}Th_{xs}$, ^{7}Be, and $^{210}Pb_{xs}$ activity profiles were compared with x-radiographs, the basal contacts of each deposit roughly corresponded to troughs in activity, where coarser sediments were found. Physically stratified sediment with detectable $^{234}Th_{xs}$ indicated that massive deposition had occurred within ~100 days of sample collection. The vertically sinuous profiles of the radionuclide activities found at most sites suggested deposition had occurred as separate events (Dail et al. 2007).

As an example, the suite of radionuclide and sedimentological data collected from a core in 80 m water depth provided evidence of event layers associated with 3 tropical cyclones that impacted the area in 2004, in addition to layers associated with Hurricanes Katrina and Rita in 2005 (Dail et al. 2007; Goñi et al. 2007). The annual sediment accumulation rate averaged over ^{210}Pb timescales in the region where this core was collected is approximately 2 cm yr^{-1} (Dail et al. 2007; Goñi et al. 2007). Thus, the amount of event-related material deposited was more than an order of magnitude greater than the average annual accumulation rate, highlighting the potential importance of tropical cyclones to off-shelf sediment transport (Dail et al. 2007). Similarly, these radio tracers provided the data to evaluate the thickness of event deposits throughout the shelf. The amount of sediment deposited following Hurricanes Katrina and Rita was determined by contouring the event thickness (described previously) at each site and multiplying the area between contours by the average thickness for that interval. The calculated amount of sediment associated with these 2 storms deposited on the margin was 1200 ± 200 × 10^6 t. This estimate is more than 5 times the average amount of material delivered to the continental shelf by the MARS annually. Although there are significant down-core variations in the geochemical compositions of the hurricane-deposited sediments (Fig. 5.5), these organic geochemical signatures are consistent with a massive redistribution of margin sediments (Goñi et al. 2007). The sediment samples collected following the storms are geochemically identical (e.g., ^{13}C and C:N) to materials collected on the continental shelf in earlier years (Goñi et al. 2007) and distinct from the particulate organic materials supplied by the river (Wang et al. 2004). Goñi et al. (2007) therefore concluded that the large waves associated with these storms resuspended and redistributed sediment primarily from shallow-water (<50 m) depths, where sediments typically accumulate as a result of the low mean wave and current energies. Thus, tropical

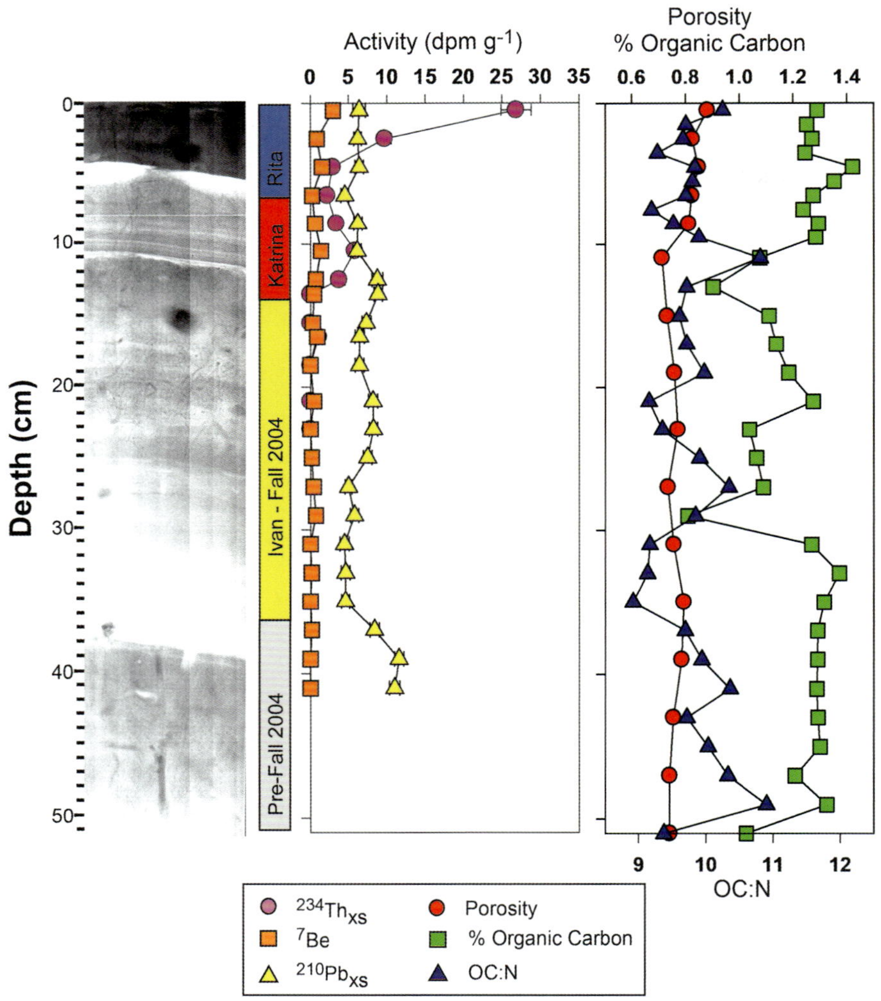

Figure 5.5. Example of down-core data analysis and x-radiography for box cores collected on the Louisiana continental shelf. Core was collected in 80 m of water west-southwest of the Mississippi Delta. X-radiograph negative (denser = lighter) portrays multiple deposits to 35 cm and bioturbated material below that depth. Down-core radiochemical data indicate new material deposited in the upper 13 cm. Excess ^{210}Pb profile reflects multiple event deposits in the upper 40 cm. Geochemical data also exhibit significant variations associated with different storm deposits.

cyclones may prove to be an important mechanism in continental shelf carbon dynamics, and changes in the frequency and intensity of large storm events at continental margins may have significant consequences geochemically and ecologically.

Decadal-Scale Sediment Accumulation

The fallout radionuclides ^{210}Pb, ^{137}Cs, and ^{239,240}Pu have been extensively used to calculate rates of sediment accumulation on decadal timescales, to provide estimates of sediment age with depth, and to establish a depositional chronology for the past 100–150 years. Developing a modern geochronology based on these radionuclides requires some basic understanding of the environment in which the samples were collected and the geochemistry of the radionuclides. However, once a chronology for a system has been determined, a wide range of sedimentological and environmental questions can be addressed, including changes in water quality, deposition history of heavy metals, or trapping of contaminants. Many authors have used geochronological techniques such as the analysis of ^{137}Cs and ^{210}Pb profiles to reconstruct contamination or sedimentation/erosion in estuaries and wetlands bordering the GOM over century timescale (e.g., Ravichandran et al. 1995a, 1995b; Santschi et al. 2001; Warnken et al. 2008; Yeager et al. 2006; Yeager et al. 2007; Oktay et al. 2000; Ravens et al. 2009; Chmura and Kosters 1994; Delaune et al. 1978; Delaune et al. 1983; Delaune et al. 1989; Milan et al. 1995; Nyman et al. 2006).

Briefly, the main processes governing excess ^{210}Pb ($^{210}Pb_{xs}$) profiles are sediment accumulation, radioactive decay, and particle mixing (Goldberg and Koide 1962; see Corbett and Walsh 2015 for a detailed description). Therefore, processes other than radioactive decay can result in an exponential decrease of $^{210}Pb_{xs}$, primarily a function of physical and biological mixing. In the simplest cores, dividing the $^{210}Pb_{xs}$ profile into 2 layers, a surface mixed layer extending to some depth in the sediment profile above a zone of no mixing, allows the sediment accumulation rate (S, cm yr^{-1}) to be calculated in the region of limited mixing via

$$A_z = A_o^{-\lambda(z/S)}, \tag{5.1}$$

where A_z is the $^{210}Pb_{excess}$ activity at depth z, A_o is the initial $^{210}Pb_{xs}$ activity upon deposition, and λ is the ^{210}Pb decay constant (0.3114 yr^{-1}). The sediment accumulation rate can then be calculated using the slope of a least-squares fit derived from the linear regression of the natural log of $^{210}Pb_{xs}$ (A_z) in relation to sample depth (z) (Appleby and Oldfield 1992). Calculating the sediment accumulation rate below the mixing zone assumes that particle reworking rates are not significant over the depth interval of calculation and that fluxes of both sediment and $^{210}Pb_{xs}$ have been constant over the age of the core (i.e., constant flux–constant sedimentation model). If deep mixing cannot be ruled out, Equation 5.1 provides an upper limit of sediment accumulation (Benninger et al. 1979; Nittrouer et al. 1984). A mass accumulation rate (g cm^{-2} yr^{-1}), accounting for variations in sediment porosity caused by the compaction of sediment after burial or during core collection, can also be calculated with this equation by plotting activities as a function of accumulated mass (g cm^{-2}) rather than linear depth (centimeters). As previously described, ^{137}Cs and ^{239,240}Pu are anthropogenic radionuclides, by-products of fission reactions, and typically used in concert with $^{210}Pb_{xs}$ dating techniques. Sediment ages are calculated by relating the initial appearance or peak in activity of these radionuclides in the strata to the years 1953 and 1963, respectively.

Following on the previous discussion of sediment dynamics in the region of the Mississippi Delta, Corbett et al. (2006) compiled ^{210}Pb data from multiple studies (e.g., Louisiana Stimulus for Excellence in Research [LaSER], Nutrient Enhanced Coastal Ocean Productivity [NECOP], Mississippi River Interdisciplinary Research [MiRIR], NSF Integrated Carbon Cycle [NSF ICC] project) conducted over approximately 3 decades (e.g., 1980s through 2005). Through evaluation of the downcore $^{210}Pb_{xs}$, Corbett et al. noted 3 depositional environments (Fig. 5.6): (1) rapid accumulation near Southwest Pass (>2.5 cm yr^{-1} or 1.8 g cm^{-2} yr^{-1}); (2) erosional surface reworked by waves and currents nearshore with no modern sediment accumulation; and (3) distal from Southwest Pass represented by a mixed surface overlying an accumulating deposit, typically <0.7 cm yr^{-1} (<0.5 g cm^{-2} yr^{-1}). A sediment budget based on the spatial distribution of accumulation rates indicates that approximately 40% of the sediment delivered annually by the Mississippi River west of the bird-foot delta is subsequently transported out of the study area (Fig. 5.6). This study focused west of the bird-foot delta since the primary distributary channel of the Mississippi River delta is Southwest Pass and delivers sediments primarily west of the delta.

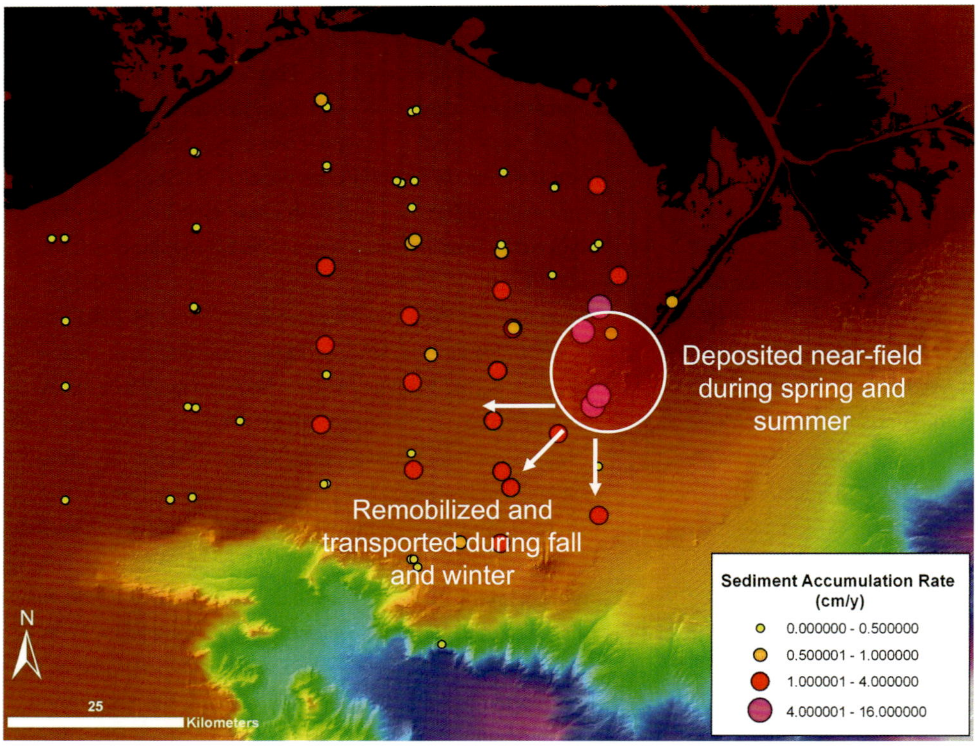

Figure 5.6. Sediment accumulation rates (cm yr^{-1}) across the deltaic region west of Southwest Pass (data from Corbett et al. 2006). Previous research using short-lived radionuclides showed fluvial sediments are initially deposited within 30 km of Southwest Pass (*white circle*) prior to redistribution and long-term accumulation.

Beyond insights into the depositional setting and magnitude of material accumulating on the shelf, ^{210}Pb is critical in establishing a sediment geochronology from cores collected throughout the GOM, thus providing invaluable information on the timing of events. For example, Osterman et al. (2005), Osterman et al. (2008), and Osterman et al. (2009) used ^{210}Pb in conjunction with foraminiferal data to conclude that the general area near the Mississippi River delta experienced hypoxia as far back as the early 1900s, before anthropogenic influence in the Louisiana Bight, and noted that more hypoxia occurred farther west by the mid-1900s. Tichenor et al. (2016) used a similar approach and concluded that sediments in the Louisiana Bight had an increased influence of hypoxic events starting around 1950.

^{210}Pb techniques have been applied beyond the continental shelf. Yeager et al. (2004) analyzed $^{210}Pb_{xs}$ in 6 cores collected from the northern GOM in water depths between 700 and 3500 m. This study was one of the first (at the time) to publish sedimentation rates in the northern GOM outside the immediate area of the Mississippi River delta. Calculated maximum sediment accumulation rates ranged from 0.04 to 0.44 cm yr^{-1} (0.02 to 0.76 g cm^{-2} yr^{-1}) but became essentially constant at ~0.08 cm yr^{-1} (~0.04 g cm^{-2} yr^{-1}) beyond the 1000 m depth contour (Yeager et al. 2004). However, these ^{210}Pb-derived rates were later shown to be mostly overestimates (by 1–2 orders of magnitude) when compared to radiocarbon dating (Santschi and Rowe 2008). Radiocarbon-derived sediment accumulation rates are more reasonable, as the ^{210}Pb-derived sediment accumulation rates would overestimate the total sediment delivery to the northern GOM also by 1–2 orders of magnitude. More recently, several sedimentation experiments have been conducted in the northern GOM related to the 2010 Deepwater Horizon (DWH) blowout event. ^{210}Pb and other short-lived radiotracers (i.e., ^{234}Th, ^{137}Cs, $^{239,240}Pu$) have been critical in establishing a down-core geochronology to interpret the influence of the DWH event recorded in the sediment column. Brooks et al. (2015) measured decadal-scale mass accumulation rates from 0.05 to 0.16 g cm^{-2} yr^{-1}, similar to previous results reported by Yeager et al. (2004). However, mass accumulation rates measured via ^{234}Th in late 2010 following the DWH event were significantly higher (0.48 to 2.40 g cm^{-2} yr^{-1}) (Brooks et al. 2015), but care should be taken when comparing rates calculated from tracers with such different half-lives (Sadler 1981; Anders et al. 1987). Regardless, evidence presented is consistent with a significant short-term (4–5 months) depositional event that deposited 0.4–1.2 cm of sediment following the DWH blowout. Other sedimentological studies have shown an ecological change associated with this event. Schwing et al. (2015) used ^{210}Pb to establish a geochronology from 2 stations and showed

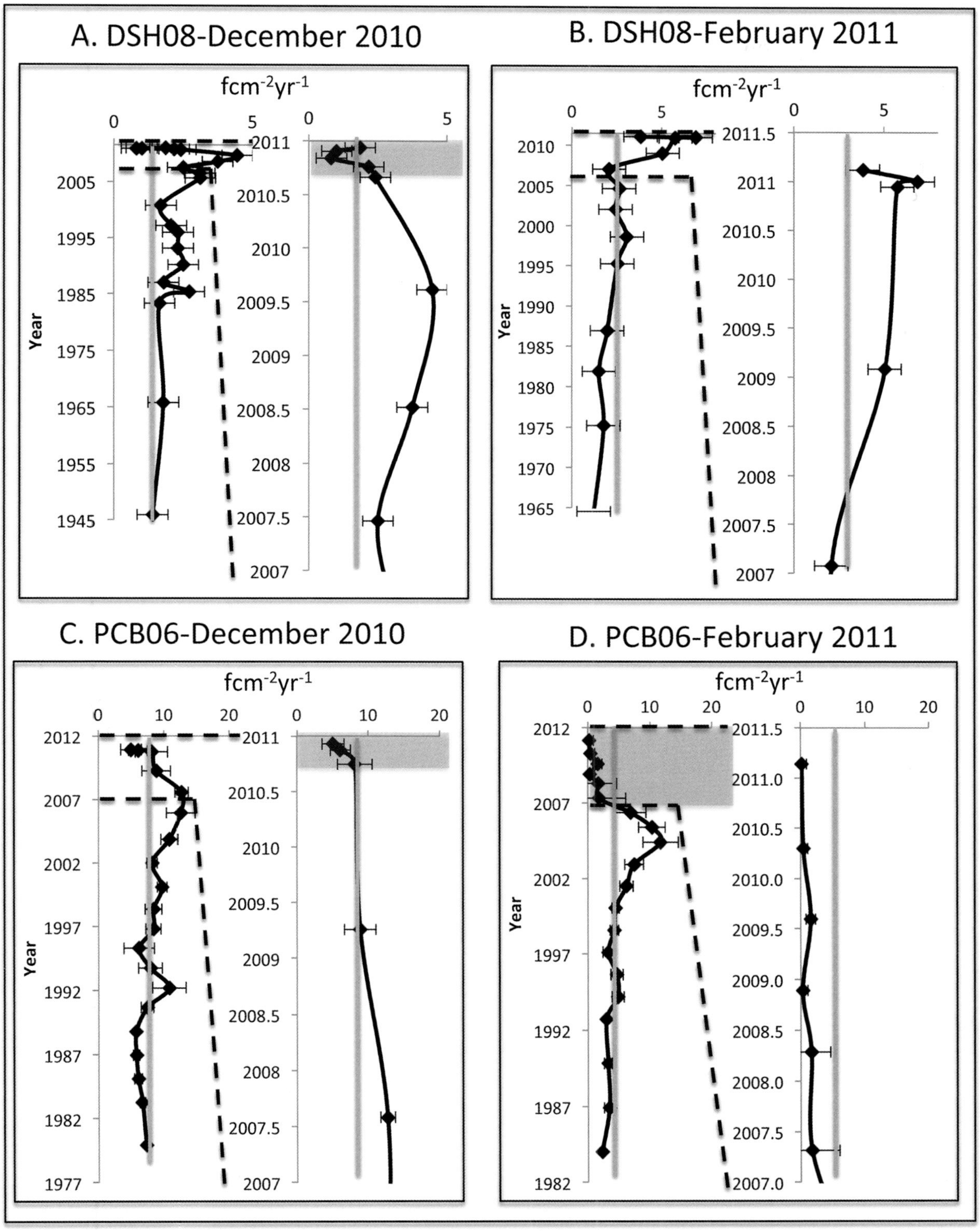

Figure 5.7. Changes in benthic foraminiferal accumulation rate records at 2 sampling sites near the Deep Water Horizon event, approximately 80 km southeast of the Mississippi Delta. Cores were collected in 2010 (*A* and *C*) and 2011 (*B* and *D*). Dates (*y*-axis) were established by modeling the down-core ^{210}Pb activity. The period from 2007 to 2011 is expanded in each section for examination of the period during and after the DWH event. The gray areas represent the decline in benthic foraminiferal accumulation rate (BFAR), and the gray lines represent the down-core mean BFAR (from Schwing et al. 2015).

a decrease in benthic foraminiferal density and benthic foraminiferal accumulation rate (Fig. 5.7) in the modern sediments relative to the down-core mean in all benthic foraminifera. These studies provide a few examples of how this particle reactive tracer has been used to increase our understanding of sedimentological processes across the continental margin and deep GOM.

Longer Timescale Sedimentary Studies with ^{14}C

Over the Quaternary Period (2.6 Myr to present), sea level has risen and fallen numerous times as ice sheets have melted and advanced. To understand sedimentation over this longer time frame, various dating methods may be employed; Walker (2005) discusses a variety of techniques, including optically stimulated luminesce (OSL), tephra chronology, and longer timescale radionuclides. Carbon-14 is one of the most widely used tools to date strata development during and since the end of the last ice age, and it is discussed further here because of its great contribution to forming our knowledge of the GOM.

Evolution of the Gulf Coast, including the Mississippi River delta and other deltas and estuaries, is important to environmental and economic concerns today and in forecasting the future (e.g., habitats and marine resources; Day et al. 2007). Building on the earlier work of Fisk (1944) and others, McFarlan (1961) and Frazier (1967) used radiocarbon dating to determine the timing of sedimentation of the Mississippi River's delta lobes (Fig. 5.8). More recent research has revised and expanded this knowledge (e.g., Coleman et al. 1998; Törnqvist et al. 2008), and studies in the GOM have provided a foundation for comparison of deltaic systems around the world (Coleman and Wright 1975; Wright and Coleman 1973; Roberts 1997; Walsh and Nittrouer 2009). Over the last few decades, many investigations have used a combination of geophysical and geologic data (coring and drilling) coupled with ^{14}C as a temporal tool to better understand the evolution of the reefs (e.g., Hine et al. 2008; Mallinson et al. 2014), rivers and deltas (e.g., Coleman et al. 1998; Blum and Törnqvist 2000; Törnqvist et al. 2004; Törnqvist et al. 2006), barriers and estuaries (e.g., Penland et al. 1988; Simms et al. 2006; Rodriguez et al. 2010), and the continental margin. See Santschi and Rowe (2008), Anderson et al. (2016), Bentley et al. (2016), and Blum et al. (2013) for recent reviews and additional references.

Climate change and coastal responses are key topics of research today, and radiocarbon is essential for developing chronologies of past environmental variability recorded in coastal and marine records. Marshes are challenged by sea-level rise, but they also record its his-

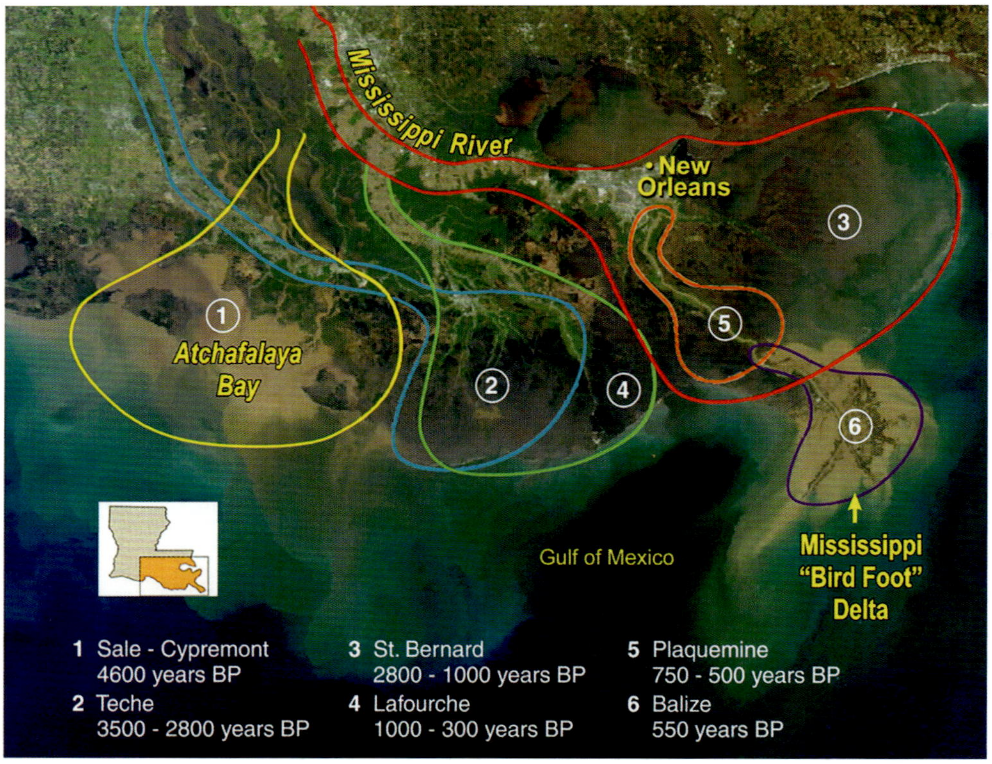

Figure 5.8. Evolution of the Mississippi Delta by lobe-switching sedimentation. The chronology of this changing depositional pattern was developed using radiocarbon data; see references in text (from Day et al. 2007).

tory and can help inform the governing processes, such as the critical processes of sediment compaction and land subsidence (e.g., Törnqvist et al. 2004; Törnqvist et al. 2006; Törnqvist et al. 2008; Blum et al. 2008; González and Törnqvist 2009; Kolker et al. 2011; Yu et al. 2012). Rapid coastal marsh losses are a significant concern, especially the best management approaches (Day et al. 2007; Allison and Meselhe 2010; Nittrouer et al. 2012). The GOM regularly experiences storms, and related deposits are captured in marshes and other coastal deposits. Several studies have used these records to build chronologies of past events (Liu and Fearn 1993, 2000; Liu et al. 2008; Wallace and Anderson 2010). Data indicate that the annual probability of hurricane landfall in the GOM is 0.4–0.5% (Wallace and Anderson 2010; Bregy et al. 2018). Strata on the continental margin and seaward are informative of climate and other changes from decades to millennia (e.g., Flower et al. 2004; Richey et al. 2007; Poore et al. 2005; Poore et al. 2009; Fig. 5.9). Data reveal abrupt events, including meltwater pulses, as well as gradual responses to global conditions (Emiliani et al. 1975; Kennett and Shackleton 1975; Leventer et al. 1982, 1983; Aharon 2003; Sionneau et al. 2010). In these many ways, radiocarbon has been used to constrain various types of ocean changes with time. Additionally, ^{14}C may be utilized to track processes, such as animal consumption of fossil carbon (Wilson et al. 2016) and oil dispersion and uptake (Chanton et al. 2012; Chanton et al. 2015; Rosenheim et al. 2016).

Another atmospherically delivered radionuclide, ^{129}I, which is derived by mainly anthropogenic but also natural processes (Moran et al. 2002), is mostly soluble in water, and its profiles could be used to date water masses in the GOM (Schink et al. 1995), alongside radiocarbon profiles (Severinghaus et al. 1996). In Galveston Bay, $^{129/127}I$ isotopic ratios were used as a novel tool for tracing terrestrial organic carbon (Schwehr et al. 2005), and Schink et al. (1995) discussed its use to date biota from the GOM. Because ^{129}I was contained in near-surface fallout dust particles from the Nevada test site, it could also be used to date Mississippi River delta sediments, along with $^{239,240}Pu$, ^{137}Cs, and ^{210}Pb (Oktay et al. 2000).

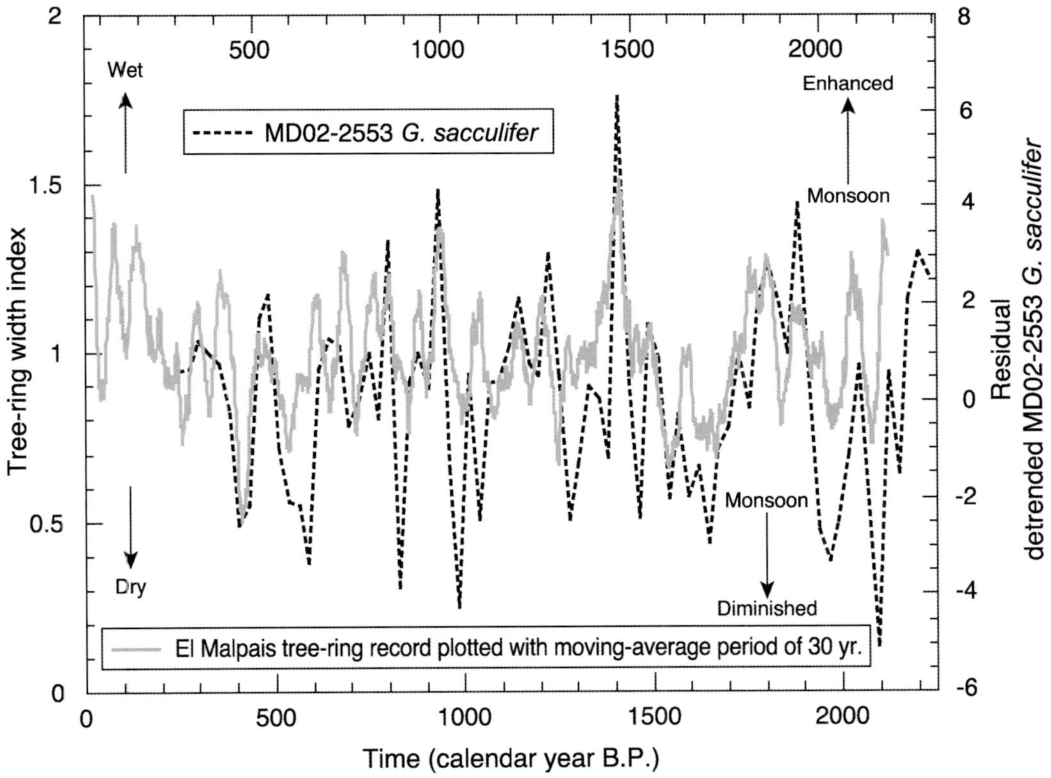

Figure 5.9. Tree-ring record from El Malpais National Monument in west-central New Mexico and foraminifera (*Globigerinoides sacculifer*) data from Pigmy Basin in northern Gulf of Mexico. Data indicate microfossils can be used as a proxy record for the monsoon in the southwestern United States and northern Mexico rainfall and demonstrate monsoon variability (from Poore et al. 2005).

Within the lower Mississippi River, $^{129}I/^{127}I$ ratios were closely related to flow rates and soil leaching rates (Oktay et al. 2001).

Particle Flux through Surface Waters of the GOM via ^{234}Th

The flux of particulate matter through the water column affects many biogeochemical processes in marine environments. Quantifying the magnitude and spatial variability of this flux from the surface to the deep ocean is important for understanding carbon and nutrient cycling and for assessing the fate and transport of contaminants introduced into ocean basins. Many particle-reactive elements are scavenged (i.e., adsorbed) by settling particles and therefore follow the pathways of the particles. Removal of radionuclides or other elements may be biologically-mediated via uptake on organic coatings or in the siliceous or calcareous parts of marine organisms. Precipitation of an oxide phase during oxidation-reduction reactions involving elements including iron and manganese can also efficiently scavenge dissolved radionuclides. Fate and transport of trace metals and pollutants can be determined by characterizing the pathways of a naturally occurring radionuclide that has similar chemical behavior (McKee et al. 1984).

U- and Th-series radionuclides have played a central role in the study of geochemical processes that regulate the composition of ocean waters. Th isotopes, in particular, have been extensively utilized as in situ tracers of particle dynamics (Bacon and Anderson 1982). Th-234 ($t_{1/2}$ = 24.1 days) is continuously produced in oceanic waters by the alpha decay of its parent nuclide ^{238}U (t_{12} = 4.5 × 10^9yr). The near-constant concentration of ^{238}U in the water column results in a constant production rate of ^{234}Th (2430 atoms min^{-1} m^{-3} at a salinity of 35‰). The particle-reactive nature of ^{234}Th often results in radioactive disequilibrium relative to its parent nuclide, which is chemically inert and remains dissolved in seawater as $UO_2(CO_3)_3^{4-}$. The extent of radioactive disequilibrium occurring between ^{234}Th and its soluble parent, ^{238}U, can be used to quantify removal rates for Th. Removal rates can be calculated for timescales that are comparable to the mean life of ^{234}Th (i.e., processes that occur on the order of days to months) (Coale and Bruland 1985).

One-dimensional scavenging models have previously been used to characterize the magnitude and variability of ^{234}Th adsorption in the water column and establish potential scavenging mechanisms. Coale and Bruland (1985, 1987) developed a model for upper-ocean cycling in which ^{234}Th was used as a tracer of the sinking flux of particulate organic carbon (POC). Th-234 residence times in the water column provide insight into the efficiency of removal processes. It has been shown that particle residence times in the euphotic zone are comparable to the recycling times of nutrients, such as phosphate and nitrate (Coale and Bruland 1985). Field data from the water column have shown that the scavenging-rate constants of dissolved particle-reactive radionuclides are controlled by suspended-matter concentrations (McKee et al. 1984). Coale and Bruland (1985) found that the scavenging of dissolved ^{234}Th was most intense in the surface mixed layer of the open ocean and that the scavenging-rate constant roughly correlated with the production rate of particles (i.e., primary production).

As part of a larger project focused on the influences of atmospheric deposition, organic complexation, and photochemical processes on the redox cycle of iron in surface waters of the GOM, Allen (2002) evaluated particle scavenging and residence times via ^{234}Th within the southern and northeastern GOM by sampling varying water masses: open Gulf, coastal waters, and within the Loop Current. This Loop Current (the focus of this discussion) is assumed to be a stable, oligotrophic water mass where lateral advection of shelf waters, input from rivers, and upwelling are minimal (Müller-Karger et al. 1991). Sampling stations were occupied in the spring and summer of 2001 to evaluate seasonal variations in chemical and biological processes. Fluorescence levels in the water column during the spring were 2–3 times higher than during the summer (Fig. 5.10). The large differences between spring and summer were attributed to the low particle concentrations in the summer waters. Also during the spring and summer cruises, samples inside the Loop Current showed distinct decreases in particle export and scavenging, with a particle residence time increasing 3–4 times in the summer (i.e., 29 ± 8 d spring; 100 ± 30 d summer). There was significant disequilibrium between ^{234}Th and ^{238}U throughout the water column during the spring (Fig. 5.10, top panel), indicating shorter particle residence times. However, during the summer, ^{234}Th was in equilibrium or in excess throughout the water column below ~80 m (Fig. 5.10, bottom panel). From the ^{234}Th profiles, it is clear that the particulate, dissolved, and total ^{234}Th activities in the summer were higher than in the spring, indicating longer particle residence times. These findings are comparable

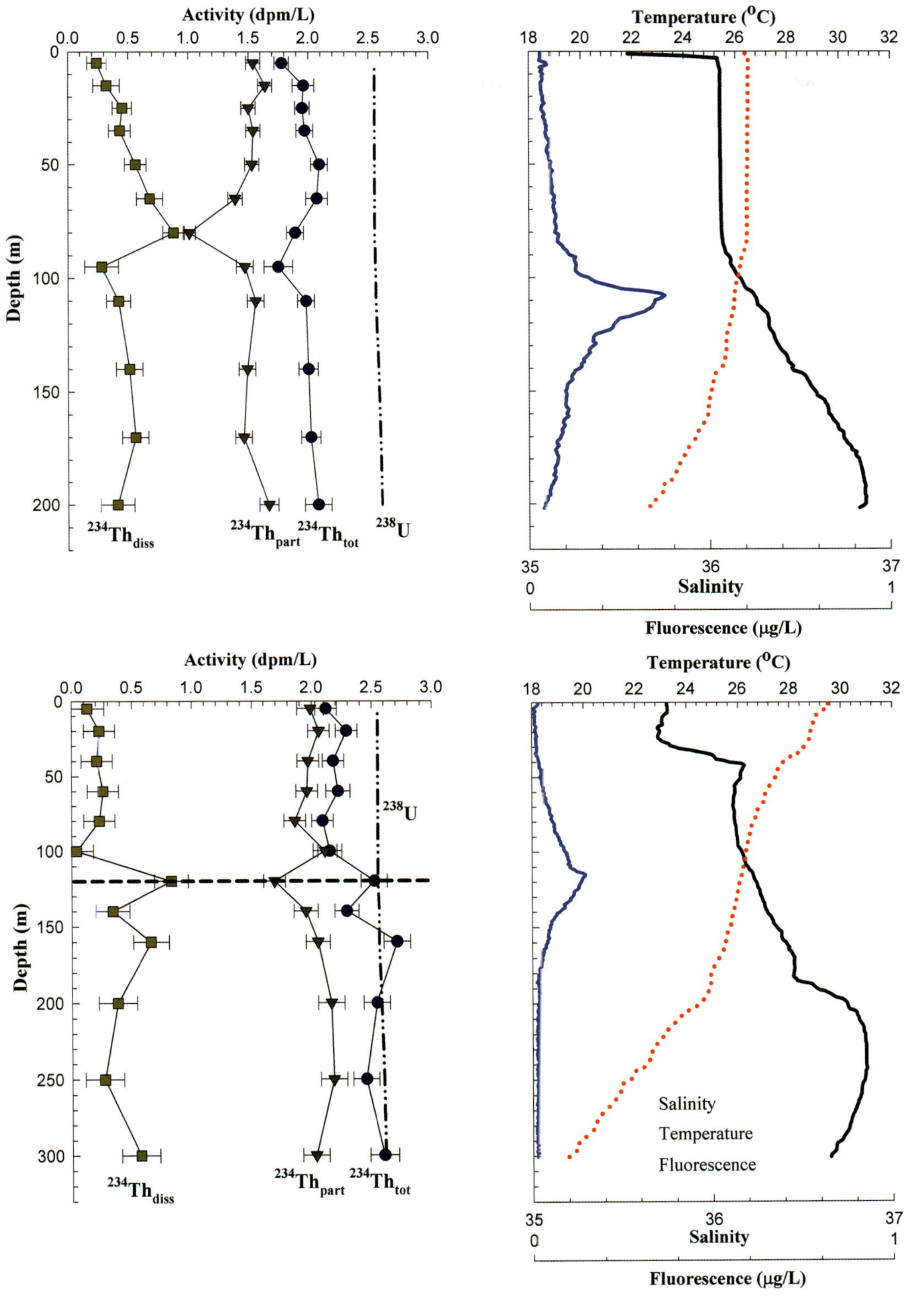

Figure 5.10. ^{234}Th is continuously produced in oceanic waters by the alpha-decay of its parent ^{238}U (2.5 dpm/L). The particle-reactive nature of ^{234}Th results in radioactive disequilibrium relative to its parent radionuclide, which is chemically inert and remains dissolved in seawater. The deficit of ^{234}Th in surface waters relative to its parent (^{234}Th:^{238}U disequilibrium) can be related to the direct export of particles into the deep ocean. ^{234}Th was used to evaluate particle flux in the central Gulf of Mexico (25°N, 86°W) in the spring (*top panel*) and summer (*bottom panel*) of 2001 (Allen 2002). Increased particle scavenging and export rates were related to in situ fluorescence (primary productivity).

to the findings of Baskaran et al. (1996), who also showed that for deeper-water stations, average particle residence times were shorter in spring than in summer.

In a more detailed study of ^{234}Th profiles within warm-core and cold-core rings in the northern GOM, where ^{234}Th concentrations in suspended particles and the dissolved phase, along with ^{234}Th fluxes in short-term sediment traps (corrected for release during the 1–2 exposures), were compared to profiles of pigments, POC, uronic acids, and total acid polysaccharides. It was then shown that ^{234}Th/POC ratios in particles were strongly related to uronic acid/POC ratios, and the vertical flux of total acid polysaccharides to ^{234}Th/POC ratios in settling particles. However, profiles of ^{234}Th in this study were similar to those of Guo et al. (2002) and Santschi et al. (2005).

Radionuclide Associations with Water-Column Colloids and Particles

Colloids are a key to understanding natural processes, such as the self-cleansing capacity of natural aquatic systems. "Colloids" are defined, according to the International Union of Pure and Applied Chemistry (IUPAC), as nanoparticles and macromolecules with dimensions between 1 nm and 1 μm. Thomas Graham (1805–69) coined the term to distinguish those materials in aqueous solution that would not pass through a membrane from those that would. The term "colloid" is derived from the Greek *Kolla*, meaning "glue." In a wider sense, colloids are one phase dispersed in another phase. Thus, the term also includes aerosols, foam, and engineered nanoparticles. Because their size is in the same range as the wavelength of visible light, colloids scatter light, and thus their concentration and size distribution can also be determined, among other methods, by using light-scattering techniques (Krishnamurti 1929). The existence of colloidal substances has been described in the literature for almost 100 years. Radiochemists were among the first to describe colloidal substances that bind low concentrations of radionuclides in the absence of a stable element as a carrier (Paneth 1914).

While the name "colloids" was for many years interpreted to represent mainly metal oxides and clay minerals, customarily by inorganic chemists and environmental chemists, only about 2 decades ago these chemists realized that in estuaries and the ocean, most of these colloidal substances consist of organic macromolecules, with only low concentrations of clays and metal oxides (e.g., Guo et al. 2000).

Colloids in the GOM and its estuaries, as in all natural aquatic systems, are then composed mainly of aquagenic (e.g., exopolymeric substances) and pedogenic (terrestrial) organic substances (Buffle 1991). These substances are macromolecular and mainly composed of polysaccharides (e.g., Santschi et al. 1998; Quigley et al. 2002; Guo et al. 2002; Xu et al. 2011), uronic acids and other acid polysaccharides (Guo et al. 2002; Xu et al. 2011), and proteins (Alvarado Quiroz et al. 2006). Minor components include hydroxamate siderophores (Chuang et al. 2015). These components contain chelating (e.g., for metal ions) and aromatic sites (e.g., for iodine; Schwehr et al. 2005) that can strongly bind selective trace elements and particle-reactive radionuclides (of Be, Pb, Po, Th, Pa, and others). At the molecular level, colloids form gels from nanofibrils (with dimensions of 1–2 nm to several micrometers; Santschi et al. 1998).

The particle reactivity of Th(IV), an A-type hard-metal ion, also leads to significant associations of ^{234}Th to particles and colloids, which makes it suitable as a tracer for deriving their residence times. Colloidal organic matter (COM) is continuously transformed, removed, and reintroduced or renewed in natural waters. When colloids are enriched to 100% polysaccharides, they have a modern radiocarbon age (Santschi et al. 1998), while apparent radiocarbon ages of COM range from modern to thousands of years, because modern radiocarbon ages cannot distinguish between turnover times of hours to years, whereas ^{234}Th can. When ^{234}Th was used in conjunction with radiocarbon to determine particle and colloid turnover rates in the northern GOM, timescales were on the order of days to weeks, depending on particle concentration (Santschi et al. 1995). In the Galveston Bay estuary, ^{234}Th-derived turnover rates of particles were considerably shorter, on the order of a few hours (Baskaran et al. 1992; Baskaran and Santschi 1993).

Analysis of radiocarbon, combined with $\delta^{13}C$, was successfully applied to colloidal macromolecular organic carbon from Galveston Bay and the GOM to delineate a multi-endmember model that included the estuary, surface, and deep Gulf (e.g., Guo et al. 2003). Normally, the substances with the highest molecular weight have the shortest residence time (Guo et al. 1997; Santschi et al. 1995). Continuous removal by aggregation of macromolecules to suspended particles via "marine snow" composed of exopolymeric substances (EPS) provides an

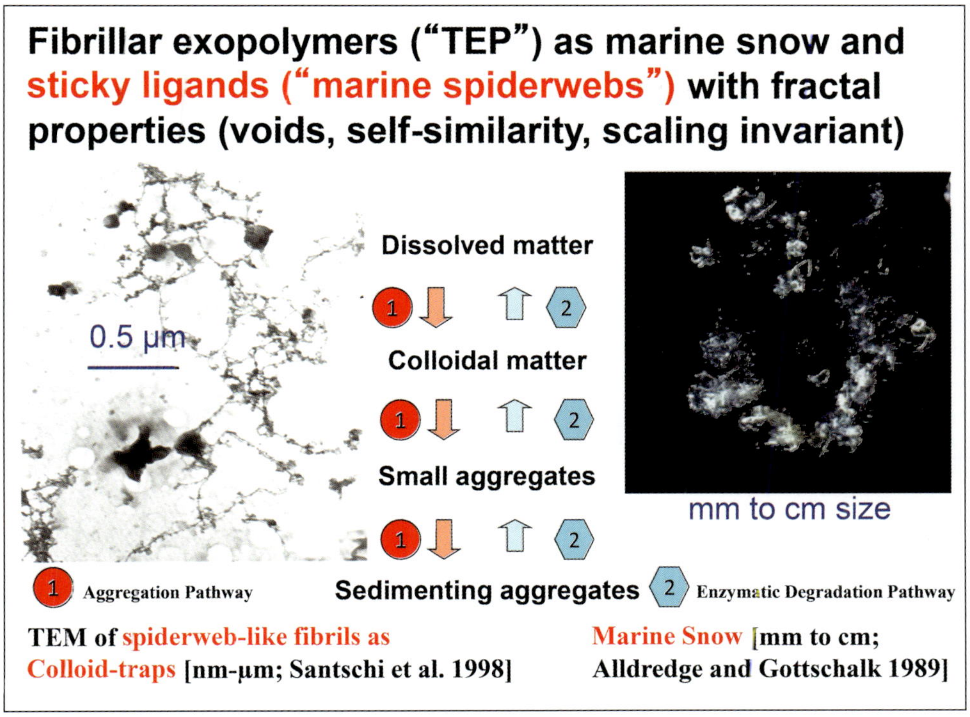

Figure 5.11. The continuous removal by aggregation of macromolecules operates opposite to the main pathway of degradation, producing marine gels (*left*) from colloidal fibrillary macromolecular organic matter and marine snow (*right*) aggregates (Santschi et al. 1998; Alldredge and Gottschalk 1989).

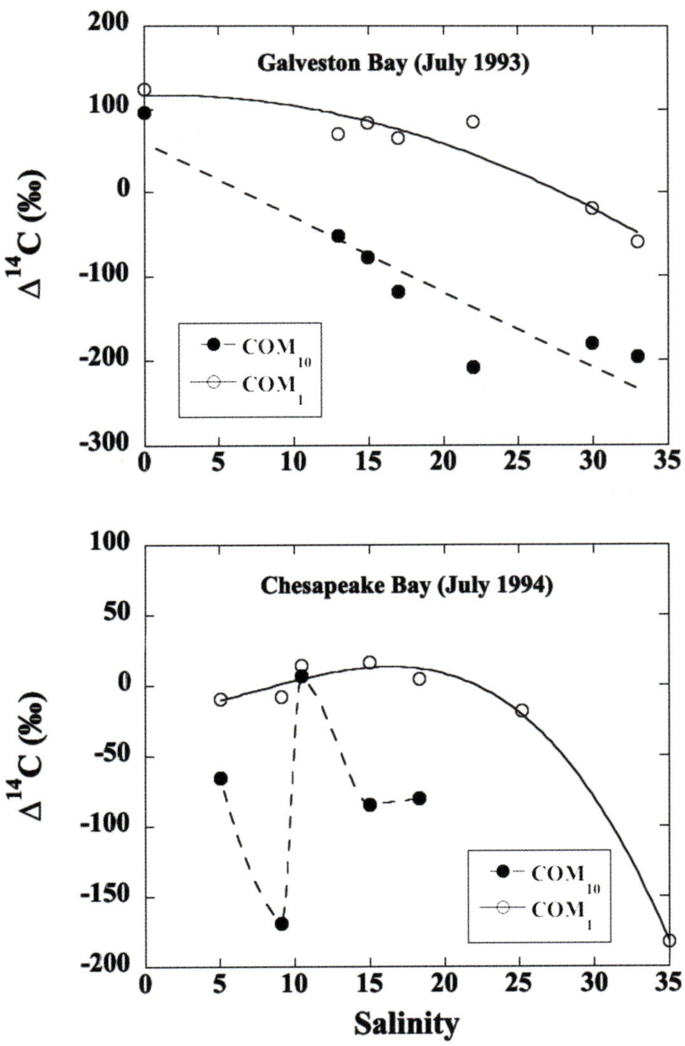

Figure 5.12. Sediment resuspension can reverse the normal pattern of increasing residence times with decreasing molecular weight of COM (COM1 ≥ 1 kDa; COM10 ≥ 10 kDa), commonly found in estuary waters without sediment resuspension (*upper graph*) (Santschi et al. 1995). Resuspension can result in HMW-NOM (COM10), having an older radiocarbon age (i.e., more negative $\Delta^{14}C$) than LMW-NOM, once again the reverse of what is commonly found estuarine waters without sediment resuspension (Guo and Santschi 1997; with permission from Marine Chemistry, Elsevier Inc.).

alternative pathway to degradation whereby it is thought that EPS can provide the "glue" to allow aggregation of particles (Alldredge and Gottschalk 1989; Santschi et al. 1998; Thornton 2004; see Fig. 5.11). In shallower systems such as Galveston Bay, sediment resuspension can reverse the normal pattern of increasing residence times with decreasing molecular weight of organic compounds, as shown by the reversed pattern when residence times are plotted against salinity (Fig. 5.12), and impart high-molecular-weight natural organic matter (HMW-NOM) with older radiocarbon ages than low-molecular-weight (LMW) NOM; Guo and Santschi 1997).

Guo and Santschi (2007) and Doucet et al. (2007) provide extensive details on methods, approaches, techniques, and results on colloidal associations of radionuclides and trace elements. These authors provide many examples from the GOM of studies on colloidal associations of NOM, including EPS, radionuclides, and trace elements. To carry out meaningful research on colloidal associations of radionuclides, however, one needs a well-calibrated ultrafiltration system. Several prime examples using this system are the calibration studies carried out by Guo and Santschi (1996), Guo et al. (2000), and Guo et al. (2001), where a permeation model was successfully applied to ultrafiltered organic matter and selected radionuclides.

The strong association between selected radionuclides (e.g., ^{234}Th) and colloids leads to a close relationship between their colloidal fractions (Fig. 5.13). Demonstrations of significant colloidal associations of selected radionuclides by Baskaran and Santschi (1993), Baskaran et al. (1992), Baskaran et al. (1996), Baskaran et al. (1997), and Stordal et al. (1996) paved the way to further verify the Brownian colloidal pumping model by Honeyman and Santschi (1989), which links adsorption to particles and colloids to the coagulation of colloids. These authors successfully predicted the frequently observed decreasing (log) partitioning coefficients (K_d), and increasing kinetic constants, with particle concentration. This dependency is due to sorption to aggregating colloids and a consequence of colloidal, macromolecular binding ligands in the filter-passing fraction; this is illustrated in Figure 5.14 by the dependency of log scavenging-rate constants (λ_s) and log particle-water partition coefficients (K_d) with log suspended-particle concentration (SPM; Baskaran et al. 1996).

Transects along salinity gradients in estuaries (e.g., Baskaran et al. 1992; Baskaran and Santschi 1993; Guo et al. 1995; Schwehr et al. 2005) and positioned relative to warm-core and cold-core rings in the GOM (e.g., Guo et al. 2002; Santschi et al. 2003; Xu et al. 2011) were carried out mainly to verify the Brownian pumping model that predicts particle-concentration effects on particle-water partition coefficients and particle or colloid residence times due to colloidal associations of Th nuclides and other radionuclides (Baskaran et al. 1992; Guo et al. 1995; Guo et al. 2002; Baskaran and Swarzenski 2007; Baskaran and Santschi 1993; Schwehr et al. 2005) and organic carbon transport rates (e.g., Baskaran et al. 1996; Guo et al. 1997; Hung et al. 2004; Hung et al. 2010; Xu et al. 2011).

Studies on the composition of colloidal macromolecular organic matter pointing to specific carrier mole-

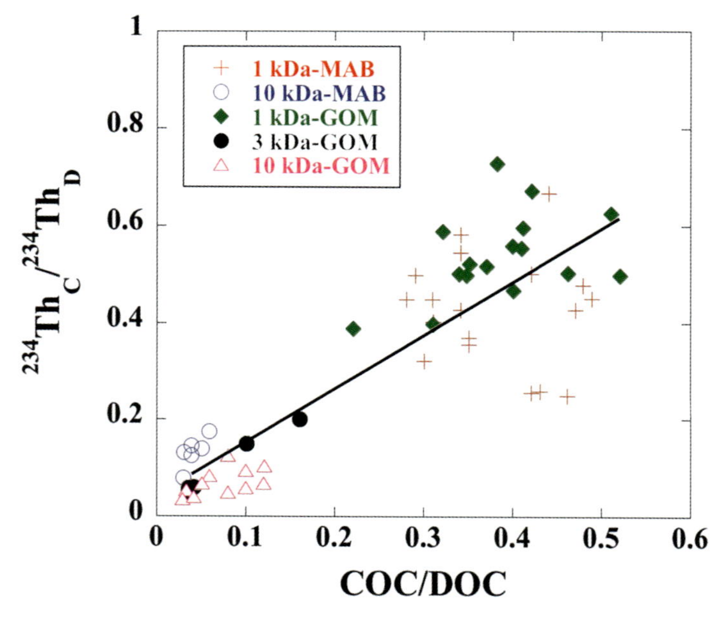

Figure 5.13. Th-234 partitions (colloidal [Th_C] to total dissolved [Th_D]) in proportion to the way colloidal macromolecular natural organic carbon (COC) partitions to total dissolved organic carbon (DOC) (Guo et al. 1997). These results indicate that the higher the proportion of colloidal organic matter, the higher the proportion of colloidal ^{234}Th (with permission from Colloids and Surfaces, A, Elsevier Inc.).

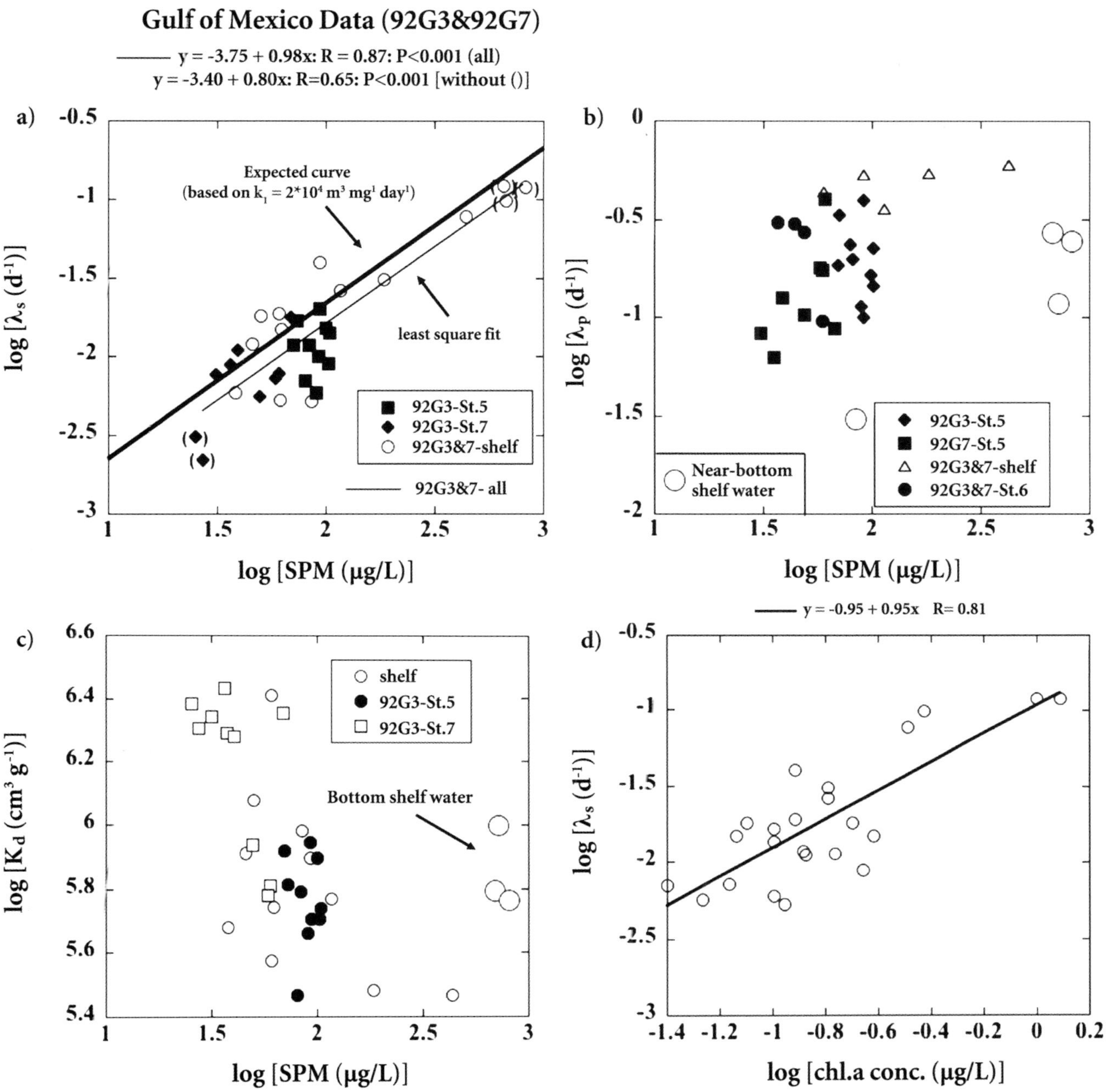

Figure 5.14. Suspended particulate matter (SPM) dependence of the ^{234}Th-derived kinetic-rate constants and particle-water partition coefficients (K_d) that confirm predictions by the Brownian colloidal pumping model of Honeyman and Santschi (1989). (*a*) Logarithm of the model-derived first-order dissolved ^{234}Th scavenging-rate constant (λ_s) vs. logarithm of the SPM concentration from all available data from the Gulf of Mexico (92G3 and 92G7); (*b*) logarithm of the model-derived first-order particulate ^{234}Th scavenging-rate constant (λ_p) vs. logarithm of the SPM concentration from all data (92G3 and 92G7); (*c*) logarithm of the distribution coefficient ($Kd = [^{234}Th_p]/([^{234}Th_d][SPM])$ vs. logarithm of the SPM concentration from all data (92G3 and 92G7); (*d*), scavenging-rate constants (λ_s) also relate to pigment (Chl-*a*) concentrations (from Baskaran et al. 1996; with permission from Continental Research, Elsevier Inc.).

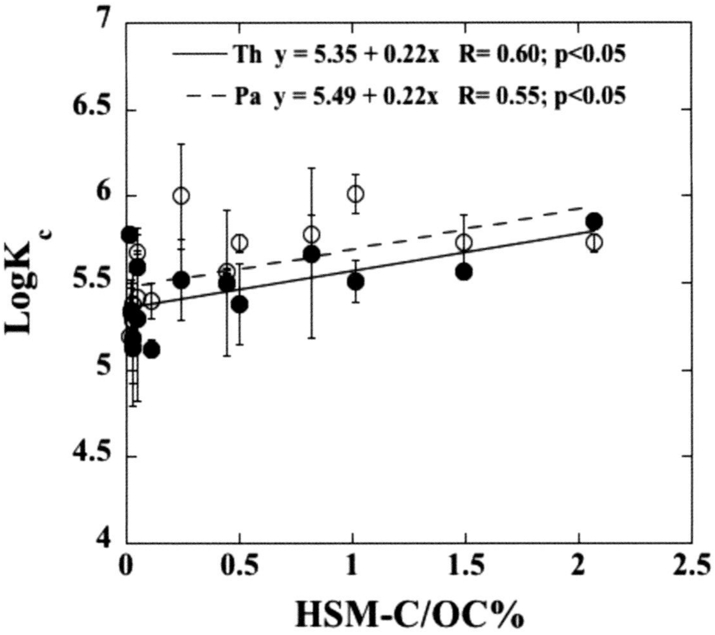

Figure 5.15. A close relationship exists between Th and Pa radionuclide particle-water partition coefficients (expressed as log K_c) and the relative concentration of hydroxamate siderophores (HSM-C/OC) in ultrafiltered (and subsequently diafiltered) colloid samples collected from the northern Gulf of Mexico and the Pacific Ocean (Chuang et al. 2015; with permission from Marine Chemistry, Elsevier Inc.).

cules for selected radionuclides include those of Alvarado Quiroz et al. (2006) and Chuang et al. (2015). Quigley et al. (2002) and Alvarado Quiroz et al. (2006) attributed potential radionuclide-binding ligands in colloidal EPS from the GOM using radiotracers to potential functional groups. When Chuang et al. (2015) compared colloids from the GOM with those collected from other oligotrophic sites, such as the Pacific Ocean, they found for the first time a close relationship between ^{234}Th bound to colloids and the presence of hydroxamate siderophore ligands in colloids. Hydroxamate siderophores are produced by microorganisms to increase the micronutrient iron availability for uptake and are normally found only in the truly dissolved LMW fraction (Fig. 5.15). Earlier accounts of what is known about Th-binding ligands have been summarized by Santschi et al. (2006).

Radionuclides and Water Exchange

Submarine Groundwater Discharge

Submarine groundwater discharge (SGD) is potentially a significant process that influences nearshore nutrient and geochemical budgets. Johannes stated, "SGD should occur anywhere that an aquifer is hydraulically connected with the sea through permeable rocks or bottom sediments and where the head is above sea level" (1980, 365). This condition is met in most coastal environments throughout the Southeast and Gulf Coast regions, and research indicates that SGD is a significant nearshore process in these regions. Beyond the transport of water, groundwater discharge has been identified as a source of new and regenerated nutrients to receiving waters and may contribute nutrient loads similar to those from rivers (e.g., Valiela et al. 1990; Slomp and Van Cappellen 2004; McCoy et al. 2007). The delivery of trace metals and organic carbon via SGD has also been recognized and suggested to be important in regional and global geochemical budgets (Duncan and Shaw 2003; Windom et al. 2006; Johannesson and Burdige 2007; Moore 2010; Johannesson et al. 2011).

Advancement in knowledge of SGD has been rapid, but the subject has been researched intensely for less than 20 years. Estimates of SGD have been calculated in many areas throughout the Southeast and Gulf Coast regions, and more recent research has been focused on characterization and quantification of SGD sources. SGD is defined as "any and all flow of water on continental margins from the sea bed to the coastal ocean, regardless of fluid composition or driving force" (Burnett et al. 2003, 3). Therefore, as defined, SGD may include both fresh and saline sources of groundwater (Fig. 5.16). Several methods are utilized to assess SGD (e.g., direct measurement, geochemical tracers, numerical modeling), which have provided opportunities to compare results and assess strengths and weaknesses associated with each method.

SGD can be traced from the land-ocean boundary across continental shelves using radium isotopes (Charette et al. 2001; Peterson et al. 2009) and radon (Corbett

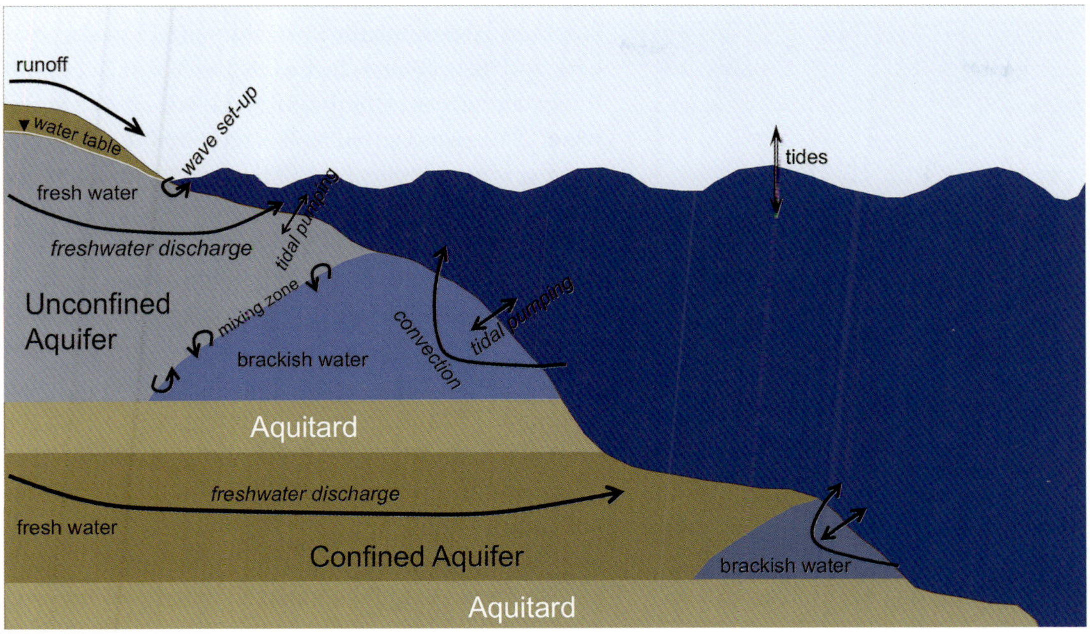

Figure 5.16. Primary sources of water and drivers of submarine groundwater discharge (SGD) in coastal aquifers (after Taniguchi et al. 2002).

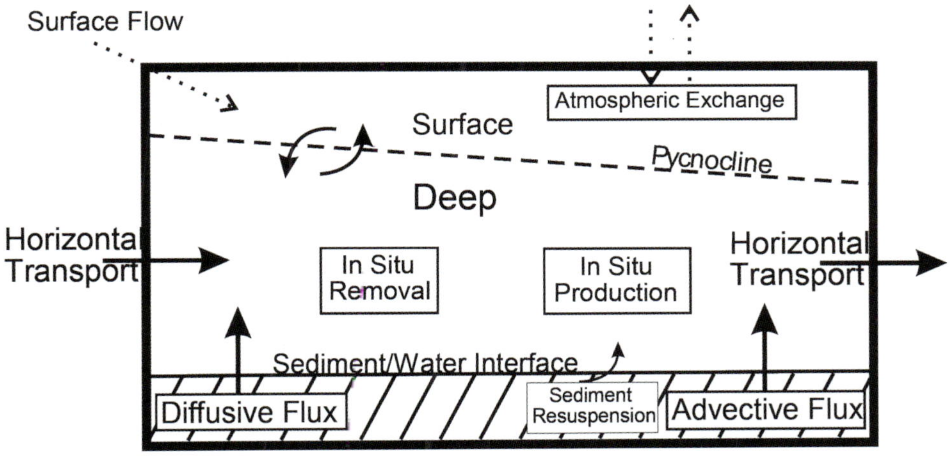

Figure 5.17. A simple box model can be used to describe the sources and sinks of natural groundwater tracers. In situ removal and production are dependent on the tracer of interest and may include desorption/adsorption, decay, and so on.

et al. 2000; Burnett et al. 2006; McCoy et al. 2007; McCoy et al. 2007). Ra and Rn (^{224}Ra $t_{12} = 3.63$ d, ^{223}Ra $t_{12} = 11.43$ d, ^{228}Ra $t_{12} = 5.75$ yr, ^{226}Ra $t_{12} = 1600$ yr, ^{222}Rn $t_{12} = 3.83$ d) are naturally occurring radionuclides that are enriched in groundwater relative to seawater (Moore 1996; Cable et al. 1996; Burnett et al. 2003). Ra is generated naturally by the decay of Th, which is present in soils, rocks, and sediments (see Fig. 5.1). Th is highly particle reactive in the marine environment, and Ra is particle reactive in freshwater environments. When Ra-enriched water enters the marine environment, Ra desorbs from particles and moves with the associated water mass. In contrast, ^{222}Rn is an inert gas produced through the decay of ^{226}Ra. Rn is typically elevated in groundwater due to the production and recoil processes originating from Ra within the aquifer matrix, thus accumulating as a dissolved gas in groundwater. The water-column standing crop of the dissolved natural tracer represents a balance between its sources and sinks. A simple box model for these tracers may be used to describe the source and sink terms in the coastal system (Fig. 5.17). In general, the tracer inventory is a balance between (1) benthic advective-diffusive exchange; (2) in situ production and loss, such as decay, desorption, and adsorption; (3) horizontal water-column advection; (4) eddy diffusion through the pycnocline; and (5) air-sea exchange. Ultimately, these tracers allow

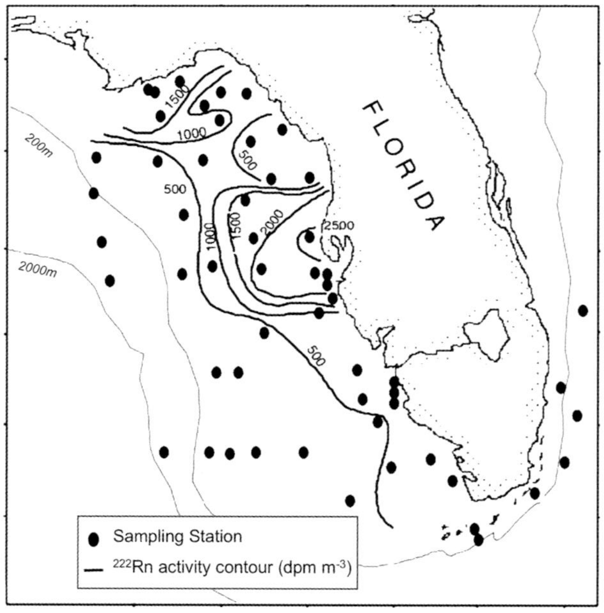

Figure 5.18. Contours of unsupported ^{222}Rn in bottom waters off the west Florida shelf (modified from Fanning et al. 1987).

for the evaluation of the volume and sometimes pathways of SGD to be identified via the spatial distribution of Rn- and Ra-enriched surface waters (Moore 1996).

Some of the earliest work linking these tracers to groundwater in the GOM was completed on the west Florida continental shelf. Fanning et al. (1987) measured unsupported ^{222}Rn (^{222}Rn minus that supported by ^{226}Ra) bottom-water concentrations on the west Florida shelf ranging from 500 to 2500 dpm m^{-3} and suggested groundwater discharge or diffusion from underlying phosphorite deposits might be responsible for these relatively high concentrations (Fig. 5.18). Cable et al. (1996) noted that the highest ^{222}Rn contours aligned with a high abundance of submarine freshwater springs along the Florida coastline. These authors went on to develop an SGD tracing tool based on Rn inventories and applied the model to the northeastern GOM. The study found that SGD rates of 2–10 cm d^{-1} were needed to maintain water-column ^{222}Rn inventories. Although the SGD velocities (cm d^{1}) may seem small in comparison to the velocity of water exiting a river, the large spatial distribution over which this process acts can provide a volumetrically important source of fluids; and Cable et al. showed that this source was comparable in discharge to that of the major rivers of Florida.

In contrast to the karst hydrogeology of the west Florida coast, the underlying geology of coastal Louisiana is dominated by discontinuous, near-surface sands, point bars, and distributary-channel deposits associated with the deltaic history of the area (Rollo 1966). Although this region has been referred to as a river-dominated ocean margin, several studies have suggested that groundwater may also play an important role in transporting water and bioactive elements to the coastal waters along this deltaic system (Krest et al. 1999; Moore and Krest 2004; Kolker et al. 2013). Krest et al. (1999) used a Ra mass-balance approach to estimated SGD on the continental shelf of Louisiana adjacent to the Mississippi River. They used ^{226}Ra and ^{228}Ra as indicators of SGD and, by process of elimination, suggested that elevated activities were most likely due to advection of Ra-rich SGD and that this discharge may enter bottom waters of the Gulf shelf between the mouth of the Mississippi and the Atchafalaya Rivers at a rate as high as 1 cm d^{-1}. Moore and Krest (2004) reported enriched ^{223}Ra and ^{224}Ra activities around the Atchafalaya and Mississippi Rivers and estimated an SGD rate of 2.5 cm d^{-1}, equivalent to 7% of average Mississippi River discharge, would be required to support the observed surface-water activities. Interestingly, McCoy et al. (2007) conducted a follow-up study using ^{3}He, ^{4}He, ^{222}Rn, and ^{226}Ra to evaluate SGD along the continental shelf west of the Mississippi River delta and found significantly lower rates of SGD (0.01–0.14 cm d^{-1}). These tracer-based rates of SGD agreed with a corresponding hydrogeologic modeling study (Thompson et al. 2007). The higher rates measured by Krest et al. (1999) and Moore and Krest (2004) across the continental shelf via Ra isotopes may be attributed to the discharge of Ra-enriched "formation water" associated with oil and gas extraction throughout the region. This formation water is a part of the oil extraction process that leads to the release of fluids to surface waters and includes the brine fluid associated with the hydrocarbon-bearing strata and any water and chemicals used as part of the extraction process. In fact, Krest et al. (1999) calculated a discharge of only 0.03 cm d^{-1} of formation water was needed to support observed elevated bottom-water radium activities without invoking groundwater sources.

Although SGD may not be a significant process delivering freshwater along the continental shelf west of the Mississippi River delta (McCoy et al. 2007; Thompson et al. 2007), other studies have used a similar radiotracer approach to demonstrate the importance of this hydrologic process in adjacent coastal bays (Kolker et al. 2013). This study estimated groundwater discharge into Barataria Bay via ^{222}Rn mass balance. The authors found that SGD varied seasonally and ranged between 280 and 4760 $m^3\ s^{-1}$, equivalent to 0.6–10 cm d^{-1} when normalized to the area of the basin (4200 m^2), or 0–15% of the total flow of the Mississippi River (Kolker et al. 2013). This large volume of water entering a coastal bay would certainly have biogeochemical implications that should be considered in future work.

Cross-Shelf Mixing

The land-ocean interface is extremely dynamic. The hydrodynamics of coastal systems are complex and unique, although physical transport processes on multiple spatial and temporal scales are often the dominant factors in controlling geochemical and biological processes. The transport of nutrients, water-residence times, river-plume dynamics, and oil-spill trajectories are but a few of the important physical processes that influence the ecology of coastal systems. Several radionuclides, particularly the Ra quartet ($^{223,\,224,\,226,\,228}$Ra), have been used extensively in coastal and continental shelf systems to better define physical processes responsible for the distribution of chemical and biological species.

Moore (2000) first described the utility of radium isotopes in quantifying fluxes of dissolved components across the continental shelf. The change in activity of the radionuclides as a function of distance from shore can be used to compute a coastal mixing rate (K_h), shown in Equation 5.2:

$$A_x = A_o\, exp\left[-x\sqrt{\frac{\lambda}{K_h}}\right] \tag{5.2}$$

where A_x is the isotopic activity (i.e., ^{223,224,228}Ra) at distance x from the coast; A_0 is the activity at the shoreline ($x = 0$); and λ is the corresponding decay constant. This model is based on the assumptions that (1) the system is at steady state, (2) the only source of Ra is a constant flux at the landward boundary (A_0), and (3) Ra losses are through mixing and decay only. This suite of Ra isotopes can also be used to calculate the water-mass age (Moore 2003), shown in Equation 5.3:

$$AR_{obs} = AR_i\left[\frac{exp(-\lambda_{224}\, t)}{exp(-\lambda_{224}\, t)}\right] \tag{5.3}$$

where AR_{obs} is the observed ^{224}Ra/^{223}Ra activity ratio at each sample location, AR_i is the ^{224}Ra/^{223}Ra activity ratio of the source area, and λ_{224} and λ_{223} are the decay constants for each radionuclide (d^{-1}). The equation can be solved for t (days) and defines the "age" of the water mass at the time of sampling. This calculation assumes that the activity ratio of the source (AR_i) is constant.

Given the assumptions of this model and the complexity of some continental shelves (e.g., Louisiana shelf adjacent to the Mississippi River delta), some coastal systems may be difficult to model due to additional offshore sources of radionuclides (e.g., particle desorption, benthic flux) and lateral advective transport. However, these models have been applied as part of several groundwater-related studies along the northeastern GOM and the west Florida shelf (Table 5.1). As an example, Smith and Swarzenski (2012) quantified the nearshore (i.e., 15 km) mixing rates via $^{223,\,224}$Ra as part of a larger effort to estimate the SGD-based nutrient fluxes to the west Florida shelf and evaluate these in terms of estimated nutrient demands of relatively large red-tide blooms (*K. brevis*). Using Equation 5.2 and a regression analysis, they obtained the horizontal mixing rates from each radionuclide and each sampling period (November 2009, June 2010; Fig. 5.19). The calculated mixing rates (14.2–73.6 $km^2\ d^{-1}$) were used to evaluate cross-shelf ^{222}Rn transport in a larger mass-balance model to quantify the advective flux of water and associated nutrients across the sediment-water interface. Smith and Robbins (2012) applied a similar approach as described previously for Ra isotopes but used instead ^{222}Rn, which has a similar half-life but must be corrected

Table 5.1. Summary of cross-shelf mixing (K_h) rates quantified via radiotracers in the Gulf of Mexico

Location	Isotopes	K_h ($km^2\ d^{-1}$)	Reference
Northeastern GOM, Turkey Point, FL	^{223}Ra, ^{224}Ra	$10^{-0.63} - 10^{0.02}$	Moore 2003
Northeastern GOM, Turkey Point, FL	^{223}Ra, ^{224}Ra	$10^{-0.10}$, $10^{-0.04}$	Santos et al. 2009
West Florida shelf, Pinellas County, FL	$^{223,\,224}$Ra	$10^{1.2} - 10^{1.9}$	Smith and Swarzenski 2012
West Florida shelf, Pinellas County, FL	^{222}Rn	$10^{-0.38} - 10^{3.4}$	Smith and Robbins 2012

for any loss to the atmosphere through evasion. Although this approach resulted in a greater variation in mixing rates, the range and average of these modeled results agree well with other mixing rates obtained by the Ra method (Table 5.1). This 1-D modeling approach has been widely used on continental shelves worldwide. Continued refinement of these models (Hancock et al. 2006; Zhang et al. 2011) will ultimately provide opportunities to use the approach more broadly in the GOM to better understand the interaction of coastal and offshore waters.

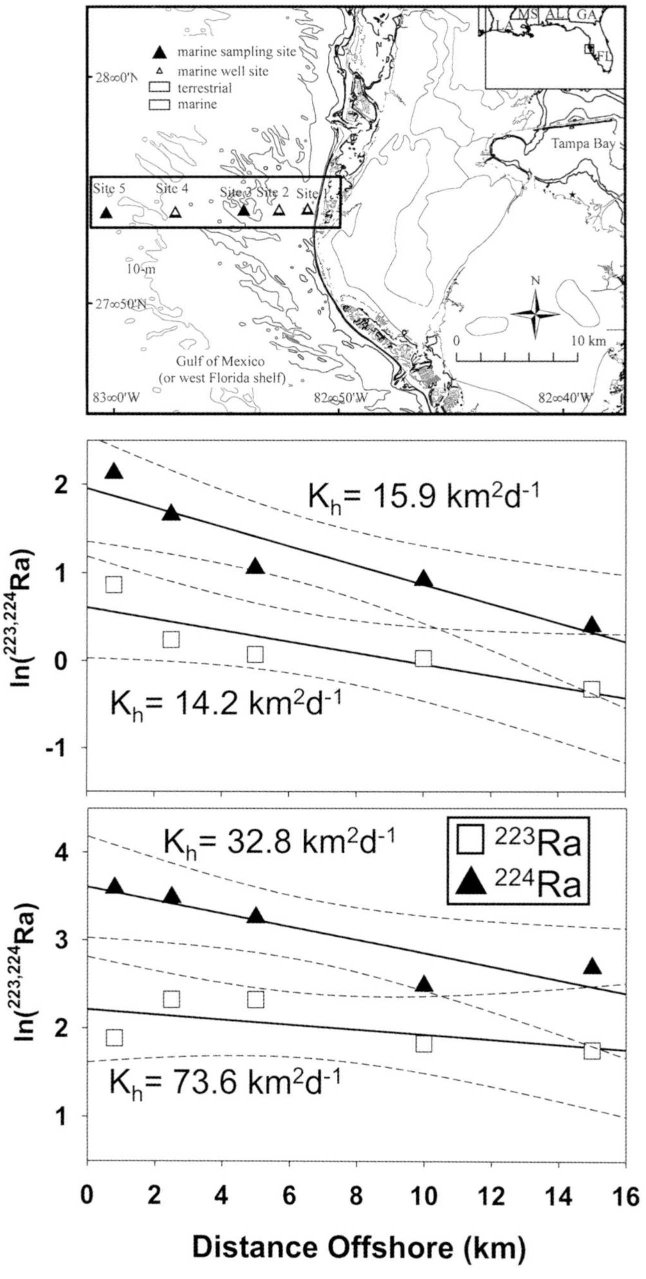

Figure 5.19. Samples collected from a shore-perpendicular transect off Pinellas County, Florida (*A*). Sampling transect extended from Tampa Bay across the peninsula to 15 km offshore near the 10 m isobath. $^{223,224}Ra$ were used to evaluate cross-shelf mixing rates (K_h) during November 2009 (*B*) and June 2010 (*C*) (modified from Smith and Swarzenski 2012).

Conclusion

Radionuclides play a critical role in chemical oceanography to help establish the magnitude and/or timescale of processes throughout the marine environment, including those in the atmosphere, water column, and sediments, as well as across all associated interfaces (e.g., atmosphere-water; benthic-pelagic; surface-deep ocean). These radiotracers have been used extensively throughout the GOM, from the coast to open ocean. Here, we explain some of the most important uses of radioisotopes to demonstrate the breadth of their application, but certainly many other excellent examples and applications were not included as studies are numerous and vary tremendously.

Fundamentally, there are 3 types of radionuclides: (1) those along primordial decay chains (e.g., ^{234}Th, ^{210}Pb); (2) those cosmogenically formed (e.g., ^{3}He, ^{7}Be); and (3) those anthropogenically created (e.g., ^{137}Cs) or enhanced (e.g., ^{14}C). This summary demonstrates how radioactive nuclides can provide insights into a multitude of processes across a wide range of temporal and spatial scales. Depending on their chemical behavior, they can be applied to understand hydrodynamic, solute, or particulate processes over short (days) to long (millennia) timescales. These microscopic tools are invaluable for quantifying natural and anthropogenic impacts and thus can help inform modeling efforts and management decisions today and into the future.

Acknowledgments

We thank John Jaeger and Kevin Yeager for their detailed and thoughtful critique of an earlier draft of this chapter. We appreciate the leadership of Tom Bianchi for spearheading this effort and providing us the opportunity to summarize a lot of excellent research conducted in the Gulf of Mexico.

References

Aarkrog, A. 1988. Radiological impact of Chernobyl debris compared with that from nuclear weapons fallout. *Journal of Environmental Radioactivity* 6(2):151–162.

Adolphi, F., and R. Muscheler. 2016. Synchronizing the Greenland ice core and radiocarbon timescales over the Holocene—Bayesian wiggle-matching of cosmogenic radionuclide records. *Climate of the Past* 12(1):15–30.

Aharon, P. 2003. Meltwater flooding events in the Gulf of Mexico revisited: Implications for rapid climate changes during the last deglaciation. *Paleoceanography* 18:1079. doi:10.1029/2002PA000840.

Alldredge, A. L., and C. G. Gottschalk. 1989. Direct observations of the mass flocculation of diatom blooms: Characteristics, settling velocities and formation of diatom aggregates. *Deep Sea Research* 36:159–171.

Allen, M. 2002. Particle scavenging and mixing rates in surface waters of the Gulf of Mexico using ^{234}Th and ^{222}Rn. Master's thesis, East Carolina University. 91 pp.

Aller, R. C., L. K. Benninger, and J. K. Cochran. 1980. Tracking particle-associated processes in nearshore environments by use of ^{234}Th/^{238}U disequilibrium. *Earth and Planetary Science Letters* 47:161–175.

Aller, R. C., and J. K. Cochran. 1976. ^{234}Th/^{238}U disequilibrium in nearshore sediment: Particle reworking and diagenetic time scales. *Earth and Planetary Science Letters* 29:37–50.

Allison, M. A., G. C. Kineke, E. S. Gordon, and M. A. Goñi. 2000. Development and reworking of a seasonal flood deposit on the inner continental shelf off the Atchafalaya River. *Continental Shelf Research* 20:2267–2294.

Allison, M. A., and E. A. Meselhe. 2010. The use of large water and sediment diversions in the lower Mississippi River (Louisiana) for coastal restoration. *Journal of Hydrology* 387:346–360.

Alvarado Quiroz, N. A., C. C. Hung, and P. H. Santschi. 2006. Binding of thorium (IV) to carboxylate, phosphate and sulfate functional groups from marine exopolymeric substances (EPS). *Marine Chemistry* 100:337–353.

Anders, M. H., S. W. Krueger, and P. M. Sadler. 1987. A new look at sedimentation rates and the completeness of the stratigraphic record. *Journal of Geology* 95:1–14.

Anderson, E. C., and W. F. Libby. 1951. World-wide distribution of natural radiocarbon. *Physical Review* 81:64.

Anderson, J. B., D. J. Wallace, A. R. Simms, A. B. Rodriguez, R. W. R. Weight, and Z. P. Taha. 2016. Recycling sediments between source and sink during a eustatic cycle: Systems of Late Quaternary northwestern Gulf of Mexico basin. *Earth Science Reviews* 153:111–138.

Appleby, P. G., and F. Oldfield. 1992. Application of Pb-210 to sedimentation studies: Uranium-series disequilibrium: Applications to earth, marine, and environmental problems. Pp. 731–778 in M. Ivanovich and R. S. Harmon (eds.), *Uranium-Series Disequilibrium: Applications to Earth, Marine, and Environmental Problems*, 2nd ed. Oxford: Clarendon Press.

Arzayus K. M., R. M. Dickhut, and E. A. Canuel. 2002. Effects of physical mixing on the attenuation of polycyclic aromatic hydrocarbons (PAHs) in estuarine sediments. *Organic Geochemistry* 33(12):1759–1769.

Bacon, M. P., and R. F. Anderson. 1982. Distribution of thorium isotopes between dissolved and particulate forms in the deep sea. *Journal of Geophysical Research* 87:2045–2056.

Baskaran, M., M. Ravichandran, and T. S. Bianchi. 1997. Cycling of ^{7}Be and ^{210}Pb in a high DOC, shallow, turbid estuary of Southeast Texas. *Estuarine, Coastal and Shelf Science* 45:165–176.

Baskaran, M., and P. H. Santschi. 1993. The role of particles and colloids in the transport of radionuclides in coastal environments of Texas. *Marine Chemistry* 43:95–114.

Baskaran, M., P. H. Santschi, G. Benoit, and B. D. Honeyman. 1992. Scavenging of Th isotopes by colloids in seawater of the Gulf of Mexico. *Geochimica et Cosmochimica Acta* 56:3375–3388.

Baskaran, M., P. H. Santschi, L. Guo, T. S. Bianchi, and C. Lambert. 1996. ^{234}Th:^{238}U disequilibria in the Gulf of Mexico: The importance of organic matter and particle concentration. *Continental Shelf Research* 16:353–380.

Baskaran, M., and P. W. Swarzenski. 2007. Seasonal variations on the residence times and partitioning of short-lived radionuclides (^{234}Th, ^{7}Be and ^{210}Pb) and depositional fluxes of ^{7}Be and ^{210}Pb in Tampa Bay, Florida. *Marine Chemistry* 104(1–2):27–42.

Begemann, F., and W. F. Libby. 1957. Continental water balance, ground water inventory and storage times, surface ocean mixing rates and world-wide water circulation patterns from cosmic-ray and bomb tritium. *Geochimica et Cosmochimica Acta* 12(4):277–296.

Benninger, L. K., R. C. Aller, J. K. Cochran, and K. K. Turekian. 1979. Effects of biological mixing on the ^{210}Pb chronology and trace metal distribution in a Long Island Sound sediment core. *Earth Planetary Science Letter* 43:241–259.

Bentley, S. J., M. D. Blum, J. Maloney, L. G. Pond, and R. L. Paulsell. 2016. The Mississippi River source-to-sink system: Perspectives on tectonic, climatic, and anthropogenic influences, Miocene to Anthropocene. *Earth Science Review* 153:139–174.

Blum, M., J. Martin, K. Milliken, and M. Garvin. 2013. Paleovalley systems: Insights from Quaternary analogs and experiments. *Earth Science Review* 116:128–169.

Blum, M. D., J. H. Tomkin, A. Purcell, and R. R. Lancaster. 2008. Ups and downs of the Mississippi Delta. *Geology* 36:675–678.

Blum, M. D., and T. E. Törnqvist. 2000. Fluvial responses to climate and sea-level change: A review and look forward. *Sedimentology* 47:2–48.

Bourdon, B., S. Turner, G. M. Henderson, and C. C. Lundstrom. 2003. Introduction to U-series geochemistry. *Reviews in Mineralogy and Geochemistry* 52:1–21. doi:10.2113/0520001.

Bregy, J. C., D. J. Wallace, R. T. Minzoni, and V. J. Cruz. 2018. 2500-year paleotempestological record of intense storms for the northern Gulf of Mexico, United States. *Marine Geology* 396:26-42.

Bronk Ramsey, C. 1995. Radiocarbon calibration and analysis of stratigraphy: The OxCal program. *Radiocarbon* 37(2):425–430.

Bronk Ramsey, C., and S. Lee. 2013. Recent and planned developments of the program OxCal. *Radiocarbon* 55(2–3):720–730.

Brooks G. R., R. A. Larson, P. T. Schwing, I. Romero, C. Moore, G.-J. Reichart, T. Jilbert, et al. 2015. Sedimentation pulse in the NE Gulf of Mexico following the 2010 DWH blowout. *PLOS ONE* 10(7):e0132341. doi:10.1371/journal.pone.0132341.

Buffle, J. 1991. *Complexation Reactions in Aquatic Systems: An Analytical Approach*. Chichester, UK: John Wiley and Sons/Ellis Horwood. 692 pp.

Burnett, W. C., P. K. Aggarwal, A. Aureli, H. Bokuniewicz, J. E. Cable, M. A. Charette, E. Kontar, et al. 2006. Quantifying submarine groundwater discharge in the coastal zone via multiple methods. *Science of the Total Environment* 367:498–543.

Burnett, W. C., H. Bokuniewicz, M. Huettel, W. S. Moore, and M. Taniguchi. 2003. Groundwater and pore water inputs to the coastal zone. *Biogeochemistry* 66:3–33.

Cable, J. E., W. C. Burnett, J. Chanton, and G. L. Weatherly. 1996. Estimating groundwater discharge into the northeastern Gulf of Mexico using radon-222. *Earth Planet Science Letter* 144:591–604.

Canuel, E., C. Martens, and L. Benninger. 1990. Seasonal variations in Be-7 activity in the sediments of Cape Lookout Bight, North Carolina. *Geochimica Cosmochimica Acta* 54:237–245.

Chanton J. P., J. Cherrier, R. M. Wilson, J. Sarkodee-Adoo, S. Bosman, A. Mickle, and W. M. Graham. 2012. Radiocarbon evidence that carbon from the Deepwater Horizon spill entered the planktonic food web of the Gulf of Mexico. *Environmental Research Letters* 7. doi:10.1088/1748-9326/7/4/045303.

Chanton J., T. Zhao, B. E. Rosenheim, S. Joye, S. Bosman, C. Brunner, K. M. Yeager, A. R. Diercks, and D. Hollander. 2015. Using natural abundance radiocarbon to trace the flux of petrocarbon to the seafloor following the Deepwater Horizon oil spill. *Environmental Science & Technology* 49(2):847–854.

Charette, M. A., K. O. Buesseler, and J. E. Andrews. 2001. Utility of radium isotopes for evaluating the input and transport of groundwater-derived nitrogen to a Cape Cod estuary. *Limnology and Oceanography* 46:465–470.

Chmura, G. L., and E. C. Kosters. 1994. Storm deposition and ^{137}Cs accumulation in fine-grained marsh sediments of the Mississippi Delta plain. *Estuarine, Coastal and Shelf Science* 39:33–34.

Chuang, C. Y., P. H. Santschi, L. S. Wen, L. D. Guo, C. Xu, S. Zhang, Y. Jiang, et al. 2015. Binding of Th, Pa, Pb, Po and Be radionuclides to marine colloidal macromolecular organic matter. *Marine Chemistry* 173:320–329.

Coale, K. H., and K. W. Bruland. 1985. Th-234/U-238 disequilibria within the California Current. *Limnology and Oceanography* 30(1):22–33.

———. 1987. Oceanic stratified euphotic zone as elucidated by ^{234}Th:^{238}U disequilibria. *Limnology and Oceanography* 31(1):189–200.

Coleman, J. M. 1988. Dynamic changes and processes in the Mississippi River delta. *Geological Society of America Bulletin* 100:999–1015.

Coleman, J. M., H. H. Roberts, and G. W. Stone. 1998. Mississippi River delta: An overview. *Journal of Coastal Research* 14(3):698–716.

Coleman, J. M., and L. D. Wright. 1975. Modern river deltas: Variability of processes and sand bodies. Pp. 99–149 in M. L. Broussard (ed.), *Deltas, Models for Exploration*. Houston: Geological Society.

Corbett, D. R., M. D. Dail, and B. A. McKee. 2007. High frequency time-series of the dynamic sedimentation processes on the western shelf of the Mississippi River delta. *Continental Shelf Research* 27:1600–1615.

Corbett, D. R., K. Dillon, W. C. Burnett, and J. Chanton. 2000. Estimating the groundwater contribution into Florida Bay via natural tracers, ^{222}Rn and CH_4. *Limnology and Oceanography* 45(7):1546–1557.

Corbett, D. R., B. McKee, and M. Allison. 2006. Nature of decadal-scale sediment accumulation in the Mississippi River deltaic region. *Continental Shelf Research* 26:2125–2140.

Corbett, D. R., B. McKee, and D. Duncan. 2004. Evaluation of short-term sediment dynamics via naturally-occurring radioisotopes in the Mississippi River deltaic region. *Marine Geology* 209:91–112.

Corbett, D. R., and J. P. Walsh. 2015. ^{210}Lead and ^{137} cesium: Establishing a chronology for the last century. Chap. 24 in I.

Shennan, A. J. Long, and B. P. Horton (eds.), *Handbook of Sea Level Research*. New York: John Wiley and Sons.

Craig, H., and D. Lal. 1961. The production rate of natural tritium. *Tellus* 13(1):85–105.

Dail, M. D., D. R. Corbett, and J. P. Walsh. 2007. Assessing the importance of a major hurricane on continental margin sedimentation in the Mississippi Delta region. *Continental Shelf Research* 27:1857–1874.

Day, J. W. J., D. F. Boesch, E. J. Clairain, G. P. Kemp, S. B. Laska, W. J. Mitsch, K. Orth, et al. 2007. Restoration of the Mississippi Delta: Lessons from Hurricanes Katrina and Rita. *Science* 315:1679–1684.

Day, J., Jr., C. Madden, R. Twilley, R. Shaw, B. McKee, M. Dagg, D. Childers, R. Raynie, and L. Rouse. 1995. The influence of the Atchafalaya River discharge on Fourleague Bay, LA. Pp. 151–160 in K. Dyer and R. Orth (eds.), *Changes in Fluxes in Estuaries*. Fredensborg, Denmark: Olsen and Olsen.

Delaune, R. D., R. H. Baumann, and J. G. Gosselink. 1983. Relationships among vertical accretion, coastal submergence, and erosion in a Louisiana Gulf Coast marsh. *Journal of Sedimentary Petrology* 53:147–157.

Delaune, R. D., W. H. Patrick Jr., and R. J. Buresh. 1978. Sedimentation rates determined by ^{137}Cs. *Nature* 275:532–533.

Delaune, R. D., J. H. Whitcomb, W. H. Patrick, J. H. Pardue, and S. R. Pezeshki. 1989. Accretion and canal impacts in a rapidly subsiding wetland. I: ^{137}Cs and ^{210}Pb techniques. *Estuaries* 12:247–259.

DeMaster, D. J., S. A. Kuehl, and C. A. Nittrouer. 1986. Effects of suspended sediments on geochemical processes near the mouth of the Amazon River: Examination of biological silica uptake and fate of particle-reactive elements. *Continental Shelf Research* 6:107–125.

Doney, S. C., W. J. Jenkins, and J. L. Bullister. 1997. A comparison of ocean tracer dating techniques on a meridional section in the eastern North Atlantic. *Deep Sea Research I: Oceanographic Research Papers* 44(4):603–626.

Doucet, F. J., J. R. Lead, and P. H. Santschi. 2007. Colloid-trace element interactions in aquatic systems: Environmental colloids and particles: Behavior, separation and characterization. Chap. 3 in K. Wilkinson and J. Lead (eds.), *International Union of Pure and Applied Chemistry (IUPAC) Series on Analytical and Physical Chemistry of Environmental Systems*. New York: John Wiley.

Duncan, T., and T. J. Shaw. 2003. The mobility of rare earth elements and redox sensitive elements in the groundwater/seawater mixing zone of a shallow coastal aquifer. *Aquatic Geochemistry* 9:233–255.

Dutkiewicz, V. A., and L. Husain. 1985. Stratospheric and tropospheric components of ^{7}Be in surface air. *Journal of Geophysical Research* 90(D3):5783–5788.

Emiliani, C., S. Gartner, B. Lidz, K. Eldridge, D. K. Elvey, T. C. Huang, J. J. Stipp, and M. F. Swanson. 1975. Paleoclimatological analysis of Late Quaternary cores from the northeastern Gulf of Mexico. *Science* 189:1083–1088.

Fanning, K. A., L. Torres, and J. Breland II. 1987. ^{226}Ra and ^{222}Rn along Florida's coasts. Pp. 203–209 in W. Properzio (ed.), *Proceedings of the Florida Natural Radiation and Technologically Enhanced Radiation Symposium*. Daytona Beach: Florida Chapter Health Physics Society.

Fisk, H. N. 1944. *Geological Investigation of the Alluvial Valley of the Lower Mississippi River*. Vicksburg: US Army Corps of Engineers, Mississippi River Commission. 78 pp.

Flower, B., D. Hastings, H. Hill, and T. Quinn. 2004. Phasing of deglacial warming and Laurentide ice sheet meltwater in the Gulf of Mexico. *Geology* 886:32.

Frazier, D. E. 1967. Recent deltaic deposits of the Mississippi River: Their development and chronology. *Gulf Coast Association of Geological Societies Transactions* 17:287–315.

Goldberg, E. D., and M. Koide. 1962. Geochronological studies of deep-sea sediments by the ionium-thorium method. *Geochimica et Cosmochimica Acta* 26:417–450.

Goñi, M., D. R. Corbett, Y. Alleau, J. P. Walsh, D. Mallinson, M. A. Allison, E. Gordon, S. Petsch, and T. M. Dellapenna. 2007. The effects of Hurricanes Katrina and Rita on the seabed of the Louisiana shelf. *Sedimentary Record* 5(1):4–9.

González, J. L., and T. E. Törnqvist. 2009. A new Late Holocene sea-level record from the Mississippi Delta: Evidence for a climate/sea level connection? *Quaternary Science Reviews* 28:1737–1749.

Green, M. A., R. C. Aller, J. K. Cochran, C. Lee, and J. Y. Aller. 2002. Bioturbation in shelf/slope sediments off Cape Hatteras, North Carolina: The use of Th-234, Chl-*a*, and Br to evaluate rates of particle and solute transport. *Deep Sea Research Part II: Tropical Studies in Oceanography* 49:4627–4644.

Guo, L., C. C. Hung, P. H. Santschi, and I. D. Walsh. 2002. ^{234}Th scavenging and its relationship to acid polysaccharide abundance in the Gulf of Mexico. *Marine Chemistry* 78(2–3):103–119.

Guo, L., B. Hunt, and P. H. Santschi. 2001. Ultrafiltration behavior of major ions (Na, Ca, Mg, F, Cl, and $SO4^{2-}$) in natural waters. *Water Research* 35(6):1500–1508.

Guo, L., and P. H. Santschi. 1996. A critical evaluation of cross-flow ultrafiltration techniques for sampling colloidal organic carbon in seawater. *Marine Chemistry* 55:113–127.

———. 1997. Isotopic and elemental characterization of

colloidal organic matter from the Chesapeake Bay and Galveston Bay. *Marine Chemistry* 59:1–15.

———. 2007. Ultrafiltration and its applications to sampling and characterisation of aquatic colloids. Chap. 4 in K. J. Wilkinson and J. R. Lead (eds.), *Environmental Colloids and Particles: Behaviour, Separation and Characterisation*, vol. 10. Chichester, UK: John Wiley and Sons.

Guo, L., P. H. Santschi, and M. Baskaran. 1997. Interactions of thorium isotopes with colloidal organic matter in oceanic environments. *Colloids and Surfaces A: Physicochemical and Engineering Aspects* 120:255–271.

Guo, L., P. H. Santschi, M. Baskaran, and A. Zindler. 1995. Distribution of dissolved and particulate ^{230}Th and ^{232}Th in seawater from the Gulf of Mexico and off Cape Hatteras, as measured by SIMS. *Earth and Planetary Science Letters* 133:117–128.

Guo, L., N. Tanaka, D. M. Schell, and P. H. Santschi. 2003. Nitrogen and carbon isotopic composition of HMW dissolved organic matter in marine environments. *Marine Ecology Progress Series* 252:51–60.

Guo, L., L.-S. Wen, D. Tang, and P. H. Santschi. 2000. Re-examination of cross-flow ultrafiltration for sampling marine colloids: Evidence from molecular probes. *Marine Chemistry* 69(1–2):75–90.

Hancock, G. J., I. T. Webster, and T. C. Stieglitz. 2006. Horizontal mixing of Great Barrier Reef waters: Offshore diffusivity determined from radium isotope distribution. *Journal of Geophysical Research* 111:C12019. doi:10.1029/2006JC003608.

Hine, A. C., B. Jarrett, S. Locker, D. Mallinson, D. Naar, and B. Donahue. 2008. Coral reefs, present and past, on the west Florida shelf and platform margin. Chap. 4 in B. M. Riegl and R. E. Dodge (eds.), *Coral Reefs of the USA*. New York: Springer.

Honeyman, B. D., and P. H. Santschi. 1989. A Brownian-pumping model for trace metal scavenging: Evidence from Th isotopes. *Journal of Marine Research* 47(4):950–995.

Hung, C. C., L. Guo, K. A. Roberts, and P. H. Santschi. 2004. Upper ocean carbon flux determined by size-fractionated ^{234}Th data and sediment traps in the Gulf of Mexico. *Geochemical Journal* 38(6):601–611.

Hung, C. C., C. Xu, P. H. Santschi, S. Zhang, K. A. Schwehr, A. Quigg, L. Guo, et al. 2010. Comparative evaluation of sediment-trap and ^{234}Th-derived POC fluxes from the upper oligotrophic waters of the Gulf of Mexico and the subtropical northwestern Pacific Ocean. *Marine Chemistry* 121:132–144.

Hutchinson, S. M. 1995. Use of magnetic and radiometric measurements to investigate erosion and sedimentation in a British upland catchment. *Earth Surface Processes and Landforms* 20:293–314.

Ivanovich, M., and R. S. Harmon (eds.). 1992. *Uranium-Series Disequilibrium: Applications to Earth, Marine, and Environmental Sciences*. 2nd ed. Oxford: Clarendon Press. 910 pp.

Johannes, R. E. 1980. The ecological significance of the submarine discharge of groundwater. *Marine Ecology Progress Series* 3:365–373.

Johannesson, K. H., and D. J. Burdige. 2007. Balancing the global oceanic neodymium budget: Evaluating the role of groundwater. *Earth and Planetary Science Letters* 253:129–142.

Johannesson K. H., D. A. Chevis, D. J. Burdige, J. E. Cable, J. B. Martin, and M. Roy. 2011. Submarine groundwater discharge is an important net source of light and middle REEs to coastal waters of the Indian River Lagoon, Florida, USA. *Geochimica et Cosmochimica Acta* 75:825–843.

Kaste, J. M., and M. Baskaran. 2011. Meteoric Be-7 and Be-10 as process tracers in the environment. Chap. 5 in M. Baskaran (ed.), *Handbook of Environmental Isotope Geochemistry*, vols. 1–2. Berlin: Springer.

Kennett, J. P., and N. J. Shackleton. 1975. Laurentide ice sheet meltwater recorded in Gulf of Mexico deep-sea cores. *Science* 188:147–150.

Kolker, A. S., M. A. Allison, and S. Hameed. 2011. An evaluation of subsidence rates and sea-level variability in the northern Gulf of Mexico. *Geophysical Research Letters* 38:L21404.

Kolker A. S., J. E. Cable, K. H. Johannesson, M. A. Allison, and L. V. Inniss. 2013. Pathways and processes associated with the transport of groundwater in deltaic systems. *Journal of Hydrology* 498:319–334.

Kolker, A. S., M. D. Miner, and H. D. Weathers. 2012. Depositional dynamics in a river diversion receiving basin: The case of the West Bay Mississippi River diversion. *Estuarine, Coastal and Shelf Science* 106:1–12.

Kovanen, D. J., and D. J. Easterbrook. 2002. Paleodeviations of radiocarbon marine reservoir values for the northeast Pacific. *Geology* 30(3):243–246.

Krest, J. M., W. S. Moore, and Rama. 1999. ^{226}Ra and ^{228}Ra in the mixing zone of the Mississippi and Atchafalaya Rivers: Indicators of groundwater input. *Marine Chemistry* 64:129–152.

Krishnamurti, K. 1929. The scattering of light in colloidal solutions and gels. *Nature* 124:690–691.

Krishnaswamy, S., D. Lal, J. M. Martin, and M. Meybeck. 1971. Geochronology of lake sediments. *Earth and Planetary Science Letters* 11:401–414.

Kuehl, S. A., D. J. DeMaster, and C. A. Nittrouer. 1986. Nature of sediment accumulation on the Amazon continental shelf. *Continental Shelf Research* 6:311–336.

Kuehl, S. A., C. A. Nittrouer, M. A. Allison, L. E. C. Faria, D. A. Dukat, J. M. Jaeger, T. D. Pacioni, A. G. Figueiredo, and E. C. Underkoffler. 1996. Sediment deposition, accumulation, and sea bed dynamics in an energetic fine-grained coastal environment. *Continental Shelf Research* 16:787–815.

Kummu, M., H. de Moel, G. Salvucci, D. Viviroli., D. J. Ward, and O. Varis. 2016. Over the hills and further away from coast: Global geospatial patterns of human and environment over the 20th–21st centuries. *Environmental Research Letters* 11(3):034010.

Leventer, A., D. F. Williams, and J. P. Kennett. 1982. Dynamics of the Laurentide ice sheet during the last deglaciation: Evidence from the Gulf of Mexico. *Earth and Planetary Science Letters* 59:11–17.

———. 1983. Relationships between anoxia, glacial meltwater and microfossil preservation in the Orca Basin, Gulf of Mexico. *Marine Geology* 53:23– 40.

Liu, K-b., and M. L. Fearn. 1993. Lake-sediment record of Late Holocene hurricane activities from coastal Alabama. *Geology* 21:793–796.

———. 2000. Reconstruction of prehistoric landfall frequencies of catastrophic hurricanes in northwestern Florida from lake sediment records. *Quaternary Research* 54:238–245.

Liu, K-b., H. Y. Lu, and C. M. Shen. 2008. A 1,200-year proxy record of hurricanes and fires from the Gulf of Mexico coast: Testing the hypothesis of hurricane and fire interactions. *Quaternary Research* 69:29–41.

Macdonald, R. W., E. C. Carmack, and D. W. R. Wallace. 1993. Tritium and radiocarbon dating of Canada basin deep waters. *Science* 259(5091):103–104.

Mallinson, D., A. Hine, D. Naar, S. Locker, and B. Donahue. 2014. New perspectives on the geology and origin of the Florida Middle Ground carbonate banks, west Florida shelf, USA. *Marine Geology* 355:54–70.

McFarlan, E., Jr. 1961. Radiocarbon dating of Late Quaternary deposits, south Louisiana. *Geological Society of American Bulletin* 72:129–158.

McCoy, C. A., D. R. Corbett, J. E. Cable, and R. K. Spruill. 2007. Hydrogeological characterization and quantification of submarine groundwater discharge in the southeast coastal plain of North Carolina. *Journal of Hydrology* 339:159–171.

McCoy, C. A., D. R. Corbett, B. A. McKee, and Z. Top. 2007. An evaluation of submarine groundwater discharge along the continental shelf of Louisiana using a multiple tracer approach. *Journal of Geophysical Research: Oceans* 112. doi:10.1029/2006JC003505.

McKee, B. A., R. C. Aller, M. A. Allison, T. S. Bianchi, and G. C. Kineke. 2004. Transport and transformation of dissolved and particulate materials on continental margins influenced by major rivers: Benthic boundary layer and seabed processes. *Continental Shelf Research* 24:899–926.

McKee, B. A., D. J. DeMaster, and C. A. Nittrouer. 1984. The use of ^{234}Th/^{238}U disequilibrium to examine the fate of particle-reactive species on the Yangtze continental shelf. *Earth and Planetary Science Letters* 68:431–442.

———. 1986. Temporal variability in the partitioning of thorium between dissolved and particulate phases on the Amazon shelf: Implications for the scavenging of particle reactive species. *Continental Shelf Research* 6:87–106.

Meade, R. H. 1996. River-sediment inputs to major deltas. Chap. 3 in J. D. Milliman and B. U. Haq (eds.), *Sea-Level Rise and Coastal Subsidence*. London: Kluwer.

Meade, R., and R. Parker. 1985. Sediment in rivers of the United States. Pp. 49–60 in *National Water Summary 1984: Hydrologic Events, Selected Water-Quality Trends, and Ground-Water Resources*, Water Supply Paper 2275. Washington DC: US Geological Survey.

Milan, C. S., E. M. Swenson, R. E. Turner, and J. M. Lee. 1995. Assessment of the ^{137}Cs method for estimating sediment accumulation rates: Louisiana salt marshes. *Journal Coastal Research* 11:296–307.

Milliman, J. D., and R. H. Meade. 1983. World-wide delivery of river sediment to the oceans. *Geology* 91:1–21.

Moore, W. S. 1996. Large groundwater inputs to coastal waters revealed by ^{226}Ra enrichments. *Nature* 380:612–614.

———. 2000. Determining coastal mixing rates using radium isotopes. *Continental Shelf Research* 20:1993–2007.

———. 2003. Sources and fluxes of submarine groundwater discharge delineated by radium isotopes. *Biogeochemistry* 66:75–93.

———. 2010. The effects of submarine groundwater discharge on the ocean. *Annual Review of Marine Science* 2:59–88.

Moore, W. S., and J. M. Krest. 2004. Distribution of ^{222}Ra and ^{226}Ra in the plumes of the Mississippi and Atchafalaya River and the Gulf of Mexico. *Marine Chemistry* 86:105–119.

Moran, J. E., S. D. Oktay, and P. H. Santschi. 2002. Sources of iodine and ^{129} iodine in rivers. *Water Resource Research* 38(8). doi:10.1029/2001WR000622.

Moran, S. B., and J. N. Smith. 2000. Th-234 as a tracer of scavenging and particle export in the Beaufort Sea. *Continental Shelf Research* 20(2):153–167.

Mossa, J. 1996. Sediment dynamics in the lowermost Mississippi River. *Engineering Geology* 45:457–479.

Müller-Karger, F. E., J. J. Walsh, R. H. Evans, and M. B. Meyers. 1991. On the seasonal phytoplankton concentration and sea surface temperature cycles of the Gulf of Mexico as determined by satellites. *Journal of Geophysical Research* 96(C7):12645–12665.

Neubauer, S. C., I. C. Anderson, J. A. Constantine, and S. A. Kuehl. 2002. Sediment deposition and accretion in a Mid-Atlantic tidal freshwater marsh. *Estuarine, Coastal and Shelf Science* 54:713–727.

Nittrouer, C. A., J. A. Austin, M. E. Field, J. H. Kravitz, J. M. P. Syvitski, and P. I. Wiberg. 2007. *Writing a Rosetta Stone: Insights into Continental-Margin Sedimentary Processes and Strata*. International Association of Sedimentologists Special Publication no. 37. Oxford: Blackwell Publishing. 549 pp.

Nittrouer, J. A., J. L. Best, C. Brantley, R. W. Cash, M. Czapiga, P. Kumar, and G. Parker. 2012. Mitigating land loss in coastal Louisiana by controlled diversion of Mississippi River sand. *Nature Geoscience* 5:534–537.

Nittrouer, C. A., D. J. DeMaster, B. A. McKee, N. H. Custhall, and L. I. Larsen. 1984. The effect of sediment mixing on ^{210}Pb accumulation rates for the Washington continental shelf. *Marine Geology* 31:297–316.

Nyman, J. A., R. J. Walters, R. D. Delaune, and W. H. Patrick Jr. 2006. Marsh vertical accretion via vegetative growth. *Estuarine Coastal Shelf Sciences* 69:370–380.

Oktay, S. D., P. H. Santschi, J. E. Moran, and P. Sharma. 2000. The ^{129} iodine bomb pulse recorded in Mississippi River delta sediments: Results from isotopes of I, Pu, Cs, Pb, and C. *Geochimica et Cosmochimica Acta* 64(6):989–996.

———. 2001. ^{129}I and ^{127}I transport in the Mississippi River. *Environmental Science & Technology* 35:4470–4476.

Oldfield, F., and P. G. Appleby. 1984. Empirical testing of ^{210}Pb-dating models. Pp. 93–124 in E. Y. Haworth and J. W. G. Lund (eds.), *Lake Sediments and Environmental History*. Leicester, UK: Leicester University Press.

Olsen, C. R., I. L. Larsen, P. D. Lowery, and N. H. Cutshall. 1986. Geochemistry and deposition of ^{7}Be in river-estuarine and coastal waters. *Journal of Geophysical Research* 91:896–908.

Osterman, L. E., R. Z. Poore, and P. Swarzenski. 2008. The last 1000 years of natural and anthropogenic low-oxygen bottom-water on the Louisiana shelf, Gulf of Mexico. *Marine Micropaleontology* 66:291–303.

Osterman, L. E., R. Z. Poore, P. Swarzenski, D. Senn, and S. DiMarco. 2009. The 20th-century development and expansion of Louisiana shelf hypoxia, Gulf of Mexico. *Geo-Mar Letters* 29(6):405.

Osterman, L. E., R. Z. Poore, P. W. Swarzenski, and R. E. Turner. 2005. Reconstructing a 180 year record of natural and anthropogenic induced low-oxygen conditions from Louisiana continental shelf sediments. *Geology* 33:329–332.

Palinkas, C. M., C. A. Nittrouer, R. A. Wheatcroft, and L. Langone. 2005. The use of ^{7}Be to identify event and seasonal sedimentation near the Po River delta, Adriatic Sea. *Marine Geology* 222–223:95–112.

Paneth, F. 1914. Colloidal solutions of radioactive substances. *Kolloid-Zeitschrift* 13:1–4.

Pearce, C., A. Varhelyi, S. Wastegard, F. Muschitielloo, N. Barrientos, M. O'Regan, T. M. Cornin, et al. 2017. The 3.6 ka Aniakchak tephra in the Arctic Ocean: A constraint on the Holocene radiocarbon reservoir age in the Chukchi Sea. *Climate of the Past* 13(4):303–316.

Penland, S., R. Boyd, and J. R. Suter. 1988. Transgressive depositional systems of the Mississippi Delta plain: A model for barrier shoreline and shelf sand development. *Journal of Sedimentary Petrology* 58:932–949.

Pennington, W., R. S. Cambray, and E. M. Fisher. 1973. Observations on lake sediments using fallout ^{137}Cs as a tracer. *Nature* 242:324–326.

Peterson, R. N., W. C. Burnett, N. Dimova, and I. R. Santos. 2009. A comparison of measurement methods for radium-226 on manganese-fiber. *Limnology and Oceanography: Methods* 7(2):196–205.

Poore, R. Z., K. L. DeLong, J. N. Richey, and T. M. Quinn. 2009. Evidence of multidecadal climate variability and the Atlantic Multidecadal Oscillation from a Gulf of Mexico sea-surface temperature-proxy record. *Geo-Marine Letters* 29. doi:10.1007/s00367-009-0154-6.

Poore R. Z., M. J. Pavich, and H. D. Grissino-Mayer. 2005. Record of the North American southwest monsoon from Gulf of Mexico sediment cores. *Geology* 33:209–212.

Presley, B. J., J. Trefry, and R. Shokes. 1980. Heavy metal inputs to Mississippi delta sediments. *Water, Air, and Soil Pollution* 13:481–494.

Quigley, M. S., P. H. Santschi, C.-C. Hung, L. Guo, and B. D. Honeyman. 2002. Importance of polysaccharides for ^{234}Th complexation to marine organic matter. *Limnology and Oceanography* 47:67–377.

Rabalais, N. N., W.-J. Cai, J. Carstensen, D. J. Conley, B. Fry, X. Quiñones-Rivera, R. Rosenberg, et al. 2014. Eutrophication-driven deoxygenation in the coastal ocean. *Oceanography* 70:123–133.

Ravens, T. M., R. C. Thomas, K. A. Roberts, and P. H. Santschi. 2009. Causes of salt marsh erosion in Galveston Bay, Texas. *Journal of Coastal Research* 25(2):265–272.

Ravichandran, M., M. Baskaran, P. H. Santschi, and T. S. Bianchi. 1995a. Geochronology of sediments of Sabine-Neches Estuary, Texas. *Chemical Geology (Isotope Geoscience)* 125:291–306.

———. 1995b. History of trace metal pollution in Sabine-Neches estuary, Beaumont, Texas. *Environmental Science & Technology* 29(6):1495–1503.

Reimer, P. J., and R. W. Reimer. 2001. A marine reservoir correction database and on-line interface. *Radiocarbon* 43(2A):461–463.

Richey, J. N., R. Z. Poore, B. P. Flower, and T. M. Quinn. 2007. 1400 year multiproxy record of climate variability from the northern Gulf of Mexico. *Geology* 35:423–426.

Roberts, H. H. 1997. Dynamic changes of the Holocene Mississippi River delta plain: The delta cycle. *Journal of Coastal Research* 13(3):605–627.

Rodriguez, A. B., A. R. Simms, and J. B. Anderson. 2010. Bay-head deltas across the northern Gulf of Mexico back step in response to the 8.2 ka cooling event. *Quaternary Science Reviews* 29:3983–3993.

Rollo, J. R. 1966. *Ground-Water Resources of the Greater New Orleans Area, Louisiana*. Bulletin no. 13. Baton Rouge: Louisiana Department of Conservation, Louisiana Geological Survey, and Louisiana Department of Public Works Water Resources. 45pp.

Romans, B. W., S. Castelltort, J. A. Covault, A. Fildani, and J. P. Walsh. 2016. Environmental signal propagation in sedimentary systems across timescales. *Earth-Science Reviews* 153:7–29.

Rosenheim, B. E., M. A. Pendergraft, G. C. Flowers, R. Carney, J. L. Sericano, R. M. Amer, J. Chanton, Z. Dincer, and T. L. Wade. 2016. Employing extant stable carbon isotope data in Gulf of Mexico sedimentary organic matter for oil spill studies. *Deep Sea Research Part II: Tropical Studies in Oceanography* 129:249–258.

Rutherford, E. 1904. The succession of changes in radioactive bodies. *Philosophical Transactions of the Royal Society A* 204:169–219.

Sadler, P. M. 1981. Sediment accumulation rates and the completeness of stratigraphic sections. *Journal of Geology* 89:569–584.

Santos, I. R., W. C. Burnett, T. Dittmar, I. G. N. A. Suryaputra, and J. Chanton. 2009. Tidal pumping drives nutrient and dissolved organic matter dynamics in a Gulf of Mexico subterranean estuary. *Geochimica et Cosmochimica Acta* 73(5):1325–1339.

Santschi, P. H., M. Allison, S. Asbill, A. B. Perlet, S. Cappellino, C. Dobbs, and L. McShea. 1999. Sediment transport and Hg recovery in Lavaca Bay, as evaluated from radionuclide and Hg distributions. *Environmental Science & Technology* 33:378–391.

Santschi, P. H., E. Balnois, K. Wilkinson, J. Zhang, J. Buffle, and L. Guo. 1998. Fibrillar polysaccharides in marine macromolecular organic matter, as imaged by atomic force microscopy and transmission electron microscopy. *Limnology and Oceanography* 43(5):896–908.

Santschi, P. H., A. B. Burd, J.-F. Gaillard, and A. A. Lazarides. 2005. Transport of materials and chemicals by nano-scale colloids and micro to macro-scale flocs in marine, freshwater and engineered systems. Chap. 9 in S. N. Liss, I. G. Droppo, G. G. Leppard, and T. G. Milligan (eds.), *Flocculation in Natural and Engineered Environmental Systems*. Boca Raton, FL: CRC Press.

Santschi, P. H., L. Guo, M. Baskaran, S. Trumbore, J. Southon, T. S. Bianchi, B. Honeyman, and L. Cifuentes. 1995. Isotopic evidence for the contemporary origin of high-molecular weight organic matter in oceanic environments. *Geochimica et Cosmochimica Acta* 59(3):625–631.

Santschi, P. H., C. C. Hung, L. Guo, J. Pinckney, G. Schultz, N. Alvarado-Quiroz, and I. Walsh. 2003. Control of acid polysaccharide production and ^{234}Th and POC export fluxes by marine organisms. *Geophysical Research Letters* 30(2). doi:10.1029/2002GL016046.

Santschi, P. H., J. W. Murray, M. Baskaran, C. R. Benitez-Nelson, L. Guo, C.-C. Hung, C. Lamborg, S. B. Moran, U. Passow, and M. Roy-Barman. 2006. Thorium speciation in seawater. *Marine Chemistry* 100:250–268.

Santschi, P. H., S. D. Oktay, and L. Cifuentes. 2007. Carbon isotopes and iodine concentrations in a Mississippi River delta core recording land use, sediment transport, and dam building in the river's drainage basin. *Marine Environmental Research* 63:278–290.

Santschi, P. H., B. J. Presley, T. L. Wade, B. Garcia-Romero, and M. Baskaran. 2001. Historical contamination of PAHs, PCBs, DDTs, and heavy metals in Mississippi River delta, Galveston Bay and Tampa Bay sediment cores. *Marine Environmental Research* 52(1):51–79.

Santschi, P. H., and G. T. Rowe. 2008. Radiocarbon-derived sedimentation rates in the Gulf of Mexico. *Deep Sea Research II: Gulf of Mexico Special Issue* 55:2572–2576.

Savoye, N., K. O. Buesseler, D. Cardinal, and F. Dehairs. 2004. ^{234}Th deficit and excess in the Southern Ocean during spring 2001: Particle export and remineralization. *Geophysical Research Letters* 31(12). doi:10.1029/2004GL019744.

Schink, D. R., P. H. Santschi, O. Corapcioglu, S. Oktay, and U. Fehn. 1995. Prospects for "iodine-129" dating of marine organic matter using AMS. *Nuclear Instruments and Methods in Physics Research Section B: Beam Interactions with Materials and Atoms* 99:524–527.

Schink, D. R., P. H. Santschi, O. Corapcioglu, P. Sharma, and U. Fehn. 1995. ^{129}I in Gulf of Mexico waters. *Earth and Planetary Science Letters* 135:131–138.

Schmidt, S., H. C. de Stigter, and T. C. E. van Weering. 2001. Enhanced short-term sediment deposition within the Nazare Canyon, North-East Atlantic. *Marine Geology* 173(1–4):55–67.

Schwehr, K. A., P. H. Santschi, and D. Elmore. 2005. The dissolved organic iodine species of the isotopic ratio of ^{129}I/^{127}I: A novel tool for tracing terrestrial organic carbon in the estuarine surface waters of Galveston Bay, Texas. *Limnology and Oceanography: Methods* 3:326–337.

Schwing, P. T., I. C. Romero, G. R. Brooks, D. W. Hastings, R. A. Larson, and D. J. Hollander. 2015. A decline in benthic foraminifera following the Deepwater Horizon event in the northeastern Gulf of Mexico. *PLOS ONE* 10(3). doi:10.1371/journal.pone.0120565.

Severinghaus, J. P., W. S. Broecker, T.-H. Peng, and G. Bonani. 1996. Transect along 24šN latitude of ^{14}C in dissolved inorganic carbon in the subtropical North Atlantic Ocean. *Radiocarbon* 38:407–414.

Siani, G., M. Paterne, E. Michel, R. Sulpizio, A. Sbrana, M. Arnold, and G. Haddad. 2001. Mediterranean Sea surface radiocarbon reservoir age changes since the last glacial maximum. *Science* 294(5548):1917–1920.

Simms, A. R., J. B. Anderson, and M. Blum. 2006. Barrier-island aggradation via inlet migration: Mustang Island, Texas. *Sedimentary Geology* 187:105–125.

Sionneau T., V. Bout-Roumazeilles, B. P. Flower, A. Bory, N. Tribovillard, C. Kissel, B. Van Vliet-Lanoe, and J. C. M Serrano. 2010. Provenance of freshwater pulses in the Gulf of Mexico during the last deglaciation. *Quaternary Research* 74(2):235–245.

Slomp, C. P., and P. Van Cappellen. 2004. Nutrient inputs to the coastal ocean through submarine groundwater discharge: Controls and potential impact. *Journal of Hydrology* 295:64–86.

Smith, C. G, and L. L. Robbins. 2012. *Surface-Water Radon-222 Distribution along the West-Central Florida Shelf.* US Geological Survey Open-File Report 2012–1212. Reston, VA: US Geological Survey. 26 pp.

Smith, C. G., and P. W. Swarzenski. 2012. An investigation of submarine groundwater borne nutrient fluxes to the west Florida shelf and recurrent harmful algal blooms. *Limnology and Oceanography* 57:471–485.

Smoak, J. M., D. J. DeMaster, S. A. Kuehl, R. H. Pope, and B. A. McKee. 1996. The behavior of particle-reactive tracers in a high turbidity environment: ^{234}Th and ^{210}Pb on the Amazon continental shelf. *Geochimica et Cosmochimica Acta* 60:2123–2137.

Sommerfield, C. K., C. A. Nittrouer, and C. R. Alexander. 1999. ^{7}Be as a tracer of flood sedimentation on the northern California continental margin. *Continental Shelf Research* 19:335–361.

Stordal, M. C., P. H. Santschi, and G. A. Gill. 1996. Colloidal pumping: Evidence for the coagulation process using natural colloids tagged with 20^{3}Hg. *Environmental Science & Technology* 30:3335–3340.

Stuiver, M., and P. J. Reimer. 1993. Extended ^{14}C data base and revised CALIB 3.0 ^{14}C age calibration program. *Radiocarbon* 35:215–230.

Stuiver, M., P. Reimer, E. Bard, W. Beck, G. Burr, K. Hughen, B. Kromer, G. McCormac, J. van der Plicht, and M. Spurk. 1998. INTCAL98 radiocarbon age calibration, 24,000 cal BP. *Radiocarbon* 40:1041–1084.

Stuiver, M., P. J. Reimer, and R. Reimer. 2013. CALIB radiocarbon calibration. *CALIB ^{14}C Calibration Program*. Belfast: Queen's University.

———. 2017. CALIB 7.1. http://calib.org.

Subt, C., K. A. Fangman, J. S. Wellner, and B. E. Rosenheim. 2016. Sediment chronology in Antarctic deglacial sediments: Reconciling organic carbon C-14 ages to carbonate C-14 ages using ramped PyrOx. *Holocene* 26(2):265–273.

Suess, H. E., 1965. Secular variations of the cosmic-ray-produced carbon 14 in the atmosphere and their interpretations. *Journal of Geophysical Research* 70(23):5937–5952.

———. 1967. Bristlecone pine calibration of the radiocarbon time scale from 4100 BC to 1500 BC. Pp. 143–151 in *Radioactive Dating and Methods of Low-Level Counting*. Vienna: International Atomic Energy Agency.

Swarzenski, P. W., D. R. Corbett, J. M. Smoak, and B. A. McKee. 2000. The use of U-Th series radionuclides and transient tracers in oceanography: An overview. Pp. 33–53 in R. E. Hester and R. M. Harrison (eds.), *Chemistry in the Marine Environment*, Issues in Environmental Science and Technology, no. 13. London: Royal Society of Chemistry.

Syvitski, J. P. M., C. J. Vörösmarty, A. J. Kettner, and P. A. Green. 2005. Impact of humans on the flux of terrestrial sediment to the global coastal ocean. *Science* 308:376–380.

Taniguchi, M., W. C. Burnett, J. E. Cable, and J. V. Turner. 2002. Investigation of submarine groundwater discharge. *Hydrological Processes* 16(2):2115–2129.

Thompson, C., L. Smith, and R. Maji. 2007. Hydrogeological modeling of submarine groundwater discharge on the continental shelf of Louisiana. *Journal of Geophysical Research* 112:C03014.

Thornton, D. C. O. 2004. Formation of transparent exopolymeric particles (TEP) from macroalgal detritus. *Marine Ecology Progress Series* 282:1–12.

Tichenor, H. R., S. J. Culver, D. R. Corbett, J. P. Walsh, and M. A. Buzas. 2016. Does the PEB index respond only to hypoxia in the Mississippi Delta, Gulf of Mexico? *Journal of Foraminiferal Research* 46(1):48–60.

Törnqvist, T. E., S. J. Bick, J. L. González, K. Van der Borg, and A. F. M. De Jong. 2004. Tracking the sea-level signature of the 8.2 ka cooling event: New constraints from the Mississippi Delta. *Geophysical Research Letters* 31:L23309.

Törnqvist, T. E., S. J. Bick, K. Van der Borg, and A. F. M. De Jong. 2006. How stable is the Mississippi Delta? *Geology* 34:697–700.

Törnqvist, T. E., J. L. González, L. A. Newsom, K. Van der Borg, A. F. M. De Jong, and C. W. Kurnik. 2004. Deciphering Holocene sea-level history on the U.S. Gulf Coast: A high-resolution record from the Mississippi Delta. *Geological Society of America Bulletin* 116:1026–1039.

Törnqvist, T. E., D. J. Wallace, J. E. A. Storms, J. Wallinga, R. L. Dam, M. Blaauw, M. S. Derksen, C. J. W. Klerks, C. Meijneken, and E. M. A. Snijders. 2008. Mississippi Delta subsidence primarily caused by compaction of Holocene strata. *Nature Geoscience* 1:173–176.

Turekian, K. K. 2003. Uranium decay series. *Reviews in Mineralogy and Geochemistry* 52(1). doi:10.2113/0520004.

UNSCEAR. 2000. *Sources and Effects of Ionizing Radiation*. Vol. 1: *Sources*. United Nations Committee on the Effects of Atomic Radiation. http://www.unscear.org/unscear/en/publications/2000_1.html.

Valiela, I., J. Costa, K. Foreman, J. M. Teal, B. Howes, and D. Aubrey. 1990. Transport of water-borne nutrients from watersheds and their effects on coastal waters. *Biogeochemistry* 10:177–198.

Walker, M. 2005. *Quaternary Dating Methods*. New York: John Wiley and Sons. 304 pp.

Wallace, D. J., and J. B. Anderson. 2010. Evidence of similar probability of intense hurricane strikes for the Gulf of Mexico over the Late Holocene. *Geology* 38:511–514.

Walsh, J. P., M. A. Allison, D. R. Corbett, S. L. Goodbred Jr., S. A. Kuehl, C. Nittrouer, and A. S. Ogston. 2013. Shelf and slope sedimentation associated with large deltaic systems. Pp. 86–117 in T. Bianchi, M. Allison, and W. J. Cai (eds.), *Biogeochemical Dynamics at Large River-Coastal Interfaces: Linkages with Global Climate Change*. Cambridge: Cambridge University Press.

Walsh, J. P., D. R. Corbett, D. Mallinson, M. Goñi, M. Dail, C. Loewy, K. Marciniak, et al. 2006. Mississippi Delta mudflow activity and 2005 Gulf hurricanes. *EOS, Transactions, American Geophysical Union* 87:179–477.

Walsh, J. P., and C. A. Nittrouer. 2009. Towards an understanding of fine-grained river-sediment dispersal on continental margins. *Marine Geology* 263:34–45.

Wang, L., M. Sarnthein, H. Erlenkeuser, J. O. Grimalt, P. Grootes, S. Heilig, E. Ivanova, M. Kienast, C. Pelejero, and U. Pflaumann. 1999. East Asian monsoon climate during the Late Pleistocene: High-resolution sediment records from the South China Sea. *Marine Geology* 156(1–4): 245–284.

Wang, X.-C., R. F. Chen, and G. B. Gardner. 2004. Sources and transport of dissolved and particulate organic carbon in the Mississippi River estuary and adjacent coastal waters of the northern Gulf of Mexico. *Marine Chemistry* 89:241–256.

Waples, J. T., C. Benitez-Nelson, N. Savoye, M. R. van der Loeff, M. Baskaran, and O. Gustafsson. 2006. An introduction to the application and future use of ^{234}Th in aquatic systems. *Marine Chemistry* 100:166–189.

Warnken, K. W., P. H. Santschi, K. A. Roberts, and G. A. Gill. 2008. The cycling and oxidation pathways of organic carbon in a shallow estuary along the Texas Gulf Coast. *Estuarine, Coastal and Shelf Science* 76:69–84.

Wilson, R. M., J. Cherrier, J. Sarkodee-Adoo, S. Bosman, A. Mickle, and J. P. Chanton. 2016. Tracing the intrusion of fossil carbon into coastal Louisiana macrofauna using natural ^{14}C and ^{13}C abundances. *Deep Sea Research Part II: Tropical Studies in Oceanography* 129:89–95.

Windom, H., W. S. Moore, L. F. Niencheski, and R. Jahnke. 2006. Submarine groundwater discharge: A large, previously unrecognized source of dissolved iron to the South Atlantic Ocean. *Marine Chemistry* 102:252–266.

Wright, L. D., and J. M. Coleman.1973. Variations in morphology of major river deltas as functions of ocean wave and river discharge regimes. *American Association of Petroleum Geology Bulletin* 57:370–398.

Wright, L. D., and C. A. Nittrouer. 1995. Dispersal of river sediments in coastal seas—6 contrasting cases. *Estuaries* 18:494–508.

Xu, C., P. H. Santschi, C.-C. Hung, S. Zhang, K. A. Schwehr, K. A. Roberts, L. D. Guo, et al. 2011. Controls of Th-234 removal from the oligotrophic ocean by polyuronic acids and modification by microbial activity. *Marine Chemistry* 123:111–126.

Yeager, K. M., P. H. Santschi, H. S. Rifai, M. P. Suarez, R. L. Brinkmeyer, C.-C. Hung, K. Schindler, M. Andres, and

E. Weaver. 2007. Dioxin chronology and fluxes in sediments of the Houston Ship Channel, Texas: Influence of non-steady state sediment transport and total organic carbon. *Environmental Science & Technology* 41:5291–5298.

Yeager, K. M., P. H. Santschi, and G. T. Rowe. 2004. Sediment accumulation and radionuclide inventories (^{239,240}Pu, ^{210}Pb and ^{234}Th) in the northern Gulf of Mexico, as influenced by organic matter and macrofaunal density. *Marine Chemistry* 91:1–14.

Yeager, K. M., P. H. Santschi, K. J. Schindler, M. J. Andres, and E. A. Weaver. 2006. The relative importance of terrestrial versus marine sediment sources to the Nueces-Corpus Christi Estuary, Texas: An isotopic approach. *Estuaries and Coasts* 29(3):443–454.

Yu, S.-Y., T. E. Törnqvist, and P. Hu. 2012. Quantifying Holocene lithospheric subsidence rates underneath the Mississippi Delta. *Earth and Planetary Science Letters* 331–332:21–30.

Zhang, L., J. Zhang, P. W. Swarzenski, and Z. Liu. 2011. Radium isotope tracers to evaluate coastal ocean mixing and residence times. Pp. 331–343 in M. Baskaran (ed.), *Handbook of Environmental Isotope Geochemistry*. Berlin: Springer.

6

Dissolved, Colloidal, and Particulate Organic Matter in the Gulf of Mexico

Christopher L. Osburn, Andrew R. Margolin, Laodong Guo, Thomas S. Bianchi, and Dennis A. Hansell

Introduction

Organic Matter in the Marine Carbon Cycle

Production, consumption, and transformation of organic matter (OM) are key controls on marine biogeochemical cycles. Production is the formation of OM largely via photosynthetic carbon fixation (primary production), whereas consumption is both organismal assimilation into biomass (anabolism) and organismal respiration (catabolism), the latter of which is dominated by aerobic respiration in the oxygenated water column. Anaerobic respiration via electron acceptors other than oxygen occurs within oxygen-depleted environments (e.g., sediment pore waters, microzones within marine particles). Transformation can be envisioned as changes to OM size, composition, and reactivity, as it is actively cycled through marine food webs and in surface waters can be subject to photochemistry.

In this chapter, we focus on 3 broad classes of OM: dissolved, colloidal, and particulate. We restrict our focus on these pools of OM to the water column, leaving a detailed discussion of the sediments to Chapter 7; however, we acknowledge the importance of the "rain" of particulate organic matter (POM) through the water column to the sediments, as well as injection of pore-water dissolved organic matter (DOM) from resuspended surface sediments into bottom waters and diffusive fluxes at the sediment-water interface as key linkages in the marine carbon cycle. Further, we focus spatially from the coastal to pelagic waters of the Gulf of Mexico (GOM). River and estuarine OM has been considered in detail elsewhere (Bianchi et al. 1999; Bauer and Bianchi 2011; Bianchi and Bauer 2011). DOM is the largest reservoir of reduced carbon in the global ocean, ca. 662 Pg C (1 Pg = 10^{15} g), approximating that of the growing atmospheric CO_2 reservoir (Hansell et al. 2009; Ciais et al. 2013). Colloidal organic matter (COM) is operationally defined as high-molecular-weight (HMW) DOM but is treated separately because of special physicochemical properties that make it react and cycle differently from low-molecular-weight (LMW) DOM (Benner and Amon 2015). POM is comparatively larger material than COM or DOM and represents both living and nonliving biomass. All 3 pools are major components of the marine carbon cycle and can be actively transformed along the OM size continuum during their transport and cycling within marine ecosystems.

As part of that cycling, OM plays a key role in modulation of the inorganic carbon cycle, especially the influence of the ocean's C buffer, via uptake of CO_2 through primary production and release of CO_2 via remineralization. We can view this cycle by first focusing on the formation of POM via primary production in surface waters as the start of the ocean's "biological pump" (Jiao et al. 2010; Fig. 6.1).

This fixed carbon constitutes the bulk of the POM that then is transported to depth. After fixation, POM can be transformed into, or be caused to released as, DOM (and COM) via several ecosystem processes involving graz-

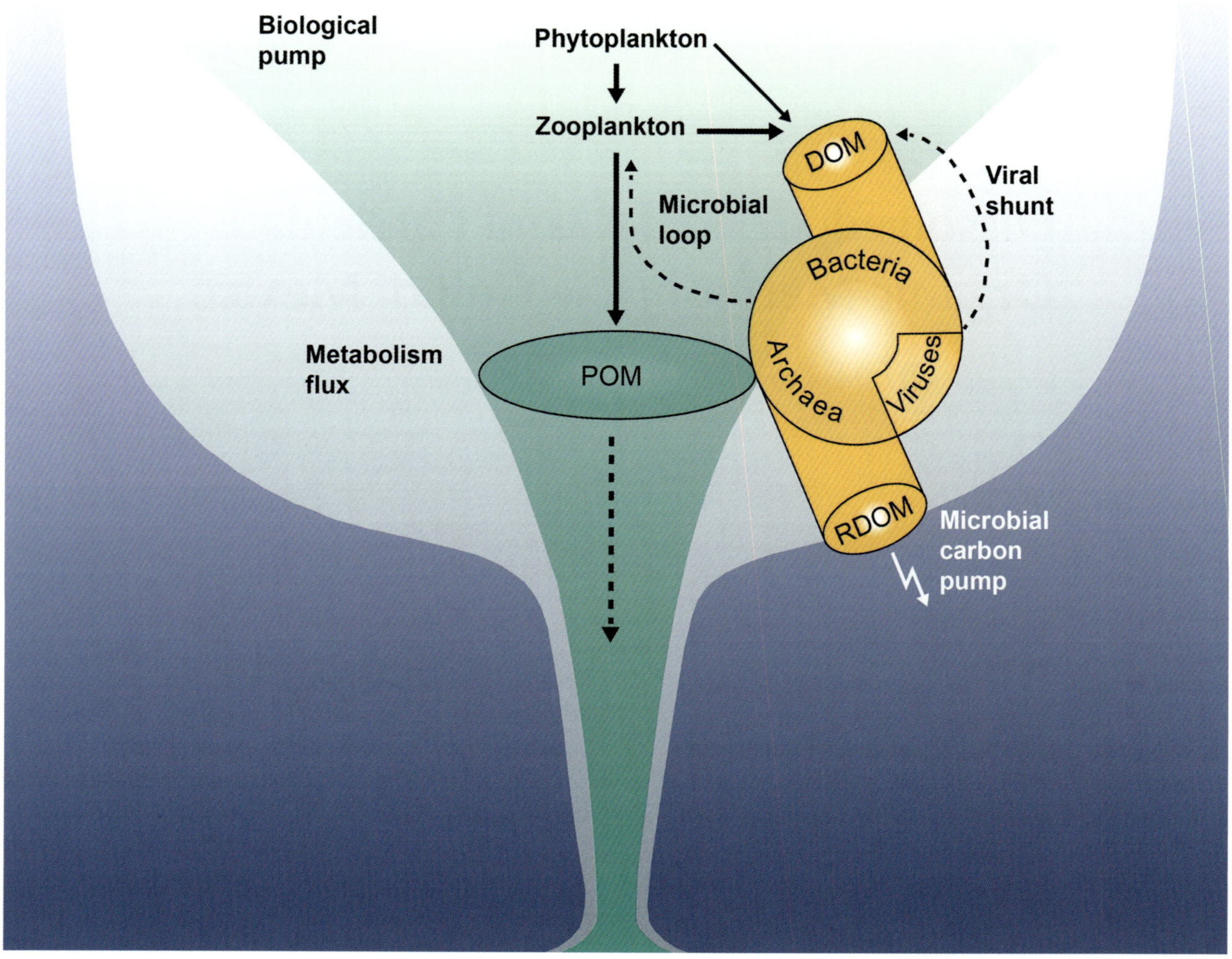

Figure 6.1. The biological pump in the Gulf of Mexico, based on the formation of organic matter from primary production in surface waters (from Jiao et al. 2010; used with permission).

ing and predation by metazoans and further processing by microbes (e.g., archaea, bacteria, and viruses); this microbial loop is thought to contribute to the production of the refractory DOM (RDOM) pool in deep-ocean waters. On the other hand, DOM can coagulate or flocculate into POM and form gels through a self-assembly process (Verdugo et al. 2004; Verdugo 2012). The DOM may then be biologically or photochemically oxidized to CO_2, although all sizes of OM can be a substrate for respiration. Therefore, OM cycling via production and degradation processes modulates the CO_2 balance of the oceans and hence has important implications for global climate (Sarmiento and Gruber 2002; Emerson and Hedges 2008; Bauer et al. 2013; Ciais et al. 2013; Le Quéré et al. 2013; Holte et al. 2017).

Terrigenous and Marine Sources of Organic Matter in the Ocean

Discussion of OM cycles in the marine environment must consider the 2 dominant sources of that material. GOM, a marine basin enclosed by the North American continent, has significant terrigenous inputs of OM, primarily by the Mississippi and Atchafalaya Rivers, as well as other small rivers, and secondarily by bordering estuaries and fringing coastal wetlands (Guo et al. 1999, and references therein). Fluvial input of OM to the GOM via rivers into estuaries and coastal waters effectively connects the Gulf to ecosystem processes occurring on land in each river's watershed. Dominant land-based sources of terrestrial OM include soils, anthropogenic inputs (urban and agricultural), and wetlands.

These fluvial inputs do not themselves act as "pipes" but rather exhibit substantial biogeochemical processing of materials depending on elevation gradients, flowpaths across and through sediment-water interfaces, hydrological storage and retention, and "microbial capacities" to transform OM in fluvial networks (Cole et al. 2007; Battin et al. 2008). For smaller rivers, export from land to ocean of terrestrial OM may be dominated by a few annual extreme storm events (Yuan et al. 2004; Inamdar and Mitchell 2006; Raymond and Saiers 2010). The importance of extreme events for importing terrestrial OM and its resulting biogeochemical processing in the GOM has recently been reported (Yuan et al. 2004; Bianchi et al. 2013). More recently, the important role of low-frequency large events, associated with rapid increases in headwater stream velocity in the transport of OM through the aquatic continuum, has been referred to as the pulse-shunt concept (PSC; Raymond et al. 2016).

Terrestrial sources reflect land-based primary production of OM, incorporation and modification into soil ecosystems, and export (with aforementioned transformation) in fluvial systems connected to the ocean (e.g., Bianchi et al. 2014). Therefore, the biochemical composition of land plants and soil organisms can be expected to exert unique geochemical fingerprints to be detected in OM pools in the GOM. In this chapter we examine the chemical properties of OM to evaluate its terrestrial origins, utilizing chemical information from the wealth of studies that have focused specifically on terrestrial inputs. However, we do not specifically deal with the riverine, estuarine, and wetland ecosystem processes that can modify those properties. Therefore, for information on riverine and estuarine OM biogeochemistry in the GOM region, see, for example, Coffin and Cifuentes (1999); Guo et al. (1994); Bianchi et al. (1999); Benner and Opsahl (2001); Guo et al. (2009); Cai et al. (2012); and Cai et al. (2015).

Planktonic sources of OM in the GOM can be thought of as the primary and secondary production in the Gulf's coastal and pelagic ecosystems. The biochemical composition of marine planktonic OM is very different from that of terrigenous OM (especially that derived from soils), providing discrete chemical signals that aid in discriminating planktonic from terrigenous OM. The vertical export of planktonic OM provides a substrate for pelagic ecosystems, actively cycling through marine food webs. In addition, it is generally considered that the rain of POM from surface to deep water (Diercks and Asper 1997; Green and Dagg 1997) that escapes consumption and mineralization in the water column is the major source of carbon and energy for sediment-dwelling organisms in benthic ecosystems. Hence the importance of planktonic material formed in the water column is expressed in benthic-pelagic coupling and movement of major bioactive elements.

We describe the characteristics of these major pools of OM in the GOM and summarize estimates of OM stocks and fluxes within the GOM. We begin with some definitions of OM physical and chemical characteristics; then we discuss the traits, distribution, and fate of 3 major categories of OM. We indicate where data are scarce and identify the important unknowns. We then summarize the key points and explore future directions for OM research in the GOM.

Characterization of Organic Matter

Because marine OM is a combination of hundreds to thousands of molecules, defining its chemical composition is critical; several excellent summaries have been published (Mopper et al. 2007; Repeta 2015). Here we review basic concepts that allow a discussion of what is known about OM geochemistry in the GOM. OM is operationally defined by size of material retained by, or passing through, filters (Fig. 6.2). In the broadest sense, filtration separates POM from DOM because POM is retained on the filter whereas DOM is contained in the filtrate. Historically, several different filter sizes and types have been used to make this separation, based on needs of the researcher as well as advances in filter technology over time. Glass microfiber filters have mesh sizes of roughly 1 µm; the most commonly used mesh size for marine studies is 0.7 µm. These filters are not only economical, but they can also be combusted at high temperature prior to use to remove any organic background. In oligotrophic or deep (>250 m) waters where the POM-to-DOM ratio is small (<1:40), filtration may not be necessary when considering the bulk OM pool, and some studies have used total organic carbon (TOC) to represent DOC to avoid filtration and possible contamination during sample filtration. To consider the sampling of POM (e.g., excluding bacteria), filters constructed from polymers (or quartz) can have more narrowly defined size ranges, affording a more distinct separation of OM size. Most marine OM studies utilize 0.2 or 0.7 µm as the POM/DOM boundary. For the size fractionation of DOM, or separating the bulk DOM into colloidal or HMW- and LMW-DOM (Guo et al. 1995; Guo and Santschi 2007; Emerson and Hedges

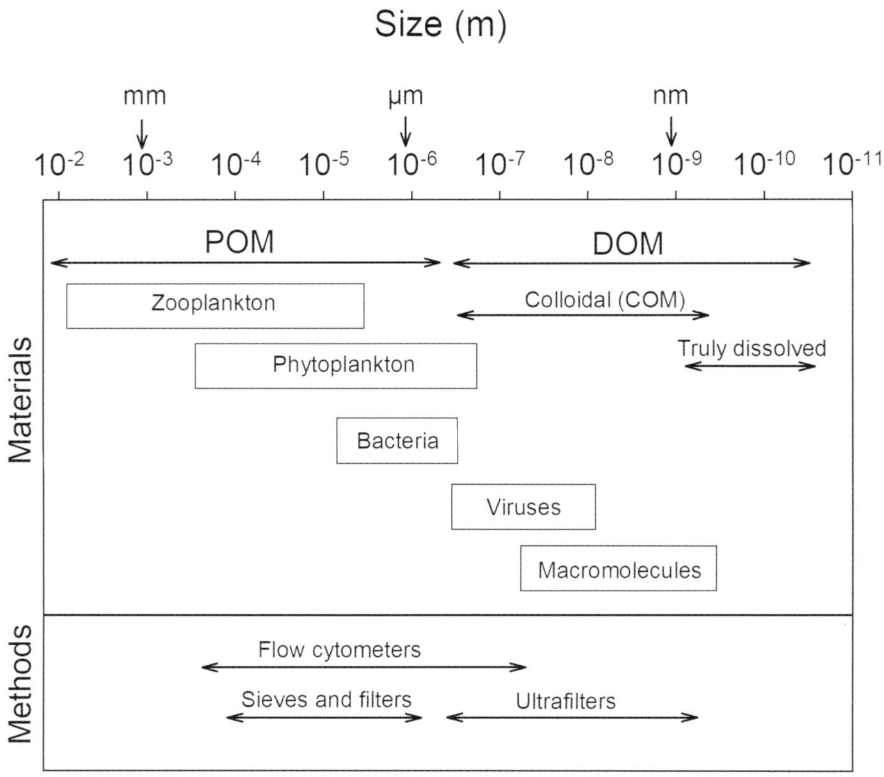

Figure 6.2. Size distribution of organic matter in seawater (adapted from Repeta 2015).

2008), filters with pore sizes ranging from 0.2 µm (to exclude bacteria) to 0.45 or 0.7 µm have been widely used (Benner et al. 1992; Guo et al. 1995; Cai and Guo 2009). However, quantitative difference in DOM quantity and quality between the 0.2 µm and 0.45 or 0.7 µm filtrate is largely unknown (Xu and Guo 2017).

Sources of OM can be defined based on their provenance (i.e., origin) with respect to the system in which they are found. "Autochthonous" sources are formed in place: for example, OM formation by marine primary and secondary production. "Allochthonous" sources are formed outside a system and transported to it. Multiple allochthonous sources of OM to the GOM occur; rivers and estuaries are the primary sources.

The literature also describes OM in the oceans as "terrestrial" and "planktonic." These terms describe the provenance of OM but also are meant to distinguish between broad compositions of chemical characteristics of these materials. Terrestrial OM refers to OM originating from terrestrial primary production and degradation processes (e.g., soil formation) and is dominated by land-plant detritus and soil OM. Planktonic OM includes primary and secondary production from bacterioplankton, phytoplankton, and zooplankton in the water column and the benthos. This definition considers estuaries, coastal wetlands (freshwater/salt marsh, mangroves), and emergent and submerged aquatic vegetation as localized inputs not important to the flux of OM across the shelf-slope–abyssal-plain transition. Sedimentary organic matter (SOM) defines the POM deposited in surface sediments but does not refer specifically to its vascular plant or planktonic origin. SOM is considered allochthonous, and processes such as erosion of open-water marshes and sediment resuspension can mobilize SOM, releasing it into the overlying water column.

Several chemical approaches are used to characterize DOM and POM. "Elemental analysis" quantifies specific elemental pools of DOM and POM. The elements most frequently measured are the major bioactive elements carbon (C), nitrogen (N), and phosphorus (P) and reported in molar units (e.g., µM) or mass units (mg L^{-1}). Organic carbon (OC) concentrations in sediments are usually expressed as a percentage of dry mass or as a mass of OC per dry mass of sediment (e.g., mg C kg^{-1}). Dissolved organic carbon (DOC) and particulate organic carbon (POC) are measured most often to quantify OM in seawater. DOC is measured via high-temperature catalytic oxidation or wet-chemical oxidation to CO_2, which is quantified relative to a series of standards. Dissolved organic nitrogen (DON) is measured indirectly

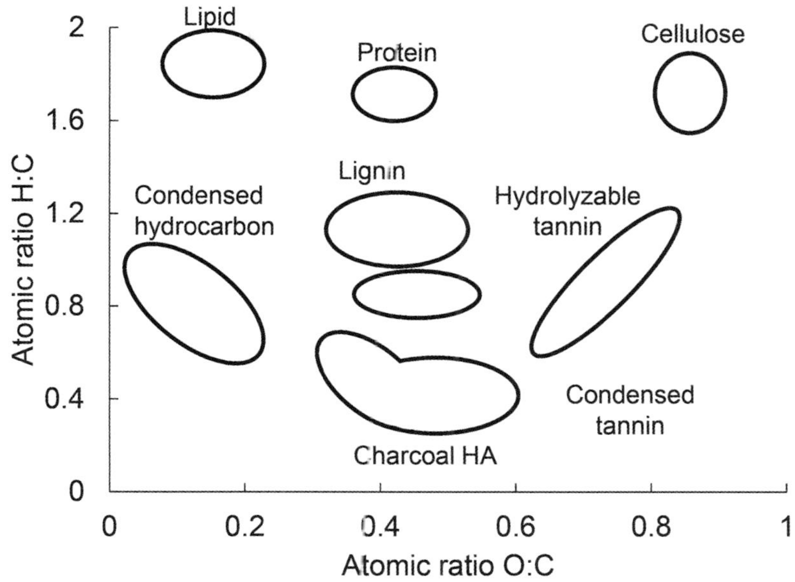

Figure 6.3. Van Krevelen plot of atomic H:C ratios vs. atomic O:C ratios obtained from molecular formulas measured by FT-ICR-MS of organic matter. Ovals indicate general regions corresponding to specific compound groups (from Mopper et al. 2007).

by difference between total nitrogen (TN) and the sum of dissolved inorganic nitrogen (DIN: nitrite + nitrate + ammonium) in samples. POC and particulate organic nitrogen (PON) are typically measured simultaneously using an elemental analyzer (EA), by combusting material collected on glass microfiber filters as described previously. Filters containing samples are vapor-acidified to remove carbonate prior to combustion with EA. Filters are combusted to CO_2 and NO_x gases at high temperature in a stream of oxygen, and the NO_x is reduced to N_2 gas over a copper-containing reduction column; each gas is measured separately after chromatographic separation. (It is important to note that the PON measurement actually is particulate nitrogen [PN], as inorganic N is not removed by vapor acidification.) Elemental analysis can also be applied to the hydrogen (H), oxygen (O), and sulfur (S) content of OM, though for consideration of sources and cycling, C and N are the most commonly measured.

Recent advances in mass spectrometry have improved the resolution of elemental analysis to discrete molecular formulas contained within a sample. The coupling of an electrospray ionization (ESI) inlet system with Fourier-transform ion cyclotron resonance mass spectrometry (FT-ICR-MS) affords a means to conduct a very-high-resolution mass spectral analysis. Size range of ions is limited to 200 to 1000 Da, but thousands of ions can be measured with enough resolution to confidently assign specific elemental formulas, though each unique formula likely represents a mixture of isomers (Repeta 2014). The technique is highly informational but not quantitative; however, the resulting formulas can be plotted as functions of their elemental ratios (e.g., H:C and O:C) in van Krevelen diagrams (Fig. 6.3). Certain regions of van Krevelen space indicate major compound classes such as lipids, carbohydrates, proteins, lignins, and condensed hydrocarbons (Koch et al. 2005; Mopper et al. 2007; Sleighter and Hatcher 2008). Multivariate statistical processing and Spearman rank correlations have been employed to relate the high-data-density output of FT-ICR-MS with physical and chemical parameters of DOM and its environment (e.g., fluorescence or watershed land use) (see Stubbins et al. 2014; Wagner et al. 2015).

Instruments used to measure DOC, POC, and PON, such as TOC analyzers and EAs, respectively, are routinely coupled to magnetic-sector isotope ratio mass spectrometers (IRMS) to measure stable C and N isotopes of these fractions of OM (e.g., Osburn and St-Jean 2007; Lalonde et al. 2014; Beaupré 2015). Stable isotope ratios (R) (e.g., ^{13}C:^{12}C or ^{15}N:^{14}N) are reported as delta values (δ) in per mil units (‰), which represent the deviation of the sample ratios from the ratios of an internationally accepted standard (the Vienna Pee Dee Belemnite standard, VPDB; Equation 6.1):

$$\delta^{N}X = \left[\frac{R_{sample}}{R_{standard}} -1\right] \times 1000‰, \tag{6.1}$$

where N is the heavy isotope of element X (e.g., ^{13}C, ^{15}N). Samples are calibrated to the VPDB scale by measurement of standards of known isotopic composition.

Radiocarbon isotope (^{14}C) measurements on DOM and POM are conducted on OM converted to CO_2 and then reduced to graphite targets, which are then measured for their ^{14}C abundance by accelerator mass spectrometry (AMS; McNichol and Aluwihare 2007; Beaupré 2015). $\Delta^{14}C$ is defined as the ratio ($\delta^{14}C = {}^{14}C{:}{}^{12}C$) of a sample to a standard, similarly to $\delta^{13}C$ (Equation 6.2):

$$\Delta^{14}C = \delta^{14}C\text{-}2\,(\delta^{13}C\text{+}25)\left[1+\frac{\delta^{14}C}{1000}\right]. \qquad (6.2)$$

Further considerations of radioactive decay, imperfect mixing of the ^{14}C isotope in the Earth system, the "Suess effect" caused by the combustion of isotopically light fossil fuels, and "reservoir effect" must be considered in the computation of these values (Ascough et al. 2005; Beaupré 2015). However, the advantage of $\Delta^{14}C$ values is their ability to be converted to radiocarbon ages that estimate the age of the original OM (Equation 6.3).

$$^{14}C\text{ age} = -\left[\frac{5568\text{ years}}{\text{In}2}\right]\text{In}\left[1+\frac{\Delta^{14}C}{1000}\right]. \qquad (6.3)$$

$\Delta^{14}C$ values can be used with $\delta^{13}C$ values in source-mixing models to apportion OM sources between allochthonous and autochthonous endmembers (Peterson et al. 1985; Goñi et al. 1998; Bauer 2002; McCallister et al. 2004; Cui et al. 2016).

A fraction of DOM absorbs light in the ultraviolet (UV) and visible regions of the electromagnetic spectrum and is termed "chromophoric dissolved organic matter" (CDOM). A fraction of CDOM re-emits absorbed light energy as fluorescence (FDOM). The absorbance and fluorescence properties derived from discrete compounds such as lignin and pigments and classes of compounds such as humic substances (of both terrestrial and marine origin) are consistent with absorption and fluorescence of rivers, estuaries, and the ocean (Del Vecchio and Blough 2004; Coble 2007; Nelson and Siegel 2013; Stedmon and Nelson 2015; Osburn et al. 2016). OM-containing aromatic rings, for example, absorb quite strongly between 250 and 280 nm. CDOM absorption or fluorescence often is strongly correlated to DOC concentrations, especially in coastal waters, with different regression slopes depending on DOM sources and composition (Osburn et al. 2016; Zhou et al. 2016).

A major advantage of measuring CDOM absorption and fluorescence is the rapidity in which each measurement can be conducted, which has allowed high resolution in situ monitoring of CDOM distributions in rivers and oceans (Chen et al. 2004; Chen and Gardner 2004; Comny et al. 2004). CDOM absorption is quantified as the Napierian absorption coefficient, which is defined as the blank-corrected absorbance, A, at a wavelength, λ, divided by the path length of the cell (L) in which the absorbance measurement is taken (in meters):

$$a_\lambda = \frac{2.303A_\lambda}{L}. \qquad (6.4)$$

CDOM absorption spectra exhibit a featureless decay with increasing wavelength. Exponential equations are used to model CDOM absorption coefficients (a_λ, in m^{-1}) at any wavelengths, λ, based on a reference wavelength , a_{λ_0} for example, a_{375}:

$$a_\lambda = a_{\lambda_0}\, e^{-S(\lambda-\lambda_0)}, \qquad (6.5)$$

where S is the spectral slope coefficient, and for most samples (especially for rivers, estuaries, and coastal waters) the model performs well between 300 and 650 nm (Stedmon et al. 2000). Helms et al. (2008) introduced the spectral slope ratio, S_R, which is specific in the ultraviolet region of the CDOM spectrum and which contains considerable information related to DOM chemistry. S_R is calculated as the ratio of the S value from 275 to 295 nm to the S value from 350 to 400 nm. This ratio has been shown to increase with salinity from river to seawater and with decreasing molecular weight of DOM. Processes such as photochemical degradation also can increase S and S_R (Spencer et al. 2009). Typically, S_R values for terrestrial CDOM are <1.0 and for seawater are >2.0.

Specific ultraviolet absorption (SUVA) at a specific wavelength is computed as the decadic absorption (Napierian divided by 2.3030) divided by the DOC concentration in mg C L^{-1} (Weishaar et al. 2003). $SUVA_{254}$ is a common index computed because of the peak absorption by aromatic carbon. $SUVA_{254}$ values range from 0 to 5 L mg C^{-1} m^{-1}; values higher than 6 likely have high iron content and should be interpreted with caution (Poulin et al. 2014).

CDOM fluorescence is measured as emission spectra collected over a series of excitation wavelengths, which most often are concatenated into an excitation-emission matrix (EEM) and visualized as a contour plot. EEMs exhibit regular features associated with known sources of OM (Table 6.1).

While CDOM absorbance and fluorescence contain no discrete chemical information, their properties are

Table 6.1. Common regions of organic matter fluorescence maxima along with source descriptions

Type	Label[a]	Description	Ex. range (nm)	Em. range (nm)
Visible	A	Terrestrial humic substances	260	400–460
Visible	M	Microbial humic substances	290–310	370–410
Visible	C	Terrestrial humic substances	320–360	420–460
UVA	N	Recent phytoplankton production	280	370
UVA	B	Tyrosine-like	275	305
UVA	T	Tryptophan-like	275	340

Source: Modified from Stedmon and Nelson (2015).
Note: Ex. = Excitation; Em. = Emission.
[a] Based on Coble (2007).

relatable to DOM chemistry (e.g., Hernes and Benner 2003; Hernes et al. 2009; Spencer et al. 2009; Osburn and Stedmon 2011; Fichot and Benner 2012). Moreover, the optical properties of CDOM allow for relating these measurements to remote-sensing information; in fact, retrievals of CDOM absorption from satellite imagery allows for quantifying DOM chemistry at greater spatiotemporal resolution (Mannino et al. 2008; Tehrani, D'Sa, and Osburn 2013; Tehrani, D'Sa, Ko, et al. 2013). Recent work has shown clear linkages between DOM's optical and chemical properties in a variety of estuarine and coastal environments, including the GOM (Fichot and Benner 2012; Cawley et al. 2014; Osburn and Bianchi 2016, and references therein).

Biomarkers are specific organic compounds or compound classes that can be used to assess the biochemical composition of OM in aquatic ecosystems (see Bianchi and Canuel 2011). Typically, biomarkers make up a much smaller fraction of OM than do elemental composition that is measured on the bulk OM. Biomarkers often are extracted from complex matrices (seawater, sediments) and can represent biosynthetic pathways grouped into primary and secondary metabolisms. Carbohydrates (mono and polysaccharides), proteins (amino acids), lipids (fatty acids and sterols), and lignin all are classes of biomarkers commonly used to characterize OM in marine environments (see Bianchi and Canuel 2011). Hydrolyzable neutral sugars and amino acids are considered the most "biolabile" material and are readily consumed in surface waters, selectively removed during composition (Benner 2002). Lipids and lignin are structurally more complex and tend to be considered "recalcitrant" due to their persistence in POM and sediments. We next provide a brief overview of these major compound classes; for more-detailed descriptions and further information, see Killops and Killops (2013); Bianchi and Canuel (2011); and Hansell and Carlson (2015).

Carbohydrates, the most abundant biomolecules on earth, represent about 20–40% of phytoplankton biomass and 75% of vascular plant biomass (Aspinall 1983; Parsons et al. 1984); thus, they make up very large fractions of POM and DOM (Pakulski and Benner 1994). Carbohydrates are measured after hydrolysis of larger biomolecules as simple sugars (monosaccharides), classified as neutral sugars, the major group of sugars; and acidic sugars, amino sugars, and *O*-methyl sugars, the minor groups (Opsahl and Benner 1999; Bergamaschi et al. 1999; Benner and Kaiser 2003; Kaiser and Benner 2009).

Proteins make up about 50% of OM and contain >80% of the organic N in marine organisms (Romankevich 1984). Amino acids, which are essential to all organisms, are used as important biomarkers for protein sources and represent an important fraction of POC (13–37%) (Cowie et al. 1992; Van Mooy et al. 2002; Kaiser and Benner 2009). Twenty amino acids are found in proteins, including 11 that can be synthesized by animals, called nonessential amino acids; and 9 must be obtained through dietary means, called essential amino acids. The essential amino acids are synthesized by plants, fungi, and bacteria. Distributions of protein amino acids show remarkable uniformity among organisms (Cowie et al. 1992); thus, it has been argued that differences in amino acid composition arise from degradation processes. In addition, nonprotein amino acids have been used as an index of degradation. Amino acids are characterized as hydrolysis products of peptides and proteins. Pools of amino acids can be measured as the total hydrolyzable amino acids (THAA) in POM and in SOM (Burdige and Martens 1988; Keil and

Kirchman 1991a, 1991b; Van Mooy et al. 2002). Dissolved combined and dissolved free amino acids (DCAA and DFAA, respectively), as well as enantiomeric amino acids, are measured in DOM. Cellular proteinaceous material that has been transformed to detritus is described as particulate combined amino acids (PCAA). Finally, hydrolysis of proteins by ecto- or extracellular enzymes to peptides and amino acids leads to another category, the enzymatically hydrolyzable amino acids (EHAA) (Dauwe et al. 1999).

Lipids are defined as organic solvent-extractable compounds that represent major energy flow pathways through ecosystems. Lipids include pigments, fats, oils, waxes, phospholipids, and steroids that can be broken into compound classes such as hydrocarbons, fatty acids, and sterols. The molecular characteristics of compounds in these classes (e.g., carbon number) can be used to fingerprint OM sources to specific organisms (Canuel and Martens 1996). Fatty acids are a versatile class of lipid biomarkers with a wide array of structural features, such as number of double bonds, functional group composition, and methyl branches, allowing fatty acids to be linked to a variety of OM sources (e.g., algae, bacteria, and higher plants) (Wakeham et al. 2002; Volkman 2006). Isoprenoid lipids include steroids, hopanoids, and triterpenoids. This diverse class of lipids is derived from a common 5-carbon isoprene unit and is structurally sophisticated. The diagenetic products of isoprenoid degradation are used to classify OM according to their source: algal (steroids), bacterial (hopanoids), or vascular plant (steroids and triterpenoids) (Ourisson and Albrecht 1992; Hayes 2001; Peters et al. 2007).

Photosynthetic pigments have been used as important tracers of POC in aquatic systems (Bianchi and Canuel 2011). The primary photosynthetic pigments used in absorbing photosynthetically active radiation (PAR) are chlorophylls, carotenoids, and phycobilins—with chlorophyll dominant (Clayton 1980). Although a greater amount of chlorophyll is found on land, 75% of the global turnover (ca. 109 Mg yr^{-1}) occurs in oceans, lakes, and rivers/estuaries (Brown et al. 1991; Jeffrey and Mantoura 1997). Chlorophyll-*a* concentrations have been routinely measured for estimating standing stock and productivity of photoautotrophic cells in freshwater and marine environments. Carotenoids are an important group of tetraterpenoids consisting of a C_{40} chain with conjugated bonds (Goodwin 1980; Frank and Cogdell 1996). These pigments are found in bacteria, algae, fungi, and higher plants and serve as the principal pigment biomarkers for class-specific taxonomic work in aquatic systems. Ocean-color data, remotely sensed from satellites, provides the best practical means of thoroughly monitoring the spatial and temporal variability of near-surface phytoplankton in broad open ocean and coastal environments (Miller et al. 2005). For example, the global distribution of chlorophyll-*a* is available daily from satellite-based ocean-color sensors, such as Sea-viewing Wide Field-of-view Sensor (SeaWiFS, active 1997–2010) and Moderate Resolution Imaging Spectroradiometer (MODIS, active 2000–present).

Lignin, perhaps the most widely used vascular plant chemical biomarker for quantifying terrigenous sources of OM (Hedges and Ertel 1982), is found in the cell walls of vascular plants and is made of complex and irregular phenylpropanoid units linked together by irregular carbon-carbon and ß-*O*-4 linkages (Sarkanen and Ludwig 1971). Lignin is a structurally complex biopolymer difficult to measure unless chemically degraded to its component phenolic units. Its structural complexity makes it difficult for most heterotrophic microbes, except fungi and some bacteria, to degrade. Lignin-derived phenols are classified in groups of aldehydes, ketones, and acids based on presence of functional groups. Vanillyl (V) phenols (vanillin, acetovanillone, vanillic acid) are present in all vascular land plants. Syringyl (S) phenols (syringealdehyde, acetosyringone, syringic acid) are present only in angiosperms. Cinnamyl (C) phenols (ferulic acid, p-coumaric acid) are enriched in nonwoody plant material such as leaves and grasses. The p-hydroxy (P) phenols (p-hydroxybenzaldehyde, p-hydroxyacetophenone, p-hydrobenzoic acid) may be derived from degradation of aromatic amino acids such as tyrosine from algae as well as from angiosperms, thus must be interpreted with care. Lignin-derived phenols are photochemically sensitive; hence sunlight is a major degradation pathway for lignin (e.g., Opsahl and Benner 1998; Spencer et al. 2009).

Size Distribution

Organic matter exists as a continuum of sizes in the ocean, including dissolved, colloidal, and particulate OM. The size distribution of OM in a steady-state ecosystem can be represented as a pyramid, with DOC being the most abundant followed by colloidal organic carbon (COC) and POC (Fig. 6.4). In a specific marine ecosystem, DOC forms the basis of food webs, supporting microbial communities and linking between organic and inorganic and between biology and chemistry. Therefore, in terms of total mass (not only size distribution), DOC has the highest inventory in the ocean, followed by COC and POC based on their relative concentrations to foodwebs (Guo et al. 1994; Benner and Amon 2015; Xu and Guo 2017). However, while inventories or mass concentrations of different organic carbon pools are important, the fluxes between carbon reservoirs and chemical and biological reactivity will determine the turnover rate and overall biogeochemical cycling of OM in marine environments. For example, while smaller in pool size, POC is the active OC pool with short residence time (Santschi et al. 1995; Guo et al. 1996) and can aggregate into sinking particles or disaggregate into dissolved or colloidal phases, showing a carbon flow direction from POM to COM and DOM.

On the other hand, COC consists of macromolecules or nanoparticles, which possess high specific surface areas and are highly dispersed in nature. They have been shown to have higher biological and chemical reactivities through coagulation, biological utilization, and dynamic transformation in the water column (Santschi et al. 1995; Amon and Benner 1996; Guo et al. 1996). Indeed, over the last few decades, a growing body of literature has suggested that chemical and biological reactivities could be related to molecular size and freshness or composition (e.g., Santschi et al. 1995; Amon and Benner 1996; Guo et al. 1996; Stolpe et al. 2010; Wang et al. 2010; Benner and Amon 2015; Zhou, Stolpe, et al. 2016). Thus, mass concentrations, size spectra, and chemical composition of OM are all important factors in regulating the fate, transport, and transformation of OM in the water column. Some of the first measurements of POC and DOC in the GOM were taken by Wilson (1963) and Fredericks and Sackett (1970).

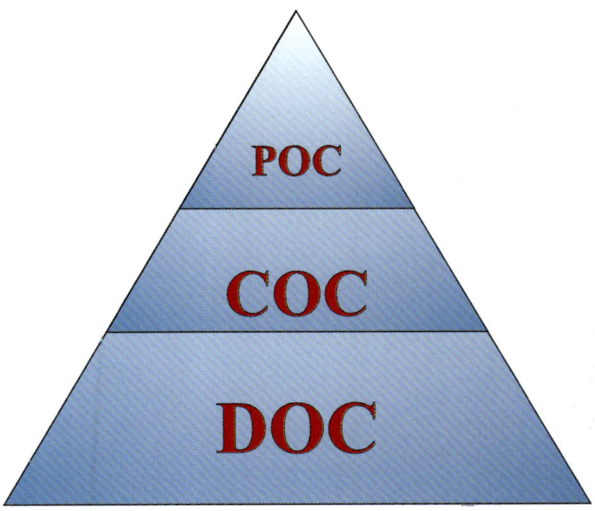

Figure 6.4. Schematic showing the relative inventory of organic matter over major size classes in the ocean water column at steady state. Note that DOC is located at the base of the marine food-web pyramid and is in the boundary between the organic and inorganic worlds, linking biology and chemistry.

As described previously, the distinction between POM and DOM is operationally defined based on filter size and the degree to which particles will settle under force of gravity. Detailed overview of the methods of filtration can be found in Grasshoff et al. (2009). However, physical definitions of POM and DOM are highly operational and arbitrary. Since different-sized DOM and POM likely have different ecological roles, physicochemical behavior, and environmental fates, it is important not to discriminate a specific size fraction in the whole OM size spectrum unless a specific study has a specific research focus. In general, POM is operationally defined as particles greater than ca. 0.4 μm in diameter and contains both living material (bacteria, phytoplankton, zooplankton) and nonliving detritus (plant and plankton fragments, fecal pellets). Sieves and filters can be used to separate POM into successive class sizes (Ho et al. 2007; Roland et al. 2009), especially useful for screening different organisms from smaller organic particles and for screening suspended and sedimentary materials into sand, silt, and clay fractions. Separation of particles by flow cytometry, coupled with mass spectrometry, also has been utilized (Minor et al. 1999). This technique allows separation of particles into subclasses based on size, shape, and spectroscopic (fluorescence) characteristics and has been adapted to distinguish between cellular POM and detrital POM via fluorescence detection of active phytoplankton cells (Minor and Nallathamby 2004). Separations by particle density have been made on suspended material and on sediments using sodium metatungstate, which separates bulk material into various density classes: for example, >2.5, 2.0–2.5, 1.6–2.0, and <1.6 g cm^{-3}(Arnarson and Keil 2001; Wakeham et al. 2009).

Separation of DOM into size classes is achieved by a variety of methods, including ultrafiltration (e.g., Guo et al. 1995; Benner et al. 1997; Guo and Santschi 2007), field-flow fractionation (e.g., Zanardi-Lamardo et al. 2002; Stolpe et al. 2010; Zhou, Stolpe, et al. 2016), and reverse osmosis coupled with electrodialysis (e.g., Green et al. 2014). Tangential-flow ultrafiltration effectively separates DOM by molecular weight into different size fractions, such as very high molecular weight (VHMW-DOM: >100 kiloDaltons, kDa), high molecular weight (HMW-DOM: 1–10 kDa), and low molecular weight (LMW-DOM: <1 kDa), depending on the specific membrane cutoff used (Guo and Santschi 2007). Recently, flow field-flow fractionation (FlFFF), a separation technique in which the retention force is provided by a cross-flow perpendicular to the channel flow, has been increasingly used to separate colloids based on their diffusion coefficients (Giddings 1993). Compared to the separation into a few size fractions, provided by cross-flow ultrafiltration techniques for DOM size fractionation, FlFFF has the advantage of providing a continuous size distribution and is capable of simultaneous separation and characterization through online detectors (e.g., Stolpe et al. 2010; Zhou and Guo 2015; Zhou et al. 2016). The material collected by these separation methods generally represents a specific colloidal size range within the bulk DOM pool. Overall, different size-fractionation techniques can be used either to determine the size distribution of bulk DOM or a specific organic component or to isolate sufficient amounts of HMW-DOM for elemental, isotopic, and molecular characterization (Guo and Santschi 2007).

Dissolved Organic Matter

Sources

The major source of DOM in the Gulf, as in any marine basin, results from either the direct or indirect production from autotrophic phytoplankton (Hansell et al. 2009). Autotrophic, or primary, production occurs near the sea surface in the euphotic zone, resulting from phytoplankton assimilating inorganic carbon and nutrients (i.e., CO_2, nitrate, and phosphate) into their biomass, as plants do on land. The most productive regions in the GOM are in coastal waters on the continental shelves, where nutrients are readily available from coastal upwelling and river input, such as near the Mississippi River delta where CDOM is abundant (Fig. 6.5), having a distribution similar to that of chlorophyll-*a* (i.e., productivity), although being primarily terrigenous in origin. Thus, the co-occurrence of peaks in phytoplankton productivity and CDOM are largely due to river inputs of nutrients and terrestrially-derived DOM, respectively. The central GOM is less productive because nutrients are limited in the euphotic zone, thereby corresponding to lower production of DOM than in coastal regions.

Globally, marine phytoplankton are responsible for fixing ~50 Pg C yr^{-1} as net primary production (Chavez et al. 2011), which not only sustains their biomass but ultimately sustains the marine food web. It is estimated that ~50% of this production is released as DOM (Nagata 2008); however, much of this pool is fleeting, as it rapidly consumed by heterotrophs. There are also bacterial sources of DOM and POM produced by both autotrophic

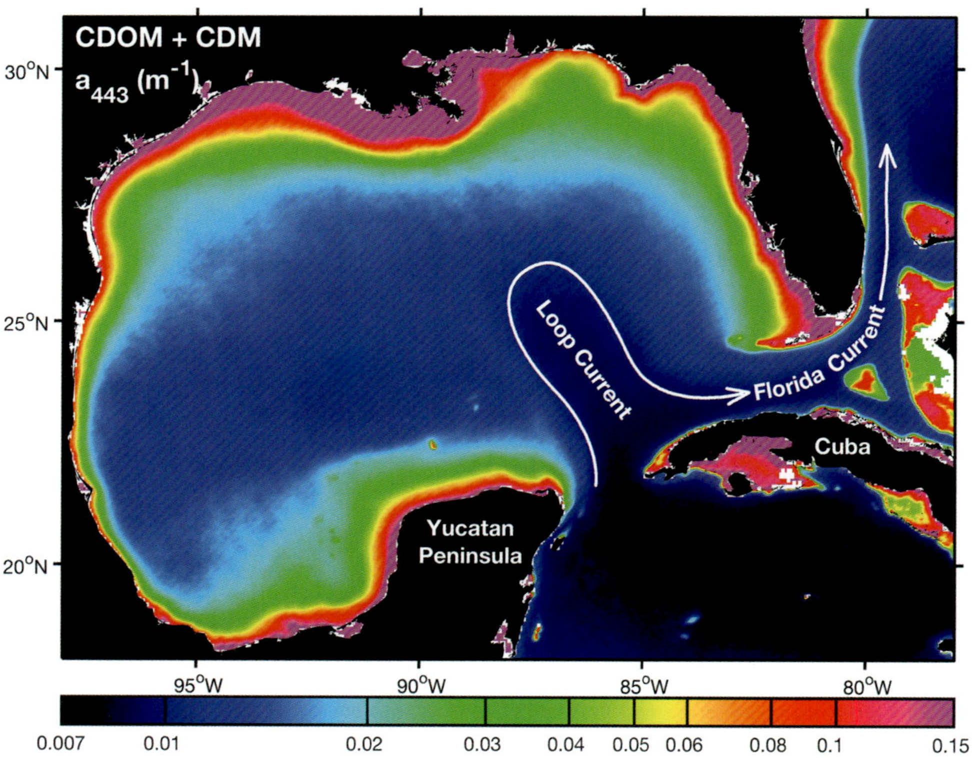

Figure 6.5. CDOM plus detrital material (CDM) from the Aqua-MODIS satellite (Werdell et al. 2013). At this wavelength, CDOM represents ~30% of the ocean's absorption, while CDM represents <10% (Nelson and Siegel 2013). Composite image from July 3, 2002, to October 1, 2016.

and heterotrophic organisms (Kaiser and Benner 2008). Net community production (NCP; i.e., net primary production minus net heterotrophic consumption) yields up to ~2 Pg C yr^{-1} globally, which can be exported to depths >100 m (Hansell et al. 2009; Hansell 2013). Numerous processes are responsible for the primary and secondary production of DOM, which are reviewed here. For further details on the production of marine DOM, see Carlson and Hansell (2015) and Nagata (2008), and references therein.

Environmental stressors such as increases in temperature and ocean acidification (both caused by anthropogenic CO_2), as well as a number of other unfavorable changes, can promote the extracellular release of DOM by phytoplankton (Myklestad 2000). Phytoplankton mortality, perhaps resulting from environmental stressors, is a major factor contributing to DOM release (Duursma 1963), while numerous studies have also found that healthy organisms release up to ~20–40% of fixed carbon as DOC (e.g., Teira et al. 2001; Marañón et al. 2005). The amount of extracellular release of DOM by both healthy and dying phytoplankton varies depending on species and specific environmental conditions (Berman-Frank and Dubinsky 1999; Nagata 2008).

In addition to the physiologically induced releases of DOM by phytoplankton, zooplankton release DOM by direct excretion, egestion, and sloppy feeding, as well as indirectly by leaching from fecal matter (Carlson and Hansell 2015). One of the earliest studies to identify the release of DOM by zooplankton was conducted by Johannes and Webb (1965), who found that the release of dissolved amino acids correlates positively with temperature. The excretion of DON by 2 copepod species was found to represent 64% and 79% of the observed concentrations of DON, with the remainder released by phytoplankton or sloppy feeding (Vincent et al. 2007). In contrast, Saba et al. (2011) found that a different species of copepod excretes 12% of ingested carbon as DOC, representing 80% of released DOC, with the remaining 20% released by sloppy feeding.

The first evidence of sloppy feeding in zooplankton came from studies that used radiocarbon-labeled algae, finding that up to 17–27% of algal carbon is released as DOC (Lampert 1978; Copping and Lorenzen 1980). It was later found that zooplankton grazing on phytoplankton can release a maximum of ~37–69% of algal carbon as a result of sloppy feeding, which depends on phytoplankton size (Strom et al. 1997; Møller and Nielsen 2001; Møller et al. 2003); if the predator-to-prey size ratio falls below ~32–55, the release of DOC via sloppy feeding becomes apparent (Møller 2005, 2007; Saba et al. 2011). Sinking fecal pellets can release up to 34% to >50% of carbon as DOC (Urban-Rich 1999; Thor et al. 2003); however, other studies have found that they release ≤6% of ingested material as DOC and DON, largely depending on the zooplankton's diet (Strom et al. 1997; Møller et al. 2003; Saba et al. 2011). Particle solubilization is another mechanism that can release notable amounts of DOM into the water column (Smith et al. 1992), specifically DOM compounds that contain carbohydrates, proteins, and lipids (Alldredge 1979; Skoog and Benner 1997; Grossart and Ploug 2001; Simon et al. 2002). Vertical migration can also play an important role in DOM release and export to depth, delivering carbon (CO_2 and DOC) and nitrogen (ammonium and DON) to depths of 300 m, equivalent to ~70% and 27% of exported POM, respectively (Steinberg et al. 2000; Steinberg et al. 2002). Bacteria and viruses can also contribute notably to the DOM pool, as the consumption by and of bacteria results in the release of DOM (Nagata 2008; Kujawinski 2011), while viral lysis also results in the release of DOM by the viral shunt (see Fig. 6.1). The viral shunt is estimated to be responsible for the release of up to 26% of annual primary production as DOM, equivalent to ≤13 Pg C yr^{-1} of DOC (Wilhelm and Suttle 1999; Chavez et al. 2011).

Chemoautotrophy is another process that results in the release of OM following CO_2 assimilation, occurring near hydrothermal vents and in hypoxic and anoxic regions (Jannasch and Mottl 1985; Jannasch et al. 1991; Taylor et al. 2001; Walsh et al. 2009). Eutrophication-driven low-oxygen events are a common occurrence in the northern GOM, resulting from accelerated OM production from nutrients into the Mississippi River, as well as terrestrial DOM in runoff, which saturates coastal waters of the GOM and creates massive "dead zones" (Malakoff 1998; Rabalais et al. 2002; Dodds 2006; Diaz and Rosenberg 2008; Bianchi et al. 2011). Other studies have also found that chemoautotrophy is evident in the oxygenated abyssal waters of the ocean (Ingalls et al. 2006; Hansman et al. 2009; Middelburg 2011).

Rivers

Globally, rivers transport ~250 Tg C of DOC to the coastal ocean and marginal seas every year (Dai et al. 2012; Raymond and Spencer 2015), and the GOM receives ~2% of this flux (4 Tg C yr^{-1}). The Mississippi is the largest river that enters the GOM in terms of discharge and presumably DOC flux as well; however, flux data are not available for all rivers that enter the basin (Table 6.2). The influence of the Mississippi on the northern GOM can clearly be seen in surface salinity and DOC measurements (Fig. 6.6). The Mississippi's discharge represents about half of the river water entering the GOM and likely half of the riverine DOC flux as well, while its neighboring Atchafalaya is also a substantial contributor; together they deliver ~70% of the river water to the GOM, along with an estimated ~75% of the riverine DOC. River discharges and fluxes reported in Table 6.2 are annual averages, since it is not uncommon for them to fluctuate by ~25% or more throughout the year, largely caused by variations in precipitation (Duan, Bianchi, and Sampere 2007; Wisser et al. 2010).

Factors that determine the amount of DOM exported from rivers to coastal waters include the volume transported, transport time, and light penetration (e.g., Bianchi et al. 2014). DOM can be added by algae if light penetration is high (Baines and Pace 1991), and it can also be added from anthropogenic sources such as wastewater (Griffith et al. 2009), which may consequently limit the penetration of light and primary production. In many rivers, transport times are short and the penetration of light can also be minimal, which limits the importance of in situ production to periods of low flow (Roach 2013) and other processes such as photo-oxidation. However, this is not always the case, especially for large rivers such as the Mississippi, which has high concentrations of nutrients that promote in situ production (Bianchi et al. 2004). Generally, rivers transport more terrigenous DOM to coastal waters when runoff and discharge are highest, such as during the spring (Guo and Macdonald 2006; Cai et al. 2015). Wetlands and floodplains can also play an important role in the transport and export of terrigenous DOM into coastal waters (Wetzel 1992; Harrison et al. 2005; Amon et al. 2012; Shen et al. 2012; Duan et al. 2017), while mangroves are also major sources (Dittmar et al. 2006). Coastal vegetation is estimated to contribute a DOC flux that is at a minimum 24% of riverine input (Raymond and Spencer 2015), equivalent to ~1 Tg C yr^{-1} for the GOM.

Human activity is also responsible for the alteration of riverine DOM composition and ultimately its delivery to the marine environment (Bernades et al. 2004; Wilson and Xenopoulos 2009; Williams et al. 2010). It is estimated that wetlands in the United States have been reduced by 6–12% since the 1780s (Dahl 1990), which could be responsible for a ~20–30% decrease in DOC export to coastal waters (Raymond and Spencer 2015). However, precipitation is expected to increase as a result of climate change (Kirtman et al. 2013), which will promote increased river discharge, so it is possible that the decrease in DOC export may be balanced by this human-induced enhancement. For example, Tropical Storm Irene resulted in the increase of DOM concentrations in streams to the extent that 40% of the average annual export occurred in the span of a few days (Yoon and Ray-

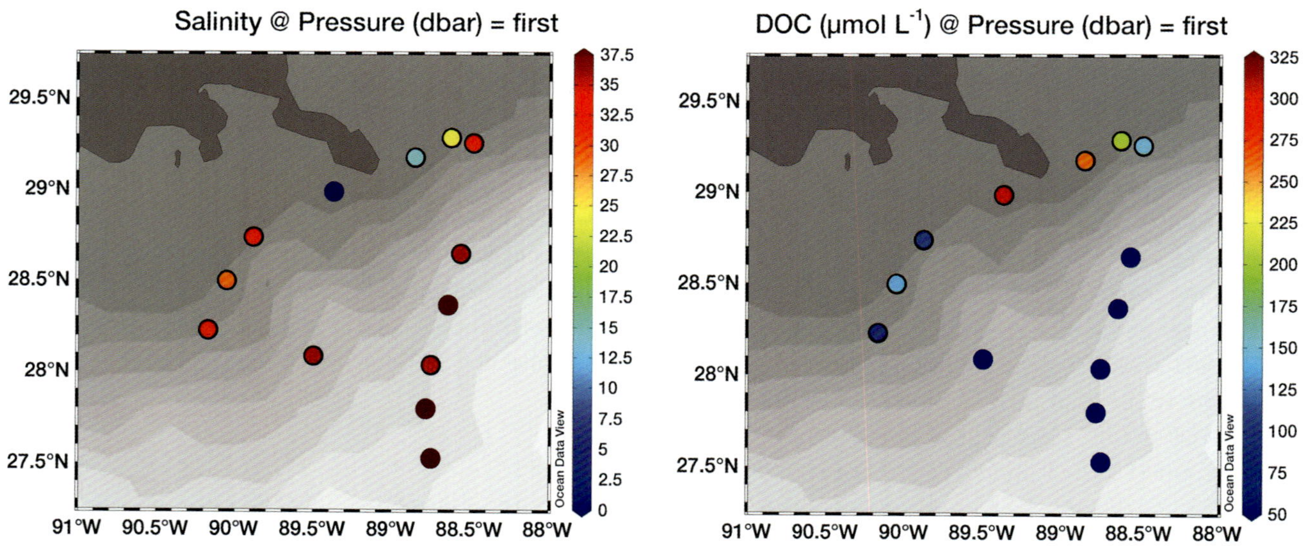

Figure 6.6. Distribution of (*A*) salinity, and (*B*) DOC concentrations in the northern Gulf of Mexico, showing the influence of the Mississippi River plume. Salinity data are reported by Kolasinski and Rosenheim (2016), and DOC data are unpublished (figure made using Ocean Data View [ODV]; Schlitzer 2014).

Table 6.2. Annual river discharge into the Gulf of Mexico

Discharge to GOM (%)	River	Country	Discharge ($km^3 yr^{-1}$)	DOC (TOC)[a] flux ($Tg C yr^{-1}$)	Mean annual DOC (TOC)[a] concentration ($\mu mol L^{-1}$)
46.824	Mississippi	USA	528.1[b]	1.75[c]–2.10[b]	331[b, d]
21.076	Atchafalaya	USA	237.7[b]	0.95[c]–1.19[b]	417[b, d]
5.941	Usumacinta	Mexico	67.0	ND	ND
4.070	Mobile	USA	45.9[b]	0.271[b]	492[d]
3.901	Papaloapán	Mexico	44.0	ND	ND
2.039	Grijalva	Mexico	23.0	ND	ND
1.951	Apalachicola	USA	22.0	ND	ND
1.951	Coatzacoalcos	Mexico	22.0	ND	ND
1.685	Panuco	Mexico	19.0	ND	ND
1.215	Sabine-Neches	USA	13.7[e]	0.087[e]	530[e]
1.099	Trinity	USA	12.4[e]	0.067[e]	420–480[e]
0.887	Pearl	USA	10.0 (7.8)	0.0452[a, f]	483[a, f]
0.754	Pascagoula	USA	8.50 (8.42)	0.0665[a, f]	658[a, f]
0.612	Tecolutla	Mexico	6.90	ND	ND
0.559	Choctawhatchee	USA	6.30 (6.28)	0.0565[a, f]	749[a, f]
0.532	Brazos	USA	6.00 (5.04)	0.0164[f]	271[e]
0.532	Suwannee	USA	6.00 (9.5)	0.406[f]	3555[f]
0.523	Escambia	USA	5.90 (6.19)	0.0458[a, f]	616[a, f]
0.523	Tonala	Mexico	5.90	ND	ND
0.319	Pontchartrain	USA	3.60[c]	0.023[e]	425–485[e]
0.257	Lavaca	USA	2.90[c]	0.03[e]	833[e]
0.248	Antigua	Mexico	2.80	ND	ND
0.231	Tuxpan	Mexico	2.60	ND	ND
0.213	Colorado	USA	2.40[e]	0.02[e]	650[e]
0.204	Nautla	Mexico	2.30	ND	ND
0.204	Tamesi	Mexico	2.30	ND	ND
0.186	Cazones	Mexico	2.10	ND	ND
0.186	Soto la Marina	Mexico	2.10	ND	ND
0.177	Candelaria	Mexico	2.00	ND	ND
0.168	Jamapa	Mexico	1.90	ND	ND

Continued

Table 6.2. Continued

Discharge to GOM (%)	River	Country	Discharge (km³ yr⁻¹)	DOC (TOC)[a] flux (Tg C yr⁻¹)	Mean annual DOC (TOC)[a] concentration (μmol L⁻¹)
0.160	Blanco	Mexico	1.80	ND	ND
0.133	Peace	USA	1.50	ND	ND
0.124	Ochlockonee	USA	1.40	ND	ND
0.098	Caloosahatchee	USA	1.10	ND	ND
0.098	Rio Grande	USA/Mexico	1.10[b]	0.00516[b]	614[b, d]
0.071	San Fernando	Mexico	0.80	ND	ND
0.066	Nueces	USA	0.75[e]	0.005–0.006[e]	560–630[e]
0.060	Champoton	Mexico	0.68	ND	ND
0.044	Actopán	Mexico	0.50	ND	ND
0.038	Misantia	Mexico	0.43	ND	ND
0.025	San Antonio–Guadalupe	USA	0.28[e]	0.001–0.002[e]	330–480[e]
0.018	Caominada	USA	0.20[e]	0.04–0.28[e]	558[e]

Note: Annual average river discharge, DOC/TOC fluxes, and concentrations into the GOM from Milliman and Farnsworth (2011) unless otherwise noted.

[a] Total organic carbon (TOC).

[b] Spencer et al. (2013).

[c] Shen et al. (2012).

[d] Concentration calculated from flux and discharge, and ND indicates no data to our knowledge. Meybeck and Ragu (1997) discharge in parentheses, used for calculating flux from concentration.

[e] Guo et al. (1999).

[f] Meybeck and Ragu (1997).

mond 2012; Raymond and Spencer 2015). Fires, either caused by the direct combustion of fossil fuels or indirectly from wildfires, result in the release of polycyclic aromatic hydrocarbon (PAH), or black carbon, which is delivered to coastal waters where concentrations are elevated from both fluvial and atmospheric inputs from land (Dittmar et al. 2012; Yang and Guo 2014). OM such as black carbon is largely resistant to degradation (Kim et al. 2006) and represents ~10% of the flux of DOM from rivers to the marine environment (Mitra et al. 2002; Jaffé et al. 2013; Raymond and Spencer 2015).

Precipitation

The abundance of OM in rainwater is highly variable; Willey et al. (2000) reported continental rain DOC concentrations of 161 μM, roughly 7-fold higher than marine rain DOC. Scant data exist for rainwater DOC in the GOM. Recent work by Mitra et al. (2017) surveyed rainwater DOC from the northeastern GOM on the Louisiana-Texas (LaTex) shelf to the Florida Keys. They found comparatively higher DOC concentrations ranging from 300 to 1600 μM, the latter approaching measurements of urban rain DOC concentrations summarized in Willey et al. (2000). Mitra et al. (2017) examined the composition of rain DOC using fluorescence and $\delta^{13}C$ values, finding signals indicative of PAHs from fossil-fuel emissions and/or continentally derived biogenic emissions. They speculate that oil and gas operations in the Gulf region may be strongly influencing the flux of airborne DOM to the coastal GOM.

Distribution of Dissolved Organic Matter

As its name suggests, marine DOM is indeed *dissolved* in seawater, so to understand its distribution in any basin, it is necessary to understand the relevant circulation pathways that transport this dissolved material. The GOM is a unique basin, connected to the western North Atlantic Ocean through the Florida Straits; the Yucatan Channel is the passageway connecting it to the Caribbean, which is in turn connected to the Atlantic by numerous passages between the Antillean islands. These passageways restrict flow below the depths of their sills (~800–2000 m), and since there is no mechanism for deepwater formation in the GOM, material such as DOM and TOC tends to be homogeneous below these depths (Fig. 6.7), where the deep waters reside for ~250 years (Rivas et al. 2005). In contrast to the deep water of the basin, the upper water column (<500 m) is dynamic due to the funneling of surface currents through the narrow Yucatan Channel and Florida Straits, as well as being influenced by river input, especially in the northern GOM.

Virtually all seawater and allochthonous marine DOM enters the GOM through the Yucatan Channel, the passage located between the Yucatan Peninsula and Cuba (see Fig. 6.5), which has a sill depth of ~2000 m. The Loop Current is responsible for transporting water northward through the Yucatan Channel into the GOM, which famously loops within the basin, turning in the east-southeast direction and occasionally shedding anticyclonic warm-core eddies, or rings, before emerging as the Florida Loop Current. The Florida Loop Current exits the GOM through the Florida Straits, which have a

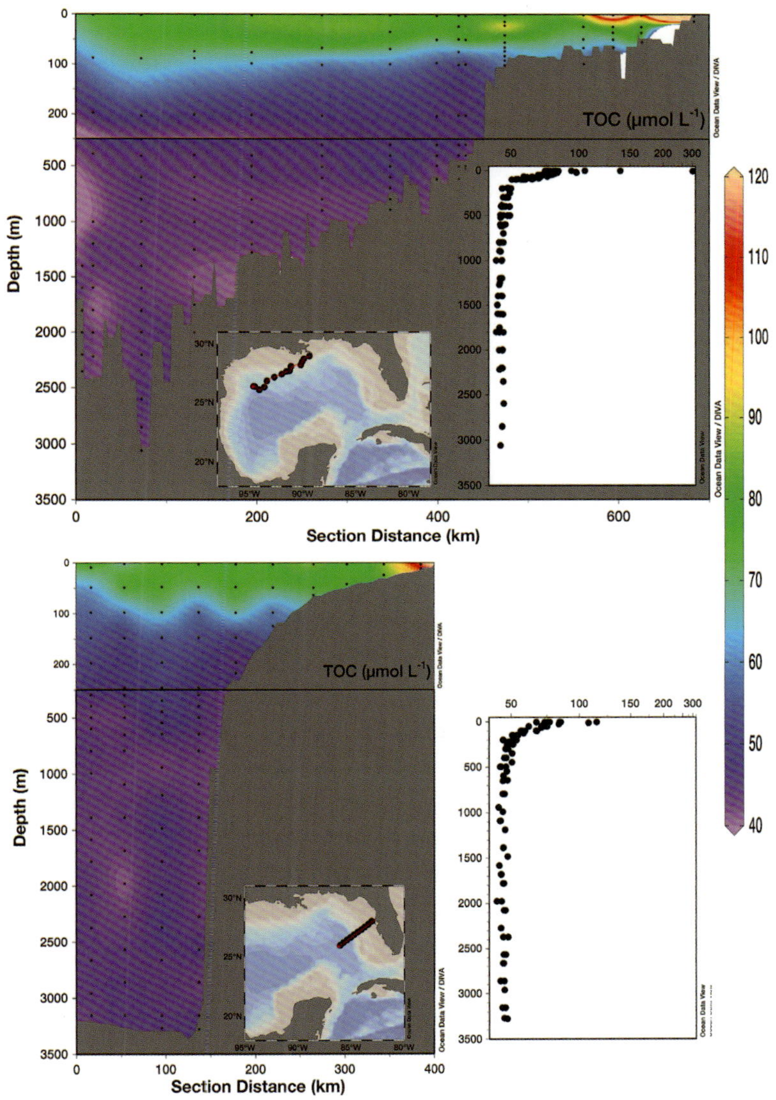

Figure 6.7. Sections and profiles of total organic carbon (TOC) in the Gulf of Mexico. The upper section and profile comprise samples collected in November 2013 (PE14-11) and July 2014 (PE15-01), respectively, while the lower section and profile comprise samples collected in July 2012 (RB12-03/GOMECC-2). The sections' upper panels are depths 0–250 m, and lower panels are 250–3500 m. Axes of profiles are nonlinear. Hydrographic data from PE14-11 and PE15-01 are reported by Young (2014) and Kolasinski and Rosenheim (2016), and TOC data are unpublished. All GOMECC-2 data are reported by Wanninkhof et al. (2014) (figure made using ODV; Schlitzer 2014).

maximum sill depth of ~800 m. While these surface features can clearly be seen in satellite measurements of sea-surface temperature, they are also apparent in satellite CDOM distributions.

DOM concentrations are highest in coastal waters due to elevated productivity and river input, decreasing from the shelf to the basin interior (Guo et al. 1995; Bianchi et al. 1997; Shen et al. 2016), which is illustrated by the distributions of sea-surface CDOM (see Fig. 6.5) and TOC (see Fig. 6.7). Concentrations are also higher in the upper water column than at depth; concentrations are ~75 μmol L^{-1} at the sea surface off the shelf and 45 μmol L^{-1} in water >500 m, as reported in previous works (Guo et al. 1995; Bianchi et al. 1997). On the continental shelf and in productive waters, it is not uncommon for concentrations to be ~100–300 μmol L^{-1} (Guo et al. 1994; Guo et al. 2009; Benner and Opsahl 2001; Hung et al. 2003) and >600 μmol L^{-1} in some of the GOM's estuaries (Guo et al. 1999; Bianchi et al. 2009; Bianchi et al. 2011; Bianchi, Grace, et al. 2014; Fichot and Benner 2011). Elevated concentrations are apparent in the shelf waters (see Fig. 6.6), with highest TOC concentrations on the sections of ~115 μmol L^{-1} near Tampa, Florida, and ~130–300 μmol L^{-1} near the Mississippi Delta.

Even in the basin interior, where productivity is minimal, DOM concentrations are highest at the surface; concentrations decrease with depth as light and biological activity become limited below the euphotic zone. Based on the GOM's volume of 2.43 × 10^6 km^3 and the deep-basin DOC concentration of 45 μmol L^{-1}, the GOM has a DOC inventory of >1.3 Pg C, or ~0.2% of the oceanic inventory (Hansell et al. 2009). Considering that the volume of the GOM is ~0.2% of the global ocean's volume (~1.3 × 10^9 km^3) and that DOC concentrations in the deep Gulf are similar to those found in the North Atlantic (Hansell et al. 2009), the GOM's DOC inventory is not surprising.

Determining the Gulf's DOM distribution and concentrations (i.e., ~45 μmol L^{-1} at depth) allows the identification of anomalies, especially in the regions of offshore oil recovery. In the deep GOM, anomalous DOC concentrations were observed following the Deepwater Horizon (DWH) oil spill, including surface samples as high as ~600 μmol L^{-1} (Zhou and Guo 2012; Zhou et al. 2013; Bianchi, Osburn, et al. 2014). Over the course of the 87-day spill (April 20–July 15, 2010), ~795 × 10^6 L of oil were released into the GOM from the Macondo wellhead at a depth of ~1600 m (Joye 2015), during which deep (1100–1400 m) DOC concentrations as high as ~500 μmol L^{-1} were observed (Zhou et al. 2013). A few months after the well was capped, deep concentrations decreased to near-background concentrations, reaching ~30–60 μmol L^{-1}; Zhou and Guo (2012) attributed the very low values to scavenging, removal via oil droplets, and/or sorption onto GF/F filters during sample collection. Higher values were attributed to oil, made apparent by optical properties, such as $SUVA_{254}$.

Optical Properties

Surface-layer CDOM (and DOM) is highest on the shelf and near the influence of rivers (Fig. 6.8), with concentrations increasing toward the mouth of the Mississippi, similar to those of DOC (see Fig. 6.6). Generally, CDOM correlated with terrigenous DOC concentrations (Mannino et al. 2008; Fichot and Benner 2011), although their relationship varies regionally and seasonally, in some cases varying by more than a factor of 25 (Del Vecchio and Blough 2004; Nelson and Siegel 2013; Stedmon and Nelson 2015). However, extensive work has been done to determine relationships between CDOM and DOC, especially in the northern GOM near the Mississippi plume (Fichot and Benner 2011, 2012, 2014; Shen et al. 2012). Specifically, Fichot and Benner (2012) identified a relationship between $S_{275\text{-}295}$ and lignin, which originates from vascular land plants; this relationship has been employed to trace the delivery and removal of terrigenous DOC in the northern GOM (Fichot and Benner 2014; Fichot et al. 2014). A similar relationship between $S_{275\text{-}295}$ and lignin was identified in the Arctic Ocean (Fichot et al. 2013), also heavily influenced by rivers; however, this relationship is not applicable for waters far from riverine sources, as $S_{275\text{-}295}$ diminishes in the sea surface because it is susceptible to photo-oxidation (Helms et al. 2008).

From the relationships between terrigenous DOC and $S_{275\text{-}295}$, Fichot et al. (2014) were able to utilize satellite reflectance data to develop an algorithm for calculating $S_{275\text{-}295}$ in the northern GOM, since $S_{275\text{-}295}$ is not reported by Aqua-MODIS. Through the reflectance-derived $S_{275\text{-}295}$, they were able to observe pulses of terrigenous DOC export from the northern GOM shelf being delivered to the basin interior. In addition to the relationship of $S_{275\text{-}295}$ with terrigenous DOC, similar relationships have been identified between terrigenous DOC and various CDOM a_λ wavelengths (Osburn et al. 2014), as well as OM molecular weight (Helms et al. 2008), aromaticity (Weishaar et al. 2003), and mineral-

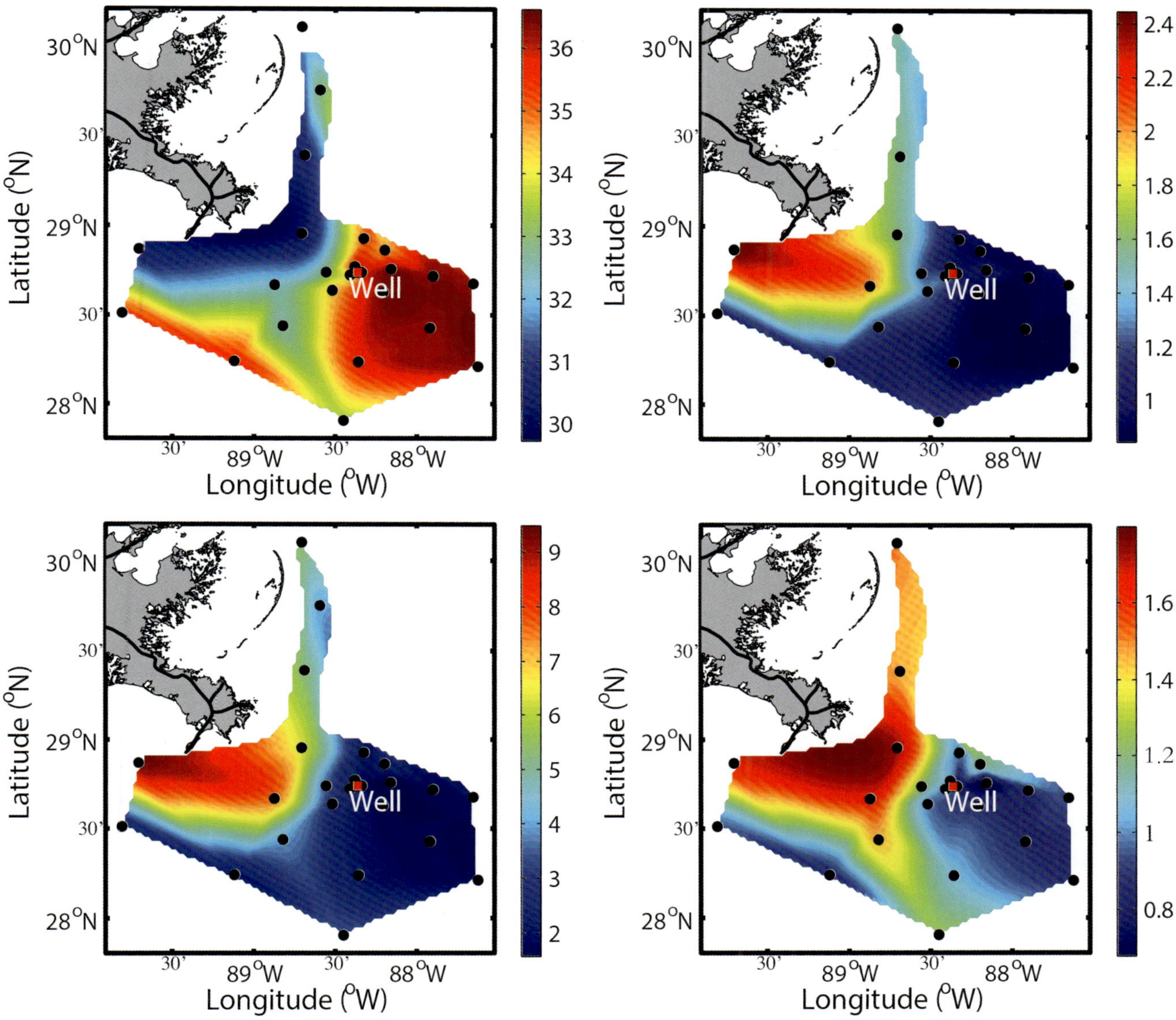

Figure 6.8. Distribution of (*A*) salinity, (*B*) DOC (in mg L^{-1}), (*C*) a254, and (*D*) SUVA254 values in the northern Gulf of Mexico, showing the influence of the Mississippi River plume and oil spill (Zhou and Guo 2012).

ization (Nelson et al. 2010). $SUVA_{254}$ has been employed as a tracer for remnant oil from the DWH spill, which was elevated 15 months after the wellhead was capped (Zhou and Guo 2012), suggesting the presence of oil-contaminated DOM or enhanced microbial activity and mineralization near the location of the wellhead. In addition to $SUVA_{254}$, CDOM a_{254} was also elevated, while DOC did not stand out from other neighboring stations (see Fig. 6.8). $SUVA_{254}$ correlates well with DOM aromaticity (Weishaar et al. 2003), suggesting the presence of oil-derived DOM near the wellhead, which would otherwise be undetectable since DOC concentrations are not anomalous. Samples collected 2 years after the oil spill at depths >1500 m ranged from background levels to ~300 µmol L^{-1} at the stations closest to the Macondo wellhead (Bianchi, Osburn, et al. 2014). Interestingly, deep samples collected off the west Florida shelf in July 2012 did not exhibit TOC concentrations higher than background values; however, these samples were collected farther from the wellhead than those reported by Bianchi, Osburn, et al. (2014).

Fate of Dissolved Organic Matter

The majority of newly produced DOM in surface waters is rapidly removed, with only a small portion resisting degradation and thus accumulating in the upper water column (Carlson and Hansell 2015; Shen et al. 2016). Recent production of DOM is enriched in biochemical compounds such as dissolved carbohydrates (CHO) present in surface waters and proportionally comparable to discrete regions of high production as indicated by chlorophyll-*a* maxima. While the upper water column, including surface water, and the chlorophyll-*a* maximum layer contains more polysaccharides, including diluted acid-hydrolyzable polysaccharides (HCl-PCHO) and diluted acid-resistant polysaccharides (HR-PCHO), bottom waters contain more monosaccharides but less polysaccharides (Fig. 6.9), mostly due to the degradation and metabolism of DOM in the deep water column (see Pakulski and Benner 1994; Lin and Guo 2015).

This accumulated material is evident in the profiles of TOC (see Fig. 6.7), where concentrations in the upper 500 m are in excess of the ~43 µmol L^{-1} deepwater concentrations. DOM accumulation results from the biological production of OM being greater than its removal (i.e., NCP), which is apparent throughout the global ocean (Hansell and Carlson 1998; Hansell et al. 2009). A consequence of this excess DOM is that heterotrophic bacteria selectively metabolize the most readily available energy sources (i.e., biolabile DOM) (Floodgate 1995; Carlson and Hansell 2015), leaving behind more recalcitrant forms that accumulate seasonally in the mixed layer. Seasonal removal of this accumulated material is clearly illustrated in the Bermuda Atlantic Time-series Study (BATS) measurements of TOC (Fig. 6.10), with the accumulated TOC signal decreasing during convective winter mixing events as it is exported to the waters below (Carlson et al. 1994; Hansell and Carlson 2001). This exported material stimulates the growth of upper mesopelagic heterotrophic prokaryotes that consume much of the exported DOM at 140 and 250 m (Carlson et al. 2004; Carlson et al. 2009; Morris et al. 2005; Letscher et al. 2013); the associated DOM mineralization can account for as much as half the oxygen utilization at these depths (Doval and Hansell 2000).

While the export of DOM is important, OM export is dominated by sinking POM, which is referred to as the biological pump, accounting for ~80% of the exported carbon to depths >100 m (Ducklow et al. 2001; Hopkinson and Vallino 2005; Carlson and Hansell 2015). Another proposed mechanism for storing carbon in seawater is the microbial carbon pump (MCP), which sequesters recalcitrant DOM based on its reactivity, independent of depth (Jiao et al. 2010; Jiao and Azam 2011). It is estimated that the MCP is responsible for sequestering 8–23 Tg C yr^{-1} globally (Benner and Herndl 2011), which is at least 1 order of magnitude less than the biological pump, having an estimated export of 700 Tg C yr^{-1} below 2000 m (Henson et al. 2012).

Marine DOM consists of innumerable compounds (Kujawinski 2011), each resisting degradation from times ranging from minutes to millennia (Nagata 2008;

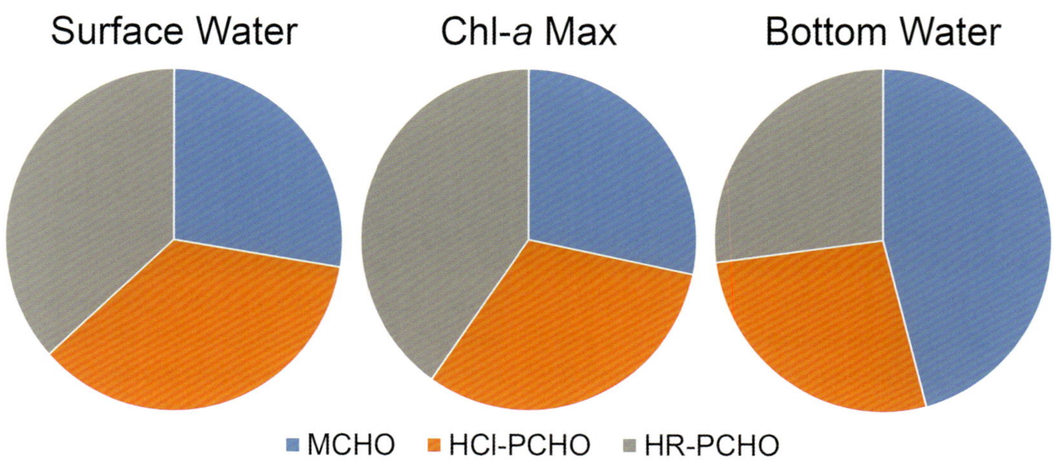

Figure 6.9. Different carbohydrate components within the total dissolved CHO pool, including monosaccharides (MCHO), diluted acid hydrolyzable polysaccharides (HCl-PCHO), and diluted acid-resistant polysaccharides (HR-PCHO) in surface water, Chl-*a* maximum layer and bottom water in the northern Gulf of Mexico (replotted from data in Lin and Guo 2015).

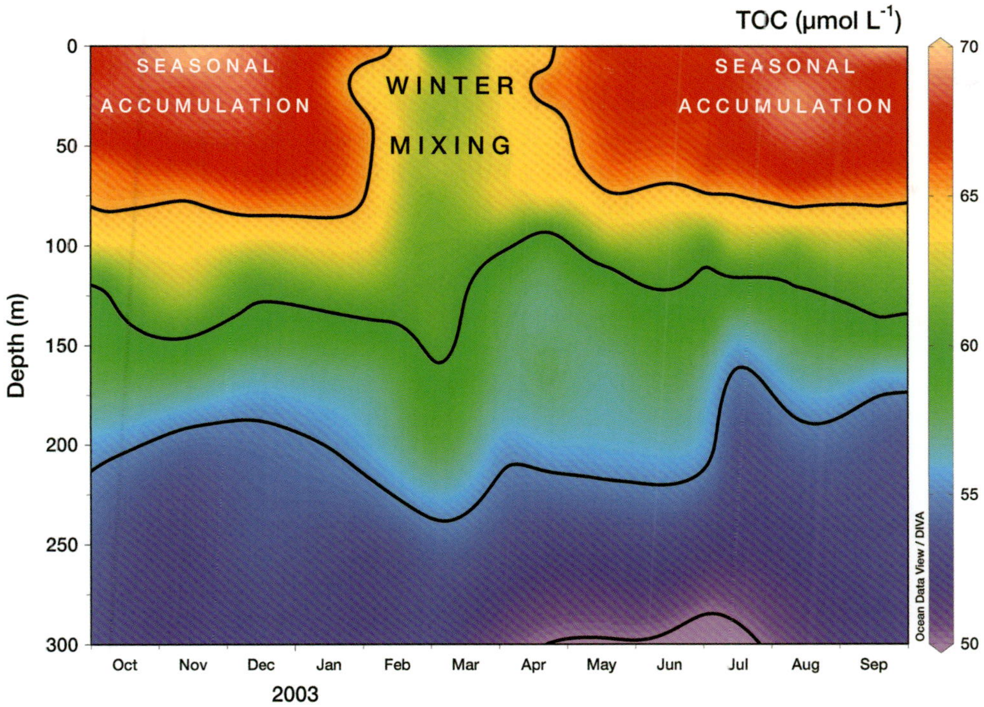

Figure 6.10. TOC concentrations from the BATS, showing the seasonal accumulation of TOC and the winter mixing that results in its seasonal export and removal. Contours in increments of 5 μmol L^{-1}, from 50 to 65 μmol L^{-1} (figure made using ODV; Schlitzer 2014).

Hansell 2013; Carlson and Hansell 2015). The most rapidly utilized pool of this material, referred to as labile DOC (LDOC), has a lifetime of minutes to days (Amon and Benner 1996; Cherrier and Bauer 2004; Carlson and Hansell 2015). Since the pool of LDOC is so reactive, its presence and removal are best determined by measurement of the bacterial carbon demand (BCD) rather than direct measurements of LDOC concentration, which is maintained at nanomolar levels (Skoog et al. 1999; Carlson and Hansell 2015). Based on BCD measurements, it is estimated that ~50% of net primary production is transformed into LDOC (Nagata 2008). Globally, the LDOM inventory is <200 Tg C and has an annual production rate 100 times higher than its inventory (i.e., ~20 Pg C yr^{-1}) (Hansell 2013). Since the majority of LDOC is rapidly turned over near the surface waters, the GOM's LDOC reservoir can be approximated based on its surface area relative to the ocean, equivalent to <0.8 Tg C. Since the GOM is a small, enclosed basin surrounded by productive shelves, these LDOC inventory estimates are likely conservative. While LDOC is an important source of carbon for the microbial community, since it is mineralized so rapidly, it has a negligible role in the biological pump; however, it is important for maintaining inorganic nutrient loads in the euphotic zone. LDOC is most abundant in productive coastal waters and follows the distribution of chlorophyll-*a* (see Fig. 6.5). This co-occurrence is due to LDOC turnover by microbes, which contributes to phytoplankton production (via nutrient cycling) and the production of LDOC from phytoplankton (via leaky cells and sloppy feeding by zooplankton).

Fractions of DOM can have intermediate reactivity times, ranging from weeks to centuries, depending on ocean-mixing timescales, such as annual winter mixing or cross-shelf mixing in summer in the northern GOM (Shen et al. 2016); these are commonly referred to as semilabile and semirefractory DOC (SLDOC and SRDOC, respectively). SLDOC is the portion of DOC that accumulates in the surface waters on seasonal timescales and is removed within weeks to years once exported to the mesopelagic (i.e., 100–500 m) by convective winter mixing (Hansell and Carlson 2001; Hansell et al. 2012). TOC measurements at BATS clearly show seasonal SLDOC accumulation prior to winter mixing and its subsequent export and mineralization (see Fig. 6.10). Following this export, the inorganic by-products of SLDOC mineralization (i.e., CO_2 and nutrients) are typically returned to the surface if this seasonal ventilation penetrates the waters below ~250 m, where they can be renewed by the enriched deep waters (Carlson et al. 1994; Doval and Hansell 2000; Hansell and Carlson 2001). It is uncommon for seasonal mixing to result in deep ventilation, however, resulting in SLDOC contributing only to local export and not being a mechanism for long-term sequestration of carbon from surface waters and air-sea exchanges (Hansell et al. 2012; Carlson and Hansell 2015). During the spring, upper-

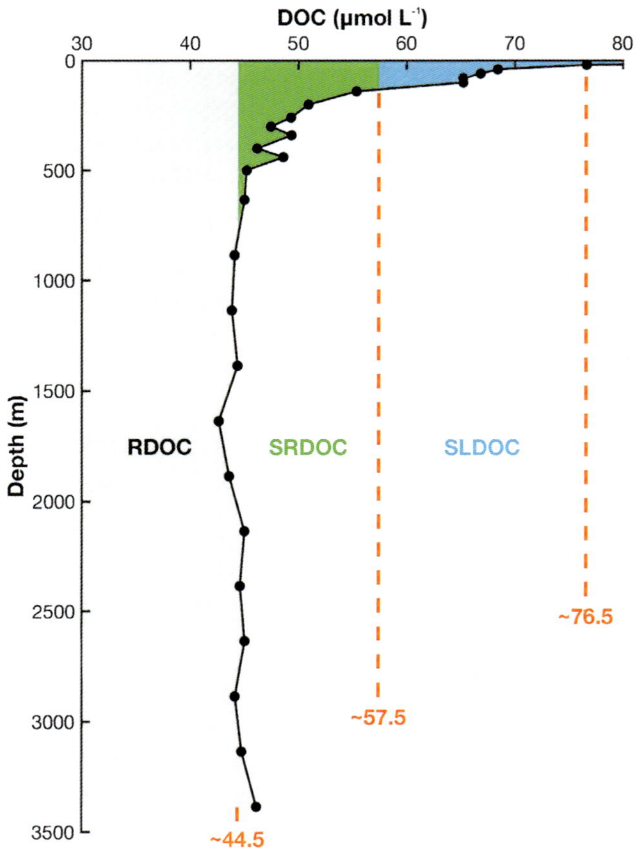

Figure 6.11. Profile of mean DOC measurements in the Gulf of Mexico, binned in increments of 20 m in the upper 500 m and increments of 250 m below, showing the fractions of recalcitrant DOC as described by Hansell (2013). Data are unpublished.

water-column TOC concentrations are restored, increasing by ~3–10 µmol L^{-1} as SLDOC accumulates in the restratified euphotic zone (see Fig. 6.10). SLDOC globally has an inventory of 6 Pg C and is renewed about every 2 years with a production rate of 3.4 Pg C yr^{-1} (Hansell et al. 2012; Hansell 2013). Based on the DOC concentration gradient of ~19 µmol L^{-1} in the upper ~140 m (Fig. 6.11), the GOM's SLDOC inventory is ~20 Tg C, with an estimated production rate of ~10 Tg C yr^{-1}.

SRDOC's lifetime is longer than SLDOC's, ranging from decades to centuries (Hansell 2013; Carlson and Hansell 2015). In the GOM, SRDOC is evident in the vertical DOC gradient in the deeper mesopelagic between ~100 and 500 m (see Fig. 6.11), within the permanent pycnocline. At higher latitudes, such as in the North Atlantic and Mediterranean, these deep gradients can extend to >1000 m because of deepwater formation (Carlson et al. 2010; Santinelli et al. 2010); however, this phenomenon does not occur in the GOM. The global SRDOC inventory is ~14 Pg C, much larger than the SLDOC inventory, but the SRDOM production rate is only ~0.34 Pg C yr^{-1} (Hansell et al. 2012). Like SLDOM, SRDOM contributes little to the vertical export of carbon, unless this material escapes below the permanent pycnocline. SRDOC accounts for ~5–20% of the apparent oxygen utilization at depth (Doval and Hansell 2000; Aristegui et al. 2002; Carlson et al. 2010); the sinking and mineralization of POM account for the remainder. Based on the DOC concentration gradient of ~13 µmol L^{-1} between ~140 and 500 m (see Fig. 6.11), the GOM's SRDOC inventory is ~27 Tg C, with an estimated production rate of ~0.66 Tg C yr^{-1}.

DOM that persists for longer than centuries is referred to as refractory DOC (RDOC), with the oldest fraction referred to as ultrarefractory DOC (URDOC). RDOC can easily be identified at depths >500 m, where the vertical gradient in DOC is minimal, having a mean concentration of ~44.5 µmol L^{-1} in the GOM (see Fig. 6.11). As in the global ocean, RDOC likely represents 90% of the total inventory (Hansell et al. 2009; Hansell et al. 2012; Hansell 2013), containing ~1.3 Pg C as RDOC in the GOM and making its bulk DOC reservoir equivalent to ~1.4 Pg C. RDOC has a mean age of 4000–6000 years (Williams and Druffel 1987; Druffel et al. 1989; Bauer et al. 1992; Bauer 2002), ranging from ~1000 to ~40,000 years and dominated by the younger fractions (Hansell 2013; Carlson and Hansell 2015). Since the GOM's RDOC pool persists for much longer than the basin's 250-year residence time (Rivas et al. 2005), it is unlikely that there would be any deep basin gradients; however, this remains unknown. For these older pools of DOC, photo-oxidation, hydrothermal circulation, and sorption onto sinking particles and/or delivery to sediments are thought to be their primary sinks in the marine environment (Dittmar and Paeng 2009; Stubbins et al. 2012; Hawkes et al. 2015).

Removal Mechanisms

A mechanism that may stimulate the removal of otherwise recalcitrant forms of DOM is the priming effect, which has long been recognized in soil science and is also possibly important in the marine environment (Guenet et al. 2010; Bianchi 2011). When water containing RDOC is mixed with water containing LDOC compounds, the LDOC may act as a substrate for the mineralization of RDOC by heterotrophic bacteria; this mechanism is a priming effect referred to as cometabolism (Horvath 1972; Guenet et al.

2010). Net mutualism is another priming-effect mechanism through which RDOC- and LDOC-metabolizing communities release enzymes for degrading both classes of compounds as well as nutrients to sustain the degradation of LDOC and RDOC (Bianchi 2011). The third possible mechanism for stimulating the priming effect occurs when the availability of LDOC promotes the heterotrophic microbial community to synthesize RDOM-degrading enzymes (Guenet et al. 2010). Coastal regions that receive a large influx of terrigenous DOM, such as in the northern GOM near the Mississippi Delta, are areas where priming may be prevalent, since terrigenous DOM is not apparent in the open ocean (Bianchi 2011). However, the priming effect has been observed in the open ocean based on incubation studies in which glucose (i.e., LDOC) and nutrients reduced DOC concentrations to below initial conditions (Carlson et al. 2002). Furthermore, priming has been proposed as a mechanism for the nonconservative behavior of RDOC in the deep Pacific (Hansell and Carlson 2013).

In coastal and surface waters, priming may be the major removal mechanism for RDOM; it may also be an important removal mechanism for terrigenous DOM in river-dominated coastal waters, such as those in the northern GOM. Another removal mechanism for terrigenous DOM is the consumption and/or transformation by heterotrophic bacteria (Søndergaard and Middelboe 1995; Tremblay and Benner 2009; Kaiser et al. 2017), and it can also be removed abiotically by photo-oxidation (Duan, Bianchi, Shiller, et al. 2007; Spencer et al. 2009), flocculation (Sholkovitz 1976; Benner and Opsahl 2001), and self assembly (Verdugo 2012; Ding et al. 2008). Recalcitrant DOM may also be removed by photo-oxidation in coastal waters as well as in the surface of the open ocean (Mopper et al. 1991; Benner and Kaiser 2011). By considering radiocarbon measurements of DOC, Cherrier et al. (1999) found heterotrophic bacteria in surface waters to be older than expected, suggesting that RDOM may be transformed into more labile forms. RDOC residing in deep waters is biologically refractory, but when it is upwelled to the surface, it can be broken down as a result of exposure to UV sunlight, resulting in either the phototransformation into labile forms and/or photo-oxidation into inorganic carbon (Mopper et al. 1991; Moran and Zepp 1997; Stubbins et al. 2012).

While photo-oxidation may represent a major removal pathway for RDOC, its role varies depending on the chemical composition of said DOM. For example, Benner and Biddanda (1998) demonstrated that bacterial production in euphotic-zone water decreased by ~75% when exposed to UV light, whereas bacterial production increased by ~40% when deep waters (150–1000 m) were exposed. Photochemical degradation may play a prominent role in RDOC removal within the euphotic zone, especially for the optically active DOM components; this removal is also apparent in the abyssal ocean where decreasing gradients (i.e., removal) have been observed (Carlson et al. 2010; Hansell et al. 2009; Hansell et al. 2012).

The circulation of water through hydrothermal vent systems and porous basalts may be a notable sink for RDOC or even URDOC, considering RDOC's slow removal rate of ~3 Tg C yr^{-1} (Hansell 2013). Multiple studies have shown that high-temperature vents are responsible for removing ~0.01–0.1 Tg C yr^{-1} (Lang et al. 2006; Hawkes et al. 2015), suggesting that hydrothermal systems represent a minor removal pathway for RDOC. These systems could, however, represent the major sink for URDOC, considering that its removal rate is ~0.001 Tg C yr^{-1} (Hansell 2013).

Particulate matter may also play an important role in removing DOM via sorption onto sinking POM. This process has become evident by radiocarbon measurements finding POM that is older than expected, suggesting that RDOC is incorporated into particulates, perhaps representing ~14% of POM in the deep Pacific (Druffel and Williams 1990). One mechanism that could explain this phenomenon is the formation of self-assembled microgels (SAGs), which are initially neutrally buoyant, but under certain conditions can be compressed into dense, sinking particles (Verdugo 2012).

Microbial degradation of DOM by organisms such as heterotrophic prokaryotes, the primary DOM consumers in the marine environment, results in either the mineralization of DOM to CO_2 and nutrients (Ducklow et al. 1986; Goldman and Dennet 2000) or the transformation of DOM to POM via the microbial loop (Azam et al. 1983; Pomeroy et al. 2007). In fact, bacteria also can be sources of DOM and POM in the ocean (Kaiser and Benner 2008). Net bacterial production ranges regionally, from 20% to 50% of primary production in productive coastal environments, such as near the Mississippi Delta, to <10% in oligotrophic regions such as in the central GOM (Cole et al. 1988; Del Giorgio and Cole 1998; Ducklow 1999). DOM is considered to be the major source of carbon to microbial communities in the marine environment, with prokaryotes rather than eukaryotic organisms dominating utilization (Carlson and Hansell 2015).

Colloidal Organic Matter

Importance of Colloidal Organic Matter in Organic Matter Cycling

DOM is a major component of the global carbon cycle and plays an important role in regulating the biogeochemical cycling of nutrients and trace elements in aquatic systems (Aiken et al. 2011; Bauer et al. 2013; Hedges 2002). Bulk DOM has been shown to be heterogeneous in size or molecular weight, composition, and chemical reactivity (Benner and Amon 2015; Guo et al. 1996; Hansell 2013). Indeed, bulk DOM is composed of mostly COM or HMW-DOM, especially in estuarine and coastal environments and the upper water column (Guo et al. 1995; Cai and Guo 2009; Guo and Santschi 2007). Colloids have been recognized as an intermediary between dissolved and particulate phases and play a critical role in regulating the fate, transport, and bioavailability of trace metals and organic pollutants in aquatic systems (Aiken et al. 2011; Guo et al. 2002; Guo et al. 2003; Lead and Wilkinson 2006). Thus, COM is important not only in the dynamic cycling of organic matter but also in the biogeochemical cycling of bioactive elements in the ocean.

The abundance and chemical composition of HMW or COM have long been reported for coastal and open-water stations in the Gulf (Maurer 1976; Benner et al. 1992; Guo et al. 1994; Guo et al. 1995; Santschi et al. 1995; Bianchi et al. 1997; Santschi et al. 1998; Passow et al. 2012) in addition to river and estuarine waters in the northern GOM (Guo and Santschi 1997; Cai and Guo 2009; Duan, Bianchi, Shiller, et al. 2007; Guo et al. 2009; Wang et al. 2010). In general, concentrations of both the >1 kDa (COC_1) and the >10 kDa COC (COC_{10}) decreased from river to coastal waters (e.g., Guo and Santschi 1997; Cai and Guo 2009; Guo et al. 2009) and from low-salinity to high-salinity waters, even within offshore waters in the GOM (Figure 6.12). For example, concentrations of the >1 kDa COC in river waters were as high as 181–187 μM in the Mississippi River (Guo et al. 2009; Cai et al. 2015), 329 μM in the Trinity River, Texas (Guo and Santschi 1997), 100–400 μM in the Pearl River, Mississippi (Duan, Bianchi, Shiller, et al. 2007), and 135–567 μM in the Jourdan River, Mississippi (Wang et al. 2010). Concentrations of COC then decreased from river water to estuarine water to coastal waters (Guo and Santschi 1997; Guo et al. 2009). Vertically, concentrations of COC also decreased from surface water to deep water in all COC size fractions, such as the >1 kDa, >3 kDa, and >10 kDa COC (Figure 6.13).

As shown in Figure 6.14, the percentage of DOC in each specific size fraction generally increased with decreasing molecular size, with the <1 kDa LMW-DOC predominant in the bulk DOC pool in both surface and deep water. In GOM surface water, for example, the percentage of COC in the bulk DOC pool was 4% in the 10 kDa–0.2 μm colloidal fraction, 12% in the 3–10 kDa COC fraction, and 22% in the 1–3 kDa COC fraction, leaving the majority of the bulk DOC (up to 62%) in the <1 kDa LMW fraction. In the deep water of the GOM, the larger-

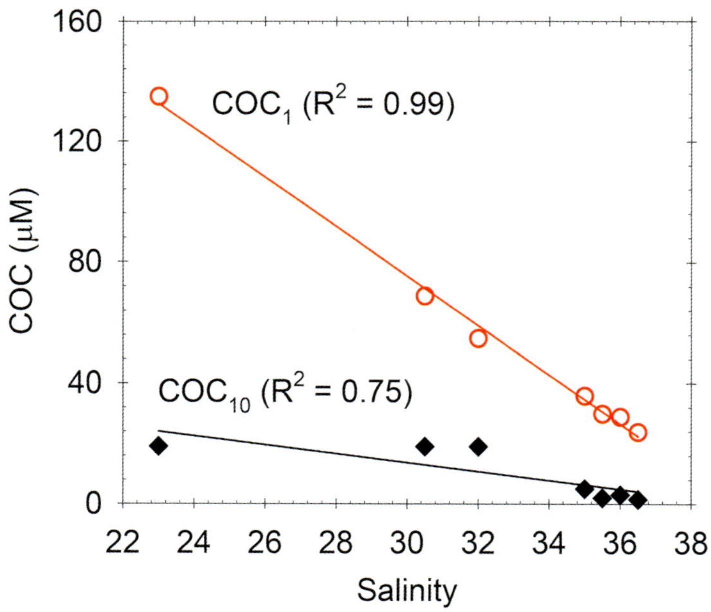

Figure 6.12. Variations in COC concentration with salinity, showing a linear decrease with increasing salinity or bulk DOC concentration in surface waters from the northern Gulf of Mexico (from Guo et al. 1995).

sized COM was reduced to 3% in the 10 kDa–0.2 μm and 7% in the 3–10 kDa colloidal fraction, while the 1–3 kDa medium-sized COM remained somewhat constant as that in surface water. In contrast, the <1 kDa LMW-DOC increased from 62% of the bulk DOC in surface waters to 67% in deep water in the GOM, as also shown for the increase in monosaccharides abundance from surface to deep water (see Fig. 6.9). In other words, deep water contains lower colloidal abundance than surface waters in the GOM, similar to the distribution of carbohydrates in the water column (Pakulski and Benner 1994; Hung et al. 2003; Lin and Guo 2015) and the freshness or higher biological/chemical reactivity of HMW-DOM in marine environments (Bianchi et al. 1995; Santschi et al. 1995). It is worthwhile to point out that colloidal abundance reported in the literature could be highly variable even for the same water sample, depending on the methods used (Buesseler et al. 1996; Guo and Santschi 2007), mainly because colloidal abundance can be overestimated by using concentration differences between the initial solution and the LMW size fraction, since the LMW-DOM can be concentrated during ultrafiltration (Guo and Santschi 1997, 2007). Truly colloidal abundance can be quantified using ultrafiltration, as shown in many previous studies (Guo and Santschi 1997; Guo et al. 2000; Cai and Guo 2009), but requires the sampling and measurements of ultrafiltration time-series samples, which is time consuming and rarely adopted in most previous studies.

Size Spectra of Colloidal Organic Matter

Size distributions of DOM in seawater among different colloidal size fractions could be obtained via ultrafiltration using membranes with different molecular weight (MW) cutoffs (Guo et al. 1995; Benner and Amon 2015). In addition to ultrafiltration, FlFFF techniques allow a continuous separation of colloids or nanoparticles across the entire colloidal size spectrum, ranging from 0.3 kDa to 0.45 or 0.7 μm based on flow dynamic separation (Baalou-

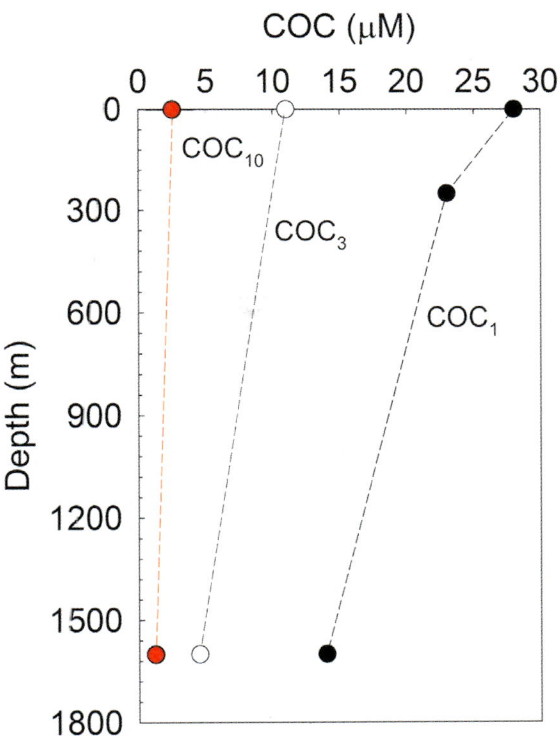

Figure 6.13. Vertical distributions of different-sized colloidal organic carbon in the Gulf of Mexico (from Guo et al. 1995). Note that concentrations of each COC size fraction (e.g., the >1 kDa COC_1; the >3 kDa COC_3; and the >10 kDa COC_{10}) decreased consistently with increasing water depth.

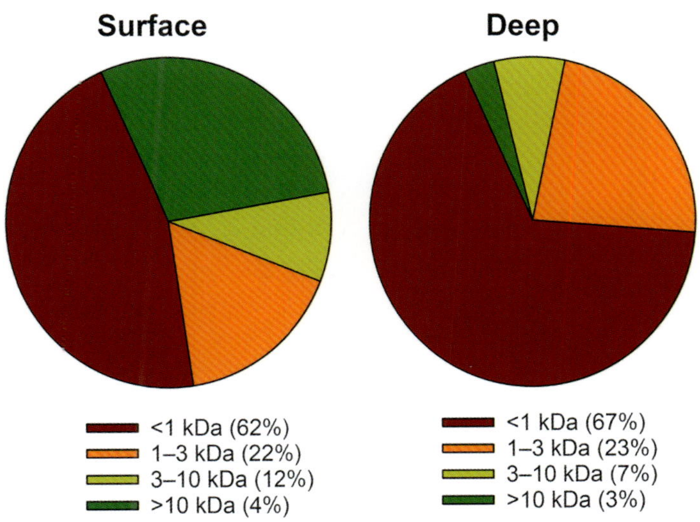

Figure 6.14. Examples showing the molecular size distribution of bulk DOC (<0.2 μm) in the <1 kDa, 1–3 kDa, 3–10 kDa, and the 10 kDa-0.2 μm size fractions in the Gulf of Mexico (replotted from Guo et al. 1995).

sha et al. 2011; Zhou and Guo 2015). Recently, colloidal size distributions of OM have been reported for seawater in the GOM, using the FlFFF techniques coupled with different online detectors, mostly for estuarine and coastal waters (e.g., Wells 2004; Zanardi-Lamardo et al. 2004; Stolpe et al. 2010; Stolpe et al. 2014; Zhou et al. 2016). Based on these applications, some results of size distributions of COM in the Gulf are summarized here.

Examples of DOM size distribution based on ultrafiltration and concentration differences are shown in Figure 6.14. Since ultrafiltration methods provide only physical separation of a specific DOM size fraction, specific characterization is done separately, depending on available COM sample sizes and analytical methods. Based on data shown in Figure 6.14, within the >1 kDa bulk COM in the Gulf, 10% of the bulk COM was measured in the 10 kDa–0.2 μm larger-sized COM, 26% were in the 3–10 kDa medium-sized COM, and 64% were in the 1–3 kDa smaller-sized COM fraction. The decrease in OC quantity with decreasing colloidal size is consistent with a nor-

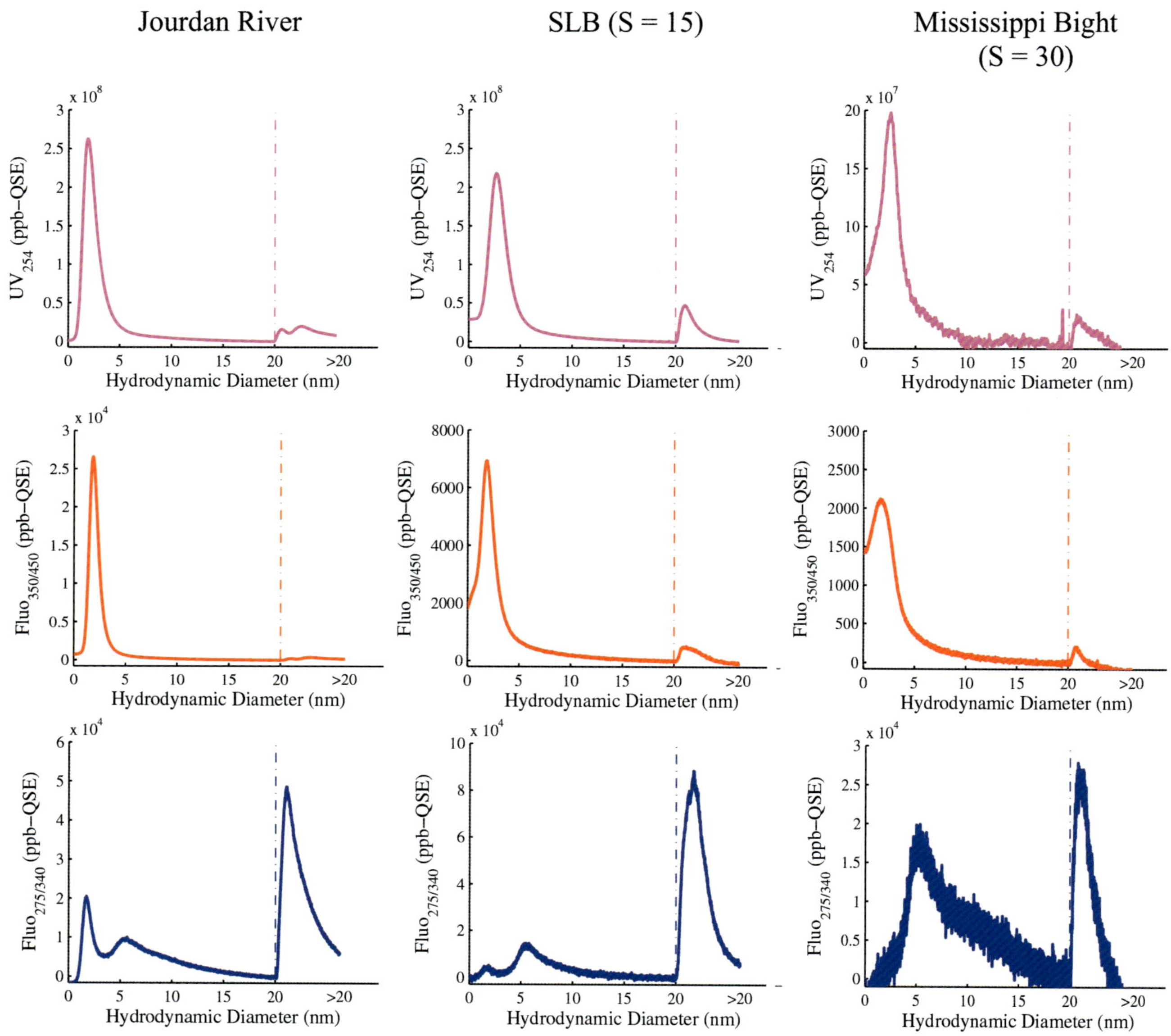

Figure 6.15. Typical FlFFF fractographs of colloidal organic matter along a salinity gradient from river water, the Pearl River, to estuarine water from the Bay St. Louis (SLB) and coastal seawater in the Mississippi Bight in the northern Gulf of Mexico. Examples shown here include the UV-absorbance (*upper panels*), humic-like (*middle panels*), and protein-like (*lower panels*) fluorescent DOM components (from Zhou, Stolpe, et al. 2016).

mal size spectrum in a steady-state ecosystem, as shown in Figure 6.4. Chemical composition or quality changes within different colloidal size fraction remain poorly understood.

Using FlFFF coupled with online detectors, such as UV-absorbance and fluorescence detectors, typical size spectra of COM in river, estuarine, and coastal seawater in the northern Gulf are shown in Figure 6.15. Examples shown here include the UV-absorbance (upper panels), humic-like (middle panels), and protein-like (lower panels) fluorescent DOM components in samples along a salinity gradient from the Jourdan River to Bay St. Louis (BSL) and the Mississippi Bight in the northern GOM. Both UV-absorbance and humic-like fluorescent components show similar size spectra with a predominant single peak in the size range between 0.5 and 4 nm and a small peak in the larger colloidal size (>20 nm). This is consistent with the fact that humic substances normally have an average molecular weight of 2–3 kDa (Aiken et al. 2011). In contrast, protein-like components had a broader size spectrum with diverse molecular sizes. In addition to smaller-sized colloids (0.5–4 nm) and medium-sized colloids (e.g., 4–8 nm), the protein-like components also had a significant portion of larger-sized colloids (>20 nm). The percentage of smaller-sized colloids in general decreased from river water to estuarine water to coastal seawater in both humic-like and protein-like components. Similarly, Stolpe et al. (2010) and Stolpe et al. (2014) also showed the same colloidal size spectra for the humic-like and protein-like organic components.

As shown in Figure 6.16, colloidal fluorescent humic-like components were partitioned mostly in the 0.5–4 nm size fraction, followed by the medium-sized (4–20 nm) and then larger-sized (>20 nm) colloids. Interestingly, the percentage of the relative abundance of colloidal fluorescent humic-like components decreased in both the smaller-sized and larger-sized colloids, but increased in the medium-sized colloids along a salinity gradient from river, estuarine, and coastal seawater in the northern GOM. This indicates that colloidal humic-like components can transfer actively between different-sized fractions during estuarine mixing, with the smaller colloids coagulating into medium-sized colloids (Zhou, Stolpe, et al. 2016). In contrast, the colloidal protein-like components had the highest abundance in the larger-sized colloids, followed by the medium-sized and smaller-sized colloids. In addition, the percentage of the relative abundance of colloidal protein-like components increased in the larger-sized but decreased in the medium-sized colloidal fraction, while the smaller-sized colloidal fraction remained somewhat similar along a salinity gradient from river, estuarine, and coastal seawater in the northern GOM. The dynamic changes in both molecular size and composition between colloidal size fractions during estuarine mixing can be revealed only by the FlFFF analysis. While the examples shown here are mostly from the results of optical properties, coupling of additional detectors with the FlFFF separation system will allow the characterization of other organic compound classes or specific organic functionalities.

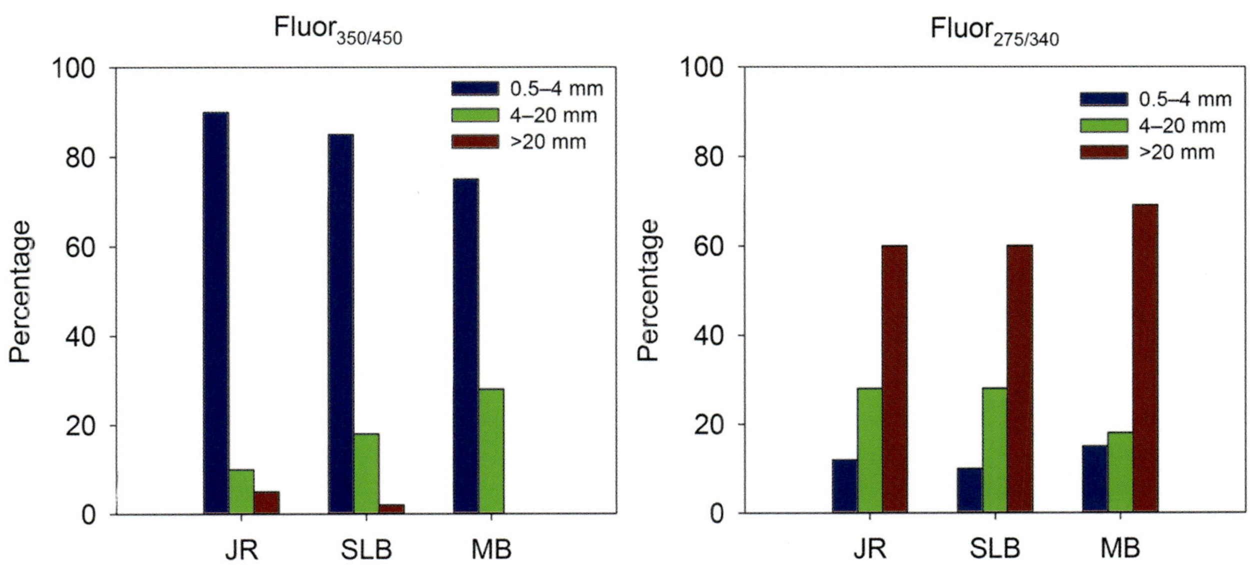

Figure 6.16. Variations in size spectra of colloidal fluorescent DOM, including humic-like and protein-like components from river, estuarine, and coastal seawater samples in the northern Gulf of Mexico (from Zhou, Stolpe, et al. 2016).

Chemical Composition of Colloidal Organic Matter

In addition to colloidal abundance and size spectra discussed previously, chemical composition of isolated COM samples has been reported, including elemental composition, isotopic signatures, and molecular composition of COM in the Gulf. Chemical and molecular characterization of DOM in the ocean has been limited because of the relatively low mass concentration of DOC in seawater and the presence of salts, which are about 35,000 times as large as the mass concentration of DOC. Therefore, isolation and purification of DOM are prerequisites before further chemical and molecular characterization (Guo and Santschi 2007), except for carbon isotope analysis (e.g., Druffel et al. 1992). Nevertheless, physical separation methods using ultrafiltration or reverse osmosis/electrodialysis only isolate, by molecular weight/size, the HMW- or colloidal DOM, which is usually in the >1 kDa fraction (Guo and Santschi 2007). Thus, these methods miss the LMW-DOM, the predominant DOM fraction in open-ocean environments (see Figure 6.14). In addition, other DOM separation methods, such as the solid-phase extraction method and XAD resin column, require chemical manipulation before DOM separation, which isolates selectively only certain portions of DOM (e.g., Chen et al. 2016). Regardless of these limitations, most previous studies are based on these separation methods and reported the chemical, isotopic, and molecular composition on a specific DOM fraction. Since DOM is highly heterogeneous in molecular size and composition (Xu and Guo 2017), chemical and molecular characterization on a specific DOM size fraction, in addition to information on the bulk DOM, should provide new insights into our better understanding of chemical/molecular heterogeneity of DOM in marine environments (e.g., Santschi et al. 1995; Guo et al. 1996; Bianchi et al. 1997).

As shown in Table 6.3, the two most common isolated COM sample sizes, including the 10 kDa–0.2 μm and 1 kDa–0.2 μm, have been characterized for elemental (C, N, and S), molecular and biomarker (lignin, CHO, protein, plant pigments), and isotopic composition ($\delta^{13}C$, $\delta^{15}N$, and $\Delta^{14}C$). The C/N ratio has been used to trace the diagenetic status of COM, showing a general increase with decreasing size of OM (e.g., Guo and Santschi 1997), while plant pigments were used to examine OM sources (e.g., Bianchi et al. 1995). In addition, lignin phenols were used to track the transport pathways of terrestrial OM along the shelf/slope/basin in the GOM (Bianchi et al. 1997; Duan, Bianchi, and Sampere 2007). Radiocarbon has been widely used to determine apparent ^{14}C ages of COM in the >10 kDa and/or the >1 kDa COM fractions (Santschi et al. 1995; Guo et al. 1996; Loh et al. 2004). It was first found that the 10 kDa–0.2 μm HMW-COM was much younger than the bulk COM pool (1 kDa–0.2 μm) in the water column of the GOM (Santschi et al. 1995). Most important, the carbohydrate component was found to contain much higher $\Delta^{14}C$ values and was much younger (~1000 years) than the >1 kDa bulk COM (Santschi et al. 1998), showing variable ^{14}C ages of COM components and indicating different sources and cycling pathways among different COM components (Santschi et al. 1998; Guo et al. 2009). At the land-ocean interface, the ^{14}C age increased with increasing depth and increasing salinity in the Mississippi River plume (Guo et al. 2009). However, the 10 kDa–0.2 μm HMW-COM was found to be older than the bulk COM pool at times in Galveston Bay as a result of sediment resuspension (e.g., Guo and Santschi 1997).

Trace-metal contents of isolated COM samples have also been reported. Most metals were in the COM concentration range between <0.1 and 50 μg/g, with a few exceptions, such as Fe, which is much higher than that of other metals (e.g., Guo et al. 2000). In addition, COM samples were found to have metal/OC ratios similar to those observed in humic substances and marine plankton samples. The metal/OC ratio was found to follow the order of Cu > Ni, Cr, Zn > Mn > Co > Pb, Cd, which is similar to the Irving-Williams order except for Mn. Thus, the binding of metals with marine colloids is dominated by their affinity for organic matter.

Biological and Chemical Reactivity of Colloidal Organic Matter

As discussed previously, the HMW-DOM or COM in seawater has been shown to contain higher carbohydrate contents and have higher apparent ^{14}C ages than those of the bulk DOM and COM (e.g., Kaiser and Benner 2009). Indeed, incubation experiments have shown higher degradation rates than those of the <1 kDa LMW-DOM counterpart (Amon and Benner 1996; Benner and Amon 2015). Results from Th-234/U-238 disequilibrium also consistently show much shorter turnover times in the HMW-COM than in the bulk COM pool (e.g., Baskaran and Santschi 1993; Santschi et al. 1995; Guo et al. 1997; Chuang et al. 2015), consistent with the higher $\Delta^{14}C$ values or younger ^{14}C ages observed for the 10 kDa–0.2 μm COM pool in the GOM. Based on the radiocarbon composition in different OM size

Table 6.3. Compilation of elemental, molecular, and isotopic composition of HMW-DOM and COM in estuarine and coastal seawater in the Gulf of Mexico

Location	COM fraction	Elemental composition	Molecular composition	Isotopic composition	Main points	References
Slope water	1 kDa–0.2 μm		^{13}C-NMR		Carbohydrates are a major component of HMW-DOM	Benner et al. 1992
Shelf/slope water	1 kDa–0.2 μm	—	Plant pigments/ biomarkers	—	Source from plankton but contents are lower than POC pool	Bianchi et al. 1995
Shelf/slope water	1 kDa–0.2 μm		Lignin, loliolides		Important for lateral transport of terrestrial OC	Bianchi et al. 1997
Shelf/slope water	>10 kDa and >1kDa COM	—	—	^{234}Th, δ^{13}C, and radiocarbon	Younger and high turnover rates in larger-sized COM	Santschi et al. 1995
Shelf/slope stations	>10 kDa and >1kDa COM	—	—	^{234}Th	Shorter turnover time for the >10 kDa COM than the >1 kDa COM	Guo et al. 1997
Galveston Bay	>1 kDa >10 kDa	C, N, and S	—	δ^{13}C and radiocarbon	C/N ratio increased with decreasing size; older COM>10 kDa suggested sources from sediment	Guo and Santschi 1997
Slope water	>1 kDa COM	C and N	Polysaccharides	δ^{13}C and Δ^{14}C	Younger ^{14}C ages in CHO components	Santschi et al. 1998
Shelf/slope waters	>1 kDa COM	Trace metals	—	—	Most metals with a concentration <0.1–50 μg/g COM, with a Me/OC ratio similar to that of humic substances and plankton	Guo et al. 2000
Galveston Bay and GOM	>1 kDa and >10 kDa COM	C and N	—	δ^{13}C and δ^{15}N	δ^{13}C increased with Δ^{14}C in deep water but decreased with Δ^{14}C in surface waters; opposite is true for δ^{15}N and Δ^{14}C relationship	Guo et al. 2003
Mississippi and Pearl Rivers	>1 kDa	—	Lignin	□	Lower contents in the Mississippi River	Duan, Bianchi, Shiller, et al. 2007
Mississippi River plume	>1 kDa	C and N	Compound class by pyrolysis GC-MS	δ^{13}C, δ^{15}N, and radiocarbon	Aromatic OM and Δ^{14}C decreased, but C/N ratio, δ^{13}C, CHO, and uronic acids increased with increasing salinity	Guo et al. 2009
Shelf/slope waters	1 kDa–0.2 μm	C and N	Proteins, THAA, polysaccharides, uronic acids, hydroxamate siderophores, hydroquinone		Binding capacity with trace elements differed among different organic functionalities	Chuang et al. 2015

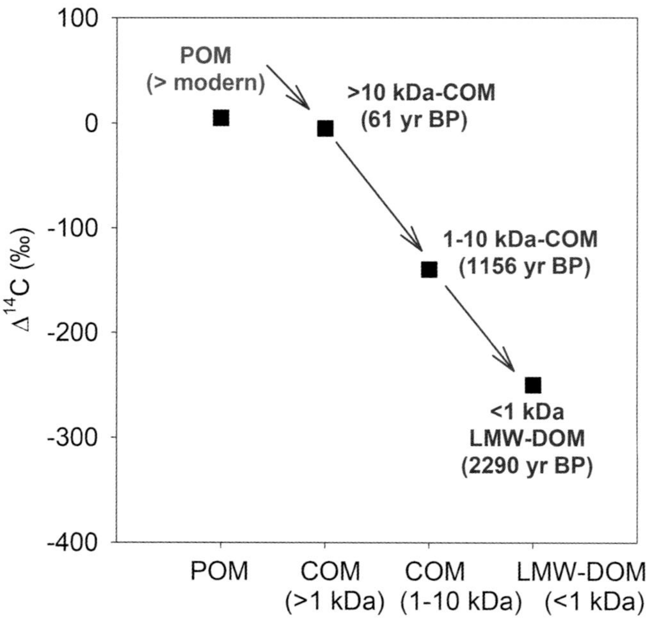

Figure 6.17. Changes in $\Delta^{14}C$ value and ^{14}C age among different OM size fractions, including particulate organic matter (POM), larger sized COM (10 kDa–0.2 μm), smaller sized COM (1–10 kDa), and the <1 kDa LMW-DOM at an offshore station in the Gulf of Mexico (data from Guo et al. 1996; replotted from Santschi et al. 2006).

fractions, including the <1 kDa DOM, 1–10 kDa COM, 10 kDa–0.2 μm COM, and POM (Guo et al. 1996), it clearly demonstrates that the $\Delta^{14}C$ values consistently decreased with decreasing molecular or particulate size and an apparent ^{14}C age increased with decreasing molecular size of DOM (Fig. 6.17). These results point to an organic carbon flow direction mainly from POM to larger-sized COM and to <1 kDa LMW-DOM, consistent with the high biological and chemical reactivity measured in the HMW or colloidal fractions of DOM (Benner and Amon 2015). High biological and chemical reactivity observed for the HMW-DOM supports the notion that marine DOM is highly heterogeneous in molecular size, composition, and reactivity. Thus, COM plays a more active ecological role and is important in regulating carbon cycling and environmental fate of bioactive elements and particle-reactive trace elements in marine environments.

Particulate Organic Matter

Sources of Particulate Organic Matter

The rain of POM is a key component of the Gulf's pelagic and sedimentary ecosystems. Major allochthonous sources of POM to the Gulf are rivers and estuaries, coastal wetlands, and atmospheric deposition, whereas autochthonous sources include primary and secondary production. Physical processes, such as resuspension, can redistribute these primary sources (Bianchi et al. 2014). Across the shelf to abyssal plain regions, primary production dominates the surface POM and constitutes the major portion of POM that is transported to depth. However, lateral transport of terrestrial POM deposited on the inner shelf across the shelf and slope, possibly as advection of POM in the benthic nepheloid layer (BNL), and sediment focusing all can mobilize this detrital OM and increase its proportional transport at depth and its deposition in the outer shelf, slope, and abyssal plain sediments (Bianchi et al. 1997; Gordon and Goñi 2003; Santschi and Rowe 2008). Much of what is known about POM in deeper waters of the GOM (>2000 m) comes from work in the early 2000s associated with the Deep Gulf of Mexico Benthos (DGoMB) program (Rowe et al. 2008).

No discussion of POM in the Gulf can take place without attention to riverine inputs, especially from the Mississippi-Atchafalaya River system (MARS). The Atchafalaya River receives roughly 30% of the flow of the Mississippi River (MR) via diversions created by the US Army Corps of Engineers. POM contributions are seasonally variable with no clear relationship to discharge (Bianchi, Wysocki, et al. 2007). The magnitude of POC flux to the GOM is about 9.3×10^8 kg yr^{-1}. This material is largely composed of C3 plant vegetation (Bianchi, Wysocki, et al. 2007). Marsh inputs appear to account for up to 30% of the POM exported from the MARS to the inner shelf (Bianchi et al. 2011). Isotopic and biomarker data from the Atchafalaya River suggested at least 2 sources of allochthonous terrestrial POM, "fresher" (less-degraded) plant debris near the river mouth and more-degraded soil-derived material deposited seaward onto the shelf (Gordon and Goñi 2003). These results indicated that C4 soil OM was a primary source of organic carbon to the shelf seaward toward the open Gulf. The dominance of OM enriched in fresh vascular plant material near the river mouth could result from hydrodynamic sorting of larger particles (McKee et al. 2004; Bianchi, Galler, et al. 2007; Bianchi, Allison, et al. 2014).

Recent work based on the DWH oil spill suggests that hydrocarbons stored beneath the seafloor can be important localized petrogenic sources of OM (Graham et al. 2010; De Gouw et al. 2011; Passow et al. 2012; Cherrier et al. 2013). For example, oil can be released naturally in hydrocarbon seeps but also released catastrophically via spills. Hydrocarbon sources of OM to the Gulf may be considered "natural" but also allochthonous.

POM in the Gulf is a mixture of multiple sources that is transformed during transport and deposition. This means that bulk concentrations, while useful, are insufficient to elucidate the complexity of POM biogeochemistry across the continental margin. Several chemical measurements are required to elucidate origins and to separate sources. Indeed, much work has been done on utilizing a series of techniques ranging from stable isotopes to biochemical markers to spectroscopy. Next, we consider the chemical composition of POM as a means to understand its sources and cycling within the GOM.

Chemical Composition of Particulate Organic Matter

C comprises the majority of POM along with other bioactive elements such as N, P, O, and S. The composition of POM is related to its sources and its transformations. However, the composition is not static; once formed and even in mixtures of multiple sources, OM composition changes from one that is easily characterized to one that is difficult to characterize. This paradigm is known as the "composition conundrum" and confounds a complete description of POM chemistry in the ocean to this day, despite advances in analytical chemistry (Hedges et al. 2000; Lee et al. 2004; Sleighter and Hatcher 2007).

At the bulk level, the ratio of C:N can be informative concerning the source of the OM (Meyers 1997; Hatcher 2004). Terrestrial OM tends to be enriched in C relative to N; hence, terrestrial OM tends to have higher C:N ratios than planktonic OM (Table 6.4; Bianchi and Bauer 2011). Many components of terrestrial plants (e.g., cellulose and lignin) do not contain nitrogen; however, protein-like material (amino acids, amino sugars, purines, nucleic acids) does. Thus, distinctions among bulk sources of OM (land plant vs. plankton) occur because of variation among its biochemical components.

Carbon-stable isotope ($\delta^{13}C$) values of organic matter in seawater also vary greatly among sources, with bulk

Table 6.4. Elemental and stable carbon isotope values for organic matter in seawater

OM source or biochemical class	C:N	$\delta^{13}C$ (‰ vs. Vienna PDB)	Comments
Terrestrial			
Leaves	100	-27 to -29	
Wood	1000	-27 to -29	
Marine vascular plants			
Spartina alterniflora	24–45	-13 to -15	Salt marsh grass
Zostera marina	17–70	-8 to -12	Seagrass
Marine plankton	6–11	-20	Bulk material
Bacteria	5.7	-20 to -25[a]	
Biochemicals			
Polysaccharide	n.d.	-18 to -20	For marine plankton
Protein	3.9	-16 to -18	For marine plankton
Lipid	n.d.	-25 to -35	For marine plankton
Chlorophyll	12.5	n.d.	
Lignin	n.d.	-25 to -32[b]	C3 plant
Lignin	n.d.	-12 to -19[b]	C4 plant

Source: Adapted from Bianchi and Bauer (2011); and Whitehead (2008).
Note: n.d. = not determined (lacks nitrogen).
[a] Range of river to coastal ocean values from Coffin and Cifuentes (1999).
[b] Goñi and Eglinton (1996).

values being a mass balance of individual biochemicals of fresh plant and plankton material mixed with detrital material (e.g., humic substances; see Table 6.2). δ^{13}C-POM values in suspended matter for the MR show relatively narrow variation from -24‰ to -26‰ (Bianchi, Wysocki, et al. 2007). This implies a relatively constant source, or rather a constant mixture of sources. Autochthonous production in the river may have influenced the enriched values (-27‰) as a result of inputs of planktonic OM from diatoms, effectively diluting the terrestrial- and soil-OM contributions, as evidenced by a lack of correlation between POC and lignin-phenol concentration, the latter being specific biomarkers for land plants (Duan and Bianchi 2006). This supposition is supported by a Lagrangian study of the lower MR in which a decrease in chlorophyll-*a* concentrations between Baton Rouge, Louisiana, and Head of Passes (with no salt wedge present) suggested that upper tributaries of the MR (e.g., Missouri and Ohio Rivers) are "seeding" phytoplankton in the lower river (Dagg et al. 2005). However, the Duan and Bianchi study provided key insight to resolve the issue of the source of POM in the lower MR because δ^{13}C values were measured on the lignin phenols themselves. Enriched δ^{13}C values (-19‰ to -24‰) for ferulic acid (which itself is present in greater abundances in nonwoody vegetation; Hedges and Mann 1979) indicate contributions of C4 plant materials and appear to suggest substantial amounts of MR POM derived from upper Midwest vegetation (Goñi et al. 1998). In contrast, δ^{13}C values of woody-plant biomarkers such as acetovanillone and syringic acid, both derived from lignin, were relatively depleted (-26‰ to -30‰) and indicative of forested vascular plant OM. Agricultural land-use practices in the MR watershed might therefore moderate the contributions from C3 land plants and influence the overall loading of POM into the Gulf via the MR (Onstad et al. 2000).

Recent work has utilized optical properties of absorbance and fluorescence commonly applied to DOM to characterize POM extracted in dilute base (BEPOM), allowing rapid characterization of bulk OM cycling (Osburn et al. 2012). In this technique, adapted from extraction of humic substances from soils and sediments, material collected on a 0.7 μm porosity glass-fiber filter is extracted in 0.1 *M* NaOH for 24 hours in the dark. The resulting extract is clarified of any filter particles and measured for its absorption and fluorescence properties, similar to DOM (e.g., Stedmon et al. 2003; Coble 2007). In GOM surface waters, BEPOM was enriched in protein-like fluorescence, which reflected primary production by phytoplankton (Brym et al. 2014). In contrast, BEPOM samples in bottom waters collected from the continental shelf near the DWH site were enriched in microbial humic fluorescence signatures suggestive of bacterial processing and had δ^{13}C values of -31‰ to -38‰, indicative of a fossil methane influence of perhaps 30–40% in the POM, on the same order of magnitude as estimates using Δ^{14}C measurements (Brym et al. 2014; Cherrier et al. 2013). These results suggest that the chemistry of POM in the Gulf is intimately linked to petrogenic carbon sources released from the subsurface as well as allochthonous and autochthonous carbon sources from terrestrial and marine environments, respectively.

Several workers have shown that the primary riverine dispersal system, the MARS, exerts influence over the POM of the inner shelf, decreasing in importance toward the shelf-slope transition (Goñi et al. 1998; Bianchi et al. 1997; Bianchi et al. 2002; Gordon and Goñi 2003; Sampere et al. 2008; Bianchi et al. 2011; Table 6.5). Some of this seaward transport is due to resuspension and advection across the shelf, often induced by storms and hurricanes (Allison et al. 2010; Sampere et al. 2011). However, 3 key results of POM chemical composition show this general offshore progression. First, C:N ratios typically decrease from values of 29 near river mouths to 8–13 over the remainder of the shelf, signals that remain constant into the offshore region (Gordon and Goñi 2003). Second, carbon-stable isotopes show enrichment in ^{13}C from values ~-28‰ inshore near the river to offshore near the shelf/slope break (-20‰). In fact, there is little change in either C:N or δ^{13}C values of SOM collected from >500 m water depth in Mississippi Canyon (Waterson and Canuel 2008). Third, yields of phenols derived from the terrestrial biomarker lignin (mg phenol 100 mg OC^{-1}) decreased with decreasing POC concentrations seaward across the shelf to the slope as the terrestrial influence waned. A notable exception was spillover from mobile muds that redistribute terrestrial POM to deeper shelf/slope regions (Bianchi et al. 2006; Sampere et al. 2008; Sampere et al. 2011).

Hydrodynamic sorting is an important mechanism for transporting river OM into the Gulf, thus strongly influencing the dominance of terrestrial sources on the chemical composition of its POM. Sand-size fractions of sediment contained as much terrestrial organic carbon as did mud fractions based on C:N elemental ratios and lignin-phenol yields (Bianchi, Galler, et al. 2007). Wakeham et al. (2009) used density fractionation to compare POM properties stored in continental margin sediments.

Table 6.5. Concentration of SPM and POC and atomic C:N ratios for the lower Mississippi River and the shelf and slope of the Gulf of Mexico

Location	Water depth (m)	SPM (mg L^{-1})	POC (μM)	C:N	Reference
Lower river	0.5[a]	107	195.4	11	Bianchi et al. (2002)
	16[b]	108	26.9	10.7	
Lower river	0.5[a]	125	228.3	10	
	8[b]	213	353.6	8.4	
Inner shelf	1[a]	23.8	43.5	8.4	
	50[b]	28.1	7.0	17.5	
Inner shelf	1[a]	35	119.1	7.1	
	26.5[b]	42	17.4	12	
Shelf	16	n/a	5.8	n/a	Bianchi et al. (1997)
Shelf/break	100	n/a	2.8	n/a	
Shelf/break	356	n/a	2.0	n/a	
Slope	1602	n/a	2.7	n/a	
Slope	1123	n/a	1.3	n/a	
Shelf/break	539	n/a	1.0	n/a	
Shelf	49	n/a	2.3	n/a	
Shelf	27	n/a	8.5	n/a	
Shelf	56	n/a	1.3	n/a	
Shelf	101	n/a	3.4	n/a	
Shelf/break	402	n/a	2.4	n/a	
Slope	1603	n/a	1.7	n/a	

Note: SPM = suspended particulate matter.

POC = particulate organic carbon.

n/a = not available.

[a] Mean of surface samples <5 m water depth.

[b] Mean of bottom samples >10 m water depth.

Samples were fractionated into density "classes" and comparisons made among sediments collected at 50 m and 540 m water depth in the GOM, representing a prodelta deposition area and sediments, which were compared to sediments collected in the lower MR. Most OC was contained in the mesodensity fraction (1.6 to 2.5 g cm^{-3}), and the least dense fractions were enriched in lignin primarily of C3 plant origin. These results were consistent with those of Bianchi et al. (2002) and demonstrated that nearshore sediment OM is made up mostly of vascular plant debris. Hydrodynamic sorting occurs in the BNL, a layer of thick mobile muds in which higher-energy environments can winnow away finer material and resuspend, sort, and transport material farther offshore (Allison et al. 2000; Allison, Bianchi, et al. 2007).

To summarize, the chemistry of POM in the Gulf is shaped to a large degree by riverine forcings in 2 ways. In the first way, the MARS system itself injects substantial amounts of POM that is derived from its extensive watershed and is dominated by soil and plant debris. These detrital materials are relatively depleted in N relative to autochthonous sources and carry $\delta^{13}C$ signatures of -25‰ to -30‰, reflective of carbon fixation via the C3 pathway. POM having enriched $\delta^{13}C$ values (-10‰ to 15‰) can be indicative of C4 plants, which has been attributed to the large agricultural production in the MR watershed but might also reflect wetland plants such as *Spartina* spp. Terrestrial POM inputs to the shelf via the MARS system are enriched in lignin. In a separate but equally important way, the MARS system delivers nutrients that fuel coastal phytoplankton production. POM derived from primary production is enriched in "labile" biochemicals such as carbohydrates and amino acids, as well as more refractory lipids. Enrichment of amino acids lowers the C:N ratio compared to that in terrestrial POM, whereas $\delta^{13}C$ values are enriched (-19‰ to -22‰) because of the different inorganic carbon source present in seawater. Lignin is absent in POM derived solely from phytoplankton. As the freshly produced material rains down through the water column, it is consumed; however, formation of marine snow that increases sinking rates can facilitate rapid transport to surface sediments, mitigating degradation in the water column.

Distribution of Particulate Organic Matter

While much of the POM loading via rivers and coastal outwelling dominates POM concentration on the inner shelf, substantial inputs of POM to the pelagic water column occur from primary production in surface waters. The MR drains nearly 40% of the coterminous United States and contains much higher concentrations of suspended sediments and POM than seawater does (Milliman and Meade 1983; Bianchi et al. 1997). The distribution of POM in the Gulf due to the MR is restricted, however, because of strong wind fields and eastward or westward along-shelf surface currents, which limit the impacts of the river to the continental shelf (Yuan et al. 2004; Walker et al. 2005).

Primary-production rates are important to consider as the initial formation of POC followed by estimates of rates of its vertical transport (fluxes). River loading of particulate (and dissolved) materials can strongly control inshore shelf conditions, but circulation features such as eddies can facilitate advection of POM formed nearshore in enhanced nutrient regimes into the interior of the Gulf. In fact, warm-core rings (WCR) and cold-core rings (CCR), as well as the Loop Current, are prominent features in the circulation of the GOM that serve to create conditions conducive for primary production of POM as well as agents of its dispersal. Thus, physical-biogeochemical coupling is very important for the distribution of POM in the Gulf. Early work has shown that diatoms, prymnesiophytes, and cyanobacteria dominated the Texas shelf and shelf/slope break waters in the winter season (~85%), with diatoms and cyanobacteria the dominant taxa in fall (~90%) (Lambert et al. 1999). In general, diatoms and cyanobacteria dominated the shelf waters while cyanobacteria, pelagophytes, and prymnesiophytes were dominant in slope waters. Interestingly, the highest concentrations of prymnesiophytes, based on 19' hexanoyloxyfucoxanthin (19-hex) concentrations, were found in slope waters in January 1993 and were likely related to the presence of a WCR at that time. Prymnesiophytes have been shown to be ubiquitous in oligotrophic, tropical waters and may have been transported into this region of the Gulf by a warm-core eddy (Biggs 1992; Claustre and Marty 1995).

Rates of primary production of POM by phytoplankton are strongly linked to sources of nutrients such as the MARS. Chen et al. (2000) calculated an annual average primary production for coastal waters of the northwestern GOM of 0.44 g C m^{-2} d^{-1}. The highest values of primary production were nearly 8-fold greater than the average (up to 3.31 g C m^{-2} d^{-1}) and occurred in coastal waters between Terrebonne Bay and Atchafalaya Bay. Riverine nutrient inputs could be driving this trend, with circulation patterns dictating where nutrients are dispersed and features such as eddies restricting water-mass movements and facilitating phytoplankton growth (Scavia et al. 2003; Dagg et al. 2007). Net primary production (NPP) in the northwestern GOM (west of 91°W) calculated from remote-sensing imagery using SeaWiFS averaged 0.4 g C m^{-2} d^{-1} (Biggs et al. 2008), which matches quite well with the in situ and experimental measurements made by Chen et al. (2000). However, average NPP calculated for the northeastern GOM stations east of 91°W from SeaWiFS imagery was nearly double, at 0.7 g C m^{-2} d^{-1}. This near doubling of mean daily NPP was interpreted to be due to eddies in the northeastern Gulf entraining and transporting MR water containing relatively higher amounts phytoplankton offshore, as evidenced by relatively higher sea-surface chlorophyll values.

This observation was important because it underscores the coupled physical-biogeochemical processes in coastal to offshore waters of the GOM that largely constrain the distribution of POM and can have wide-ranging effects on marine biodiversity and fisheries. In addition to distributing nutrients, advection can be a key means for distributing OM formed inshore toward the open Gulf. Toner et al. (2003) found that pervasive cyclonic and anticyclonic eddies in the Gulf generated advective paths that connected the inshore and remote shelf regions. The paths were modeled using a Lagrangian approach and aligned remarkably well with the plume events recorded with the chlorophyll-*a* ocean-color products from SeaWiFS. Two such events were studied, demonstrating interactions between a cyclonic eddy and the Loop Current and between cyclonic and anticyclonic eddies. Intereddy advection was therefore seen as a key means of moving OM formed inshore over the shelf and into the Gulf interior. The linkage of the physical with the biological in eddies is strengthened in CCR in which nutrients are upwelled and abundant, unlike in the nutrient-poorer WCR.

Given the strength of the MR flow (at times >20,000 m^3 s^{-1}) and the nutrients associated with it, the MR's plume (MRP) also exhibits important physical-biogeochemical linkages. Walker and Rabalais (2006) investigated the surface circulation and structure of the MRP using satellite measurements of chlorophyll-*a*, suspended sediment, and temperature. They found that during high river discharge the plume's extent distributed freshwater, sediment, and carbon more than 2700 km^2 and more than 20 km southwest across the shelf. POM was not quantified, but concentrations of total suspended solids (TSS) approached 360 mg L^{-1}. Winds are important drivers as dispersal mechanisms for suspended materials in river plumes, and seasonal variations in wind fields as a consequence strongly influence the distribution of POM across the continental shelf. Along-shore easterly winds connect the eastern (Louisiana, Mississippi, Alabama) shelf with the western Louisiana-Texas shelf via strong (40–90 cm s^{-1}) westward-flowing currents, which can join with the flow out of Southwest Pass, pushing the MRP farther west. Easterly winds trap river discharge and nutrients, effectively stalling the westward flow and promoting phytoplankton primary production on the shelf. In the summer, southerly to southwesterly winds can reverse direction of the MRP to the east for 1–2 months. This effect can entrain river water and its load of nutrients and OM over the continental slope via slope eddies and the Loop Current.

Aside from delivering nutrients, the MARS system is a primary source of POM to the northern Gulf. The annual flux of POC from the MR to the GOM was estimated at 0.8 × 10^9 kg C yr^{-1} (Duan and Bianchi 2006; Bianchi, Galler, et al. 2007). Mean POC concentration in the lower MR, representing its discharge to the Gulf, was 1.56 mg L^{-1}. During a study of POM loads to the Louisiana shelf after Hurricane Lili in 2002, Goñi et al. (2006) derived a relationship between TSS and POC: POC = TSS × 0.0206 −0.309 (R^2 = 0.70; N = 99). Using mean TSS values from Bianchi, Galler, et al. (2007), we computed a mean POC concentration for the lower MR of 2.00 mg L^{-1}, quite similar to the mean POC from that study. This estimate suggests that linkages between riverine export of POC and its distribution on the shelf are strong.

The mean areal extent of the MRP under moderate flow is 1814 km^2 (Walker et al. 2005). Therefore, we were able to compute a mean daily flux of POC from the river to this region of the shelf of 1.30 mg C m^{-2} d^{-1}. This value is a first-order estimate of the POM contribution from the MR to the GOM shelf carbon cycle, nearly double estimates of primary production previously discussed. Inventories of POM have been studied across the continental shelf and across the continental slope. Higher concentrations of POC were found on the inner shelf but decreased dramatically across the shelf break and slope (Table 6.6). Maximum POC values were in the surface waters (0.5–2 m) and ranged from 43.5 to 353.6 μM in the lower river and the inner shelf and then declined to 2.3–8.5 μM on the outer shelf and in deeper water (Bianchi et al. 2002; Bianchi, Wysocki, et al. 2007). Shelf-break and slope POC values were lower still (1.0–2.8 μM). Further, POC concentrations decreased with depth (Fig. 6.18). Other work in the MPR region has shown that chlorophyll and degradation products (indicative of zooplankton graz-

Table 6.6. Comparison of POC flux and SCOC rates for the northern Gulf of Mexico

Site	POC flux rate	SCOC rate	Interpretation of difference
Head of Mississippi Canyon	77	36.5	Positive imbalance and net burial of OM
Midslope	15.8	34.3	Negative balance; detrital OM from resuspension supporting benthic communities
Mississippi Fan	4.2	15.7	Negative balance; detrital OM from resuspension supporting benthic communities
Abyssal plain	4	3.9	Ecosystem in balance

Source: Data compiled from Rowe et al. (2008).

Note: SCOC = sediment community oxygen consumption; values are means of multiple observations and in units of mg C m^{-2} d^{-1}.

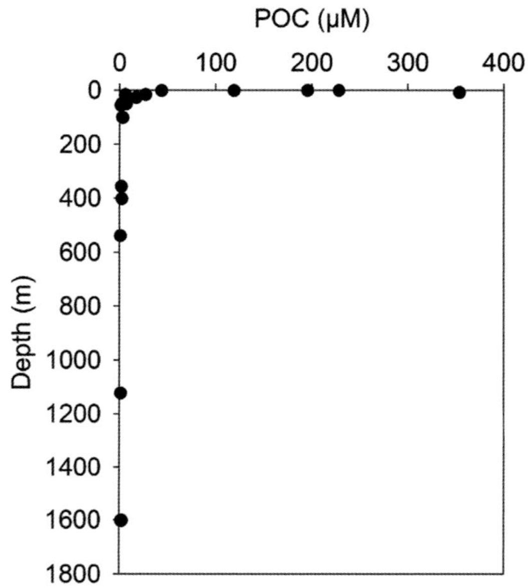

Figure 6.18. POC concentrations plotted as a function of depth for the continental shelf and slope (data compiled from Bianchi et al. 1997 and Bianchi et al. 2002).

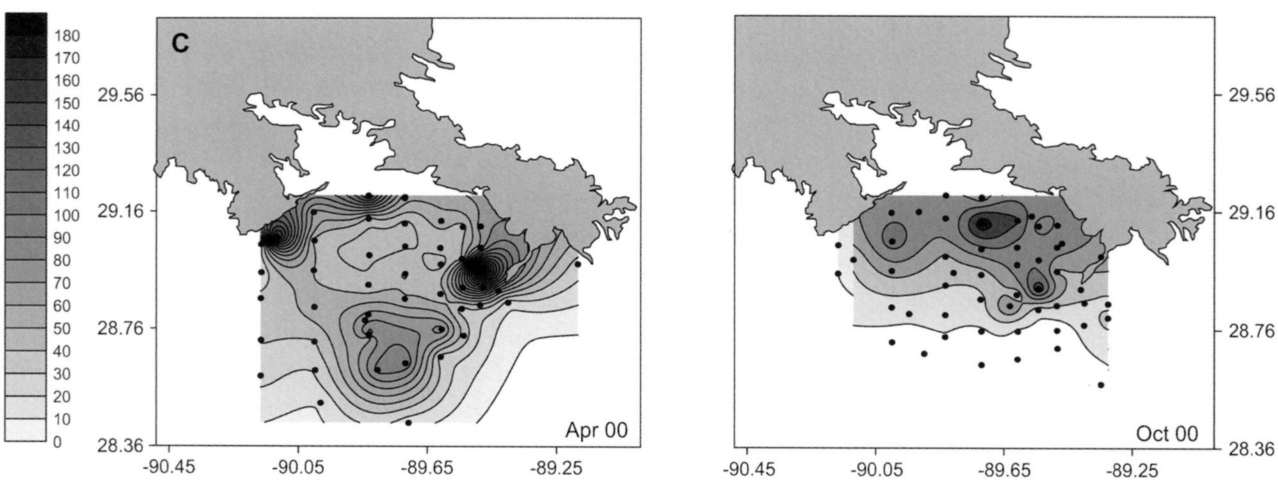

Figure 6.19. Spatial distribution of POC concentrations in μM from surface waters on the Louisiana shelf during April and October 2000. Black dots represent the station locations for each sampling period. Bathymetric curves for 50 m, 100 m, 150 m, and 200 m are shown (Wysocki et al. 2006).

ing) observed in surface sediment indicated a transfer of autochthonous carbon from the highly productive photic zone to the sediment, coupling phytoplankton-derived POC in surface waters with organic-carbon deposition in surface sediment (Wysocki et al. 2006). Moreover, cross-shelf changes in chlorophyll indicated a westward transport of phytoplankton that was directly and indirectly linked with river discharge and pigment decay dynamics. Although POC was not correlated with salinity, the greater distribution of POC across the Louisiana shelf during high flow in April clearly reflects the actual transport of river-derived POC and river-derived nutrients linked with marine POC (Fig. 6.19).

C:N ratios of the inner-shelf POM suggested that surface waters were influenced by primary production and deeper waters by terrestrial inputs. Lignin-phenol compositions of surface sediments collected from sites where POM samples were taken suggested a strong influence of woody terrestrial (angiosperm) vegetation, as sediment samples were enriched in syringyl lignin relative to vanillyl phenols and depleted in cinnamyl phenols. Hydrodynamic sorting of large wood fragments was invoked to explain these patterns, resulting in higher estimates of woody angiosperm OM in the river's primary dispersal system.

However, recent work has revealed that wetlands also may be important sources of POM to the Gulf. Bianchi et al. (2011) used a combination of stable isotopes and biomarkers to examine relative abundances of riverine, marine, and wetland OM to the Louisiana shelf. Marine sources supplied 55–61% of OC inputs to the shelf during the spring, while riverine OC inputs were highest in the fall (63%), attributed to lower marine inputs, perhaps limited by nutrients and sunlight. OM inputs from coastal marshes accounted for 19–34% of the OC in the spring and 3–15% in the fall. While lower than values for riverine or marine sources, the marsh values were determined to be significantly more important to the inner shelf than previously realized. That realization appears to be consistent across the GOM irrespective of coastal wetland type. Mangrove wetlands, for example, those fringing the Shark River estuary on the west coast of Florida, also may be important contributors of POM to the Gulf. Using a suite of lipid biomarkers to constrain a 3-endmember mixing model of freshwater, marine, and mangrove wetland sources, He et al. (2014) found that mangroves were the dominant source of POM to the Shark River estuary. Tidal pumping was determined to be the major factor for the importance of mangrove material, via resuspension of surface sediments during incoming tide and export to the shelf during outgoing tide. Rosenheim et al. (2013) used ramped pyrolysis, heating OM in an inert atmosphere and burning off increasingly less volatile compounds as temperature is increased, to examine the $\delta^{13}C$ and $\Delta^{14}C$ values of increasingly "recalcitrant" fractions of OM. Higher Mississippi and Atchafalya River discharges contained comparatively older POC. The implication of this study was that materials stored in the floodplain and perhaps wetlands, as well as petrogenic OC, can be mobilized during higher river flows and transported to the GOM.

At depth, currents can cause resuspension of surface sediments across the shelf, increase particle concentrations in bottom waters, and transport particles downslope. A mechanism for the distribution of POM is the BNL, also described as mobile muds (cf. Aller 1998). The BNL acts like a conduit by distributing materials across the shelf in tight linkage with shelf dynamics. Nearshore, high, localized inputs of sediment from rivers and bays and high-energy waves (perhaps caused by frontal systems) resuspend and redistribute materials, whereas near the shelf edge, near-inertial waves and shelf waves provide energy to resuspend and transport fine sediment (Sahl et al. 1987). Also of importance, the nearshore BNL can act as a highly efficient biogeochemical reactor in which substantial OC remineralization occurs (Aller 1998). Seasonal variability in river and coastal wetland sources of OM will influence the extent of mobile muds across the shelf environment as well as the quality of the material contained within them (Allison et al. 2000; Corbett et al. 2004; Corbett et al. 2006). For example, in the Bianchi et al. (2011) study, sediments and POC from the lower MR exhibited very similar lignin phenol and $\delta^{13}C$ values, indicative of relatively unreactive material being deposited to coastal sediments. Prior work on the nature of POM exported by the MR has suggested that OM near the river mouth is less dense material dominated by woody plant fragments, whereas deposition farther offshore is largely C4 plants derived from the river's watershed (Goñi et al. 1998; Onstad et al. 2000; Bianchi, Wysocki, et al. 2007). Bianchi, Wysocki, et al. (2007) have theorized that much of the predegradation of the POM exported by the river is caused by seasonal storage of POM in the BNL, which has settled out of suspension during low-flow conditions. Nevertheless, a gradient of lignin from marshes in Barataria Bay across the inner shelf strongly suggests the export of marsh detritus (Sampere et al. 2008). This means that local inputs of material to the BNL from coastal wetlands fringing the Gulf could be an unrealized and important source of POM.

Finally, storms can play a role in movement of material from coastal rivers and estuaries offshore, affecting the distribution of POM. Yuan et al. (2004) demonstrated that storms can disrupt the localized along-shore current patterns that constrain the extent of the MR's influence on the open GOM. Using satellite imagery, these workers examined sea-surface chlorophyll (SSC) distributions after passage of 2 storms: Hurricane Lili and Tropical Storm Barry. Eddies formed by each storm injected nutrient-rich water into the open Gulf, where the light limitation from turbid plume waters was diminished, and promoted phytoplankton blooms. The blooms were visible in SeaWiFS satellite imagery after the storm. The "tongue" extending out of the MR delta transitioned into a ring and persisted for several days.

Winter storms also can disrupt persistent stratification that occurs across the LaTex shelf (intensifying the periods of shelf hypoxia) as a result of the strong Louisiana Coastal Current (Wiseman et al. 1997). Vertical mixing can resuspend materials in the BNL and surface sediments during such events, which is an effective means of redistributing POM from shallow-water depths (<10 m) inshore to deeper waters (~20 m) offshore and across the continental shelf (Allison et al. 2000). Goñi et al. (2006) found a post-storm deposit from Hurricane Lili (2002) near the Atchafalaya River mouth across the inner shelf that ranged from 0.5–20 cm in thickness, with a mean range of 3–10 cm. Silt and sands dominated the storm deposits and contained elevated OC contents (1–2% OC). POC concentrations in the water column ranged from 0.2 to 5.0 mg L^{-1}. After Hurricane Ivan, Sampere et al. (2008) found increased %OC in surface sediments of the shelf out to the Mississippi Canyon that contained relatively more river-dominated woody material than pre-Ivan sediments. In the deep part of the canyon (>500 m) fresher signatures of biomarkers indicated the redistribution of "fresh" nearshore surface sediment into the deeper waters of Mississippi Canyon. Recent work has shown that a BNL in Mississippi Canyon presumed to be formed after Hurricane Isaac is enriched in macroaggregates that appeared to be a mixture of lithogenic materials and OM from surface sediments, which compared to OM-poor material collected in sediment traps was hypothesized to be due to resuspension and redistribution of sediments from the shelf to the canyon (Ziervogel and Arnosti 2016). The Sampere et al. (2008) study also suggested the importance of cyclonic storm events such as hurricanes for resuspending marsh-derived POM deposited in wetland basins across the GOM into overlying waters where transport and redistribution to the inner shelf can occur. Despite marshes being important potential sources of POM exported to the Gulf during storms, these workers also suggest that marsh-derived material is likely diluted by riverine POM and sediment and rapidly oxidized in shallow shelf waters.

Fate of Particulate Organic Matter

Once formed, POM may be consumed and degraded to smaller particles, reaching sizes <0.2 μm, at which point OM is considered truly dissolved. Vertical transport of POM in the water column, from surface water to sediments, represents a major pathway for the fate of OM, leading to its consumption or burial in sediments. Microbial degradation of POM constitutes its primary removal mechanism in the water column and results in a rather consistent change in its biochemical composition during sinking (Wakeham et al. 1997; Lee et al. 2004). However, microbes also can add to and modify the composition of POM via aggregation (Biddanda and Pomeroy 1988; Alldredge and Silver 1988; Simon et al. 2002; Kaiser and Benner 2008). The fate of POM in the ocean is linked to the depth of the water column, since a deeper water column allows a longer time for POM degradation to occur. Depth profiles of POC concentrations reflect the removal of POC by grazing and mineralization in that roughly 90% of surface primary production is consumed in an otherwise well-oxygenated water column (Meyers 1997). Therefore, POM deposition to surface sediments should decrease with increasing water depth (Suess 1980). The fate of POM in surface sediments is considered in Chapter 7.

Flux estimates for POC transport from surface waters to deep waters in the GOM have been based on 3 primary observations: sediment traps, radioisotope tracers, and models based on remote-sensing observations. Most of these estimates are done at steady state. Sediment-trap estimates of total particulate carbon flux range from 0.2 to 1.8 g C m^{-2} d^{-1}, varying temporally between the MRP and adjacent shelf region (Redalje et al. 1994). The highest vertical POC flux occurred in May (1.80 g C m^{-2} d^{-1}) for the MRP, roughly 4.5 times greater than on the shelf (0.4 g C m^{-2} d^{-1}). In mid- to late summer, vertical POC fluxes for the MRP and the shelf were much more similar, 0.29 and 0.18 g C m^{-2} d^{-1}, respectively. This distinction was sensible given the higher primary production in the plume during the spring. Diatoms appear to be key sources of planktonic POM raining from surface waters to the sediments (Dagg et al. 2005).

Seaward of the MRP and inner-shelf regions of the GOM, vertical POM fluxes decrease substantially and are strongly influenced by coupled physical-biological processes. Control of rates of primary production exert the most influence. For example, Biggs (1992) calculated primary production rates across a mesocyclonic eddy in the western GOM; within the warm ring of the eddy, where nitrate concentrations were near limits of detection, primary production averaged 160 mg C m^{-2} d^{-1}, translating to vertical POC fluxes (at 50 and 100 m water depths) of 4–12 mg C m^{-2} d^{-1} (assuming 50% C by mass of the daily mass flux that was calculated). The range of primary production values were consistent with those found in earlier surveys of the region (100–200 mg C m^{-2} d^{-1}). However, in the cold core, with higher nutrient concentrations, primary production was substantially greater at ~250 mg C m^{-2} d^{-1}. Moreover, vertical POC fluxes estimated from mass fluxes, as described previously, were 200 and 132 mg C m^{-2} d^{-1}, at 50 and 100 m, respectively.

Sinking particles were collected in sediment traps deployed in WCRs and CCRs in separate north-south transects along 94°W in 2000 and 2001 (Fig. 6.20; Hung et al. 2003). These transects extended from coastal shelf regions near Galveston, Texas, offshore to the oligotrophic GOM over the continental slope. CCRs are nutrient rich and higher in primary production than WCRs, so differences in vertical POC fluxes were expected based on these nutrient regimes. The higher productivity was evident in depth profiles of POC concentrations in the CCRs, which were generally double those in the WCRs. Likewise, POC flux measurements, which decreased with depth from 65 to 120 m, were roughly double in the CCRs than in the WCRs.

It should be noted that sampling protocols have perhaps influenced observations about POC mass fluxes. Utilizing ^{234}Th as a radiochemical tracer, Hung et al. (2004) found good agreement in POC flux values (39–148 mg C m^{-2} d^{-1}) computed directly and measured from sediment traps at 120 m water depth, compared with POC flux values (38–163 mg C m^{-2} d^{-1}) for particles >53 µm, using the ^{234}Th deficit multiplied by the POC/^{234}Th ratio (averaged from 0 to 120 m) for suspended particles >53 µm. Using the ^{234}Th deficit multiplied by the POC/^{234}Th ratio for sinking particles collected by sediment traps at 120 m produced much higher POC sinking rates of 52–450 mg C m^{-2} d^{-1}. Vertical POC fluxes computed by ^{234}Th disequilibria on small (1–10 µm) and intermediate (10–53 µm) particle size ranges produced even higher estimates: 505–2554 mg C m^{-2} d^{-1} and 19–399 mg C m^{-2} d^{-1}, respectively. POC fluxes computed from sediment traps may underestimate fluxes, possibly due to particles being degraded in traps during collection and undertrapping due to effects of currents and swimmers. However, the matchup between the direct sediment-trap vertical POC flux and the ^{234}Th radiotracer approach applied to the >53 µm size fraction did produce the most consistent results.

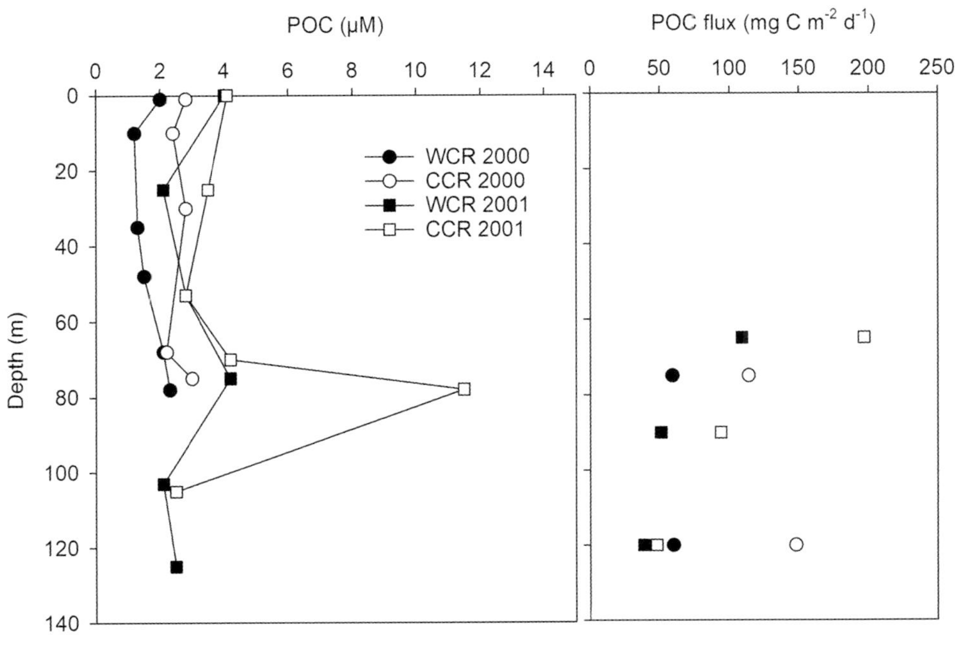

Figure 6.20. POC concentrations (*left*, in µM) and vertical POC flux rates (*right*) measured from sediment traps in nutrient-poor warm-core rings (WCR) and nutrient-enriched cold-core rings (CCR) sampled in the northern Gulf of Mexico in 2000 and 2001 (from Hung et al. 2003).

Later, Hung et al. (2010) compared POC flux rates among particle size classes computed as the POC/^{234}Th ratio times the ^{234}Th flux and found that rates for particle sizes of 10–50 μm were most similar to POC fluxes computed from sediment-trap studies. Measurements of ^{234}Th activity and POC concentration were done on the same filter rather than in separate filters as done in previous studies. Filtration may disaggregate larger particles loosely held as macroaggregates, and thus the fate of POM may be more closely linked with larger aggregates than those ultimately measured on filters. This study also observed that POC fluxes were highly correlated to uronic acid, while the correlation with bacterial biomass was negative (Santschi et al. 2003). The positive correlation with uronic acid suggested that this material acted as a scavenger for ^{234}Th. Uronic acid is a component of extracellular material in all microorganisms and is thought to form gels that hold aggregates together. The negative correlation of POC with bacterial biomass suggested that bacteria play an important role in the disaggregation process. Bacterial enzymes are key mediators of OM degradation and therefore contribute to decreased POC fluxes, controlling the fate of this material in the biological pump (Arnosti 2011).

Remote-sensing information from satellite platforms, such as SeaWiFS and MODIS, have been used to estimate SSC concentrations from which POC fluxes can be estimated. Based on SSC, the Vertically Generalized Production Model (VGPM) is used to estimate a depth-integrated NPP, which is finally used in an exponential decay model to compute POC fluxes (Müller-Karger et al. 2005; Biggs et al. 2008). Distinct differences in POC fluxes were determined for the northwestern and northeastern GOM shelf by comparing 3-year averages from 1998 to 2001 (Biggs et al. 2008). The northwestern GOM (west of 91°W) had a mean daily POC flux estimated at 9.4 mg C m^{-2} d^{-1}, while the northeastern GOM (east of 91°W) had a mean daily POC flux estimated at 17.9 mg C m^{-2} d^{-1}. In the northeastern region, maxima occurred during November to February and during summer when production was high. Warm slope eddies were thought to have advected low-salinity “green water” from the MRP offshore, which would explain the discrepancy. Both regional shelf estimates were far larger than that near the edge of the continental margin at the shelf/slope break. South of 26°N, primary production was much lower than in the northeastern and northwestern regions, which translated into a seasonal mean POC flux estimate of 2.9 mg C m^{-2} d^{-1}.

El Niño–Southern Oscillation events have also been shown to impact POC distribution and abundance in the GOM. For example, chlorophyll concentrations, during the winter to spring transition in 1998 in regions of the northeastern GOM off the Florida coast, were up to 4 times higher than during the same periods in 1999 and 2000. This was attributed to higher freshwater discharge during the 1997–98 winter-spring transition, linked with unusually strong upwelling observed along the coast in spring 1998 during an El Niño–Southern Oscillation event.

Low vertical flux rates across the shelf/slope break and into the oligotrophic open GOM in some cases have led to a supply of autochthonous OM insufficient to support benthic ecosystems. The DGoMB program examined vertical POM fluxes and associated benthic-pelagic coupling with surface benthic ecosystems via the sum of sediment community oxygen consumption (SCOC) and burial (Rowe et al. 2008). Comparisons between SCOC estimates and POC fluxes computed from remote-sensing information from the SeaWiFS satellite (Biggs et al. 2008) made for sites at the head of Mississippi Canyon, down the slope, and onto the abyssal plain show clear differences in all but the abyssal plain site (see Table 6.6). The negative imbalances were interpreted to indicate that sources of OM other than the rain of POC must be supporting benthic communities, most likely detrital material transported via resuspension processes as described previously.

Making sense of direct and indirect means of computing vertical export fluxes of POM from the surface ocean to bottom sediments requires comparative studies, few of which have been conducted. However, recent work by Maiti et al. (2016) has attempted to make comparisons for the northern Gulf. They focus on the oligotrophic GOM to preclude interference of POC exports from the MARS system but also to avoid interference in satellite estimates of chlorophyll that may be due to CDOM. This study employed a second radiochemical tracer for POC flux using the disequilibrium between ^{210}Po and ^{210}Pb, which provides estimates on a timescale of weeks to months, compared to ^{234}Th, which is on a timescale of days to weeks. Sediment traps were deployed for 2 days in regions east of the MRP at times coinciding with high primary production near, north of, and south of the DWH spill site (Biggs et al. 2008). The ^{210}Po fluxes at 150 m were 23–30 mg C m^{-2} d^{-1} and decreased by about 50% at 350 m to 10–14 mg C m^{-2} d^{-1}. The inference here was that roughly 50% of the POC was being mineralized over that depth range. The ^{210}Po tracer compared favorably to POC flux measurements made at 150 m using the ^{234}Th tracer (27–40 mg C m^{-2} d^{-1}), as well as direct measurements from sediment traps (29–32 mg C m^{-2} d^{-1}). These results were consistent between 2012 and 2013 sampling years.

Satellite-based estimates of NPP and temperature were used in 3 different models to estimate vertical POC exports (Laws, Dunne, and Henson; see Maiti et al. 2016 for details on each model) to compare with indirect measurements from radiochemical tracers and direct measurements from sediment traps. Model estimates based on VGPM using the Laws and Dunne models were 2-fold greater than in situ observations, whereas estimates using the Henson model were similar to, or slightly lower than, the in situ observations. The discrepancy highlighted the fact that global productivity models based on temperature may omit other biogeochemical and scaling factors required to accurately predict vertical POC export fluxes across oceanic biomes.

Vertical export flux transitions the fate of POM to sedimentary environments, as it becomes a component of the SOM pool, which is considered in Chapter 7. However, we provide here a few key observations to contrast with aforementioned vertical POM flux measurements. After correcting for carbonate concentrations, Morse and Beazley (2008) found that between water depths of 200 and 3500 m in Mississippi Canyon, sedimentary OC concentrations decreased from 2% to 0.4% OC by weight (wt%). Further, decreases in the OC to surface-area ratio (OC/SA) (0.52–0.11 mg m^{-2}) with increasing water depth in that region supported the notion that POC decreased as a function of particle size, which is consistent with the concept that degradation processes effectively reduce the size of POM in addition to its overall concentrations. In sharp contrast, across the Texas shelf of the western GOM to Alaminos Canyon and into the submarine canyons, the influence of particle transport on the fate of POM in the GOM was apparent. The %OC was noticeably higher, ranging from 0.63 to 0.95 wt%. OC/SA values ranged from 0.14 to 0.44 mg m^{-2}, and δ^{13}C values of the sediment OC ranged from -20.6‰ to -33.5‰. The main implication of this distinction was that westward currents had deposited sediments transported by the MR to the region, though the δ^{13}C values -30‰ did suggest some possible contribution from hydrocarbons. In fact, near the MR itself, in waters known for high primary production, sediment OC was 1.1 wt%, which was greater than the mean for the 3 locations studied (OC = 0.8 wt%). OC/SA values were restricted to a narrower range than those at the other sites (0.21–0.39 mg m^{-2}), suggesting a riverine input, though δ^{13}C values of sediment OC indicated a marine origin (-21.1‰ to -23.6‰). It is clear from the examination of surface sediments that the fate of particles in the GOM is controlled by transport pathways either from surface to bottom or via advective and lateral transport by currents across the GOM's extensive shelf areas.

However, much of POM's fate also is a function of aggregate dynamics. Coagulation of aggregates occurs via collision and attachment processes between detritus, living cells, fecal pellets, and inorganic materials, the rate of which is controlled by particle size, concentration, and adhesion ("stickiness"; Burd and Jackson 2009). Aggregates exhibit a transient nature in which shear forces of tides and currents, as well grazing, can all serve to reduce their lifetimes (Passow 2016). Marine snow aggregates that settle rapidly from the water column and escape these disaggregation processes are the primary transport mechanism for POM deposition from the surface to bottom waters. It has long been known that bacteria are primary agents of disaggregation via enzymatic hydrolysis (Smith et al. 1992). Settling speeds of marine snow in the GOM have been measured at 10–85 m d^{-1} (mean 34 m d^{-1}) for aggregates ranging in size from 0.5 to 3.5 mm (Diercks and Asper 1997). The wide range in settling speeds of aggregates is consistent with the rather fragile nature of larger aggregates, which can perhaps disaggregate as the result of shear forces during vertical transport. For example, medium-sized aggregates may actually have sunk as larger loosely aggregated particles (Hung et al. 2010). These workers opined that particle sizes are dynamic and that particles form and then disaggregate, perhaps continuously, while settling from the water column.

In the aftermath of the DWH spill in 2010 a substantial fraction of the oil aggregated into marine snow and rapidly sedimented from the water column. Passow et al. (2012) characterized oil-produced aggregates, observing rather large sizes and material rich in mucus. Sinking properties of oil aggregates mimicked those of diatoms, and the composition appeared heterogeneous and made up of a variety of compounds. They theorize that 3 mechanisms were responsible for the formation and cycling of oil aggregates:

1. production of mucous webs through the activities of bacterial oil degraders associated with the floating oil layer;

2. production of oily particulate matter through interactions of oil components with suspended matter and their coagulation; and
3. coagulation of phytoplankton with oil droplets incorporated into aggregates.

It thus appears that oil is an important contributor to marine snow formation and transport. Oil aggregates are anticipated to be important vectors for transfer of fossil carbon released in seeps and catastrophically through disasters such as the DWH spill into surface sediments. Further, these aggregates likely experience rapid and intense cycling, possibly due to enrichment of hydrolytically labile "particle glue," such as emulsifying carbohydrate-rich exopolymeric substances (EPS). In a study of enzyme activities within and outside the deep hydrocarbon plume associated with DWH, Ziervogel and Arnosti (2016) discovered significantly higher peptidase and glucosidase activities, as well as higher rates of polysaccharide hydrolysis. Release of EPS and other products by oil-degrading bacterial communities may in fact "prime" further POM degradation by bacterial communities (cf. Bianchi 2011).

A final consideration on the fate of POM in shallow coastal environments is its photodissolution to DOM, a process that is receiving increasing attention (Kieber et al. 2006; Mayer, Schick, Bianchi, et al. 2009). The mechanism causing this process is sunlight degradation of large organic molecules to smaller molecules that effectively are considered pieces of the large macromolecular particle but by operational definition would be <1 μm, hence DOM (Shank et al. 2011). Obviously, this process is more important for organic-rich sediments because abundant chromophores need to be present to initiate photochemical reactions that either directly photolyze bonds causing production of LMW-OM or promote secondary photoreactions (perhaps formation of reactive oxygen species) that dissociate bonds and degrade POM to DOM. A 7 wt% OC sediment from Florida Bay at particle concentrations <40 mg L^{-1} increased DOC concentrations by 1 mg C L^{-1} under exposure in a solar simulator when compared to dark controls in which no increase in DOC was measured. However, it is possible the photodissolved DOM originates from sunlight-driven desorption of comparatively smaller molecules bound to inorganic particle surfaces. Fluorescence measurements of photodissolved DOM from Florida Bay sediments determined that most of the released material was humic-like and probably soil OM sorbed onto particles and transported to coastal waters (Shank et al. 2011). This photodissolution of OM from the particle phase was consistent with other studies of coastal sediments and algal detritus (Kieber et al. 2006; Mayer et al. 2006).

Biomarkers

Chemical compounds specific to organisms, sources, and degradation pathways are termed "biomarkers" and supply information into the reactivity and cycling of OM. Biomarkers are measured using molecular-level techniques (see the section on chemical composition of OM). In the GOM, OM classification into riverine, marine, and wetland sources has been achieved via a combination of measurements of bulk characteristics (C:N ratio, stable isotope composition) and biomarkers such as carbohydrates or lignin. Lipid profiles of OM (mostly POM) have been used to assess trophic level changes in addition to organismal sources.

Pakulski and Benner (1994) showed that total carbohydrate concentrations in the GOM near the shelf break decreased with depth from 35 μM to 10 μM. Surface concentrations were slightly higher during the day than at night (32.5 μM vs. 24.5 μM). The authors found no significant difference in carbohydrate concentrations between filtered and unfiltered samples, which indicated that most of the carbohydrate was in the DOM. The distributions suggested that polysaccharides dominate in the surface waters, representing 10–34% of the DOC. With increasing depth, monosaccharides become more abundant and also represent more of the DOC. Acid polysaccharide (APS) compounds such as uronic acid may indicate rapid degradation of carbohydrates in POM by heterotrophic communities and release from photosynthetic cyanobacteria (Hung et al. 2003).

Benner and Opsahl (2001) undertook a study of carbohydrate and lignin composition of ultrafiltered DOM (UDOM) in the MRP waters over different seasons. Over the salinity gradient of 0–36 in the MRP, neutral sugars exhibited nonconservative distributions clustered over a salinity range of 15–30, with greater-than-expected concentrations based on mixing of river water and seawater endmember concentrations. The pattern in neutral sugars was seasonally consistent with positive deviations from conservative mixing with respect to DOC concentrations and coincided with high planktonic biomass and high rates of autotrophic and heterotrophic processes. This led the authors to consider that nutrients delivered by the river were fueling fresh production of POM, which was rapidly cycled to DOM. In contrast, lignin-phenol concentrations declined nonconservatively with salinity, but with lower concentrations than expected at higher salinities, the implication was that solar photo-oxidation of lignin was occurring. Key to this latter interpretation was the selective removal of syringyl phenols, which have been shown to be particularly susceptible to photo-oxidation (Opsahl and Benner 1998). This study provided important insight into DOC cycling that focused not just at the bulk level, where production and removal processes could be obscured or perhaps canceled out, but rather demonstrated separate pools of organic matter cycling co-occurring and driven by a range of physicochemical processes. In particular, removal of DOM via the combined effects of microbial and photochemical oxidation constitutes important control on the terrigenous flux of OM to the oceans (Fichot and Benner 2014).

Lipid biomarkers have been measured in POM and SOM to trace the flux of materials from land to sea and from surface waters to surface sediments. Lipid extracts are fractioned into neutral lipids and fatty acids, and combinations of various individual compounds within each grouping provide useful source discriminations. For example, the terrestrial to aquatic fatty acid ratio (TAR_{FA}) is the sum of long-chain even-numbered saturated fatty acids, which indicates land plants, over the sum of short-chain even-numbered fatty acids, which indicates planktonic OM (Meyers 1997; Equation 6.6):

$$TAR_{FA} = \left[\frac{(C_{24} + C_{26} + C_{28})}{(C_{12} + C_{14} + C_{16})} \right] \tag{6.6}$$

Waterson and Canuel (2008) examined fatty acid and sterol distributions in SOM collected from the MR, Barataria Bay estuary, a coastal marsh, and across the shelf-slope transition to Mississippi Canyon. Mean TAR_{FA} ratios decreased from the river (0.6) to the shelf (0.2), slope (0.3), and canyon (0.2). TAR_{FA} in the estuary was lower than in the river (0.4), but the ratio in the marsh SOM was higher (0.9). These results are consistent with $\delta^{13}C$ and C:N results showing that terrestrial OM is deposited throughout the shelf-to-slope region into the canyon region. Further evaluation of fatty acid and sterol distributions were conducted to estimate relative fractions of autochthonous and allochthonous sources of SOM, comparing these 2 lipid-based approaches to an approach based on $\delta^{13}C$ values (Table 6.7). This approach allowed them to conclude that overall autochthonous sources of OM dominated the surface sediments away from the river mouth and marsh regions where vascular plant material dominated. The importance of autochthonous sources increased offshore, and these sources were consistent with higher chlorophyll pigment fluorescence measured in the estuary and shelf regions. While results of these 3 modeling approaches were overall corroborative, distinct differences were observed. Another important finding of this modeling exercise was the variability in estimates using these techniques, leading the authors to

Table 6.7. Mass balance and mixing model results estimating the mean proportion of autochthonous and allochthonous sources of SOM in the northern Gulf of Mexico

Region	$\delta^{13}C$[a]		Fatty acids[b]		Sterols[c]	
	% Autochthonous	% Allochthonous	% Autochthonous	% Allochthonous	% Autochthonous	% Allochthonous
River	20	80	62	39	53	48
Marsh	51	49	50	50	28	72
Estuary	60	40	67	33	62	38
Shelf	60	40	78	23	67	33
Slope	67	33	71	29	68	32
Canyon	71	29	83	17	67	34

Source: After Waterson and Canuel (2008).

[a] Autochthonous and allochthonous percentages calculated by isotopic mass balance. In the marsh region, % autochthonous refers to the percentage of C4 marsh plants.

[b] Autochthonous is defined as the even- and odd-numbered short-chain fatty acids, while allochthonous is defined as the even- and odd-numbered long-chain fatty acids, each normalized to the total saturated fatty acids.

[c] Autochthonous was the fraction of sterols that identify specific phytoplankton groups (e.g., diatoms) and zooplankton, whereas allochthonous was the fraction of sterols specific to higher plants and freshwater algae.

advocate for a multitracer approach to overcome the pros and cons of each individual approach.

Hydrodynamic sorting was previously described as an important control on the composition of POM and SOM, influencing the amount and nature of organic materials exported from rivers into the GOM. The paradigm here is that larger coarse material is enriched in terrigenous OM (modern woody-plant fragments) and is deposited inshore along with silt and sand (Bianchi et al. 2002; Bianchi, Wysocki, et al. 2007; Bianchi, Galler, et al. 2007; Gordon and Goñi 2003, 2004; Bianchi, Grace, et al. 2014). Smaller, finer material is deposited farther offshore and contains lower amounts of terrigenous OM, which is itself highly degraded and dominated by soils eroded from northwestern grasslands of the MR watershed. This latter material is hydrodynamically sorted in the BNL by waves, currents, and storms, which selectively transport this material farther offshore. Wakeham et al. (2009) conducted density fractionations on 3 sediment samples from the Waterson and Canuel (2008) study and examined OM provenance, age, and degradation state with an aim to understand the potential relationships between density and size in terms of OM properties deposited on continental shelves. They found a relationship between size and density within the context of hydrodynamic sorting of particles that explained results from measurements of stable and radiocarbon isotopes, lignin, and fatty acid compositions of several density fractions compared to bulk properties. For example, the lowest-density coarse

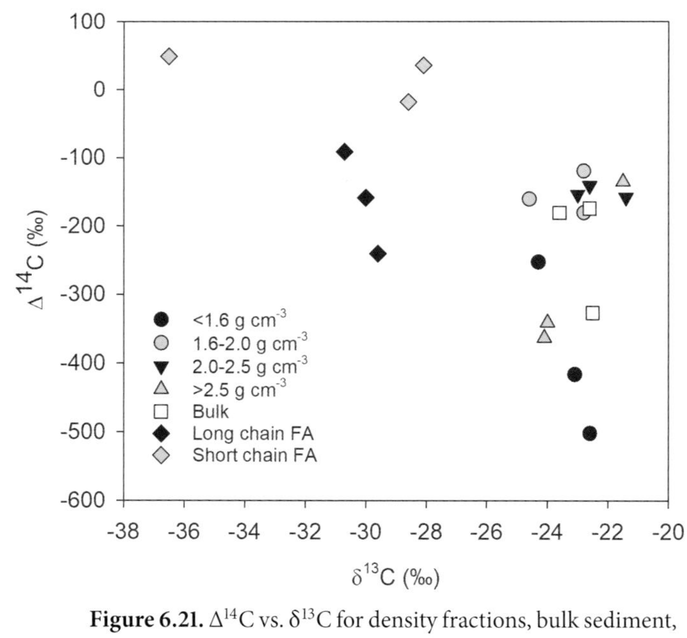

Figure 6.21. $\Delta^{14}C$ vs. $\delta^{13}C$ for density fractions, bulk sediment, and fatty acid extracts (adapted from Wakeham et al. 2009).

fraction found had the highest lignin concentration with the lowest decay index, suggestive of woody debris rapidly deposited on the inner shelf. In contrast, the highest-density fine fraction (clays) had the highest lignin decay index due to increased transport potential farther offshore allowing for greater decay.

The lowest-density fractions (<1.6 g cm^{-3}) invariably were enriched in terrestrial OM, but by mass, the so-called mesodensity fraction (1.6–2.5 g m^{-3}) contained the most OM. While containing the highest percentage of OC, the <1.6 g cm^{-3} fraction also contained the C:N ratios and the highest proportional amount of the vascular plant biomarker, lignin. Within lignin-phenol distributions, it was observed that the ratio of vanillic acid to vanillin (Ad/Al_v) increased with increasing density. Larger values of this ratio indicate oxidative degradation and suggested that the SOM in the highest-density fractions was most degraded. Radiocarbon evidence of each fraction supported this notion, and a cross-plot of $\delta^{13}C$ and $\Delta^{14}C$ values revealed a gradient showing enriched $\delta^{13}C$ values and depleted $\Delta^{14}C$ values in the MR and GOM shelf sediments (Fig. 6.21).

These results were corroborated by radiocarbon measurements on composites of long- and short-chain fatty acid (FA) composites used to compute TAR_{FA}. The long-chain FA homologs are thought to derive primarily from the waxy coatings of vascular land plants, whereas the short-chain FA homologs derive predominantly from marine plankton. Enrichment of ^{14}C in the short-chain FA indicated a largely modern and recently produced marine component compared to an older, pre-aged (aged on land) terrestrial component. Together these results suggested that the highest-density fraction of SOM is primarily composed of OC-poor material, which is older and more degraded than the lowest-density fraction. Generally, it was observed that low-density coastal sediments contain higher proportions of OM (largely from woody debris) than did high-density coastal sediments enriched in low-OM mineral-rich material deposited farther offshore. The results from this study based on density indicated that OC-rich C3 plant material settles out in the inner shelf, consistent with previous measurements of bulk POM properties in suspended material and in sandy surface sediments (Onstad et al. 2000; Bianchi et al. 2002; Bianchi, Wysocki, et al. 2007; Bianchi, Galler, et al. 2007; Bianchi et al. 2011; Bianchi, Grace, et al. 2014).

Photodissolution of sediments has important implications for the fate of organic biomarkers deposited in shallow marine coastal environments, especially photoreactive lignin and phytodetritus. Like concentrations of TOC, lignin concentrations decreased significantly after exposure of MR and coastal GOM sediments to sunlight (Mayer, Schick, Bianchi, et al. 2009). The solid-phase photodegradation reactions mimicked those occurring for dissolved lignin in which syringyl groups were relatively more photolabile than cinnamyl groups (Opsahl and Benner 1998). Furthermore, solid-phase lignin acid:aldehyde ratios increased with sunlight exposure, indicating that surface-sediment lignin, like dissolved lignin, is oxidized by sunlight. Phytodetritus also was quite susceptible to photodegradation and modification of bulk materials but also at the molecular level (Mayer, Schick, Hardy, et al. 2009). Lipid peroxides were a key result of phytodetritus photosensitization reactions (especially in oxygenated solutions), and while this process would alter lipid-based biomarkers of precursory material, lipid peroxides also can promote cross-linking and condensation reactions that bind other lipids as well as amino acid dimers (Mayer, Schick, Hardy, et al. 2009). DOM desorbed from soil particles is roughly 50% bioavailable (Butman et al. 2007), which suggests this pathway is likely an important secondary control on the fate of POM in coastal waters, though clearly more work needs to be done to constrain the rates of desorption as well as the photochemical and microbial fate of the desorbed material (Shank et al. 2011; Liu and Shank 2015).

Extracellular Enzymes

We include in this section some recent work on POM, although not strictly a biomarker—especially on aggregates formed during transport through the "twilight zone" (marine snow) (Alldredge 1979; Lee 2004)—using extracellular enzyme activity (EEA) assays. EEA assays target specific biochemical classes of OM and can provide information on its reactivity, as aggregation shapes POM degradation by providing substrate for microorganisms to colonize (Biddanda and Pomeroy 1988; Arnosti 2011). Common enzymes used to assess the quality of OM are glucosidases, peptidases, and lipases, which serve as macromolecular degradation markers for carbohydrates, proteins, and lipids, respectively. Polysaccharides are a dominant class of carbohydrates in seawater OM, and thus their reactivity is important information about OM reactivity generally (Benner 2002). Steen et al. (2012) examined hydrolysis rates in the northern GOM for several polysaccharides along with bacterial production and glucose mineralization. They found wide variation in patterns of potential hydrolysis rates in seawater consistent with the interpretation that heterotrophic microbial communities are restricted in their ability to access specific polysaccharides. Total maximum hydrolysis rates (sum of laminarin, xylan, chondroitin, fucoidan, pullulan, and arabinogalactan) were roughly 100 nm L^{-1} hr^{-1} near the MR and similar between surface and 35 m water depth. Offshore rates were about 75 nm L^{-1} hr^{-1} in surface waters and decreased dramatically to about 15 nm L^{-1} hr^{-1} with depth (to 905 m). Among specific polysaccharides, laminarin and xylan hydrolysis rates dominated in shore surface waters near the MR, whereas chondroitin dominated pelagic waters. The rates were smaller than decreases in bacterial production and glucose metabolism; the authors interpret this to indicate that microbial communities do retain their potential to degrade polysaccharides even when their activities are low. However, the microbial communities were able to increase enzyme activity upon delivery of fresh OM, which may mean that macromolecular substrates promote deepwater microbial metabolism.

Enzyme studies were conducted extensively after the DWH disaster in 2010 after observation of massive amounts of marine snow formation (Passow 2016). The larger amounts of marine snow translated into greatly enhanced degradation of POM, as evidenced by bacterial enzyme activities. Hydrolysis rates of carbohydrates and peptides were significantly higher (300% and 50% higher,

Table 6.8. Enzymatic hydrolysis rates (pmol monomer cell-1 h-1) after a 1-day incubation

	Pullulan	Laminarin	Xylan	Fucoidan	Arabinogalactan	Chondroitin
Ambient water	2.1	1.2	4.7	4.5	0	3.5
Oil water	3.0	4.8	4.8	2.4	0.13	0.19
Oil aggregates	5.6	7.6	0.8	0	0.04	0.57

Source: From Arnosti et al. (2016).

Note: The results show the effects of oil aggregate presence on activation of certain enzyme systems.

respectively) in deepwater hydrocarbon plume samples than in nonplume samples (Ziervogel and Arnosti 2016). Hydrocarbons in the plume were confirmed with fluorescence measurements. Higher bacterial abundances in the deepwater plume samples confirmed the higher bacterial activity measured by the hydrolytic enzyme rate assays. Examination of discrete polysaccharide enzyme activities supported these observations. Arnosti et al. (2016) compared hydrolysis rates of a suite of monomers, and after 1 day of incubation found discrete responses with respect to specific substrates. Laminarin, pullulan, and arabinogalactan were "activated" first in oil water and oil aggregates compared to ambient seawater (Table 6.8). Other monomers, such as fucoidin and chondroitin, exhibited lower hydrolysis rates in oil treatments than in ambient seawater. Thus, select hydrolytic enzyme activities are enhanced on POM aggregates, and despite the presence of oil, the pattern was consistent with prior work on enzyme activity on OM in the Gulf (Steen et al. 2012).

Ziervogel, Osburn, et al. (2016) recently documented a close linkage between POM and bacterial activities by combining enzyme assays measured on natural heterotrophic communities with BEPOM fluorescence analysis, which constituted a substantial fraction of the POM pool. High glucosidase and peptidase activities were observed where BEPOM fluorescence was enriched in protein-like signatures, and chlorophyll-*a* concentrations also were high. However, fresh POM was not the only biologically reactive material found in subsurface waters. In a study of resuspended material in a BNL formed after Hurricane Isaac, possibly the result of enhanced discharge of particle-rich MR water, Ziervogel, Osburn, et al. (2016) discovered that the resuspended sediments rapidly coagulate, forming macroaggregates. Most striking from this study was the reactivity of this material. Major enhancements of bacterial activity in terms of enzymatic rates, abundance, and production were measured in the BNL and compared favorably to roller bottle experiments. These results indicated that storms could periodically transport sedimented POM into the overlying water, where it is rapidly degraded.

As described earlier in the section on the fate of POM, oil is an important vector for aggregate formation and distribution. Studies of microbial enzyme activities demonstrate the role of microorganisms in the formation and fate of marine snow. Experiments over 21 days on oil released from the DWH spill (collected from the surface slick) mixed with seawater in rolling glass (roller) bottles showed enhanced enzymatic degradation rates after a lag period in which bacteria colonized the oil-formed aggregates (Ziervogel et al. 2012). An increase in lipase activity was associated with an increase in bacterial abundance in oil + seawater compared to controls. Moreover, the study found rapid conversion of oil into highly fluorescent water-soluble DOM, and the authors suggested this process to be important for altering carbon-cycling dynamics via creation of "hot spots" of intense biogeochemical cycling in the mucus-rich aggregates surrounding oil. Indeed, there appeared to be persistence of oil-derived material in the marine DOM of the Gulf at least 2 years after the DWH spill (Bianchi et al. 2013). Thus, the DWH spill resulted in a cascade of particle-related, microbially mediated reactions that appear to have altered the microbial carbon pump in the Gulf of Mexico.

Summary and Future Directions

In this chapter we provide an overview of dissolved, colloidal, and particulate OM (DOM, COM, and POM) in the Gulf from the inner shelf across the continental margin to the basin. Away from substantial riverine inputs (e.g., the MR) or estuarine inputs, the primary source of OM to the Gulf is primary production in surface waters. This material is biochemically labile, especially the HMW-DOM components, meaning that it is generally rapidly consumed by bacteria within the water column, supporting their respiration. DOM exhibits consistent cross-shelf concentrations and is generally low in the open Gulf and deep waters. Similarly for POM, cross-shelf transport and deposition-redeposition cycles transform and modify organic particles. However, direct export of POM from the surface ocean to the mesopelagic is a key part of the biological pump that captures CO_2 and sequesters it as OM at depth, until it either is deposited on the seafloor or respired to CO_2. Thus, surface-ocean production is intimately linked with SSC cycling. Major events such as tropical storms and hurricanes and hydrocarbon releases exert specific and unique pressures that, in their own way, alter and modulate carbon cycling within the GOM.

As has been mentioned earlier, a paucity of data on the inventory of OM in the GOM currently exists and is a clear future research need. A recent coastal carbon synthesis activity coordinated by the Ocean Carbon and Biogeochemistry network produced estimates of coastal carbon fluxes in which atmospheric fluxes are poorly known. For example, the annual loading of OC in rainfall to the open Gulf is not known and could be strongly influenced by anthropogenic and climatic effects, particularly extreme events such as terrestrial squalls, and tropical storms as seen in the GOM and other coastal environments (e.g., Bianchi et al. 2011; Mitra et al. 2013). Indeed, tropical storms might be key vectors for transporting terrestrial OM into the coastal ocean and modulating its distribution and ultimately its deposition (Allison et al. 2010). The delivery of terrestrial OM to coastal waters of the GOM and its ultimate fate should be investigated with respect to rates of decomposition that contribute to oxygen deficits and the widespread hypoxia in the northern GOM (Green et al. 2006; Dagg et al. 2008; Bianchi et al. 2008; Bianchi et al. 2010).

The other important aspect is the heterogeneity of the molecular size, chemical composition, biological and chemical reactivity, ecological role, and environmental fate of the DOM pool during transport across the land-ocean interface and from the upper to lower water column. In addition, the transformation processes among dissolved/colloidal/particulate phases and between organic and inorganic carbon remain poorly constrained and require further investigation. However, elucidating these reaction pathways remains difficult due to the overall low DOC concentration and high salt content in seawater, presenting challenges to separate the bulk DOC into different size fractions for further detailed characterization and incubation experiments. The small size and dilute concentrations of DOC in the ocean likely influence its reactivity (Arrieta et al. 2015).

Recent focus on the role of "blue carbon" loss from vegetated coastal wetlands is important in the context of DOM and POM cycling in the GOM. Widespread subsidence caused by modification of the landscape has rendered much of the Louisiana coast subject to some of the highest rates of relative sea-level rise in North America (Bianchi et al. 1999). While the export of OM from the MARS system, especially from the MR itself, has been well studied and is fairly well known (e.g., Tian et al. 2015), contributions of POM and DOM from eroding coastal marshes to the GOM shelf ecosystem is not well known but could be important (Ren et al. 2016). The fate of this material countermands the highly valuable ecosystem service that salt marshes provide for long-term CO_2 storage in their soils and sediments. What is also unknown is how that erosive loss might itself be mitigated by climate change in the form of mangrove encroachment. These uncertainties create critical knowledge gaps in our general understanding of the GOM ecosystem but also create specific unknowns in terms of rates of OM fluxes and their ultimate fates. As this chapter demonstrates, combinations of organic geochemistry, observations, and modeling, including remote sensing, will answer much of this uncertainty.

Operation of the Gulf's biological pump remains unclear. While very recent work has attempted to synthesize OM production and export rates among measured sediment trap archives, radioactive tracers, and satellite remote-sensing imagery, this work was constrained to the northeastern GOM at 3 sites near the DWH spill site (Maiti et al. 2016). Transport of POM into the interior of the GOM and flux to the deep waters remains poorly constrained, and satellite-based models of POM fluxes perhaps require tuning with respect to a specific ocean biome. Given the emerging understanding that hydrocarbons facilitate formation of marine snow with enough ballast to result in fast sinking rates suggests that synergy between surface productivity and oil released from seeps—and especially catastrophically—may result in "extreme rain" events of POC deposition to the surface sediments. The effects of oil-aggregate deposition for GOM carbon cycling requires further investigation.

Acknowledgments

Karl Kaiser and Robert G. M. Spencer are thanked for their helpful reviews of this manuscript. Christopher L. Osburn acknowledges support from NASA award NNX14A043G.

References

Aiken, G. R., H. Hsu-Kim, and J. N. Ryan. 2011. Influence of dissolved organic matter on the environmental fate of metals, nanoparticles, and colloids. *Environmental Science & Technology* 45:3196–3201.

Alldredge, A. L. 1979. The chemical composition of macroscopic aggregates in two neretic seas. *Limnology and Oceanography* 24:855–866.

Alldredge, A. L., and M. W. Silver. 1988. Characteristics, dynamics and significance of marine snow. *Progress in Oceanography* 20:41–82.

Aller, R. C. 1998. Mobile deltaic and continental shelf muds as suboxic, fluidized bed reactors. *Marine Chemistry* 61:143–155.

Allison, M. A., T. S. Bianchi, B. A. McKee, and T. P. Sampere. 2007. Carbon burial on river-dominated continental shelves: Impact of historical changes in sediment loading adjacent to the Mississippi River. *Geophysical Research Letters* 34:L01606. doi:10.1029/2006GL028362.

Allison, M. A., T. M. Dellapenna, E. S. Gordon, S. Mitra, and S. T. Petsch. 2010. Impact of Hurricane Katrina (2005) on shelf organic carbon burial and deltaic evolution. *Geophysical Research Letters* 37:L21605. doi:10.1029/2010GL044547.

Allison, M. A., G. C. Kineke, E. S. Gordon, and M. A. Goñi. 2000. Development and reworking of a seasonal flood deposit on the inner continental shelf off the Atchafalaya River. *Continental Shelf Research* 20:2267–2294.

Amon, R. M. W., and R. Benner. 1996. Bacterial utilization of different size classes of dissolved organic matter. *Limnology and Oceanography* 41:41–51.

Amon, R. M. W., A. J. Rinehart, S. Duan, P. Louchouarn, A. Prokushkin, G. Guggenberger, D. Bauch, et al. 2012. Dissolved organic matter sources in large Arctic rivers. *Geochimica et Cosmochimica Acta* 94:217–237.

Arnarson, T. S., and R. G. Keil. 2001. Organic-mineral interactions in marine sediments studied using density fractionation and X-ray photoelectron spectroscopy. *Organic Geochemistry* 32:1401–1415.

Arístegui, J., C. M. Duarte, S. Agustí, M. Doval, X. A. Álvarez-Salgado, and D. A. Hansell. 2002. Dissolved organic carbon support of respiration in the dark ocean. *Science* 298:1967.

Arnosti, C. 2011. Microbial extracellular enzymes and the marine carbon cycle. *Annual Review of Marine Science* 3:401–425.

Arnosti, C., K. Ziervogel, T. Yang, and A. Teske. 2016. Oil-derived marine aggregates—hot spots of polysaccharide degradation by specialized bacterial communities. *Deep Sea Research Part II: Tropical Studies in Oceanography* 129:179–186.

Arrieta, J. M., E. Mayol, R. L. Hansman, G. J. Herndl, T. Dittmar, and C. M. Duarte. 2015. Dilution limits dissolved organic carbon utilization in the deep ocean. *Science* 348:331–333.

Ascough, P., G. Cook, and A. Dugmore. 2005. Methodological approaches to determining the marine radiocarbon reservoir effect. *Progress in Physical Geography* 29:532–547.

Aspinall, G. O. 1983. Classification of polysaccharides. Chap. 1 in G. O. Aspinall (ed.), *The Polysaccharides*. New York: Academic Press. 228 pp.

Azam, F., T. Fenchel, J. G. Field, J. S. Gray, L. A. Meyer-Reil, and F. Thingstad. 1983. The ecological role of water-column microbes in the sea. *Marine Ecology Progress Series* 10:257–263.

Baalousha, M., B. Stolpe, and J. R. Lead. 2011. Flow field-flow fractionation for the analysis and characterization of natural colloids and manufactured nanoparticles in environmental systems: A critical review. *Journal of Chromatography A* 1218:4078–4103.

Baines, S. B., and M. L. Pace. 1991. The production of dissolved organic matter by phytoplankton and its importance to bacteria: Patterns across marine and freshwater systems. *Limnology and Oceanography* 36:1078–1090.

Baskaran, M., and P. H. Santschi. 1993. The role of particles and colloids in the transport of radionuclides in coastal environments of Texas. *Marine Chemistry* 43:95–114.

Battin, T. J., L. A. Kaplan, S. Findlay, C. S. Hopkinson, E. Marti, A. I. Packman, J. D. Newbold, and F. Sabater. 2008. *Nature Geoscience* 1:95–100.

Bauer, J. E. 2002. Carbon isotopic composition of DOM. Chap. 8 in D. A. Hansell and C. A. Carlson (eds.), *Biogeochemistry of Marine Dissolved Organic Matter*. San Diego: Academic Press. 774 pp.

Bauer, J. E., and T. S. Bianchi. 2011. Dissolved organic carbon cycling and transformation. Chap. 2 in E. Wolanski and D. S. McLusky (eds.), *Treatise on Estuarine and Coastal Science*, vol. 5. Waltham, MA: Academic Press. 362 pp.

Bauer, J. E., W. J. Cai, P. A. Raymond, T. S. Bianchi, C. S. Hopkinson, and P. A. Regnier. 2013. The changing carbon cycle of the coastal ocean. *Nature* 504:61–70.

Bauer, J. E., P. M. Williams, and E. R. M. Druffel. 1992. ^{14}C activity of dissolved organic carbon fractions in the north-central Pacific and Sargasso Sea. *Nature* 357:667–670.

Beaupré, S. R. 2015. The carbon isotopic composition of marine DOC. Chap. 6 in D. A. Hansell and C. A. Carlson (eds.), *Biogeochemistry of Marine Dissolved Organic Matter*, 2nd ed. London: Academic Press. 712 pp.

Benner, R. 2002. Chemical composition and reactivity. Chap. 3 in D. A. Hansell and C. A. Carlson (eds.), *Biogeochemistry of Marine Dissolved Organic Matter*. San Diego: Academic Press. 774 pp.

Benner, R., and R. M. W. Amon. 2015. The size-reactivity continuum of major bioelements in the ocean. *Annual Review of Marine Science* 7:185–205.

Benner, R., and B. Biddanda. 1998. Photochemical transformations of surface and deep marine dissolved organic matter: Effects on bacterial growth. *Limnology and Oceanography* 43:1373–1378.

Benner, R., B. Biddanda, B. Black, and M. McCarthy. 1997. Abundance, size distribution, and stable carbon and nitrogen isotopic compositions of marine organic matter isolated by tangential-flow ultrafiltration. *Marine Chemistry* 57:243–263.

Benner, R., and G. J. Herndl. 2011. Bacterially derived dissolved organic matter in the microbial carbon pump. Chap. 2 in N. Jiao, F. Azam, and S. Sanders (eds.), *Microbial Carbon Pump in the Ocean*. Washington, DC: American Association for the Advancement of Science. 72 pp.

Benner, R., and K. Kaiser. 2003. Abundance of amino sugars and peptidoglycan in marine particulate and dissolved organic matter. *Limnology and Oceanography* 48:118–128.

———. 2011. Biological and photochemical transformations of amino acids and lignin phenols in riverine dissolved organic matter. *Biogeochemistry* 102:209–222.

Benner, R., and S. Opsahl. 2001. Molecular indicators of the sources and transformations of dissolved organic matter in the Mississippi River plume. *Organic Geochemistry* 32:597–611.

Benner, R., J. D. Pakulski, M. McCarthy, J. I. Hedges, and P. G. Hatcher. 1992. Bulk chemical characteristics of dissolved organic matter in the ocean. *Science* 255:1561–1564.

Bergamaschi, B. A., J. S. Walters, and J. I. Hedges. 1999. Distributions of uronic acids and O-methyl sugars in sinking and sedimentary particles in two coastal marine environments. *Geochimica et Cosmochimica Acta* 63:413–425.

Berman-Frank, I., and Z. Dubinsky. 1999. Balanced growth in aquatic plants: Myth or reality? Phytoplankton use the imbalance between carbon assimilation and biomass production to their strategic advantage. *BioScience* 49:29–37.

Bernardes, M. C., L. A. Martinelli, A. V. Krusche, J. Gudeman, M. Moreira, R. L. Victoria, J. P. H. B. Ometto, et al. 2004. Riverine organic matter composition as a function of land use changes, southwest Amazon. *Ecological Applications* 14:S263–S279.

Bianchi, T. S. 2011. The role of terrestrially-derived organic carbon in the coastal ocean: A changing paradigm and the priming effect. *Proceedings of the National Academy of Sciences* 108:19473–19481.

Bianchi, T. S., M. A. Allison, and W. J. Cai (eds.). 2014. *Biogeochemical Dynamics at Major River-Coastal Interfaces: Linkages with Global Change*. Cambridge: Cambridge University Press. 658 pp.

Bianchi, T. S., M. A. Allison, E. A. Canuel, D. R. Corbett, B. A. McKee, T. P. Sampere, S. G. Wakeham, and E. Waterson. 2006. Rapid export of organic matter to the Mississippi Canyon. *Eos, Transactions, American Geophysical Union* 87:565–573.

Bianchi, T. S., and J. E. Bauer. 2011. Particulate organic carbon cycling and transformation. Chap. 3 in E. Wolanski and D. S. McLusky (eds.), *Treatise on Estuarine and Coastal Science*, vol. 5. Waltham, MA: Academic Press. 362 pp.

Bianchi, T. S., and E. A. Canuel. 2011. *Chemical Biomarkers in Aquatic Ecosystems*. Princeton, NJ: Princeton University Press. 392 pp.

Bianchi, T. S., R. L. Cook, E. M. Perdue, P. E. Kolic, N. Green, Y. Zhang, R. W. Smith, et al. 2011. *Marine Environmental Research* 72:248–257.

Bianchi, T. S., S. F. Dimarco, M. A. Allison, P. Chapman, J. H. Cowan, R. D. Hetland, J. W. Morse, and G. Rowe. 2008. Controlling hypoxia on the US Louisiana shelf: Beyond the nutrient-centric view. *Eos, Transactions, American Geophysical Union* 89:236–237.

Bianchi, T. S., S. F. DiMarco, J. H. Cowan Jr., R. D. Hetland, P. Chapman, J. W. Day, and M. A. Allison. 2010. The science of hypoxia in the northern Gulf of Mexico: A review. *Science of the Total Environment* 408:1471–1484.

Bianchi, T. S., S. F. DiMarco, R. W. Smith, and K. M. Schreiner. 2009. A gradient of dissolved organic carbon and lignin from Terrebonne-Timbalier Bay estuary to the Louisiana shelf (USA). *Marine Chemistry* 117:32–41.

Bianchi, T. S., T. Filley, K. Dria, and P. G. Hatcher. 2004. Temporal variability in sources of dissolved organic carbon in the lower Mississippi River. *Geochimica et Cosmochimica Acta* 68:959–967.

Bianchi, T. S., J. J. Galler, and M. A. Allison. 2007. Hydrodynamic sorting and transport of terrestrially-derived organic carbon in sediments of the Mississippi and Atchafalaya Rivers. *Estuarine, Coastal and Shelf Science* 73:211–222.

Bianchi, T. S., F. Garcia-Tigreros, S. A. Yvon-Lewis, M. Shields, H. J. Mills, D. Butman, C. Osburn, et al. 2013. Enhanced transfer of terrestrially-derived carbon to the atmosphere in a flooding event. *Geophysical Research Letters* 40:116–122.

Bianchi, T.S., B. L. Grace, K. R. Carman, and I. Maulana. 2014. Amino acid cycling in the Mississippi River plume and effects from the passage of Hurricane Isadore and Lili. *Journal of Marine Systems* 136:10–21.

Bianchi, T. S., C. Lambert, P. H. Santschi, M. Baskaran, and L. Guo. 1995. Plant pigments as biomarkers of high-molecular-weight dissolved organic carbon. *Limnology and Oceanography* 40:422–428.

Bianchi, T. S., C. D. Lambert, P. H. Santschi, and L. Guo. 1997. Sources and transport of land-derived particulate and dissolved organic matter in the Gulf of Mexico (Texas shelf/slope): The use of lignin-phenols and loliolides as biomarkers. *Organic Geochemistry* 27:65–78.

Bianchi, T. S., S. Mitra, and B. A. McKee. 2002. Sources of terrestrially-derived organic carbon in lower Mississippi River and Louisiana shelf sediments: Implications for differential sedimentation and transport at the coastal margin. *Marine Chemistry* 77:211–223.

Bianchi, T. S., C. Osburn, M. Shields, S. Yvon-Lewis, J. Young, L. Guo, and Z. Zhou. 2014. Deepwater Horizon oil in Gulf of Mexico waters after 2 years: Transformation into the dissolved organic matter pool. *Environmental Science & Technology* 48:9288–9297.

Bianchi, T. S., J. R. Pennock, and R. R. Twilley (eds.). 1999. *Biogeochemistry of Gulf of Mexico Estuaries*. New York: John Wiley and Sons. 448 pp.

Bianchi, T. S., L. A. Wysocki, M. Stewart, T. R. Filley, and B. A. McKee. 2007. Temporal variability in terrestrially-derived sources of particulate organic carbon in the lower Mississippi River and its upper tributaries. *Geochimica et Cosmochimica Acta* 71(18):4425–4437.

Biddanda, B. A., and L. R. Pomeroy. 1988. Microbial aggregation and degradation of phytoplankton-derived detritus in seawater. I: Microbial succession. *Marine Ecology Progress Series* 42:79–88.

Biggs, D. C. 1992. Nutrients, plankton, and productivity in a warm-core ring in the western Gulf of Mexico. *Journal of Geophysical Research: Oceans* 97:2143–2154.

Biggs, D. C., C. Hu, and F. E. Müller-Karger. 2008. Remotely sensed sea-surface chlorophyll and POC flux at deep Gulf of Mexico benthos sampling stations. *Deep Sea Research Part II: Tropical Studies in Oceanography* 55:2555–2562.

Brown, S. B., J. D. Houghton, and G. A. F. Hendry. 1991. Chlorophyll breakdown. Pp. 465–489 in H. Scheer (ed.), *Chlorophylls*. Boca Raton, FL: CRC Press.

Brym, A., H. W. Paerl, M. T. Montgomery, L. T. Handsel, K. Ziervogel, and C. L. Osburn. 2014. Optical and chemical characterization of base-extracted particulate organic matter in coastal marine environments. *Marine Chemistry* 162:96–113.

Buesseler, K. O., J. E. Bauer, R. F. Chen, T. I. Eglinton, O. Gustafsson, W. Landing, K. Mopper, et al. 1996. An intercomparison of cross-flow filtration techniques used for sampling marine colloids: Overview and organic carbon results. *Marine Chemistry* 55:1–31.

Burd, A. B., and G. A. Jackson. 2009. Particle aggregation. *Annual Review of Marine Science* 1:65–90.

Burdige, D. J., and C. S. Martens. 1988. Biogeochemical cycling in an organic-rich coastal marine basin. 10: The role of amino acids in sedimentary carbon and nitrogen cycling. *Geochimica et Cosmochimica Acta* 52:1571–1584.

Butman, D., P. A. Raymond, N. H. Oh, and K. Mull. 2007. Quantity, ^{14}C age and lability of desorbed soil organic carbon in fresh water and seawater. *Organic Geochemistry* 38:1547–1557.

Cai, Y., and L. Guo. 2009. Abundance and variation of colloidal organic phosphorus in riverine, estuarine, and coastal waters in the northern Gulf of Mexico. *Limnology and Oceanography* 54:1393–1402.

Cai, Y., L. Guo, X. Wang, and G. R. Aiken. 2015. Abundance, stable isotopic composition, and export fluxes of DOC, POC, and DIC from the lower Mississippi River during 2006–2008. *Journal of Geophysical Research: Biogeosciences* 120:2273–2288.

Cai, Y., L. Guo, X. Wang, A. K. Mojzis, and D. G. Redalje. 2012. The source and distribution of dissolved and particulate organic matter in the Bay of St. Louis, northern Gulf of Mexico. *Estuarine, Coastal and Shelf Science* 96:96–104.

Canuel, E. A., and C. S. Martens. 1996. Reactivity of recently deposited organic matter: Degradation of lipid compounds near the sediment-water interface. *Geochimica et Cosmochimica Acta* 60:1793–1806.

Carlson, C. A., H. W. Ducklow, and A. F. Michaels. 1994. Annual flux of dissolved organic carbon from the euphotic zone in the northwest Sargasso Sea. *Nature* 371:405–408.

Carlson, C. A., S. J. Giovannoni, D. A. Hansell, S. J. Goldberg, R. Parsons, M. P. Otero, K. Vergin, and B. R. Wheeler. 2002. Effect of nutrient amendments on bacterioplankton production, community structure, and DOC utilization in the northwestern Sargasso Sea. *Aquatic Microbial Ecology* 30:19–36.

Carlson, C.A., S. J. Giovannoni, D. A. Hansell, S. J. Goldberg, R. Parsons, and K. Vergin. 2004. Interactions among dissolved organic carbon, microbial processes, and community structure in the mesopelagic zone of the northwestern Sargasso Sea. *Limnology and Oceanography* 49:1073–1083.

Carlson, C. A., and D. A. Hansell. 2015. DOM sources, sinks, reactivity, and budgets. Chap. 3 in D. A. Hansell and C. A. Carlson (eds.), *Biogeochemistry of Marine Dissolved Organic Matter*, 2nd ed. London: Academic Press.

Carlson, C. A., D. A. Hansell, N. B. Nelson, D. A. Siegel, W. M. Smethie, S. Khatiwala, M. M. Meyers, and E. Halewood. 2010. Dissolved organic carbon export and subsequent remineralization in the mesopelagic and bathypelagic realms of the North Atlantic basin. *Deep Sea Research Part II: Tropical Studies in Oceanography* 57:1433–1445.

Carlson, C. A., R. Morris, R. Parsons, A. H. Treusch, S. J. Giovannoni, and K. Vergin. 2009. Seasonal dynamics of SAR11 populations in the euphotic and mesopelagic zones of the northwestern Sargasso Sea. *International Society for Microbial Ecology* 3:283–295.

Cawley, K. M., Y. Yamashita, N. Maie, and R. Jaffé. 2014. Using optical properties to quantify fringe mangrove inputs to the dissolved organic matter (DOM) pool in a subtropical estuary. *Estuaries and Coasts* 37:399–410.

Chavez, F. P., M. Messié, and J. T. Pennington. 2011. Marine primary production in relation to climate variability and change. *Annual Review of Marine Science* 3:227–260.

Chen, M., S. Kim, J. E. Park, H. J. Jung, and J. Hur. 2016. Structural and compositional changes of dissolved organic matter upon solid-phase extraction tracked by multiple analytical tools. *Analytical and Bioanalytical Chemistry* 408:6249–6258.

Chen, R. F., P. Bissett, P. Coble, R. Conmy, G. B. Gardner, M. A. Moran, X. Wang, M. L. Wells, P. Whelan, and R. G. Zepp. 2004. Chromophoric dissolved organic matter (CDOM) source characterization in the Louisiana Bight. *Marine Chemistry* 89:257–272.

Chen, R. F., and G. B. Gardner. 2004. High-resolution measurements of chromophoric dissolved organic matter in the Mississippi and Atchafalaya River plume regions. *Marine Chemistry* 89:103–125.

Chen, X., S. E. Lohrenz, and D. A. Wiesenburg. 2000. Distribution and controlling mechanisms of primary production on the Louisiana-Texas continental shelf. *Journal of Marine Systems* 25:179–207.

Cherrier, J., and J. E. Bauer. 2004. Bacterial utilization of transient plankton-derived dissolved organic carbon and nitrogen inputs in surface ocean waters. *Aquatic Microbial Ecology* 35:229–241.

Cherrier, J., J. E. Bauer, E. R. M. Druffel, R. B. Coffin, and J. P. Chanton. 1999. Radiocarbon in marine bacteria: Evidence for the ages of assimilated carbon. *Limnology and Oceanography* 44:730–736.

Cherrier, J., J. Sarkodee-Adoo, T. P. Guilderson, and J. P. Chanton. 2013. Fossil carbon in particulate organic matter in the Gulf of Mexico following the Deepwater Horizon event. *Environmental Science & Technology Letters* 1:108–112.

Chuang, C. Y., P. H. Santschi, L. S. Wen, L. Guo, C. Xu, S. Zhang, Y. Jiang, et al. 2015. Binding of Th, Pa, Pb, Po and Be radionuclides to marine colloidal macromolecular organic matter. *Marine Chemistry* 173:320–329.

Ciais, P., A. J. Dolman, A. Bombelli, R. Duren, A. Peregon, P. J. Rayner, C. Miller, et al. 2013. Current systematic carbon cycle observations and needs for implementing a policy-relevant carbon observing system. *Biogeosciences Discussion* 10:11447–11581.

Claustre, H., and J. C. Marty. 1995. Specific phytoplankton biomasses and their relation to primary production in the tropical North Atlantic. *Deep Sea Research Part I: Oceanographic Research Papers* 42:1475–1493.

Clayton, R. K. 1980. *Photosynthesis: Physical Mechanisms and Chemical Patterns*. Cambridge: Cambridge University Press. 296 pp.

Coble, P. G. 2007. Marine optical biogeochemistry: The chemistry of ocean color. *Chemical Reviews* 107:402–418.

Coffin, R. B., and L. A. Cifuentes. 1999. Stable isotope analysis of carbon cycling in the Perdido Estuary, Florida. *Estuaries* 22:917–926.

Cole, J. J., S. Findlay, and M. L. Pace. 1988. Bacterial production in fresh and saltwater ecosystems: A cross-system overview. *Marine Ecology Progress Series* 43:1–10.

Cole, J. J., Y. T. Prairie, N. F. Caraco, W. H. McDowell, L. J. Tranvik, R. G. Striegl, R. G., C. M. Duarte, et al. 2007. Plumbing the global carbon cycle: Integrating inland waters into the terrestrial carbon budget. *Ecosystems* 10:172–185.

Conmy, R. N., P. G. Coble, R. F. Chen, and G. B. Gardner. 2004. Optical properties of colored dissolved organic matter in the northern Gulf of Mexico. *Marine Chemistry* 89:127–144.

Copping, A. E., and C. J. Lorenzen. 1980. Carbon budget of a marine phytoplankton-herbivore system with carbon-14 as a tracer. *Limnology and Oceanography* 25:873–882.

Corbett, D. R., B. A. McKee, and M. A. Allison. 2006. Nature of decadal-scale sediment accumulation on the western shelf of the Mississippi River delta. *Continental Shelf Research* 26:2125–2140.

Corbett, D. R., B. A. McKee, and D. Duncan. 2004. An evaluation of mobile mud dynamics in the Mississippi River deltaic region. *Marine Geology* 209:91–112.

Cowie, G. L., J. I. Hedges, and S. E. Calvert. 1992. Sources and relative reactivities of amino acids, neutral sugars, and lignin in an intermittently anoxic marine environment. *Geochimica et Cosmochimica Acta* 56:1963–1978.

Cui, X., T. S. Bianchi, J. A. Hutchings, C. Savage, and J. H. Curtis. 2016. Hydrodynamic sorting of organic carbon in surface sediments of Fiordland, New Zealand. *Journal of Geophysical Research: Biogeosciences* 121:1016–1031.

Dagg, M. J., J. W. Ammerman, R. M. W. Amon, W. S. Gardner, R. E. Green, and S. E. Lohrenz. 2007. A review of water column processes influencing hypoxia in the northern Gulf of Mexico. *Estuaries and Coasts* 30:735–752.

Dagg, M. J., T. S. Bianchi, G. A. Breed, W. J. Cai, S. Duan, H. Liu, B. A. McKee, R. T. Powell, and C. M. Stewart. 2005. Biogeochemical characteristics of the lower Mississippi River, USA, during June 2003. *Estuaries* 28:664–674.

Dagg, M. J., T. S. Bianchi, B. McKee, and R. Powell. 2008. Fates of dissolved and particulate materials from the Mississippi River immediately after discharge into the northern Gulf of Mexico, USA, during a period of low wind stress. *Continental Shelf Research* 28:1443–1450.

Dahl, T. E. 1990. *Wetlands: Losses in the United States: 1780's to 1980's*. Washington, DC: US Department of the Interior, Fish and Wildlife Service. 13 pp.

Dai, M., Z. Yin, F. Meng, Q. Liu, and W.-J. Cai. 2012. Spatial distribution of riverine DOC inputs to the ocean: An updated global synthesis. *Current Opinion in Environmental Sustainability* 4:170–178.

Dauwe, B., J. J. Middelburg, P. M. J. Herman, and C. H. R. Heip. 1999. Linking diagenetic alteration of amino acids and bulk organic matter reactivity. *Limnology and Oceanography* 44:1809–1814.

De Gouw, J. A., A. M. Middlebrook, C. Warneke, R. Ahmadov, E. L. Atlas, R. Bahreini, D. R. Blake, et al. 2011. Organic aerosol formation downwind from the Deepwater Horizon oil spill. *Science* 331:1295–1299.

Del Giorgio, P. A., and J. J. Cole. 1998. Bacterial growth efficiency in natural aquatic systems. *Annual Reviews of Ecology and Systematics* 29:503–541.

Del Vecchio, R., and N. V. Blough. 2004. Spatial and seasonal distribution of chromophoric dissolved organic matter and dissolved organic carbon in the Middle Atlantic Bight. *Marine Chemistry* 89:169–187.

Diaz, R. J., and R. Rosenberg. 2008. Spreading dead zones and consequences for marine ecosystems. *Science* 321:926–929.

Diercks, A. R., and V. L. Asper. 1997. In situ settling speeds of marine snow aggregates below the mixed layer: Black Sea and Gulf of Mexico. *Deep Sea Research Part I: Oceanographic Research Papers* 44:385–398.

Ding, Y. X., W. C. Chin, A. Rodriguez, C. C. Hung, P. H. Santschi, and P. Verdugo. 2008. Amphiphilic exopolymers from *Sagittula stellata* induce DOM self-assembly and formation of marine microgels. *Marine Chemistry* 112:11–19.

Dittmar, T., N. Hertkorn, G. Kattner, and R. J. Lara. 2006. Mangroves, a major source of dissolved organic carbon to the oceans. *Global Biogeochemical Cycles* 20:1–7.

Dittmar, T. and J. Paeng. 2009. A heat-induced molecular signature in marine dissolved organic matter. *Nature Geoscience* 2:175–179.

Dittmar, T., J. Paeng, T. M. Gihring, I. G. N. A. Suryaputra, and M. Huettel. 2012. Discharge of dissolved black carbon from a fire-affected intertidal system. *Limnology and Oceanography* 54:1171–1181.

Dodds, W. K., 2006. Nutrients and the "dead zone": The link between nutrient ratios and dissolved oxygen in the northern Gulf of Mexico. *Frontiers in Ecology and the Environment* 4:211–217.

Doval, M. D., and D. A. Hansell. 2000. Organic carbon and apparent oxygen utilization in the western South Pacific and the central Indian Oceans. *Marine Chemistry* 68:249–264.

Druffel, E. R. M., and P. M. Williams. 1990. Identification of a deep marine source of particulate organic carbon using bomb ^{14}C. *Nature* 347:172–174.

Druffel, E. R. M., P. M. Williams, J. E. Bauer, and J. R. Ertel. 1992. Cycling of dissolved and particulate organic matter in the open ocean. *Journal of Geophysical Research: Oceans* 97:15639–15659.

Druffel, E. R. M., P. M. Williams, and Y. Suzuki. 1989. Concentrations and radiocarbon signatures of dissolved organic matter in the Pacific Ocean. *Geophysical Research Letters* 16:991–994.

Duan, S., and T. S. Bianchi. 2006. Seasonal changes in the abundance and composition of plant pigments in particulate organic carbon in the lower Mississippi and Pearl Rivers. *Estuaries and Coasts* 29:427–442.

Duan, S., T. S. Bianchi, and T. P. Sampere. 2007. Temporal variability in the composition and abundance of terrestrially-derived dissolved organic matter in the lower Mississippi and Pearl Rivers. *Marine Chemistry* 103:172–184.

Duan, S., T. S. Bianchi, A. M. Shiller, K. Dria, P. G. Hatcher, and K. R. Carman. 2007. Variability in the bulk composition and abundance of dissolved organic matter in the lower Mississippi and Pearl Rivers. *Journal of Geophysical Research* 112:1–12.

Duan, S., Y. He, S. S. Kaushal, T. S. Bianchi, N. D. Ward, and L. Guo. 2017. Impact of wetland decline on decreasing dissolved organic carbon concentrations along the Mississippi River continuum. *Frontiers in Marine Science* 3:280. doi:10.3389/fmars.2016.00280.

Ducklow, H. W. 1999. The bacterial component of the oceanic euphotic zone. *FEMS Microbiology Ecology* 30:1–10.

Ducklow, H. W., D. A. Purdie, P. J. L. B. Williams, and J. M. Davies. 1986. Bacterioplankton: A sink for carbon in a coastal marine plankton community. *Science* 232:865–867.

Ducklow, H. W., D. K. Steinberg, and K. O. Buessler. 2001. Upper ocean carbon export and the biological pump. *Oceanography* 14:50–58.

Duursma, E. K., 1963. The production of dissolved organic matter in the sea, as related to the primary gross production of organic matter. *Netherlands Journal of Sea Research* 2:85–94.

Emerson, S. R., and J. I. Hedges (eds.). 2008. *Chemical Oceanography and the Marine Carbon Cycle*. Cambridge: Cambridge University Press. 470 pp.

Fichot, C. G., and R. Benner. 2011. A novel method to estimate DOC concentrations from CDOM absorption coefficients in coastal waters. *Geophysical Research Letters* 38:1–5.

———. 2012. The spectral slope coefficient of chromophoric dissolved organic matter ($S_{275-295}$) as a tracer of terrigenous dissolved organic carbon in river-influenced ocean margins. *Limnology and Oceanography* 57:1453–1466.

———. 2014. The fate of terrigenous dissolved organic carbon in a river-influenced ocean margin. *Global Biogeochemical Cycles* 28:300–318.

Fichot, C. G., K. Kaiser, S. B. Hooker, R. M. W. Amon, M. Babin, S. Bélanger, S. A. Walker, and R. Benner. 2013. Pan-Arctic distributions of continental runoff in the Arctic Ocean. *Scientific Reports* 3:1–6.

Fichot, C. G., S. E. Lohrenz, and R. Benner. 2014. Pulsed, cross-shelf export of terrigenous dissolved organic carbon to the Gulf of Mexico. *Journal of Geophysical Research: Oceans* 119:1–19.

Floodgate, G. D. 1995. Some environmental aspects of marine hydrocarbon bacteriology. *Aquatic Microbial Ecology* 9:3–11.

Frank, H. A., and R. J. Cogdell. 1996. Carotenoids in photosynthesis. *Photochemistry and Photobiology* 63:257–264.

Fredericks, A. D., and W. M. Sackett. 1970. Organic carbon in the Gulf of Mexico. *Journal of Geophysical Research* 75:2199–2206.

Giddings, J. C. 1993. Field-flow fractionation: Analysis of macromolecular, colloidal, and particulate materials. *Science* 260:1456–1465.

Goldman, J. C., and M. R. Dennett. 2000. Growth of marine bacteria in batch and continuous culture under carbon and nitrogen limitation. *Limnology and Oceanography* 45:789–800.

Goñi, M. A., and T. I. Eglinton. 1996. Stable carbon isotopic analyses of lignin-derived CuO oxidation products by isotope ratio monitoring-gas chromatography-mass spectrometry (irm-GC-MS). *Organic Geochemistry* 24:601–615.

Goñi, M. A., E. S. Gordon, N. M. Monacci, R. Clinton, R. Gisewhite, M. A. Allison, and G. Kineke. 2006. The effect of Hurricane Lili on the distribution of organic matter along the inner Louisiana shelf (Gulf of Mexico, USA). *Continental Shelf Research* 26:2260–2280.

Goñi, M. A., K. C. Ruttenberg, and T. I. Eglinton. 1998. A reassessment of the sources and importance of land-derived organic matter in surface sediments from the Gulf of Mexico. *Geochimica et Cosmochimica Acta* 62:3055–3075.

Goodwin, T. W. 1980. *The Biochemistry of the Carotenoids*. 2nd ed. London: Chapman and Hall. 224 pp.

Gordon, E. S., and M. A. Goñi. 2003. Sources and distribution of terrigenous organic matter delivered by the Atchafalaya River to sediments in the northern Gulf of Mexico. *Geochimica et Cosmochimica Acta* 67:2359–2375.

———. 2004. Controls on the distribution and accumulation of terrigenous organic matter in sediments from the Mississippi and Atchafalaya river margin. *Marine Chemistry* 92:331–352.

Graham, W. M., R. H. Condon, R. H. Carmichael, I. D'Ambra, H. K. Patterson, L. J. Linn, and F. J. Hernandez Jr. 2010. Oil carbon entered the coastal planktonic food web during the Deepwater Horizon oil spill. *Environmental Research Letters* 5:045301.

Grasshoff, K., K. Kremling, and M. Ehrhardt (eds.). 2009. *Methods of Seawater Analysis*. Weinheim, Germany: Wiley-VCH. 632 pp.

Green, E. P., and M. J. Dagg. 1997. Mesozooplankton associations with medium to large marine snow aggregates in the northern Gulf of Mexico. *Journal of Plankton Research* 19:435–447.

Green, N. W., E. M. Perdue, G. R. Aiken, K. D. Butler, H. Chen, T. Dittmar, J. Niggemann, and A. Stubbins. 2014. An intercomparison of three methods for the large-scale isolation of oceanic dissolved organic matter. *Marine Chemistry* 161:14–19.

Green, R. E., T. S. Bianchi, M. J. Dagg, N. D. Walker, and G. A. Breed. 2006. An organic carbon budget for the Mississippi River turbidity plume and plume contributions to air-sea CO_2 fluxes and bottom water hypoxia. *Estuaries and Coasts* 29:579–597.

Griffith, D. R., R. T. Barnes, and P. A. Raymond. 2009. Inputs of fossil carbon from wastewater treatment plants to U.S. rivers and oceans. *Environmental Science & Technology* 43:5647–5651.

Grossart, H.-P., and H. Ploug. 2001. Microbial degradation of organic carbon and nitrogen on diatom aggregates. *Limnology and Oceanography* 46:267–277.

Guenet, B., M. Danger, L. Abbadie, and G. Lacroix. 2010. Priming effect: Bridging the gap between terrestrial and aquatic ecology. *Ecology* 91:2850–2861.

Guo, L., C. H. Coleman, and P. H. Santschi. 1994. The distribution of colloidal and dissolved organic carbon in the Gulf of Mexico. *Marine Chemistry* 45:105–119.

Guo, L., and R. W. Macdonald. 2006. Source and transport of terrigenous organic matter in the upper Yukon River: Evidence from isotope ($\delta^{13}C$, $\Delta^{14}C$, and $\delta^{15}N$) composition of dissolved, colloidal, and particulate phases. *Global Biogeochemical Cycles* 20:1–12.

Guo, L., and P. H. Santschi. 1997. Isotopic and elemental characterization of colloidal organic matter from the Chesapeake Bay and Galveston Bay. *Marine Chemistry* 59:1–15.

———. 2007. Ultrafiltration and its applications to sampling and characterisation of aquatic colloids. Chap. 4 in K. J. Wilkinson and J. R. Lead (eds.), *Environmental Colloids and Particles: Behaviour, Separation and Characterisation*. Chichester, UK: John Wiley and Sons.

Guo, L., P. H. Santschi, and M. Baskaran. 1997. Interaction of thorium isotopes with colloidal organic matter in oceanic environments. *Colloids and Surfaces A: Physicochemical and Engineering Aspects* 120:255–271.

Guo, L., P. H. Santschi, and T. S. Bianchi. 1999. Dissolved organic matter in estuaries of the Gulf of Mexico. Chap. 9 in T. S. Bianchi, J. R. Pennock, and R. R. Twilley (eds.), *Biogeochemistry of Gulf of Mexico Estuaries*. Hoboken, NJ: John Wiley and Sons.

Guo, L., P. H. Santschi, L. A. Cifuentes, S. Trumbore, and J. Southon. 1996. Cycling of high molecular weight dissolved organic matter in the Middle Atlantic Bight as revealed by carbon isotopic (^{13}C and ^{14}C) signatures. *Limnology and Oceanography* 41:1242–1252.

Guo, L., P. H. Santschi, and K. W. Warnken. 1995. Dynamics of dissolved organic carbon (DOC) in oceanic environments. *Limnology and Oceanography* 40:1392–1403.

———. 2000. Trace metal composition of colloidal organic material in marine environments. *Marine Chemistry* 70:257–275.

Guo, L., N. Tanaka, D. M. Schell, and P. H. Santschi. 2003. Nitrogen and carbon isotopic composition of high-molecular-weight dissolved organic matter in marine environments. *Marine Ecology Progress Series* 252:51–60.

Guo, L., D. M. White, C. Xu, and P. H. Santschi. 2009. Chemical and isotopic composition of HMW-DOM from the Mississippi River plume. *Marine Chemistry* 114:63–71.

Hansell, D. A. 2013. Recalcitrant dissolved organic carbon fractions. *Annual Review of Marine Science* 5:421–445.

Hansell, D. A., and C. A. Carlson. 1998. Deep-ocean gradients in the concentration of dissolved organic carbon. *Nature* 395:263–266.

———. 2001. Biogeochemistry of total organic carbon and nitrogen in the Sargasso Sea: Control by convective overturn. *Deep Sea Research II: Tropical Studies in Oceanography* 48:1649–1667.

———. (eds.). 2015. *Biogeochemistry of Marine Dissolved Organic Matter*. 2nd ed. Cambridge, MA: Academic Press. 712 pp.

Hansell, D. A., C. A. Carlson, D. J. Repeta, and R. Schlitzer. 2009. Dissolved organic matter in the ocean: A controversy stimulates new insights. *Oceanography* 22:202–211.

Hansell, D. A., C. A. Carlson, and R. Schlitzer. 2012. Net removal of major marine dissolved organic carbon fractions in the subsurface ocean. *Global Biogeochemical Cycles* 26:1–9.

Hansman, R. L., S. Griffin, J. T. Watson, E. R. M. Druffel, A. E. Ingalls, A. Pearson, and L. I. Aluwihare. 2009. The radiocarbon signature of microorganisms in the mesopelagic ocean. *Proceedings of the National Academy of Sciences* 106:6513–6518.

Harrison, J. A., N. Caraco, and S. P. Seitzinger. 2005. Global patterns and sources of dissolved organic matter export to the coastal zone: Results from a spatially explicit, global model. *Global Biogeochemical Cycles* 19:1–18.

Hatcher, P. G. 2004. Intercomparisons of some new approaches for investigating the molecular weight distribution of dissolved organic matter. Pp. 241–243 in L. Martin-Neto, D. M. B. P. Milori, and W. T. L da Silva (eds.), *Humic Substances and Soil and Water Environment*. São Paulo, Brazil: Embrapa Instrumentação Agropecuária.

Hawkes, J. A., P. E. Rossel, A. Stubbins, D. Butterfield, D. P. Connelly, E. P. Achterberg, A. Koshinsky, et al. 2015. Efficient removal of recalcitrant deep-ocean dissolved organic matter during hydrothermal circulation. *Nature Geoscience* 8:856–861.

Hayes, J. M. 2001. Fractionation of carbon and hydrogen isotopes in biosynthetic processes. *Reviews in Mineralogy and Geochemistry* 43:225–277.

He, D., R. N. Mead, L. Belicka, O. Pisani, and R. Jaffé. 2014. Assessing source contributions to particulate organic matter in a subtropical estuary: A biomarker approach. *Organic Geochemistry* 75:129–139.

Hedges, J. I. 2002. Why dissolved organic matter? Chap. 1 in D. A. Hansell and C. A. Carlson (eds.), *Biogeochemistry of Marine Dissolved Organic Matter*. San Diego: Academic Press.

Hedges, J. I., G. Eglinton, P. G. Hatcher, D. L. Kirchman, C. Arnosti, S. Derenne, R. P. Evershed, et al. 2000. The molecularly-uncharacterized component of nonliving organic matter in natural environments. *Organic Geochemistry* 31:945–958.

Hedges, J. I., and J. R. Ertel. 1982. Characterization of lignin by gas capillary chromatography of cupric oxide oxidation products. *Analytical Chemistry* 54:174–178.

Hedges, J. I., and D. C. Mann. 1979. The characterization of plant tissues by their lignin oxidation products. *Geochimica et Cosmochimica Acta* 43:1803–1807.

Hedges, J. I., and K. Whitehead. 2008. Marine organic geochemistry. Chap. 8 in S. Emerson and J. Hedges (eds.), *Chemical Oceanography and the Marine Carbon Cycle*. Cambridge: Cambridge University Press.

Helms, J. R., A. Stubbins, J. D. Ritchie, E. C. Minor, D. J. Kieber, and K. Mopper. 2008. Absorption spectral slopes and slope ratios as indicators of molecular weight, source, and photobleaching of chromophoric dissolved organic matter. *Limnology and Oceanography* 53:955–969.

Henson, S. A., R. Sanders, and E. Madsen. 2012. Global patterns in efficiency of particulate organic carbon export and transfer to the deep ocean. *Global Biogeochemical Cycles* 26:1–14.

Hernes, P. J., and R. Benner. 2003. Photochemical and microbial degradation of dissolved lignin phenols: Implications for the fate of terrigenous dissolved organic matter in marine environments. *Journal of Geophysical Research: Oceans* 108:3291. doi:10.1029/2002JC001421, C9.

Hernes, P. J., B. A. Bergamaschi, R. S. Eckard, and R. G. M. Spencer. 2009. Fluorescence-based proxies for lignin in freshwater dissolved organic matter. *Journal of Geophysical Research: Biogeosciences* 114:G00F03. doi:10.1029/2009JG000938.

Ho, T. Y., L. S. Wen, C. F. You, and D. C. Lee. 2007. The trace metal composition of size-fractionated plankton in the South China Sea: Biotic versus abiotic sources. *Limnology and Oceanography* 52:1776–1788.

Holte, J., L. Talley, J. Gilson, and D. Roemmich. 2017. An Argo mixed layer climatology and data base. *Geophysical Research Letters* 44:5618–5626.

Hopkinson, C. S., Jr., and J. J. Vallino. 2005. Efficient export of carbon to the deep ocean through dissolved organic matter. *Nature* 433:142–145.

Horvath, R. S. 1972. Microbial co-metabolism and the degradation of organic compounds in nature. *Bacteriological Reviews* 36:146–155.

Hung, C.-C., L. D. Guo, K. A. Roberts, and P. H. Santschi. 2004. Upper ocean carbon flux determined by size-fractionated ^{234}Th data and sediment traps in the Gulf of Mexico. *Geochemical Journal* 38:601–611.

Hung, C.-C., L. Guo, P. H. Santschi, N. Alvarado-Quiroz, and J. M. Haye. 2003. Distributions of carbohydrate species in the Gulf of Mexico. *Marine Chemistry* 81:118–135.

Hung, C.-C., C. Xu, P. H. Santschi, S. J. Zhang, K. A. Schwehr, A. Quigg, L. Guo, et al. 2010. Comparative evaluation of sediment trap and ^{234}Th-derived POC fluxes from the upper oligotrophic waters of the Gulf of Mexico and the subtropical northwestern Pacific Ocean. *Marine Chemistry* 121:132–144.

Inamdar, S. P., and M. J. Mitchell. 2006. Hydrologic and topographic controls on storm-event exports of dissolved organic carbon (DOC) and nitrate across catchment scales. *Water Resources Research* 42:W03421. doi:10.1029/2005WR004212.

Ingalls, A. E., S. R. Shah, R. L. Hansman, L. I. Aluwihare, G. M. Santos, E. R. M. Druffel, and A. Pearson. 2006. Quantifying archaeal community autotrophy in the mesopelagic ocean using natural radiocarbon. *Proceedings of the National Academy of Sciences* 103:6442–6447.

Jaffé, R., Y. Ding, J. Niggemann, A. V. Vähätalo, A. Stubbins, R. G. M. Spencer, J. Campbell, and T. Dittmar. 2013. Global charcoal mobilization from soils via dissolution and riverine transport to the oceans. *Science* 340:345–347.

Jannasch, H. W., and M. J. Mottl. 1985. Geomicrobiology of deep-sea hydrothermal vents. *Science* 229:717–725.

Jannasch, H. W., C. O. Wirsen, and S. J. Molyneaux. 1991. Chemoautotrophic sulfur-oxidizing bacteria from the Black Sea. *Deep Sea Research* 38:S1105–S1120.

Jeffrey, S. W., and R. F. C. Mantoura. 1997. Development of pigment methods for oceanography: SCOR-supported working groups and objectives. Chap. 1 in S. W. Jeffrey, R. F. C. Mantoura, and S. W. Wright (eds.), *Phytoplankton Pigments in Oceanography*. Paris: UNESCO.

Jiao, N., and F. Azam. 2011. Microbial carbon pump and its significance for carbon sequestration in the ocean. Chap. 2 in N. Jiao, F. Azam, and S. Sanders (eds.), *Microbial Carbon Pump in the Ocean*. Washington, DC: American Association for the Advancement of Science.

Jiao, N., G. Herndl, D. A. Hansell, R. Benner, G. Kattner, S. W. Wilhelm, D. L. Kirchman, et al. 2010. Microbial production of recalcitrant dissolved organic matter: Long-term carbon storage in the global ocean. *Nature Reviews Microbiology* 8:593–599.

Johannes, R. E., and K. L. Webb. 1965. Release of dissolved amino acids by marine zooplankton. *Science* 150:76–77.

Joye, S. B. 2015. Deepwater Horizon, 5 years on: Baseline environmental data are crucial for understanding the impacts of oil spills. *Science* 349:592–593.

Kaiser, K., and R. Benner. 2008. Major bacterial contribution to the ocean reservoir of detrital organic carbon and nitrogen. *Limnology and Oceanography* 53:99–112.

———. 2009. Biochemical composition and size distribution of organic matter at the Pacific and Atlantic time-series stations. *Marine Chemistry* 113:63–77.

Kaiser, K., M. Canedo-Oropeza, R. McMahon, and R. M. W. Amon. 2017. Origins and transformations of dissolved organic matter in large Arctic rivers. *Nature Scientific Reports* 7:1–11.

Keil, R. G., and D. L. Kirchman. 1991a. Contribution of dissolved free amino acids and ammonium to the nitrogen requirements of heterotrophic bacterioplankton. *Marine Ecology Progress Series* 73:1–10.

———. 1991b. Dissolved combined amino acids in marine waters as determined by a vapor-phase hydrolysis method. *Marine Chemistry* 33:243–259.

Kieber, R. J., R. F. Whitehead, and S. A. Skrabal. 2006. Photochemical production of dissolved organic carbon from resuspended sediments. *Limnology and Oceanography* 51:2187–2195.

Killops, S. D., and V. J. Killops. 2013. *Introduction to Organic Geochemistry*. 2nd ed. Malden, MA: John Wiley and Sons. 393 pp.

Kim, S., L. A. Kaplan, and P. G. Hatcher. 2006. Biodegradable dissolved organic matter in a temperate and tropical stream determined from ultra-high resolution mass spectrometry. *Limnology and Oceanography* 51:1054–1063.

Kirtman, B., S. B. Power, J. A. Adedoyin, G. J. Boer, R. Bojariu, I. Camilloni, F. J. Doblas-Reyes, et al. 2013. Near-term climate change: Projections and predictability. Chap. 11 in IPCC (eds.), *Climate Change 2013: The Physical Science Basis. IPCC Working Group I Contribution to AR5*. Cambridge: Cambridge University Press.

Koch, B. P., M. Witt, R. Engbrodt, T. Dittmar, and G. Kattner. 2005. Molecular formulae of marine and terrigenous dissolved organic matter detected by electrospray ionization Fourier transform ion cyclotron resonance mass spectrometry. *Geochimica et Cosmochimica Acta* 69:3299–3308.

Kolasinski, J., and B. E. Rosenheim. 2016. Pelagic cruise CTD casts and ship-track velocity and weather data near the Mississippi Delta, July 2014. Consortium for Advanced Research on Transport of Hydrocarbon in the Environment (CARTHE), Gulf of Mexico Research Initiative. https://data.gulfresearchinitiative.org/data/R1.x134.073:0021.

Kujawinski, E. B. 2011. The impact of microbial metabolism on marine dissolved organic matter. *Annual Review of Marine Science* 3:567–599.

Lalonde, K., P. Middlestead, and Y. Gélinas. 2014. Automation of $^{13}C/^{12}C$ ratio measurement for freshwater and seawater DOC using high temperature combustion. *Limnology and Oceanography: Methods* 12:816–829.

Lambert, C., T. S. Bianchi, and P. R. Santschi. 1999. Cross-shelf changes in phytoplankton community composition in the Gulf of Mexico (Texas Shelf/Slope). *Continental Shelf Research* 19:1–21.

Lampert, W. 1978. Release of dissolved organic carbon by grazing zooplankton. *Limnology and Oceanography* 23:831–834.

Lang, S. Q., D. A. Butterfield, M. D. Lilley, H. P. Johnson, and J. I. Hedges. 2006. Dissolved organic carbon in ridge-axis and ridge-flank hydrothermal systems. *Geochimica et Cosmochimica Acta* 70:3830–3842.

Lead, J. R., and K. J. Wilkinson. 2006. Aquatic colloids and nanoparticles: Current knowledge and future trends. *Environmental Chemistry* 3:159–171.

Lee, C. 2004. Transformations in the "Twilight Zone" and beyond. *Marine Chemistry* 92:87–90.

Lee, C., S. Wakeham, and C. Arnosti. 2004. Particulate organic matter in the sea: The composition conundrum. *AMBIO: A Journal of the Human Environment* 33:565–575.

Le Quéré, C., R. J. Andres, T. Boden, T. Conway, R. A. Houghton, J. I. House, G. Marland, et al. 2013. The global carbon budget 1959–2011. *Earth System Science Data Discussions* 5:165–185.

Letscher, R. T., D. A. Hansell, C. A. Carlson, R. Lumpkin, and A. N. Knapp. 2013. Dissolved organic nitrogen in the global surface ocean: Distribution and fate. *Global Biogeochemical Cycles* 27:141–153.

Lin, P., and L. Guo. 2015. Spatial and vertical variations of dissolved carbohydrates in the northern Gulf of Mexico following the Deepwater Horizon oil spill. *Marine Chemistry* 174:13–25.

Liu, Q., and G. C. Shank. 2015. Solar radiation-enhanced dissolution (photodissolution) of particulate organic matter in Texas estuaries. *Estuaries and Coasts* 38:2172–2184.

Loh, A. N., J. E. Bauer, and E. R. M. Druffel. 2004. Variable ageing and storage of dissolved organic components in the open ocean. *Nature* 430:877–881.

Maiti, K., S. Bosu, E. J. D'Sa, P. L. Adhikari, M. Sutor, and K. Longnecker. 2016. Export fluxes in northern Gulf of Mexico: Comparative evaluation of direct, indirect and satellite-based estimates. *Marine Chemistry* 184:60–77.

Malakoff, D. 1998. Death by suffocation in the Gulf of Mexico. *Science* 281:190–192.

Mannino, A., M. E. Russ, and S. B. Hooker. 2008. Algorithm developed and validation for satellite-derived distributions of DOC and CDOM in the U.S. Middle Atlantic Bight. *Journal of Geophysical Research* 113:1–19.

Marañón, E., P. Cermeno, and V. Perez. 2005. Continuity in the photosynthetic production of dissolved organic carbon from eutrophic to oligotrophic waters. *Marine Ecology Progress Series* 299:7–17.

Maurer, L. G. 1976. Organic polymers in seawater: Changes with depth in the Gulf of Mexico. *Deep Sea Research* 23:1059–1064.

Mayer, L. M., L. L. Schick, T. S. Bianchi, and L. A. Wysocki. 2009. Photochemical changes in chemical markers of sedimentary organic matter source and age. *Marine Chemistry* 113:123–128.

Mayer, L. M., L. L. Schick, K. R. Hardy, and M. L. Estapa. 2009. Photodissolution and other photochemical changes upon irradiation of algal detritus. *Limnology and Oceanography* 54:1688–1698.

Mayer, L. M., L. L. Schick, K. Skorko, and E. Boss. 2006.

Photodissolution of particulate organic matter from sediments. *Limnology and Oceanography* 51:1064–1071.

McCallister, S. L., J. E. Bauer, J. E. Cherrier, and H. W. Ducklow. 2004. Assessing sources and ages of organic matter supporting river and estuarine bacterial production: A multiple-isotope ($\Delta^{14}C$, $\delta^{13}C$, and $\delta^{15}N$) approach. *Limnology and Oceanography* 49:1687–1702.

McKee, B. A., R. C. Aller, M. A. Allison, T. S. Bianchi, and G. C. Kineke. 2004. Transport and transformation of dissolved and particulate materials on continental margins influenced by major rivers: Benthic boundary layer and seabed processes. *Continental Shelf Research* 24:899–926.

McNichol, A. P., and L. I. Aluwihare. 2007. The power of radiocarbon in biogeochemical studies of the marine carbon cycle: Insights from studies of dissolved and particulate organic carbon (DOC and POC). *Chemical Reviews* 107:443–466.

Meybeck, M., and A. Ragu. 1997. Presenting the GEMS-GLORI, a compendium of world river discharge to the oceans. In *Freshwater Contamination, Proceedings of Rabat Symposium S4, IAHS* publ. no. 243. Bremerhaven: PANGAEA. http://hdl.handle.net/10013/epic.34684.

Meyers, P. A. 1997. Organic geochemical proxies of paleoceanographic, paleolimnologic, and paleoclimatic processes. *Organic Geochemistry* 27:213–250.

Middelburg, J. J. 2011. Chemoautotrophy in the ocean. *Geophysical Research Letters* 38:1–4.

Miller, R. L., C. E. Del Castillo, and B. A. McKee (eds.). 2005. *Remote Sensing of Coastal Aquatic Environments*. Dordrecht, Netherlands: Springer. 346 pp.

Milliman, J. D., and K. L. Farnsworth. 2011. *River Discharge to the Coastal Ocean: A Global Synthesis*. Cambridge: Cambridge University Press. 384 pp.

Milliman, J. D., and R. H. Meade. 1983. World-wide delivery of river sediment to the oceans. *Journal of Geology* 91:1–21.

Minor, E. C., T. I. Eglinton, J. J. Boon, and R. Olson. 1999. Protocol for the characterization of oceanic particles via flow cytometric sorting and direct temperature-resolved mass spectrometry. *Analytical Chemistry* 71:2003–2013.

Minor, E. C., and P. S. Nallathamby. 2004. "Cellular" vs. "detrital" POM: A preliminary study using fluorescent stains, flow cytometry, and mass spectrometry. *Marine Chemistry* 92:9–21.

Mitra, S., T. S. Bianchi, B. A. Mckee, and M. Sutula. 2002. Black carbon from the Mississippi River: Quantities, sources and potential implications for the global carbon cycle. *Environmental Science & Technology* 36:2296–2302.

Mitra, S., C. L. Osburn, and A. S. Wozniak. 2017. A preliminary assessment of fossil fuel and terrigenous influences to rainwater organic matter in summertime in the northern Gulf of Mexico. *Aquatic Geochemistry* 23:217–231.

Mitra, S., A. S. Wozniak, R. Miller, P. G. Hatcher, C. Buonassissi, and M. Brown. 2013. Multiproxy probing of rainwater dissolved organic matter (DOM) composition in coastal storms as a function of trajectory. *Marine Chemistry* 154:67–76.

Møller, E. F. 2005. Sloppy feeding in marine copepods: Prey-size-dependent production of dissolved organic carbon. *Journal of Plankton Research* 27:27–35.

———. 2007. Production of dissolved organic carbon by sloppy feeding in copepods *Acartia tonsa*, *Centropages typicus*, and *Temora longicornis*. *Limnology and Oceanography* 52:79–84.

Møller, E. F., and T. G. Nielsen. 2001. Production of bacteria substrate by marine copepods: Effect of phytoplankton biomass and cell size. *Journal of Plankton Research* 23:527–536.

Møller, E. F., P. Thor, and T. G. Nielsen. 2003. Production of DOC by *Calanus finmarchicus*, *C. glacialis* and *C. hyperboreus* through sloppy feeding and leakage from fecal pellets. *Marine Ecology Progress Series* 262:185–191.

Mopper, K., A. Stubbins, J. D. Ritchie, H. M. Bialk, and P. G. Hatcher. 2007. Advanced instrumental approaches for characterization of marine dissolved organic matter: Extraction techniques, mass spectrometry, and nuclear magnetic resonance spectroscopy. *Chemical Reviews* 107:419–442.

Mopper, K., X. Zhou, R. J. Kieber, D. J. Kieber, R. J. Sikorski, and R. D. Jones. 1991. Photochemical degradation of dissolved organic carbon and its impact on the oceanic carbon cycle. *Nature* 353:60–62.

Moran, M. A., and R. G. Zepp. 1997. Role of photoreactions in the formation of biologically labile compounds from dissolved organic matter. *Limnology and Oceanography* 42:1307–1316.

Morris, R. M., K. L. Vergin, J.-C. Cho, M. S. Rappé, C. A. Carlson, and S. J. Giovannoni. 2005. Temporal and spatial response of bacterioplankton lineages to annual convective overturn at the Bermuda Atlantic Time-series Study site. *Limnology and Oceanography* 50:1687–1696.

Morse, J. W., and M. J. Beazley. 2008. Organic matter in deepwater sediments of the northern Gulf of Mexico and its relationship to the distribution of benthic organisms. *Deep Sea Research Part II: Tropical Studies in Oceanography* 55:2563–2571.

Müller-Karger, F. E., R. Varela, R. Thunell, R. Luerssen, C. Hu, and J. J. Walsh. 2005. The importance of continental margins in the global carbon cycle. *Geophysical Research Letters* 32:L01602. doi:10.1029/2004GL021346.

Myklestad, S., 2000. Dissolved organic carbon from phytoplankton. In P. J. Wangersky (ed.), *Marine Chemistry: The Handbook of Environmental Chemistry*, vol. 5D. Heidelberg: Springer-Verlag.

Nagata, T. 2008. Organic matter-bacteria interactions in seawater. Chap. 7 in D. L. Kirchman (ed.), *Microbial Ecology of the Oceans*, 2nd ed. Hoboken, NJ: John Wiley and Sons.

Nelson, N. B., and D. A. Siegel. 2013. The global distribution and dynamics of chromophoric dissolved organic matter. *Annual Review of Marine Science* 5:447–476.

Nelson, N. B., D. A. Siegel, C. A. Carlson, and C. M. Swan. 2010. Tracing global biogeochemical cycles and meridional overturning circulation using chromophoric dissolved organic matter. *Geophysical Research Letters* 37:1–5.

Onstad, G. D., D. E. Canfield, P. D. Quay, and J. I. Hedges. 2000. Sources of particulate organic matter in rivers from the continental USA: Lignin phenol and stable carbon isotope compositions. *Geochimica et Cosmochimica Acta* 64:3539–3546.

Opsahl, S., and R. Benner. 1998. Photochemical reactivity of dissolved lignin in river and ocean waters. *Limnology and Oceanography* 43:1297–1304.

———. 1999. Characterization of carbohydrates during early diagenesis of five vascular plant tissues. *Organic Geochemistry* 30:83–94.

Osburn, C. L., and T. S. Bianchi. 2016. Editorial: Linking optical and chemical properties of dissolved organic matter in natural waters. *Frontiers in Marine Science* 3:223. doi:10.3389/fmars.2016.00223.

Osburn, C. L., T. J. Boyd, M. T. Montgomery, T. S. Bianchi, R. B. Coffin, and H. W. Paerl. 2016. Optical proxies for terrestrial dissolved organic matter in estuaries and coastal waters. *Frontiers in Marine Science* 2:127. doi:10.3389/fmars.2015.00127.

Osburn, C. L., R. Del Vecchio, and T. J. Boyd. 2014. Physicochemical effects on dissolved organic matter fluorescence in natural waters. Chap. 7 in P. G. Coble, J. Lead, A. Baker, D. M. Reynolds, and R. G. M. Spencer (eds.), *Aquatic Organic Matter Fluorescence*. New York: Cambridge University Press.

Osburn, C. L., L. T. Handsel, M. P. Mikan, H. W. Paerl, and M. T. Montgomery. 2012. Fluorescence tracking of dissolved and particulate organic matter quality in a river-dominated estuary. *Environmental Science & Technology* 46:8628–8636.

Osburn, C. L., and C. A. Stedmon. 2011. Linking the chemical and optical properties of dissolved organic matter in the Baltic–North Sea transition zone to differentiate three allochthonous inputs. *Marine Chemistry* 126:281–294.

Osburn, C. L., and G. St-Jean. 2007. The use of wet chemical oxidation with high-amplification isotope ratio mass spectrometry (WCO-IRMS) to measure stable isotope values of dissolved organic carbon in seawater. *Limnology and Oceanography: Methods* 5:296–308.

Ourisson, G., and P. Albrecht. 1992. Hopanoids. 1: Geohopanoids: The most abundant natural products on earth? *Accounts of Chemical Research* 25:398–402.

Pakulski, J. D., and R. Benner. 1994. Abundance and distribution of carbohydrates in the ocean. *Limnology and Oceanography* 39:930–940.

Parsons, T. R., Y. Maita, and C. M. Lalli. 1984. *A Manual of Biological and Chemical Methods for Seawater Analysis*. Oxford: Pergamon Press. 174 pp.

Passow, U. 2016. Formation of rapidly-sinking, oil-associated marine snow. *Deep Sea Research Part II: Tropical Studies in Oceanography* 129:232–240.

Passow, U., K. Ziervogel, V. Asper, and A. Diercks. 2012. Marine snow formation in the aftermath of the Deepwater Horizon oil spill in the Gulf of Mexico. *Environmental Research Letters* 7:035301. iopscience.iop.org/article/10.1088/1748-9326/7/3/035301/pdf.

Peters, K. E., C. C. Walters, and J. M. Moldowan. 2007. *The Biomarker Guide*, Vol. 1. Cambridge: Cambridge University Press. 474 pp.

Peterson, B. J., R. W. Howarth, and R. H. Garritt. 1985. Multiple stable isotopes used to trace the flow of organic matter in estuarine food webs. *Science* 227:1361–1363.

Pomeroy, L. R., P. J. L. B. Williams, F. Azam, and J. E. Hobbie. 2007. The microbial loop. *Oceanography* 20:28–33.

Poulin, B. A., J. N. Ryan, and G. R. Aiken. 2014. Effects of iron on optical properties of dissolved organic matter. *Environmental Science & Technology* 48:10098–10106.

Rabalais, N. N., R. E. Turner, and W. J. Wiseman Jr. 2002. Gulf of Mexico hypoxia, a.k.a. "the dead zone." *Annual Reviews of Ecology and Systematics* 33:235–263.

Raymond, P. A., and J. E. Saiers. 2010. Event controlled DOC export from forested watersheds. *Biogeochemistry* 100:197–209.

Raymond, P. A., J. E. Saiers, and W. V. Sobczak. 2016. Hydrological and biogeochemical controls on watershed dissolved organic matter transport: Pulse-shunt concept. *Ecology* 97:5–16.

Raymond, P. A., and R. G. M. Spencer. 2015. Riverine DOM. Chap. 11 in D. A. Hansell and C. A. Carlson (eds.), *Biogeochemistry of Marine Dissolved Organic Matter*, 2nd ed. London: Academic Press.

Redalje, D. G., S. E. Lohrenz, and G. L. Fahnenstiel. 1994. The relationship between primary production and the vertical export of particulate organic matter in a river-impacted coastal ecosystem. *Estuaries* 17:829–838.

Ren, W., H. Tian, W. J. Cai, S. E. Lohrenz, C. S. Hopkinson, W. J. Huang, J. Yang, B. Tao, S. Pan, and R. He. 2016. Century-long increasing trend and variability of dissolved organic carbon export from the Mississippi River basin driven by natural and anthropogenic forcing. *Global Biogeochemical Cycles* 30:1288–1299.

Repeta, D. J. 2015. Chemical characterization and cycling of dissolved organic matter. Chap. 2 in D. A. Hansell and C. A. Carlson (eds.), *Biogeochemistry of Marine Dissolved Organic Matter*, 2nd ed. London: Academic Press.

Rivas, D., A. Badan, and J. Ochoa. 2005. The ventilation of the deep Gulf of Mexico. *Journal of Physical Oceanography* 35:1763–1781.

Roach, K. A. 2013. Environmental factors affecting incorporation of terrestrial material into large river food webs. *Freshwater Science* 32:283–298.

Roland, L. A., M. D. McCarthy, T. D. Peterson, and B. D. Walker. 2009. A large-volume microfiltration system for isolating suspended particulate organic matter: Fabrication and assessment versus GFF filters in central North Pacific. *Limnology and Oceanography: Methods* 7:64–80.

Romankevich, E. A. 1984. *Geochemistry of Organic Matter in the Ocean*. Berlin: Springer-Verlag. 334 pp.

Rosenheim, B. E., K. M. Roe, B. J. Roberts, A. S. Kolker, M. A. Allison, and K. H. Johannesson. 2013. River discharge influences on particulate organic carbon age structure in the Mississippi/Atchafalaya River system. *Global Biogeochemical Cycles* 27:154–166.

Rowe, G. T., C. Wei, C. Nunnally, R. Haedrich, P. Montagna, J. G. Baguley, J. M. Bernhard, et al. 2008. Comparative biomass structure and estimated carbon flow in food webs in the deep Gulf of Mexico. *Deep Sea Research Part II: Tropical Studies in Oceanography* 55:2699–2711.

Saba, G. K., D. K. Steinberg, and D. A. Bronk. 2011. The relative importance of sloppy feeding, excretion, and fecal pellet leaching in the release of dissolved carbon and nitrogen by *Acartia tonsa* copepods. *Journal of Experimental Marine Biology and Ecology* 404:47–56.

Sahl, L. E., W. J. Merrell, D. W. McGrail, and J. A. Webb. 1987. Transport of mud on continental shelves: Evidence from the Texas shelf. *Marine Geology* 76:33–43.

Sampere, T. P., T. S. Bianchi, M. A. Allison, and B. A. McKee. 2011. Burial and degradation of organic carbon in Louisiana shelf/slope sediments. *Estuarine, Coastal and Shelf Science* 95:232–244.

Sampere, T. P., T. S. Bianchi, S. G. Wakeham, and M. A. Allison. 2008. Sources of organic matter in surface sediments of the Louisiana continental margin: Effects of major depositional/transport pathways and Hurricane Ivan. *Continental Shelf Research* 28:2472–2487.

Santinelli, C., L. Nannicini, and A. Seritti. 2010. DOC dynamics in the meso and bathypelagic layers of the Mediterranean Sea. *Deep Sea Research II: Tropical Studies in Oceanography* 57:1446–1459.

Santschi, P. H., E. Balnois, K. Wilkinson, J. Buffle, and L. D. Guo. 1998. Fibrillar polysaccharides in marine macromolecular organic matter as imaged by atomic force microscopy and transmission electron microscopy. *Limnology Oceanography* 43:896–908.

Santschi, P. H., L. Guo, M. Baskaran, S. Trumbore, J. Southon, T. S. Bianchi, B. Honeyman, and L. Cifuentes. 1995. Isotopic evidence for the contemporary origin of high-molecular weight organic matter in oceanic environments. *Geochimica et Cosmochimica Acta* 59:625–631.

Santschi, P. H., C. C. Hung, G. Schultz, N. Alvarado-Quiroz, L. Guo, J. L. Pinckney, and I. Walsh. 2003. Control of acid polysaccharide production and ^{234}Th and POC export fluxes by marine organisms. *Geophysical Research Letters* 30:1044. doi:10.1029/2002GL016046.

Santschi, P. H., J. W. Murray, M. Baskaran, C. R. Benitez-Nelson, L. D. Guo, C. C. Hung, C. Lamborg, S. B. Moran, U. Passow, and M. Roy-Barman. 2006. Thorium speciation in seawater. *Marine Chemistry* 100:250–268.

Santschi, P. H., and G. T. Rowe. 2008. Radiocarbon-derived sedimentation rates in the Gulf of Mexico. *Deep Sea Research Part II: Tropical Studies in Oceanography* 55:2572–2576.

Sarkanen, K. V., and C. H. Ludwig. 1971. *Lignins: Occurrence, Formation, Structure, and Reactions*. New York: Wiley-Interscience. 934 pp.

Sarmiento, J. L., and N. Gruber. 2002. Sinks for anthropogenic carbon. *Physics Today* 55:30–36.

Scavia, D., N. N. Rabalais, R. E. Turner, D. Justić, and W. J. Wiseman. 2003. Predicting the response of Gulf of Mexico hypoxia to variations in Mississippi River nitrogen load. *Limnology and Oceanography* 48:951–956.

Schlitzer, R., 2014. Ocean Data View. odv.awi.de.

Shank, G. C., A. Evans, Y. Yamashita, and R. Jaffé. 2011. Solar radiation–enhanced dissolution of particulate organic matter from coastal marine sediments. *Limnology and Oceanography* 56:577–588.

Shen, Y., C. G. Fichot, and R. Benner. 2012. Floodplain influence on dissolved organic matter composition and export from the Mississippi-Atchafalaya River system to the Gulf of Mexico. *Limnology and Oceanography* 57:1149–1160.

Shen, Y., C. G. Fichot, S.-K. Liang, and R. Benner. 2016. Biological hot spots and the accumulation of marine dissolved organic matter in a highly productive ocean margin. *Limnology and Oceanography* 61:1287–1300.

Sholkovitz, E. R. 1976. Flocculation of dissolved organic and inorganic matter during the mixing of river water and seawater. *Geochimica et Cosmochimica Acta* 40:831–845.

Simon, M., H.-P. Grossart, B. Schweitzer, and H. Ploug. 2002. Microbial ecology of organic aggregates in aquatic ecosystems. *Aquatic Microbial Ecology* 28:175–211.

Skoog, A., and R. Benner. 1997. Aldoses in various size fractions of marine organic matter: Implications for carbon cycling. *Limnology and Oceanography* 42:1803–1813.

Skoog, A., B. Biddanda, and R. Benner. 1999. Bacterial utilization of dissolved glucose in the upper water column of the Gulf of Mexico. *Limnology and Oceanography* 44:1625–1633.

Sleighter, R. L., and P. G. Hatcher. 2007. The application of electrospray ionization coupled to ultrahigh resolution mass spectrometry for the molecular characterization of natural organic matter. *Journal of Mass Spectrometry* 42:559–574.

———. 2008. Molecular characterization of dissolved organic matter (DOM) along a river to ocean transect of the lower Chesapeake Bay by ultrahigh resolution electrospray ionization Fourier transform ion cyclotron resonance mass spectrometry. *Marine Chemistry* 110:140–152.

Smith, D. C., M. Simon, A. L. Alldredge, and F. Azam. 1992. Intense hydrolytic enzyme activity on marine aggregates and implications for particle dissolution. *Nature* 359:139–142.

Søndergaard, M., and M. Middelboe. 1995. A cross-system analysis of labile dissolved organic carbon. *Marine Ecology Progress Series* 118:283–294.

Spencer, R. G. M., G. R. Aiken, M. M. Dornblaser, K. D. Butler, R. M. Holmes, G. Fiske, P. J. Mann, and A. Stubbins. 2013. Chromophoric dissolved organic matter export from U.S. rivers. *Geophysical Research Letters* 40:1575–1579.

Spencer, R. G. M., A. Stubbins, P. J. Hernes, A. Baker, K. Mopper, A. K. Aufdenkampe, R. Y. Dyda, et al. 2009. Photochemical degradation of dissolved organic matter and dissolved lignin phenols from the Congo River. *Journal of Geophysical Research* 114:1–12.

Stedmon, C. A., S. Markager, and R. Bro. 2003. Tracing dissolved organic matter in aquatic environments using a new approach to fluorescence spectroscopy. *Marine Chemistry* 82:239–254.

Stedmon, C. A., S. Markager, and H. Kaas. 2000. Optical properties and signatures of chromophoric dissolved organic matter (CDOM) in Danish coastal waters. *Estuarine Coastal and Shelf Science* 51:267–278.

Stedmon, C. A., and N. B. Nelson. 2015. The optical properties of DOM in the ocean. Chap. 10 in D. A. Hansell and C. A. Carlson (eds.), *Biogeochemistry of Marine Dissolved Organic Matter*, 2nd ed. London: Academic Press.

Steen, A. D., K. Ziervogel, S. Ghobrial, and C. Arnosti. 2012. Functional variation among polysaccharide-hydrolyzing microbial communities in the Gulf of Mexico. *Marine Chemistry* 138:13–20.

Steinberg, D. K., C. A. Carlson, N. R. Bates, S. A. Goldthwait, L. P. Madin, and A. F. Michaels. 2000. Zooplankton vertical migration and the active transport of dissolved organic and inorganic carbon in the Sargasso Sea. *Deep Sea Research I: Oceanographic Research Papers* 47:137–158.

Steinberg, D. K., S. A. Goldthwait, and D. A. Hansell. 2002. Zooplankton vertical migration and the active transport of dissolved organic and inorganic nitrogen in the Sargasso Sea. *Deep Sea Research I: Oceanographic Research Papers* 49:1445–1461.

Stolpe, B., L. Guo, A. Shiller, and M. Hassellöv. 2010. Size and composition of colloidal organic matter and trace elements in the Mississippi River, Pearl River and the northern Gulf of Mexico as characterized by flow field-flow fractionation. *Marine Chemistry* 118:119–129.

Stolpe, B., Z. Zhou, L. Guo, and A. Shiller. 2014. Colloidal size distributions of humic- and protein-like fluorescent organic matter in the northern Gulf of Mexico. *Marine Chemistry* 164:25–37.

Strom, S. L., R. Benner, S. Ziegler, and M. J. Dagg. 1997. Planktonic grazers are a potentially important source of marine dissolved organic carbon. *Limnology and Oceanography* 42:1364–1374.

Stubbins, A., J. F. Lapierre, M. Berggren, Y. T. Prairie, T. Dittmar, and P. A. del Giorgio. 2014. What's in an EEM? Molecular signatures associated with dissolved organic fluorescence in boreal Canada. *Environmental Science & Technology* 48:10598–10606.

Stubbins, A., J. Niggemann, and T. Dittmar. 2012. Photo-lability of deep ocean dissolved black carbon. *Biogeosciences* 9:1661–1670.

Suess, E. 1980. Particulate organic carbon flux in the oceans—surface productivity and oxygen utilization. *Nature* 288:260–263.

Taylor, G. T., M. Iabiachella, T.-Y. Ho, M. I. Scranton, R. C. Thunell, F. Muller-Karger, and R. Varela. 2001. Chemoautotrophy in the redox transition zone of the

Cariaco Basin: A significant midwater source of organic carbon production. *Limnology and Oceanography* 46:148–163.

Tehrani, N. C., E. J. D'sa, D. S. Ko, N. D. Walker, C. L. Osburn, and R. F. Chen. 2013. Colored dissolved organic matter dynamics in the northern Gulf of Mexico from ocean color and numerical model results. *Journal of Coastal Research* 30:800–814.

Tehrani, N. C., E. J. D'Sa, C. L. Osburn, T. S. Bianchi, and B. A. Schaeffer. 2013. Chromophoric dissolved organic matter and dissolved organic carbon from sea-viewing wide field-of-view sensor (SeaWiFS), moderate resolution imaging spectroradiometer (MODIS) and MERIS sensors: Case study for the northern Gulf of Mexico. *Remote Sensing* 5:1439–1464.

Teira, E., M. J. Pazo, P. Serret, P. and E. Fernandez. 2001. Dissolved organic carbon production by microbial populations in the Atlantic Ocean. *Limnology and Oceanography* 46:1370–1377.

Thor, P., H. G. Dam, and D. R. Rogers. 2003. Fate of organic carbon released from decomposing copepod fecal pellets in relation to bacterial production and ectoenzymatic activity. *Aquatic Microbial Ecology* 33:279–288.

Tian, H., W. Ren, J. Yang, B. Tao, W. J. Cai, S. E. Lohrenz, C. S. Hopkinson, et al. 2015. Climate extremes dominating seasonal and interannual variations in carbon export from the Mississippi River basin. *Global Biogeochemical Cycles* 29:1333–1347.

Toner, M., A. D. Kirwan, A. C. Poje, L. H. Kantha, F. E. Müller-Karger, and C. K. R. T. Jones. 2003. Chlorophyll dispersal by eddy-eddy interactions in the Gulf of Mexico. *Journal of Geophysical Research: Oceans* 108:3105. doi:10.1029/2002JC001499.

Tremblay, L., and R. Benner. 2009. Organic matter diagenesis and bacterial contributions to detrital carbon and nitrogen in the Amazon River system. *Limnology and Oceanography* 54:681–691.

Urban-Rich, J. 1999. Release of dissolved organic carbon from copepod fecal pellets in the Greenland Sea. *Journal of Experimental Marine Biology and Ecology* 232:107–124.

Van Mooy, B. A., R. G. Keil, and A. H. Devol. 2002. Impact of suboxia on sinking particulate organic carbon: Enhanced carbon flux and preferential degradation of amino acids via denitrification. *Geochimica et Cosmochimica Acta* 66:457–465.

Verdugo, P. 2012. Marine microgels. *Annual Review of Marine Science* 4:375–400.

Verdugo, P., A. L. Alldredge, F. Azam, D. L. Kirchman, U. Passow, and P. H. Santschi. 2004. The oceanic gel phase: A bridge in the DOM–POM continuum. *Marine Chemistry* 92:67–85.

Vincent, D., G. Slawyk, S. L'Helguen, G. Sarthou, M. Gallinari, L. Seuront, B. Sautour, and O. Ragueneau. 2007. Net and gross incorporation of nitrogen by marine copepods fed on ^{15}N-labelled diatoms: Methodology and trophic studies. *Journal of Experimental Marine Biology and Ecology* 353:295–305.

Volkman, J. K. 2006. Lipid markers for marine organic matter. Chap. 2 in J. K. Volkman (ed.), *Marine Organic Matter: Biomarkers, Isotopes and DNA*. Berlin: Springer.

Wagner, S., R. Jaffé, K. Cawley, T. Dittmar, and A. Stubbins. 2015. Associations between the molecular and optical properties of dissolved organic matter in the Florida Everglades, a model coastal wetland system. *Frontiers in Chemistry* 3:66. doi:10.3389/fchem.2015.00066.

Wakeham, S. G., E. A. Canuel, E. J. Lerberg, P. Mason, T. P. Sampere, and T. S. Bianchi. 2009. Partitioning of organic matter in continental margin sediments among density fractions. *Marine Chemistry* 115:211–225.

Wakeham, S. G., C. Lee, J. I. Hedges, P. J. Hernes, and M. J. Peterson. 1997. Molecular indicators of diagenetic status in marine organic matter. *Geochimica et Cosmochimica Acta* 61:5363–5369.

Wakeham, S. G., M. L. Peterson, J. I. Hedges, and C. Lee. 2002. Lipid biomarker fluxes in the Arabian Sea, with a comparison to the equatorial Pacific Ocean. *Deep Sea Research Part II: Tropical Studies in Oceanography* 49:2265–2301.

Walker, N. D., and N. N. Rabalais. 2006. Relationships among satellite chlorophyll-*a*, river inputs, and hypoxia on the Louisiana continental shelf, Gulf of Mexico. *Estuaries and Coasts* 29:1081–1093.

Walker, N. D., W. J. Wiseman Jr., L. J. Rouse Jr., and A. Babin. 2005. Effects of river discharge, wind stress, and slope eddies on circulation and the satellite-observed structure of the Mississippi River plume. *Journal of Coastal Research* 21:1228–1244.

Walsh, D. A., E. Zaikova, C. G. Howes, Y. C. Song, J. J. Wright, S. G. Tringe, P. D. Tortell, and S. J. Hallam. 2009. Metagenome of a versatile chemolithoautotroph from expanding oceanic dead zones. *Science* 326:578–582.

Wang, X., Y. Cai, and L. Guo. 2010. Preferential removal of dissolved carbohydrates during estuarine mixing in the Bay of Saint Louis in the northern Gulf of Mexico. *Marine Chemistry* 119:130–138.

———. 2013. Variations in abundance and size distribution of carbohydrates in the lower Mississippi River, Pearl River and the Bay of St. Louis. *Estuarine, Coastal and Shelf Science* 126:61–69.

Wanninkhof, R., M. Wood, and L. Barbero. 2014. *GOMECC-2: Gulf of Mexico East Coast Carbon Cruise (RB-12-03), July 21, 2012–August 13, 2012*. Atlantic Oceanographic and Meteorological Laboratory, National Oceanic and Atmospheric Administration. https://cdiac.ess-dive.lbl.gov/ftp/oceans/GOMECC2/Cruise_Report_June2014.pdf.

Waterson, E. J., and E. A. Canuel. 2008. Sources of sedimentary organic matter in the Mississippi River and adjacent Gulf of Mexico as revealed by lipid biomarker and δ^{13}CTOC analyses. *Organic Geochemistry* 39:422–439.

Weishaar, J. L., G. R. Aiken, B. A. Bergamaschi, M. S. Fram, R. Fujii, and K. Mopper. 2003. Evaluation of specific ultraviolet absorbance as an indicator of the chemical composition and reactivity of dissolved organic carbon. *Environmental Science & Technology* 37:4702–4708.

Wells, M. L. 2004. The colloidal size spectrum of CDOM in the coastal region of the Mississippi plume using flow field-flow fractionation. *Marine Chemistry* 89:89–102.

Werdell, P. J., B. A. Franz, S. W. Bailey, G. C. Feldman, E. Boss, V. E. Brando, M. Dowell, et al. 2013. Generalized ocean color inversion model for retrieving marine inherent optical properties. *Applied Optics* 52:2019–2037.

Wetzel, R. G. 1992. Gradient-dominated ecosystems: Sources and regulatory functions of dissolved organic matter in freshwater ecosystems. *Hydrobiologia* 229:181–198.

Wilhelm, S. W., and C. A. Suttle. 1999. Viruses and nutrient cycles in the sea: Viruses play critical roles in the structure and function of aquatic food webs. *BioSciences* 49:781–788.

Willey, J. D., R. J. Kieber, M. S. Eyman, and G. B. Avery. 2000. Rainwater dissolved organic carbon: Concentrations and global flux. *Global Biogeochemical Cycles* 14:139–148.

Williams, C. J., Y. Yamashita, H. F. Wilson, R. Jaffé, and M. A. Xenopoulos. 2010. Unraveling the role of land use and microbial activity in shaping dissolved organic matter characteristics in stream ecosystems. *Limnology and Oceanography* 55:1159–1171.

Williams, P. M., and E. R. M. Druffel. 1987. Radiocarbon in dissolved organic matter in the central North Pacific Ocean. *Nature* 330:246–248.

Wilson, H. F., and M. A. Xenopoulos. 2009. Effects of agricultural land use on the composition of fluvial dissolved organic matter. *Nature Geoscience* 2:37–41.

Wilson, J. T. 1963. Hypothesis of Earth's behaviour. *Nature* 198:925–929.

Wiseman, W. J., N. N. Rabalais, R. E. Turner, S. P. Dinnel, and A. N. D. A. MacNaughton. 1997. Seasonal and interannual variability within the Louisiana coastal current: Stratification and hypoxia. *Journal of Marine Systems* 12:237–248.

Wisser, D., B. M. Fekete, C. J. Vörösmarty, and A. H. Schumann. 2010. Reconstructing 20th century global hydrography: A contribution to the Global Terrestrial Network-Hydrography (GTN-H). *Hydrology and Earth System Sciences* 14:1–24.

Wysocki, L. A., T. S. Bianchi, R. T. Powell, and N. Reuss. 2006. Spatial variability in the coupling of organic carbon, nutrients, and phytoplankton pigments in surface waters and sediments of the Mississippi River plume. *Estuarine, Coastal and Shelf Science* 69:47–63.

Xu, H., and L. Guo. 2017. Molecular size-dependent abundance and composition of dissolved organic matter in river, lake and sea waters. *Water Research* 117:115–126.

Yang, W., and L. Guo. 2014. Abundance, distribution, and isotopic composition of particulate black carbon in the northern Gulf of Mexico. *Geophysical Research Letters* 41:7619–7625.

Yoon, B., and P. A. Raymond. 2012. Dissolved organic matter export from a forested watershed during Hurricane Irene. *Geophysical Research Letters* 39:1–6.

Young, C. M. 2014. Deployment: PE14-11. Biological and Chemical Oceanography Data Management Office. http://www.bco-dmo.org/deployment/521436.

Yuan, J., R. L. Miller, R. T. Powell, and M. J. Dagg. 2004. Storm-induced injection of the Mississippi River plume into the open Gulf of Mexico. *Geophysical Research Letters* 31. doi:10.1029/2003GL019335.

Zanardi-Lamardo, E., C. D. Clark, C. A. Moore, and R. G. Zika. 2002. Comparison of the molecular mass and optical properties of colored dissolved organic material in two rivers and coastal waters by flow field-flow fractionation. *Environmental Science & Technology* 36:2806–2814.

———. 2004. Seasonal variation in molecular mass and optical properties of chromophoric dissolved organic material in coastal waters of southwest Florida. *Marine Chemistry* 89:37–54.

Zhou, Z., and L. Guo. 2012. Evolution of the optical properties of seawater influenced by the DWH oil spill in the Gulf of Mexico. *Environmental Research Letters* 7:025301. doi:10.1088/1748-9326/7/2/025301.

———. 2015. A critical evaluation of an asymmetrical flow field-flow fractionation system for colloidal size characterization of natural organic matter. *Journal of Chromatography A* 1399:53–64.

Zhou, Z., L. Guo, and E. C. Minor. 2016. Characterization of bulk and chromophoric dissolved organic matter in the Laurentian Great Lakes during summer 2013. *Journal of Great Lakes Research* 42:789–801.

Zhou, Z., L. Guo, A. M. Shiller, S. E. Lohrenz, V. L. Asper, and

C. L. Osburn. 2013. Characterization of oil components from the Deepwater Horizon oil spill in the Gulf of Mexico using fluorescence EEM and PARAFAC techniques. *Marine Chemistry* 148:10–21.

Zhou, Z., B. Stolpe, L. Guo, and A. M. Shiller. 2016. Colloidal size spectra, composition and estuarine mixing behavior of DOM in river and estuarine waters of the northern Gulf of Mexico. *Geochimica Cosmochimica Acta* 181:1–17.

Ziervogel, K., and C. Arnosti. 2016. Enhanced protein and carbohydrate hydrolyses in plume-associated deepwaters initially sampled during the early stages of the Deepwater Horizon oil spill. *Deep Sea Research Part II: Tropical Studies in Oceanography* 129:368–373.

Ziervogel, K., C. Dike, V. Asper, J. Montoya, J. Battles, U. Passow, A. Diercks, M. Esch, S. Joye, C. Dewald, and C. Arnosti. 2016. Enhanced particle fluxes and heterotrophic bacterial activities in Gulf of Mexico bottom waters following storm-induced sediment resuspension. *Deep Sea Research Part II: Tropical Studies in Oceanography* 129:77–88.

Ziervogel, K., L. McKay, B. Rhodes, C. L. Osburn, J. Dickson-Brown, C. Arnosti, and A. Teske. 2012. Microbial activities and dissolved organic matter dynamics in oil-contaminated surface seawater from the Deepwater Horizon oil spill site. *PLOS ONE* 7:e34816. doi:10.1371/journal.pone.0034816.

Ziervogel, K., C. Osburn, A. Brym, J. Battles, S. Joye, N. D'souza, J. Montoya, U. Passow, and C. Arnosti. 2016. Linking heterotrophic microbial activities with particle characteristics in waters of the Mississippi River delta in the aftermath of Hurricane Isaac. *Frontiers in Marine Science* 3:8. doi:10.3389/fmars.2016.00008.

7

Marine Sediment Chemistry

Jeffrey Chanton, Laura Lapham, Thomas S. Bianchi, Kelsey Rogers, David Hollander, and Samantha Joye

Introduction

The Gulf of Mexico (GOM) receives drainage from two-thirds of the continental United States, one-third of Mexico, and parts of Canada, Cuba, and Central America. The largest single draining river is Old Muddy, the Mississippi, which receives contributions from the Arkansas, Red, Missouri, Yellowstone, Ohio, and Tennessee Rivers (Fig. 7.1).

At the confluence of the Mississippi and Red Rivers in central Louisiana begins the Atchafalaya River, a distributary of the Mississippi. This large riverine system particularly influences the northern GOM and delivers inorganic constituents and terrestrial organic materials to the sedi-

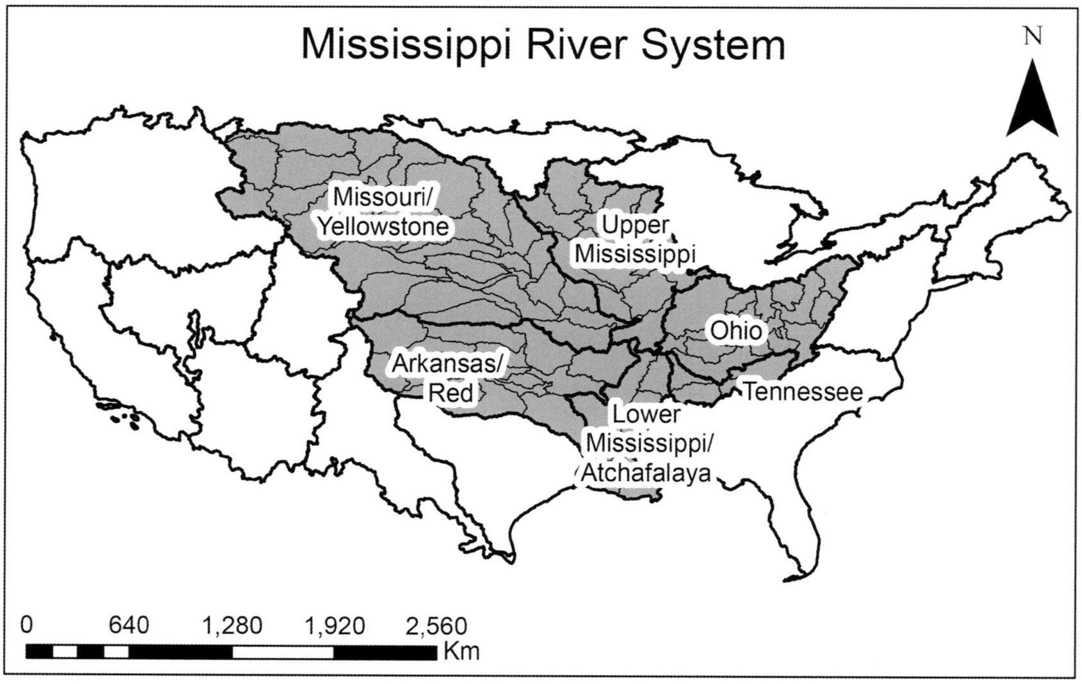

Figure 7.1. Rivers and drainage basins contributing to the Mississippi River (map drawn from data sets at https://nhd.usgs.gov/wbd.html and https://nrcs.app.box.com/v/huc/folder/39640323180).

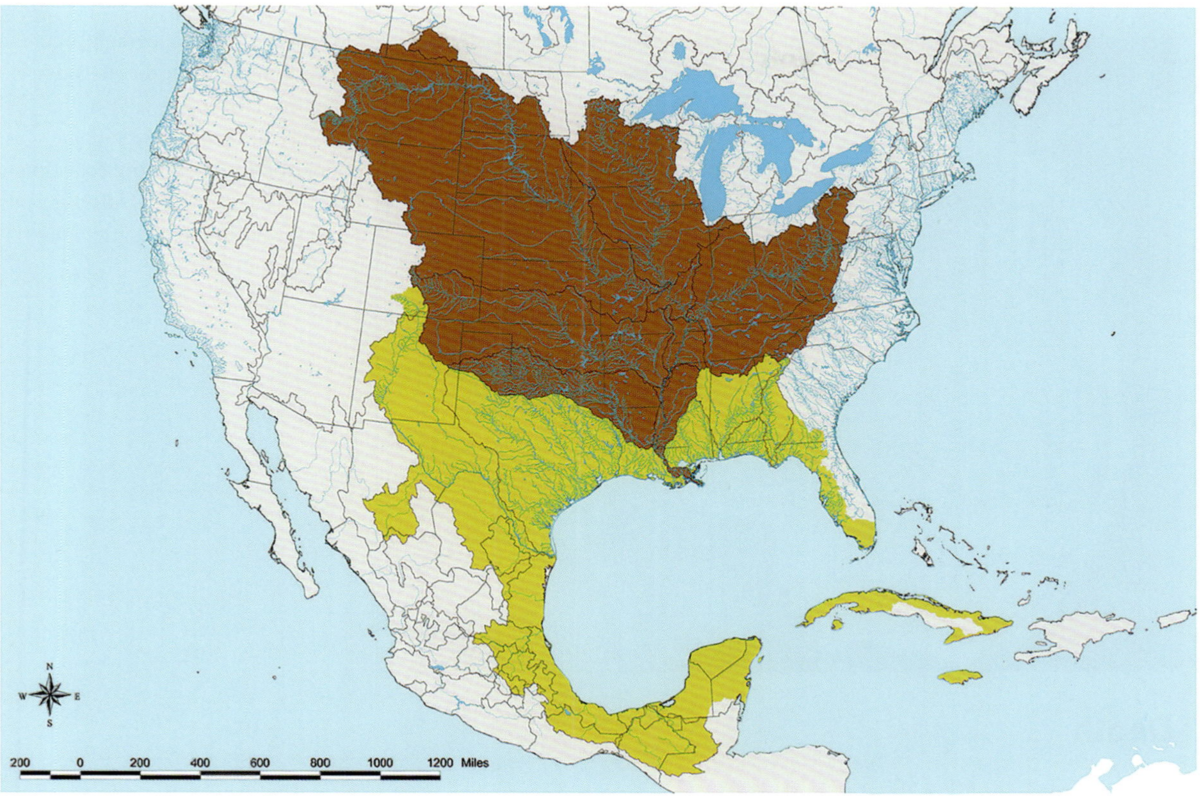

Figure 7.2. The Gulf of Mexico receives drainage from large areas in addition to the Mississippi Valley (used with permission of the Gulf of Mexico Program, US Environmental Protection Agency, Gulfport, MS, http://flowergarden.noaa.gov/image_library/maps/gomwatershed.jpg).

ments. In addition to the Mississippi, large portions of the southeastern United States, Texas, and Mexico drain into the Gulf directly or through smaller rivers (Fig. 7.2).

Two geologic features result in the advection of fluids through permeable sediments and sedimentary rock. First, the perimeter of the Gulf is bounded in 2 areas by steep carbonate escarpments, west Florida and the Yucatan Peninsula (Fig. 7.3). The karst carbonate geology of these areas results in significant submarine groundwater delivery to the Gulf (Kohout 1966; Cable, Bugna, et al. 1996; Cable, Burnett, et al. 1996; Burnett et al. 2003; Santos, Burnett, Chanton, et al. 2009; Santos, Burnett, Dittmar, et al. 2009; Santos, Dimova, et al. 2009), and deep seepage brings reduced energy-bearing fluids to the base of the Florida Escarpment (Chanton et al. 1991; Martens et al. 1991; Paull et al. 1991). Across the northern GOM, and in portions of the southern GOM, salt diapirs play an important role in deforming sediments and generate deeply sourced faults that serve as conduits, allowing seepage inputs of natural gas, petroleum, and even asphalt into the seafloor, altering the sediments and their geochemical cycles (Bernard et al. 1976; Sackett et al. 1979; Joye et al. 2004; Joye et al. 2009; MacDonald et al. 2003; MacDonald et al. 2004). This chapter describes the unique setting of the Gulf in terms of sediment deposition and the fluid flux that modulates a variety of key biogeochemical processes within sediments.

Sediment Type, Source, Transport, and Distribution

Along the northern Gulf, the De Soto Canyon (29°00'00" N, 87°30'00" W; see Fig. 7.3), about 100 km south of Pensacola, Florida, separates the northern Gulf seafloor into 2 distinct regions. West of the canyon, the seafloor is dominated by terrestrial sediments altered by salt tectonics. There, the canyon is bordered by 5 salt diapirs (Harbison 1968), and sediments are dominated by Mississippi River lithogenic inputs (Gould and Stewart 1956; Doyle and Sparks 1980). East of the De Soto Canyon, the region is carbonate-dominated biogenic sediments that grade from coarse to fine from onshore to offshore (Doyle and Sparks 1980). Similarly in the southern Gulf, the Campeche Canyon separates terrigenous clastic sediments on the west

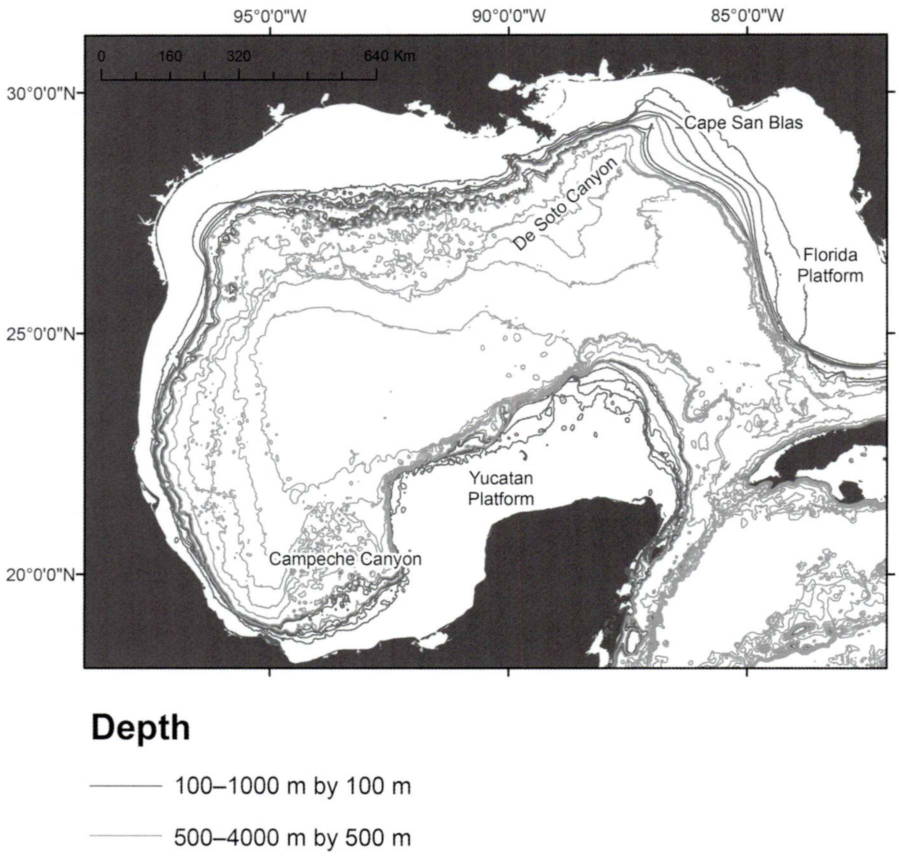

Figure 7.3. The topography of the Gulf seafloor. Note the rugged nature of the topography of the Louisiana/Texas slope and the slope of the Bay of Campeche (map produced from data at http://gcoos.tamu.edu/products/topography/SRTM30PLUS.html).

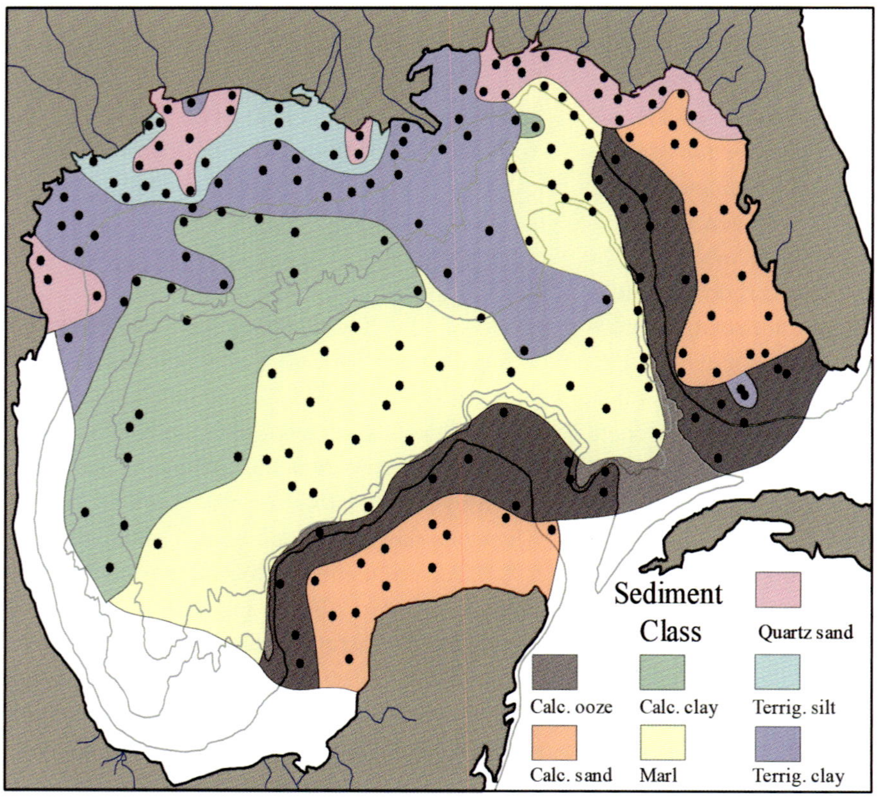

Figure 7.4. Sediment types across the Gulf of Mexico (from Balsam and Beeson 2003; used with permission).

from the biogenic carbonate deposits of the Yucatan shelf (see Fig. 7.3) to the east (Schroeder and Wiseman 1999). The GOM is bounded on its shoreline by a series of estuaries that vary from barrier bar-built lagoons to well-developed drowned river valley systems (Davis 2017). In the north is the Mississippi-Atchafalaya River system and an extensive marsh-estuarine regime (Schroeder and Wiseman 1999).

Balsam and Beeson (2003) boldly extrapolated 186 surface sediment samples to characterize the sediment distribution across the entire Gulf (Fig. 7.4). As they do on the west Florida shelf, calcareous sands dominate the Yucatan Platform, with calcareous oozes in deeper water. The shelves of Texas, Louisiana, Mississippi, Alabama, and the Florida's western panhandle are dominated by terrigenous sediment, quartz sand close to shore, with grain size decreasing to clays offshore. Terrigenous clays and silt are found offshore of the Mississippi Delta and to the west of the delta and offshore Texas and the northern Mexican Gulf Coast. The bottom of the deep Gulf is dominated by marl, which is a mixture of pelagic carbonates and terrigenous clays from the Mississippi (Balsam and Beeson 2003).

Focusing on the nearshore of the northern GOM, Doyle and Sparks (1980) drew a boundary at Florida's Cape San Blas (85.3°W), which lies 200 km northeast of the De Soto Canyon (see Fig. 7.3). They describe the clastic sand body, the MAFLA (Mississippi, Alabama, Florida) sand sheet that grades into the muds of the Mississippi Delta, to the west of the cape. They describe the west Florida margin, carbonate and evaporate rocks that have been isolated from major clastic sediment inputs since the Jurassic, to the east of the cape. The surface sediments of the west Florida shelf (Florida Platform) are a "patchy veneer of shell hash and foraminiferal, algal and even oolitic sands" (Doyle and Sparks 1980, 906). Seaward of the carbonate sands the slope sediments are planktonic forams and coccoliths. On either side of the De Soto Canyon, clay inputs differ; the Mississippi discharge is mostly the clay mineral smectite, while the Apalachicola River sediments are dominated by kaolinite (Doyle and Sparks 1980; Balsam and Beeson 2003).

Morse and Beazley (2008) reported total and organic carbon concentrations for the northern GOM, which varied from 9.1% to 1.4% by weight for inorganic plus organic carbon, 1.3% to 0.37% for organic carbon, and 1.4% to 0.55% for organic carbon on a carbonate-free basis. These authors state that the "carbonate-free" term is the better way for reporting such information, as calcium carbonate ($CaCO_3$) has been found to widely act as a "dilutant" to other sedimentary parameters. In the 28 sediment samples they analyzed, $CaCO_3$ varied from 5% to 72%. These values were greatest in the eastern Gulf and lowest near the Mississippi River mouth. The sedimentary organic carbon fraction decreased with increasing water depth to 4 km. Macrofaunal and meiofaunal abundance decreased with water depth and the associated decreasing organic carbon content. Benthic fluxes, organic matter burial rates, and electron acceptors vary strongly as a function of water depth in the Gulf (Morse and Rowe 1999).

The degree of pyritization is the ratio of pyrite iron to (pyrite iron + reactive iron) and is related to the depositional environment of the sediment. In Gulf sediments the degree of pyritization and iron sulfide content follows the same trends as those described in the previous paragraph and are high in Texas/Louisiana shelf sediments and decline offshore (Lin and Morse 1991). In typical northern Gulf deepwater sediments the degree of pyritization and sulfur content is low, unless the area is impacted by hydrocarbon seepage (Morse et al. 2002). At seep sites found at the base of the west Florida Escarpment, where methane- and ammonia-rich brines seep out onto the seafloor (Chanton et al. 1991; Martens et al. 1991), sediments can be as high as 20% sulfur by weight, predominantly as pyrite, with the degree of pyritization as high as 0.9 (Commeau et al. 1987; Chanton et al. 1993).

An excellent resource for additional information about GOM sediments, including those of its bays and estuaries, can be found in Davis (2017). This resource describes the mineralogy, grain size, and sedimentation patterns in different provinces of the Gulf in detail.

Sources and Transport of Organic Matter

The main sources of organic material to the Gulf sediments are river inputs, surface and benthic primary production, inputs of organic matter associated with oil and gas, and seafloor seepage–supported chemosynthetic production. Essentially organic matter enters Gulf sediments from either the top down (riverine, primary production) or from the bottom up (seepage), and this material is subsequently distributed through the system via physical circulation.

Terrestrial/Riverine Sources

The Mississippi and Atchafalaya Rivers deliver 60% of the suspended matter (2.1×10^{14} g) and 66% of the dissolved materials (3.1×10^{12} g of organic carbon) transported from the conterminous United States to the ocean. As the Mississippi River dominates so much of the sedimentation in the GOM, one might expect it to dominate the sources of organic matter to the Gulf (Bianchi et al. 1997). In fact, the input of terrestrial organic matter to the oceans is a difficult input to quantify, and there is considerable uncertainty regarding the extent of terrestrial organic matter inputs to both the GOM and the global ocean (Burdige 2005; Bianchi 2011). The issue is complicated because the available source tracers are not straightforward. Stable carbon isotope tracing is somewhat confounded because a mixture of C3 ($\delta^{13}C\sim$-26‰) and C4 ($\delta^{13}C\sim$-14‰) plants can generate what appears to be organic matter of marine origin ($\delta^{13}C\sim$-20‰). Lignin-derived CuO oxidation products, which are uniquely derived from terrestrial land plants (e.g., Hedges and Ertel, 1982), have been used extensively to understand the sources and distribution of terrestrial organic carbon (OC_{terr}) along the Mississippi/Atchafalaya margins (e.g., Hedges and Ertel, 1976; Goñi et al., 1997, 1998; Gordon and Goñi, 2003, 2004; Bianchi et al., 2006; Sampere et al., 2008). These studies show that the contributions of lignin phenols to overall sedimentary OC are highest in the inner shelf regions and decrease with distance from land. Sorting and differential sedimentation occur across the nearshore to the offshore (Bianchi et al. 2002). For example, larger lignin-rich C3 material settles close to shore while degraded C4 and lignin-poor material on the surface of clays may be carried to deeper waters (Gordon and Goñi 2003, 2004).

Burdige (2005) synthesized a large amount of data and surmised that terrestrial organic matter makes up 44 ± 13% of global continental margin sediments and 36 ± 11% of all marine sediments. Deltaic sediments have greater terrestrial content, and the northern Gulf certainly falls into this category. Burdige pointed out that the more refractory nature of terrestrial organic matter might reduce its impact on element cycling relative to marine organic matter. Such observations are consistent with findings from Goñi et al. (1997, 1998), where data indicate that the terrestrial organic matter that is delivered offshore is rather degraded.

Initially, the role of terrestrial organic matter in the sediments of the northern GOM presented a relatively straightforward picture. Hedges and Parker (1976) used lignin concentrations, an indicator of terrestrial organic matter, and stable carbon isotopic ratios to describe a fairly rapid decrease in terrestrially-derived organic material from onshore to offshore. They defined lignin by the parameter lambda (λ), which is the total weight of vanillyl and syringyl index phenols produced from the mild oxidation of sedimentary organic carbon. They examined 10 transects ringing the Gulf from offshore Louisiana west of the Mississippi River mouth to south of Tuxpan, Mexico (20.95°N). In general, λ declined by 80% in sediment samples from the coast to sediments collected at 150 m water depth, depending on local river input and the width of the continental shelf. The $\delta^{13}C$ of organic matter was for the most part linearly related to Λ_8 and changed from onshore values of -23‰ to -24‰ to a marine endmember ranging from -19‰ to -20‰ (Hedges and Parker 1976). Land plants (C3) and freshwater sediments were reported to have an average $\delta^{13}C$ value of -26‰, with a λ value near 9. Sediments from the outer shelf generally contained less than 10% terrestrial organic matter.

In a series of papers, Goñi et al. (1997, 1998) reassessed the importance of terrestrially-derived organic material to the Gulf from 5 to 2000 m water depth. They suggested that the low λ and the relatively enriched $\delta^{13}C$ values of Gulf offshore sedimentary organic matter could be explained by a terrestrial source made up of highly degraded C4 organic matter eroded from the (formerly) extensive grasslands of the Mississippi River drainage basin. They argued that even in grasslands converted to agriculture, the soil would carry remnant geochemical signatures of former grasslands that were eroded and transported into the fluvial systems and delivered to the Gulf. Goñi et al. (1997, 1998) found that changes in the cinnamyl:vanillyl ratio of sedimentary organic matter as one moves offshore were consistent with the increasing importance of lignin derived from grasses (rather than leaves) in offshore sediments. Compound-specific isotopic signatures of lignin phenols provided a signal of land-derived organic matter and also became ^{13}C enriched onshore to offshore in a manner consistent with 2 distinct pools of terrestrial organic matter. One of these pools was hypothesized to be significantly degraded and sorbed to clay particles. This pool was derived from the soils of grassland (C4) and deposited in deeper water. A second pool comprised fresher plant detritus and carried a stronger gymnosperm, C3 signature; these materials sedimented out closer to shore. Goñi et al. (1998) concluded that most of the terrestrial carbon in offshore sediments has a highly degraded fraction derived from C4 plants. They suggested that this source might be important for other marine basins as well. While they did suggest the importance of terrestrial inputs to the Gulf, in contrast to Hedges and Parker (1976), they did not specify a percentage contribution. Instead, they stated that the input of C4-derived organic material modifies the isotopic value of terrestrial organic carbon in offshore Gulf sediments such that it is indistinguishable from marine-derived organic carbon.

The findings of Goñi et al. (1997, 1998) were confirmed by Gordon and Goñi (2003, 2004). Using a 3-endmember mixing model, these authors reported that terrigenous organic matter accounts for about 65–80% of the organics deposited within the 25 m isobath west of the Mississippi River and offshore the mouth of the Atchafalaya River.

Additional work has further revealed that particulate organic carbon (POC) introduced by the Mississippi to nearshore waters is augmented by organic matter fixed by phytoplankton in the water column of the Louisiana shelf. This material is carried along-shelf (east to west), decomposed, buried, or transported to deeper regions in the GOM primarily via the Mississippi Canyon, which is southwest of the Mississippi River delta where it protrudes into the Gulf (see Fig. 7.3). Initially upon exiting the river channel, Mississippi River–borne particulates and sediments are deposited on the continental shelf in 20–100 m of water within 30 km of the river mouth to the west of the Mississippi River's Southwest Pass (89.4°W), albeit somewhat temporarily. Remobilization of this material and its export are driven by resuspension, mass movement along the bottom, and wind-driven (downwelling) transport of benthic boundary layers (BBL) during high-energy events in winter months or hurricane activity in the summer (Baker and Hickey 1986; DiMarco and Reid 1998; Allison et al. 2000; Allison et al. 2010; Walsh et al. 2006). Estimates suggest that 40–50% of sediment initially deposited is mobilized in this manner (Corbett et al. 2004; Corbett et al. 2006).

The particle-reactive radionuclides ^{7}Be and ^{210}Pb are robust tracers of recent river-derived inputs to the coastal zone (Olsen et al. 1986; Sommerfield et al. 1999; Allison et al. 2000; Corbett et al. 2004; Corbett et al. 2006). Changes in the spatial distribution of the ^{7}Be inventory during a high-flow period in April and a low-flow period in October 2000 demonstrated both the importance of the river as the ^{7}Be source and the potential effects of sediment remobilization during the winter (Corbett et al. 2004). Episodic mass movements are also well known along the Southwest Pass delta front (Coleman et al. 1998) and have been suggested to be a major mechanism for spin-off turbidity currents and nepheloid layers observed along the continental margin (Mulder and Alexander 2001). These processes of deposition and subsequent transport further sort river-derived organic material with woody C3 tissues preferentially sinking out of the water column in the lower river basin and along the inner shelf (Bianchi et al. 2002).

The mobilized material is transported into deeper water via the Mississippi Canyon, which lies farther to the west at 89.9°W. As the sediments are transported across the shelf, the quality of the organic matter carried to the canyon is enhanced by shelf primary production, albeit to a decreasing extent with increasing water depth (Rowe et al. 2008). Much of the in situ plume productivity supporting carbon flux is made up of diatoms (Lohrenz et al. 1999), as reflected in pigment concentrations of surface sediments below the plume (Wysocki et al. 2006). Vertical fluxes of POC in the Mississippi plume can be high (Redalje et al. 1994).

Anthropogenic impacts on the Mississippi River system have simultaneously reduced the mineral content of exported particles (Allison et al. 2007) and increased organic content. Reduced suspended inorganic sediment load results from trapping of mineral particles behind water-control structures. An increased nutrient load due to agricultural practices (fertilizer application) in the Mississippi Valley has increased phytoplankton production in the river and on the shelf and increased the export of organic detritus to deeper waters. Inputs from Barataria Bay and local marshes may also be important to offshore waters (Sampere et al. 2008).

Using a different approach, lipid biomarkers, Waterson and Canuel (2008) assessed the input of terrestrial organic matter to the Gulf. They proportioned the organic material into autochthonous and terrigenous categories and found that the deep sediments of the slope and canyon are 66–73% autochthonous and 27–34% terrestrial, while shelf sediments are 64% autochthonous and 36% terrestrial. Lignin-phenol biomarkers suggested that a fraction of total offshore organic material is represented by terrestrial components, presumably due to selective removal of the more labile fractions during transport (e.g., Mead and Goñi 2006). Examination of the content and composition of lignin products among samples of different size, surface area, and density (e.g., Gordon and Goñi 2003;

Sampere et al. 2008) show there are significant contrasts in the lignin content and compositional signatures among particles with different hydrodynamic properties. The sorting of these particles during offshore transport contributes to the spatial patterns of lignin distribution, which is also affected by degradation. Hence, for example, coarse vascular plant detritus, enriched in lignin phenols, contributes to the high lignin contents observed in the lower Mississippi River and inner-shelf regions (e.g., Bianchi et al. 2002; Gordon and Goñi 2003). Low-density, organic-rich particles with a marsh-lignin signature appear to be degraded significantly during transport (e.g., Sampere et al. 2008), whereas lithogenic-rich fine particles containing lignin derived from aged soil sources are distributed farther offshore with much less alteration (e.g., Goñi et al. 1998; Gordon and Goñi 2004; Sampere et al. 2008). The importance of a marsh component (Bianchi 2011) was not considered in the conceptual models of offshore transport suggested by Goñi et al. but would complicate matters further due to their C4 isotopic signature, which is similar to that of grassland carbon.

Primary Production of Organic Matter

Surface primary production in the northern Gulf is fueled by nutrients from the Mississippi River and can be on the order of 300–3500 g C m^{-2} yr^{-1} with a large seasonal range (Lohrenz et al. 1999; Lohrenz et al. 2014). Primary production ranges from 85 to 160 g C m^{-2} yr^{-1} on the Mexican and Texas shelves and from 30 to 475 g C m^{-2} yr^{-1} on the west Florida shelf (Vargo et al. 1987; Benway and Coble 2014). Of course, much of this production is grazed, which has a significant impact on the vertical carbon flux (Dagg et al. 2007). Estimates of carbon flux from sediment traps in the Mississippi River plume reach as high as 657 g C m^{-2} yr^{-1} in spring but are lower during other seasons (110–345 g C m^{-2} yr^{-1}) (Redalje et al. 1994). Away from the plume, rates are 66–146 g C m^{-2} d^{-1} (Qureshi 1995, cited in Dagg et al. 2007) and 5–40 g C m^{-2} yr^{-1} (Yan et al. 2016).

High primary production coupled with high freshwater inputs and summer stratification of the water column result in annual formation of a hypoxic water body in nearshore Louisiana waters west of the Mississippi River mouth (Dybas 2005). While this feature appeared early in the twentieth century or even earlier, it has become more prevalent since the 1950s as the nitrogen (N) flux from the river has tripled. The hypoxic water column has resulted in an oxygen-starved benthic community and increased anoxic conditions in underlying sediments (Rabalais et al. 2002). Increased organic carbon loading to the sediments has increased oxygen demand, exacerbating the dead zone effect. In fact, sediment oxygen demand may now be greater than water-column oxygen demand (Turner et al. 2008). In addition, the sediment forms a long-term organic carbon reservoir that buffers the continued persistence of the hypoxic zone. The associated benthic-to-water column nutrient flux from the decay of sedimentary organic matter contributes to the persistence of the hypoxic zone and will increase the difficulty in finding a remedy for the situation.

Benthic production estimates are fairly limited but were reported by Lohrenz et al. (2014) to be as high as 100–150 g C m^{-2} yr^{-1}. Moncreiff et al. (1992) estimated rates varying from 7.9 to 10,000 g C m^{-2} yr^{-1} for benthic production in the Mississippi Sound from winter to summer. Fourqurean et al. (2001) estimated rates of 18 to 1200 g C m^{-2} yr^{-1} for Florida Bay. These estimates include production by seagrass, seagrass epiphytes, and benthic microalgae. The areal extent of this production is uncertain, but there are large shallow areas with clear water in the Florida Big Bend, the large embayment to the west of Cape San Blas (see Figure 7.3; Iverson and Bittaker 1986) and on the west Florida shelf (Huettel et al. 2014). The seagrass ecosystem in the Florida Big Bend is a highly productive coastal habitat extending more than 3000 km^2 (Iverson and Bittaker 1986; Zieman and Zieman 1989) and provides nursery habitat for many of the ecologically and economically important species on the west Florida shelf. Permeable sandy sediments, which are common in the eastern Gulf, support elevated rates of benthic primary production as a result of mixing of water column and pore waters, supported by pressure gradients generated by wave and current action (Jahnke et al. 2005; Jahnke et al. 2008; Huettel et al. 2014). To the south, Florida Bay and surrounding areas contain 5500 km^2 of seagrass dominated by *Thalassia testudinum* but also contain *Halodule wrightii* and *Syringodium filiforme* (Iverson and Bittaker 1986; Zieman et al. 1989).

Inputs of Organic Matter from Seepage

The seafloor of the northern GOM is peppered with hydrocarbon springs, or cold seeps, as they were originally called, to contrast them from somewhat similar chemosynthetic communities at the hot vents discovered at spreading centers (Paull et al. 1984; Kennicutt et al. 1988; MacDonald et al. 1989; Sassen et al. 1993; Roberts and Aharon 1994;

Roberts and Carney 1997; Fisher et al. 2007). An excellent source of seep maps and their interpretation is available at the Bureau of Ocean Energy websites (https://www.boem.gov/Seismic-Water-Bottom-Anomalies-Map-Gallery/ and https://www.boem.gov/GOM-Seismic-Seafloor-Amplitude-Anomaly-Distribution/). At the core of these seeps lies the Louann salt formation, a widely distributed evaporite layer that precipitated on the Gulf floor during the Jurassic Period when the North and South American Plates separated. It has been postulated that there was a shallow sill that led to the isolation of the Gulf as a semi-enclosed basin (Salvador 1987, 1991; Hine 2013). In its early stages of formation, the Gulf was considerably larger. The basin and its flooring evaporite layer extended as far north as Oklahoma (Hine 2013). Over time, the evaporate layer was overlain with thick layers of sediment from the erosion of the Appalachian Mountains and later from the Laramide Orogeny and subsequent Rocky

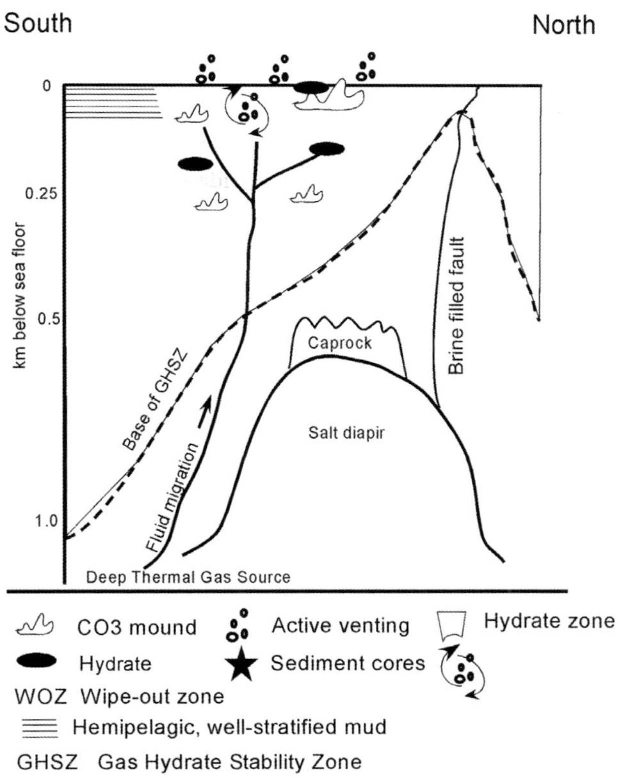

Figure 7.5. Faults in the rock and sediments overlying salt domes allow the upward movement of oil and gas toward the seafloor (Lapham 2007). The wipeout zone is the area indicated by the circular arrows around the areas of active venting, shown by the bubble-like symbol. Sediment cores were collected across these sites.

Mountains uplift. On the coastal plain and the Louisiana shelf and slope the sedimentary layers compressed the salt layers, causing them to create diapirs, vertical salt tongues that rise up through these sedimentary strata. These vertical movements created faults in the overburden that allow trapped oil and gas, formed at depth from the thermal degradation of sedimentary organic matter under high pressure, to flow upward (Fig. 7.5; Sassen et al. 1993; Sassen et al. 2001; Roberts and Carney 1997; Roberts 2001). This process has led to the rugged topography observed on the Louisiana and Texas shelves and slopes and in the Campeche basin (see Fig. 7.3). These areas have a plethora of seep expressions on the seafloor. The locations of the natural seeps have been tied to surface slicks, using images from satellites and remotely operated vehicles (MacDonald et al. 1993; MacDonald et al. 2015).

At seep sites the upwelling hydrocarbon-rich gases and fluids result in microbial plaques on the seafloor that are surrounded by chemosynthetic megafaunal communities, the most abundant being tubeworms (*Lamellibrachia* sp.) and mussels (Lloyd et al. 2010; Orcutt et al. 2010). These tubeworms appear to be ~200 years old, and their slow growth rates contrast remarkably with the rapid growth rate of tubeworms at the hydrothermal vents (Bergquist et al. 2000). Exposed gas hydrates are frequently observed breaching the seafloor at these sites, and they can break away and float upward (MacDonald et al. 1994). Exposed hydrates are stable under the conditions of seafloor temperature and pressure within the hydrate stability zone but are out of equilibrium with dissolved methane (the water bathing them is undersaturated in methane), so they slowly dissolve rather than dissociate into free gas (Lapham, Chanton, Martens, Higley, et al. 2008; Lapham et al. 2010; Lapham et al. 2014). Where salt domes intersect the seafloor, brine pools form salty lakes on the seafloor that are surrounded by apparently beachgoing mussels, which densely inhabit the edges of the brine pools (Cary et al. 1988; MacDonald et al. 1990; Joye et al. 2009).

Due to the high hydrocarbon flux, anaerobic respiration in the sediments is a very important process, particularly sulfate reduction (Boetius et al. 2000; Joye et al. 2004; Joye et al. 2009; Joye et al. 2010; Orcutt et al. 2005; Orcutt et al. 2010). Sulfate reduction generates alkalinity and the subsequent precipitation of copious amounts of authigenic carbonate (Sassen et al. 2004; Roberts et al. 2010). The $\delta^{13}C$ of these carbonates varies from -15‰ to -53‰ and follows the trend of methane $\delta^{13}C$ across multiple sites (sites are denoted on the *x*-axis in Fig. 7.6; Roberts et al. 2010). As

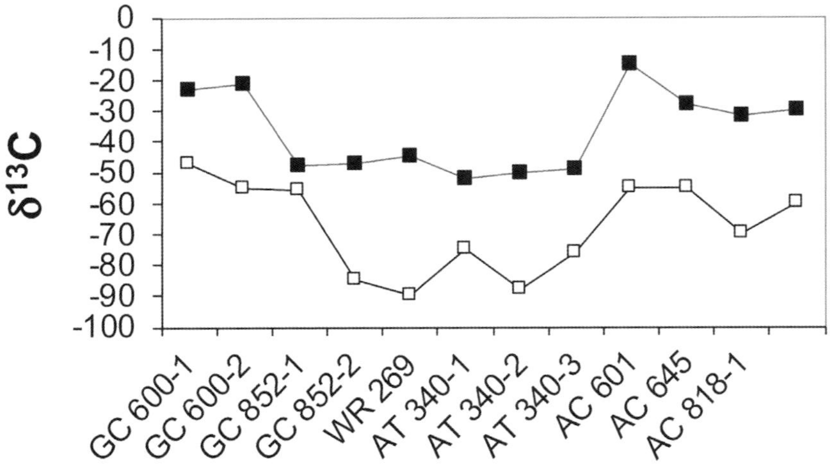

Figure 7.6. The $\delta^{13}C$ of authigenic carbonates (*filled squares*) follows the trend of methane (*open squares*) $\delta^{13}C$ across multiple seep sites, denoted on the *x*-axis (redrawn with permission from Roberts et al. 2010).

seeps age, these carbonates become more pronounced and result in hardgrounds surrounded by soft sediments. These hard-ground "islands" are colonized by coral communities (Fisher et al. 2007; Cordes et al. 2008; Cordes et al. 2009; Lessard-Pilon et al. 2010).

Sedimentary Biogeochemical Processes at Seeps

The presence of oil and gas seeping up from below enhances the organic carbon concentration of seep sediments, which can be as high as 11% (Lapham 2007). Organic matter can consist of oil or oil/methane hydrocarbons fixed into organic material by microbes and reprocessed along with organic material raining from above. This mixture is then available for microbial degradation through classic biogeochemical processes (Froelich et al. 1979). At seep sites most carbon remineralization is driven mainly by sulfate reduction and methanogenesis (Joye et al. 2004; Joye et al. 2010). Joye et al. (2010) characterized seep-influenced sedimentary habitats as (1) sites with no evidence of seepage; (2) sites of recent seepage with carbonate rock outcrops and chemosynthetic communities; (3) brine flows; (4) clam or mussel beds; (5) microbial mats, generally *Beggiatoa*; (6) sites with an abundance of oil; and (7) sites dominated by tubeworm communities. The greatest rates of sulfate reduction were found in sediments underlying the microbial mats, where rates were enhanced up to 1000-fold relative to background slope sediments (Joye et al. 2004; Joye et al. 2010; Orcutt et al. 2005; Orcutt et al. 2010).

Anaerobic oxidation of methane (AOM) has generated considerable interest, particularly at cold seeps (Reeburgh 1980; Alperin and Reeburgh 1984; Iversen and Jørgensen 1985; Valentine and Reeburgh 2000). Decades ago, Martens and Berner (1974) observed that the presence of sulfate inhibited the production of methane in marine sediments and later suggested that methane could be consumed in the sulfate-reducing zone (Barnes and Goldberg 1976; Martens and Berner 1977). However, at that time, it was generally believed that sulfate-reducing bacteria were quite limited in the types of substrates that they could utilize (Iversen and Blackburn 1981). But reports of AOM were undeniable, and Reeburgh (1980) and Iversen and Jørgensen (1985) linked methane oxidation to sulfate reduction and first reported the classic subsurface maxima in methane oxidation rates and sulfate reduction rates at the sulfate-methane transition zone.

As is generally the case in environmental science, geochemists led microbiologists to the discovery of a microbe or set of microbes that could conduct the reaction. Even now, the organism(s) responsible for AOM has not been isolated in pure culture. Hoehler et al. (1994) first suggested that the bulk of sulfate-dependent methane oxidation could be conducted by a methanogen-sulfate reducer consortium via hydrogen syntrophy. Boetius et al. (2000) then provided the first microscopic evidence for such an arrangement, observing dense clusters of methanogen-like archaea surrounded by sulfate-reducing bacteria. Recognition of the importance of this process has led to the interface of sulfate and methane in marine sediments being termed the "sulfate-methane transition zone" (SMTZ). This interface serves as a biofilter that reduces methane flux across the SMTZ (Reeburgh 2007).

Generally, it initially appeared that sulfate reduction rates in GOM sediments were decoupled from methane oxidation rates; that is, sulfate-reduction rates appeared considerably greater than rates of methane consumption (Orcutt et al. 2005; Joye et al. 2005; Joye et al. 2010; Bowles, Samarkin, Bowles, et al. 2011). This difference may result because incubations were conducted in seagoing laboratories at 1 atmosphere pressure while seafloor pressures are considerably greater. To the extent that AOM is dependent on methane concentrations, increasing the partial pressure of methane may be expected to increase methane oxidation rates. Methane has low solubility in water at atmospheric pressure, but at 500 m depth, the solubility will be 50 times greater. For a first-order process, this would increase rates 50-fold. Indeed, when pressurized incubation techniques were developed and applied and in situ methane concentrations could be replicated during rate assays, rates of AOM increased significantly (Bowles, Samarkin, and Joye 2011).

Seep Site Evolution and Development

Roberts and Carney (1997) first devised a seep evolution conceptual model for the GOM based on surficial sediments and seafloor features. Lapham, Chanton, Martens, Sleeper, et al. (2008) extended this picture of seepage rate and seafloor features by adding in geochemical profiles (Fig. 7.7). This model described high-fluid-flux areas as "mud-prone" vents that are typically associated with mud volcanoes, bacterial mats and clams, and other chemosynthetic communities and slightly biodegraded hydrocarbons (Fig. 7.7B). As the upward flux decreases, the mud-prone vents transition to active seeps. Microbial and macrofaunal activity increases, and extensive bacterial mats and gas hydrates form (Fig. 7.7C). Authigenic carbonate precipitation results from enhanced sulfate reduction. The physical formation of the hydrate and carbonate can then lead to the mineral-prone stage of the seep where the solids seal off or reduce the high-fluid-flow path, termed "the self-sealing nature of marine seeps" (Fig. 7.7D; Hovland 2002, 2390). These later-stage mineral-prone seeps thus are characterized by carbonate mounds, dead shells, and few bacterial mats, but the resultant hardgrounds may be colonized by coral communities (Fisher et al. 2007).

Lapham, Chanton, Martens, Sleeper, et al. (2008) conducted their work at Woolsey Mound, a seep site in Mississippi Canyon Lease Block 118 (MC118, 28.86°N, 88.49°W) in 850 m water, which is typical of seeps throughout the northern GOM (Macelloni et al. 2014). The site is a hydrate-carbonate mound complex that formed above a shallowly buried salt dome. The structure below the mounded area has been described through several seismic data sets (Macelloni et al. 2011; Macelloni et al. 2012; Simonetti et al. 2013) and reveals an extensive system of joints and faults associated with the deeply extended salt dome. The site is characterized by geophysical wipeout zones—blank zones in seismic interpretations where seismic signals have been reflected by either chaotic sediments or gas bubbles (Wilson et al. 2014).

Lapham, Chanton, Martens, Sleeper, et al. (2008) compared pore-water and solid phase geochemistry in 30 cores collected across Woolsey Mound within and outside wipeout zones and characterized them as a function of microbial activity. Outside wipeout zones at Woolsey Mound, shallow (~1 m) cores exhibited very low microbial activity, as evidenced by sulfate profiles remaining around seawater values down the core (Fig. 7.8). These cores could be considered typical of Gulf sediments at these water depths where no seepage occurs, as represented in Figure 7.7A. Cores collected overlying wipeout zones were classified as exhibiting moderate (Fig. 7.9) or elevated (Fig. 7.10) microbial activity and are shown in Figures 7.7B and 7.7C. Pore-water profiles with moderate activity exhibited sulfate depletion as low as 15 mM, with methane concentration as high as 56 μM. Cores with high microbial activity exhibited rapid sulfate depletion, and the formation of methane exceeded 0.5 mM. Organic carbon concentrations increased down core from 0.5% to 1.5%, consistent with a deep petroleum source. Similar high rates of sulfate reduction, anaerobic methane oxidation, and methane production have been reported in GOP seeps across the northern slope, as discussed previously.

Methane $\delta^{13}C$ isotopic composition varies, as shown in Figures 7.8 to 7.10, because of the comingling of sulfate and methane in a particular horizon, as AOM preferentially consumes the lighter isotope $^{12}CH_4$, leaving residual methane ^{13}C enriched (Alperin et al. 1988). At the depth of the SMTZ shoals with increasing microbial activity (see Figures 7.8 to 7.10), methane becomes increasingly ^{13}C depleted as a result of less AOM. Dissolved inorganic carbon (DIC) $\delta^{13}C$ varies according to the relative importance of methanogenesis, which causes carbon isotopic fractionation, in contrast to nonfractionating processes such as oxic or sulfate respiration. Methane production distributes the lighter isotope ^{12}C into methane, leaving residual CO_2 or DIC enriched in $\delta^{13}C$, while the nonfractionating respiration mechanisms produce CO_2 with an isotopic composition similar to that of the parent organic material (Corbett et al. 2013). A similar distribution of profiles as described in Figures 7.8 to 7.10 was noted across sites at the west Florida Escarpment (Chanton et al. 1991, 1993). Here, sulfate-depleted brine-rich fluids charged with ancient biogenic methane and ammonia

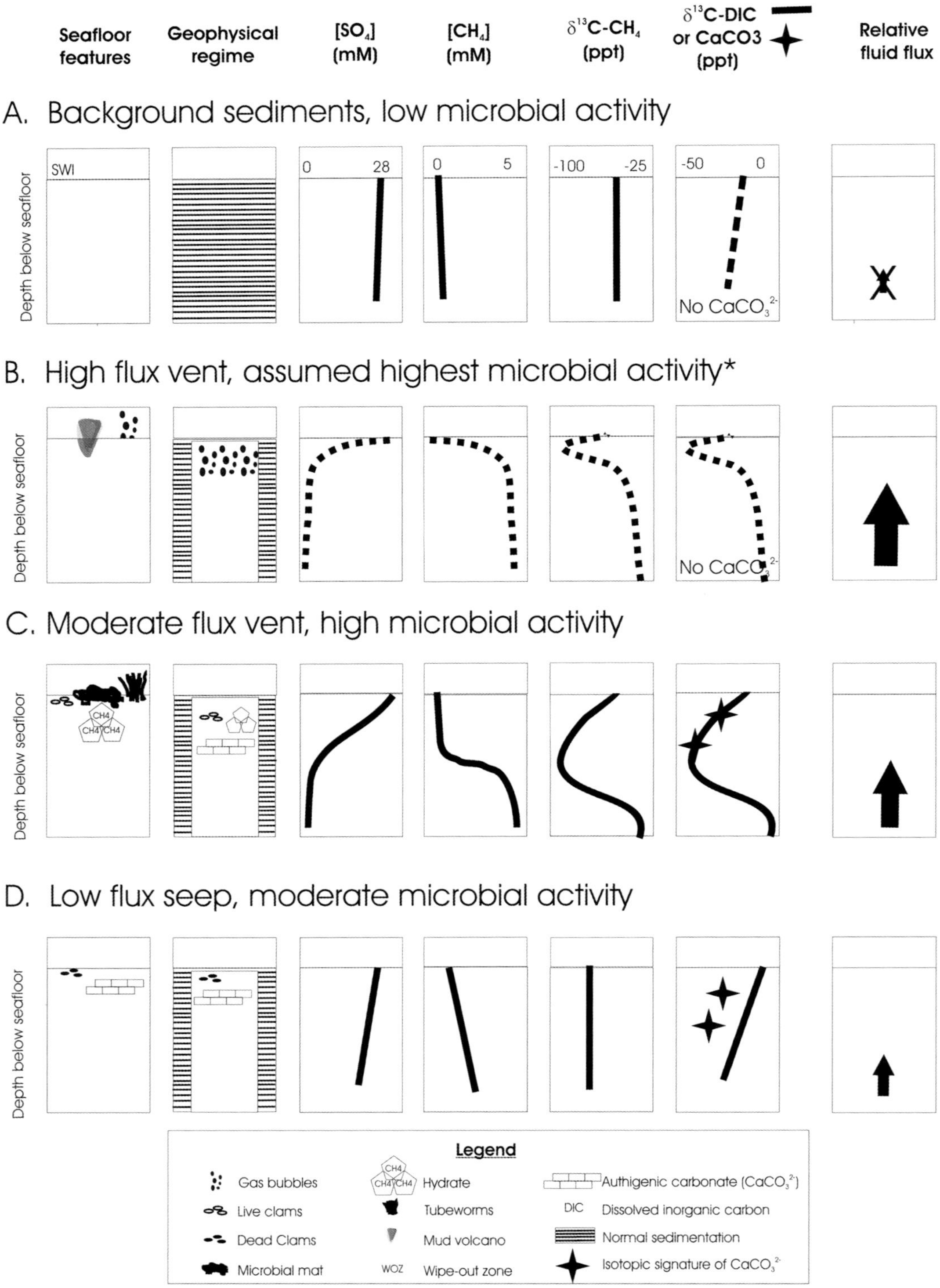

Figure 7.7. Seep-site evolution and corresponding evolution of geochemical profiles (Lapham, Chanton, Martens, Sleeper, et al. 2008). Stage *A* is background unaffected seafloor sediments, with low organic matter and vertical geochemical profiles. Stage *B* represents the highest fluid flux, a "mud volcano" with high rates of microbial activity due to high inputs of oil and gas from below. In stage *C*, the rate of fluid flux is attenuated by plugging of conduits by gas hydrates and carbonates. High microbial activity, microbial mats, and chemosynthetic communities are in evidence. In stage *D*, fluid flux is further attenuated, geochemical profiles become more vertical, and authigenic carbonates provide hardgrounds for the attachments of deep-sea corals. Wipeout zones are associated with gas bubbles. Numerals on the *x*-axis refer to the parameters horizontally across the top of the figure.

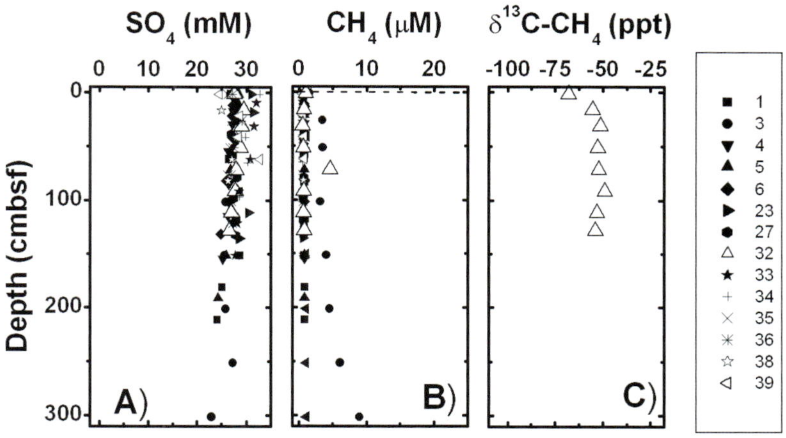

Figure 7.8. Geochemical profiles at low-microbial-activity sites typical of background locations in the Gulf not containing seeps (see Fig. 7.7A; Lapham, Chanton, Martens, Sleeper, et al. 2008). The overlapping symbols indicate similarity of concentrations at these sites.

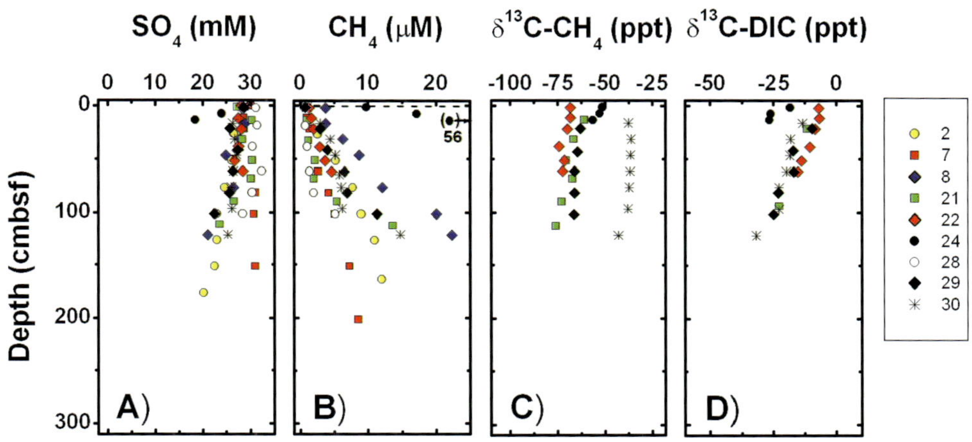

Figure 7.9. Geochemical profiles at Gulf of Mexico seep sites of moderate microbial activity sites (see Figure 7.7D; Lapham, Chanton, Martens, Sleeper, et al. 2008).

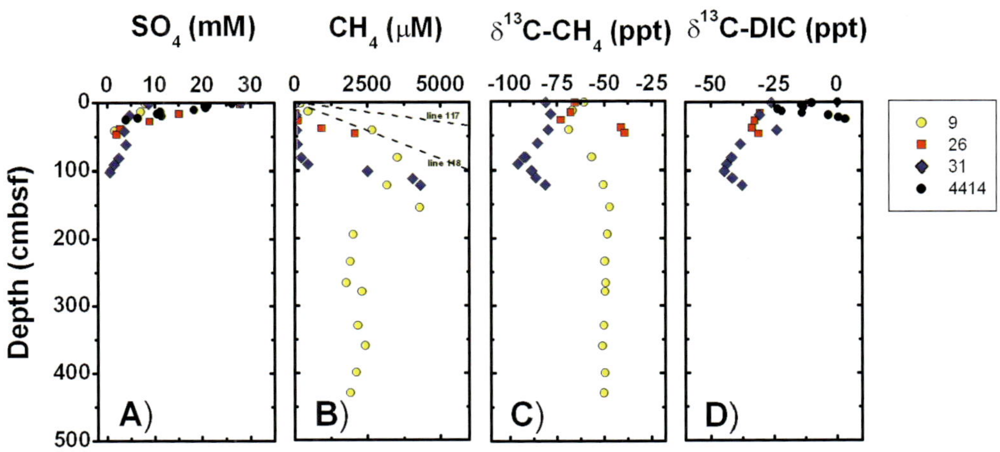

Figure 7.10. Geochemical profiles at Gulf of Mexico seep sites of high microbial activity sites (see Figure 7.7C; Lapham, Chanton, Martens, Sleeper, et al. 2008).

seep out onto the seafloor, fostering the growth of chemosynthetic communities and organic-rich and mineral-rich deposits (Paull et al. 1984; Paull et al. 1991; Commeau et al. 1987; Martens et al. 1991).

Conclusion

The seafloor of the GOM hosts an incredibly diverse range of benthic habitats supported by riverine deposits, sunlight, submarine seepage of water and hydrocarbons and is inhabited by a wide range of organisms of differing metabolisms. Sediments span the gamut from terrestrial domination to carbonate domination to hydrocarbon-seep-affected deposits. Currently attention is focused on determining the effects of the 2010 Deepwater Horizon oil spill (28.74°N, 88.37°W) on the ecosystems, seafloor, and water column of the Gulf (Joye et al. 2016). Determining the relative importance of carbon inputs by primary production, seepage, and river discharge is of great interest. The oil spill was a massive episodic input of fossil carbon that was qualitatively similar to that supplied by "natural" seepage (Chanton et al. 2015). Having a good baseline for petroleum carbon inputs and their ecosystem consequences would aid in the assessment of future extraordinary inputs if and when they occur.

Acknowledgments

We thank David Burdige and Gil Rowe for helpful reviews of this chapter. Samantha Bosman worked to get permissions for figure reproduction. Several of the authors have been supported by the Gulf of Mexico Research Initiative through its consortia over the years: Ecosystem Impacts of Oil & Gas Inputs to the Gulf (ECOGIG), Center for the Integrated Modeling and Analysis of the Gulf Ecosystem (CIMAGE), and Deep-C.

References

Allison, M. A., T. S. Bianchi, B. A McKee, and T. P. Sampere. 2007. Carbon burial on river-dominated continental shelves: Impact of historical changes in sediment loading adjacent to the Mississippi River. *Geophysical Research Letters* 34:1407–1410.

Allison, M. A., T. M. Dellapenna, E. S. Gordon, S. Mitra, and S. T. Petsch. 2010. Impact of Hurricane Katrina (2005) on shelf organic carbon burial and deltaic evolution. *Geophysical Research Letters* 37:L21605. doi:10.1029/2010GL044547.

Allison, M. A., G. C. Kineke, E. S. Gordon, and M. A. Goñi. 2000. Development and reworking of an annual flood deposit on the inner continental shelf off the Atchafalaya River. *Continental Shelf Research* 20:2267–2294.

Alperin, M. J., and W. S. Reeburgh 1984. Geochemical observations supporting anaerobic methane oxidation. Pp. 282–289 in R. Crawford and R. Hanson (eds.), *Microbial Growth on C-1 Compounds*. Washington, DC: American Society for Microbiology Press.

Alperin, M. J., W. S. Reeburgh, and M. J. Whiticar. 1988. Carbon and hydrogen isotope fractionation resulting from anaerobic methane oxidation. *Global Biogeochemical Cycles* 2:279–288.

Baker, E. T., and B. M. Hickey. 1986. Contemporary sedimentation processes in and around an active west coast submarine canyon. *Marine Geology* 71:15–34.

Balsam, W. L., and J. P. Beeson. 2003. Sea-floor sediment distribution in the Gulf of Mexico. *Deep-Sea Research I: Oceanographic Research Papers* 50:1421–1444.

Barnes, R. O., and E. D. Goldberg. 1976. Methane production and consumption in anoxic marine sediments. *Geology* 4:297–300.

Benway, H. M., and P. G. Coble (eds.). 2014. *Report of the US Gulf of Mexico Carbon Cycle Synthesis Workshop*. Woods Hole, MA: Ocean Carbon and Biogeochemistry Program and North American Carbon Program. 67 pp.

Bergquist, D. C., F. M. Williams, and C. R. Fisher. 2000. Longevity record for deep-sea invertebrate. *Nature* 403(6769):499–500.

Bernard, B. B., J. M. Brooks, and W. M. Sackett. 1976. Natural gas seepage in the Gulf of Mexico. *Earth and Planetary Science Letters* 31(1):48–54.

Bianchi, T. S. 2011. The role of terrestrially-derived organic carbon in the coastal ocean: A changing paradigm and the priming effect. *Proceedings of the National Academy of Sciences* 198:19473–19481.

Bianchi, T. S., C. D. Lambert, P. H. Santschi, and L. Guo. 1997. Sources and transport of land-derived particulate and dissolved organic matter in the Gulf of Mexico (Texas shelf/slope): The use of lignin-phenols and loliolides as biomarkers. *Organic Geochemistry* 27:65–78.

Bianchi, T. S., S. Mitra, and B. A. McKee. 2002. Sources of terrestrially-derived organic carbon in lower Mississippi River and Louisiana shelf sediments: Implications for differential sedimentation and transport at the coastal margin. Marine Chemistry 77:211–223.

Boetius, A., K. Ravenschlag, C. J. Schubert, D. Rickert, F. Widdel, A. Gieseke, R. Amann, B. B. Jorgensen, U. Witte, and O. Pfannkuche. 2000. A marine microbial consortium apparently mediating anaerobic oxidation of methane. *Nature* 407:623–626.

Bowles, M. W., V. A. Samarkin, K. M. Bowles, and S. B. Joye. 2011. Weak coupling between sulfate reduction and the anaerobic oxidation of methane in methane-rich seafloor sediments during ex situ incubation. *Geochimica et Cosmochimica Acta* 75:500–519.

Bowles, M. W., V. A. Samarkin, and S. B. Joye. 2011. Improved measurement of microbial activity in deep-sea sediments at in situ pressure and methane concentration. *Limnology Oceanography Methods* 9:499–506. doi:10.4319/lom.2011.9.499.

Burdige, D. J. 2005. Burial of terrestrial organic matter in marine sediments: A re-assessment. *Global Biogeochemical Cycles* 19:GB4011. doi:10.1029/2004GB002368.

Burnett, W. C., H. Bokuniewicz, and M. Huettel. 2003. Groundwater and pore water inputs to the coastal zone. *Biogeochemistry* 66:3. doi:10.1023/B:BIOG.0000006066.21240.53.

Cable, J. E., G. C. Bugna, W. C. Burnett, and J. P. Chanton. 1996. Application of ^{222}Rn and CH_4 for assessment of groundwater discharge to the coastal ocean. *Limnology and Oceanography* 41:1347–1353.

Cable, J., W. Burnett, J. Chanton, and G. A. Weatherly. 1996. Estimating groundwater discharge into the NE Gulf of Mexico using ^{222}Rn. *Earth and Planetary Science Letters* 144:591–604.

Cary, S. C., C. R. Fisher, and H. Felbeck. 1988. Mussel growth supported by methane as sole carbon and energy source. *Science* 240(4848):78–80.

Chanton, J. P., C. S. Martens, and C. K. Paull. 1991. Control of pore-water chemistry at the base of the Florida Escarpment by processes within the platform. *Nature* 349:229–231.

Chanton, J. P., C. S. Martens, C. K. Paull, and J. A. Coston. 1993. Sulfur isotope and porewater geochemistry of Florida Escarpment seep sediments. *Geochimica et Cosmochimica Acta* 57:1257–1266.

Chanton, J. P., T. Zhao, B. Rosenheim, S. B. Joye, S. Bosman, C. Brunner, K. Yeager, A. R. Diercks, and D. Hollander. 2015. Radiocarbon tracing of the flux of petrocarbon to the sea floor associated with the Deep Water Horizon event. *Environmental Science & Technology* 49:847–854.

Coleman, J. M., H. H. Roberts, and G. W. Stone. 1998. Mississippi River delta: An overview. *Journal of Coastal Research* 14(3):698–716.

Commeau, R. F., C. K. Paull, J. A. Commeau, and L. J. Poppe. 1987. Chemistry and mineralogy of pyrite-enriched sediments at a passive margin sulfide brine seep: Abyssal Gulf of Mexico. *Earth Planet Science Letters* 82:62–74.

Corbett, D. R., B. A. McKee, and M. A. Allison. 2006. Nature of decadal-scale sediment accumulation in the Mississippi River deltaic region. *Continental Shelf Research* 26:2125–2140.

Corbett, D. R., B. A. McKee, and D. Duncan. 2004. An evaluation of mobile mud dynamics in the Mississippi River deltaic region. *Marine Geology* 209:91–112.

Corbett, J. E., M. M. Tfaily, D. J. Burdige, W. T. Cooper, P. H. Glaser, and J. P. Chanton. 2013. Partitioning peat respiration with stable carbon isotopes. *Biogeochemistry* 114:327–340.

Cordes, E. E., D. C. Bergquist, and C. R. Fisher. 2009. Macro-ecology of Gulf of Mexico cold seeps. *Annual Review of Marine Science* 1:143–168.

Cordes, E. E., M. P. McGinley, E. L. Podowski, E. L. Becker, S. Lessard-Pilon, S. T. Viada, and C. R. Fisher. 2008. Coral communities of the deep Gulf of Mexico. *Deep Sea Research I: Oceanographic Research Papers* 55(6):777–787.

Dagg, M. J., J. W. Ammerman, R. M. W. Amon, W. S. Gardner, R. E. Green, and S. E. Lohrenz. 2007. A review of water column processes influencing hypoxia in the northern Gulf of Mexico. *Estuaries and Coasts* 30:735–752.

Davis, R. A. 2017. Sediments of the Gulf of Mexico. Pp. 165–215 in C. H. Ward (ed.), *Habitats and Biota of the Gulf of Mexico: Before the Deepwater Horizon Oil Spill*. New York: Springer-Verlag. doi:10.1007/978-1-4939-3447-8_3.

DiMarco, S. F., and R. O. Reid. 1998. Characterization of the principal tidal current constituents on the Texas-Louisiana shelf. *Journal of Geophysical Research* 103(2):3093–3110.

Doyle, L. J., and T. N. Sparks. 1980. Sediments of the Mississippi, Alabama and Florida (MAFLA) continental shelf. *Journal of Sedimentary Petrology* 50:905–916.

Dybas, C. L. 2005. Dead zones spreading in world oceans. *AIBS Bulletin* 55(7):552–557.

Fisher, C. R., H. H. Roberts, E. Cordes, and B. Bernard. 2007. Cold seeps and associated communities of the Gulf of Mexico. *Oceanography* 20:68–79.

Fourqurean, J., W. A. Willsie, C. D. Rose, and L. M. Rutten. 2001. Spatial and temporal pattern in seagrass community composition and productivity in South Florida. *Marine Biology* 138(2):341–354.

Froelich, P., G. P. Klinkhammer, M. A. Bender, N. A. Luedtke, G. R. Heath, D. Cullen, and V. Maynard. 1979. Early oxidation of organic matter in pelagic sediments of the eastern equatorial Atlantic: Suboxic diagenesis. *Geochimica et Cosmochimica Acta* 43(7):1075–1090.

Goñi, M. A., K. C. Ruttenberg, and T. I. Eglinton. 1997. Source and contribution of terrigenous organic carbon to surface sediments in the Gulf of Mexico. *Nature* 389:275–278.

———. 1998. A reassessment of the sources and importance of land-derived organic matter in surface sediments from the Gulf of Mexico. *Geochimica et Cosmochimica Acta* 62(18):3055–3075.

Gordon, E. S., and M. A. Goñi. 2003. Sources and distribution of terrigenous organic matter delivered by the Atchafalaya River to sediments in the northern Gulf of Mexico. *Geochimica et Cosmochimica Acta* 67:16–26.

———. 2004. Controls on the distribution and accumulation of terrigenous organic matter in sediments from the Mississippi and Atchafalaya River margin. *Marine Chemistry* 92:22–31.

Gould, H. R., and R. H. Stewart. 1956. Continental terrace sediments in the northeastern Gulf of Mexico. *Society* of *Economic Paleontologists and Mineralogists Special Publication* 3:2–19.

Harbison, R. N. 1968. Geology of De Soto Canyon. *Journal of Geophysical Research* 73:5175–5185.

Hedges, J. I., and P. L. Parker. 1976. Land-derived organic matter in surface sediments from the Gulf of Mexico. *Geochimica et Cosmochimica Acta* 40:1019–1029.

Hine, A. C. 2013. *Geologic History of Florida—Major Events That Formed the Sunshine State*. Gainesville: University Press of Florida. 229 pp.

Hoehler, T. M., M. J. Alperin, D. B. Albert, and C. S. Martens. 1994. Field and laboratory studies of methane oxidation in an anoxic marine sediment—evidence for a methanogen-sulfate reducer consortium. *Global Biogeochemical Cycles* 8:451–463.

Hovland, M. 2002. On the self-sealing nature of marine seeps. *Continental Shelf Research* 22:2387–2394.

Huettel, M., P. Berg, and J. E. Kostka. 2014. Benthic exchange in biogeochemical cycling and permeable sediments. *Annual Review of Marine Science* 6:23–51. doi:10.1146/annurev-marine-051413-012706.

Iverson, R. L., and H. Bittaker. 1986. Seagrass distribution and abundance in eastern Gulf of Mexico coastal waters. *Estuarine, Coastal and Shelf Science* 22:577–602.

Iversen, N., and T. H. Blackburn. 1981. Seasonal rates of methane oxidation in anoxic marine sediments. *Applied and Environmental Microbiology* 41(6):1295–1300.

Iversen, N., and B. B. Jørgensen. 1985. Anaerobic methane oxidation rates at the sulfate-methane transition in marine sediments from Kattegat and Skagerrak (Denmark). *Limnology and Oceanography* 30(5):944–955.

Jahnke, R. A., J. R. Nelson, M. E. Richards, C. Y. Robertson, A. M. F. Rao, and D. B. Jahnke. 2008. Benthic primary productivity on the Georgia midcontinental shelf: Benthic flux measurements and high-resolution, continuous in situ PAR records. *Journal of Geophysical Research* 113:80–94.

Jahnke, R. A., M. E. Richards, J. R. Nelson, C. Y. Robertson, A. M. F. Rao, and D. Jahnke. 2005. Organic matter remineralization and porewater exchange rates in permeable South Atlantic Bight continental shelf sediments. *Continental Shelf Research* 25:1433–1452.

Joye, S. B., A. Boetius, B. N. Orcutt, J. P. Montoya, H. N. Schulz, M. Erickson, and S. K Lugo. 2004. The anaerobic oxidation of methane and sulfate reduction in sediments from Gulf of Mexico cold seeps. *Chemical Geology* 205:219–238.

Joye, S. B., M. W. Bowles, V. A. Samarkin, K. S. Hunter, and H. Niemann. 2010. Biogeochemical signatures and microbial activity of different cold seep habitats along the Gulf of Mexico lower slope. *Deep Sea Research II: Tropical Studies in Oceanography* 57:1990–2001.

Joye, S. B., A. Bracco, T. Özgökmen, J. P. Chanton, M. Grosell, I. MacDonald, E. E. Cordes, J. P. Montoya, and U. Passow. 2016. The Gulf of Mexico ecosystem, six years after the Macondo oil well blowout. *Deep Sea Research II: Tropical Studies in Oceanography* 129:4–19.

Joye, S. B., V. A. Samarkin, I. R. MacDonald, K. U. Hinrichs, M. Elvert, A. P. Teske, and C. D. Meile. 2009. Metabolic variability in seafloor brines revealed by carbon and sulphur dynamics. *Nature Geoscience* 2(5):349–354.

Kennicutt, M. C., J. M. Brooks, R. R. Bidigare, and G. J. Denoux. 1988. Gulf of Mexico hydrocarbon seep communities. I: Regional distribution of hydrocarbon seepage and associated fauna. *Deep Sea Research Part I: Oceanographic Research Papers* 35:1639–1651.

Kohout, F. A. 1966. Submarine springs: A neglected phenomenon of coastal hydrology. *Hydrology* 26:391–413.

Lapham, L. L. 2007. In situ measurements of methane cycling in cold seep sediments containing gas hydrates and brines. PhD diss., University of North Carolina, Chapel Hill. 391 pp.

Lapham, L. L., J. P. Chanton, R. Chapman, and C. S. Martens. 2010. Methane under-saturated fluids in deep-sea sediments: Implications for gas hydrate stability. *Earth and Planetary Science Letters* 298:275–285.

Lapham, L. L., J. P. Chanton, C. S. Martens, P. D. Higley, H. W. Jannasch, and J. R. Woolsey. 2008. Measuring long term changes in dissolved ion and gas concentrations and stable isotopes at a hydrate site: Mississippi Canyon 118, Gulf of Mexico. *Environmental Science & Technology* 42:7367–7373.

Lapham, L. L., J. P. Chanton, C. S. Martens, K. Sleeper, and J. R. Woolsey. 2008. Microbial activity in surficial sediments overlying acoustic wipe-out zones at a Gulf of Mexico cold seep. *Geochemistry, Geophysics, Geosystems* 9:Q06001. doi:10.1029/2008GC001944.

Lapham, L. L., R. M. Wilson, I. R. MacDonald, and J. P. Chanton. 2014. Gas hydrate dissolution rates quantified with laboratory and seafloor experiments. *Geochimica et Cosmochimica Acta* 125:492–503. http://dx.doi.org/10.1016/j.gca.2013.10.030.

Lessard-Pilon, S. A., E. L. Podowski, E. E. Cordes, and C. R. Fisher. 2010. Megafauna community composition associated with *Lophelia pertusa* colonies in the Gulf of Mexico. *Deep Sea Research Part II: Tropical Studies in Oceanography* 57:1882–1890.

Lin, S., and J. W. Morse. 1991. Sulfate reduction and iron sulfide mineral formation in Gulf of Mexico anoxic sediments. *American Journal of Science* 291:55–89.

Lloyd, K. G., D. B. Lloyd, J. F. Albert, J. K. Biddle, J. P. Chanton, and A. Teske. 2010. Spatial structure and activity of sedimentary microbial communities underlying a *Beggiatoa* spp. mat in a Gulf of Mexico hydrocarbon seep. *PLOS ONE* 5:e8738.

Lohrenz, S., S. Chakraborty, M. Huettel, J. Herrera Silveira, K. Gundersen, D. Redalje, J. Wiggert, B. E. Denton, and J. Lehrter. 2014. Primary production. Pp. 28–38 in H. M. Benway and P. G. Coble (eds.), *Report of The US Gulf of Mexico Carbon Cycle Synthesis Workshop*. Woods Hole, MA: Ocean Carbon and Biogeochemistry Program and North American Carbon Program.

Lohrenz, S. E., G. L. Fahnenstiel, D. G. Redalje, G. A. Lang, X. Chen, and M. J. Dagg. 1999. Variations in primary production of northern Gulf of Mexico continental shelf waters linked to nutrient inputs from the Mississippi River. *Marine Ecology Progress Series* 155:45–54.

MacDonald, I. R., G. Bohrmann, E. Escobar, F. Abegg, P. Blanchon, V. Blinova, W. Brückmann, et al. 2004. Asphalt volcanism and chemosynthetic life in the Campeche Knolls, Gulf of Mexico. *Science* 304:999–1002.

MacDonald, I. R., G. S. Boland, J. S. Baker, J. M. Brooks, M. C. Kennicutt II, and R. Bidigare. 1989. Gulf of Mexico hydrocarbon seep communities. *Marine Biology* 101(2):235–247.

MacDonald, I. R., O. Garcia-Pineda, A. Beet, S. Daneshgar Asl, L. Feng, G. Graettinger, D. French-McCay, et al. 2015. Natural and unnatural oil slicks in the Gulf of Mexico. *Journal of Geophysical Research Oceans* 120:8364–8380. doi:10.1002/2015JC011062.

MacDonald, I. R., N. Guinasso, S. Acchleson, J. F. Amos, R. Duckworth, R. Sassen, and J. M. Brooks. 1993. Natural oil slicks in the Gulf of Mexico visible from space. *Journal of Geophysical Research* 98:16351–16364.

MacDonald, I. R., N. L. Guinasso, R. Sassen, J. M. Brooks, L. Lee, and K. T. Scott. 1994. Gas hydrate that breaches the sea floor on the continental slope of the Gulf of Mexico. *Geology* 22:699–702.

MacDonald, I. R., J. F. Reilly, N. L. Guinasso, J. M. Brooks, R. S. Carney, and W. A. Bryant. 1990. Chemosynthetic mussels at a brine-filled pockmark in the northern Gulf of Mexico. *Science* 248:1096.

MacDonald, I. R., W. W. Sager, and M. B. Peccini. 2003. Gas hydrate and chemosynthetic biota in mounded bathymetry at mid-slope hydrocarbon seeps: Northern Gulf of Mexico. *Marine Geology* 198:133–158.

Macelloni, L., B. M. Battista, and C. C. Knapp. 2011. Optimal filtering high-resolution seismic reflection data using a weighted-mode empirical mode decomposition operator. *Journal of Applied Geophysics* 75:603–614. doi:10.1016/j.jappgeo.2011.09.018.

Macelloni, L., C. B. Lutken, S. Garg, A. Simonetti, M. D'Emidio, R. M. Wilson, K. Sleeper, et al. 2014. Heat-flow regimes and the hydrate stability zone of a transient, thermogenic, fault-controlled hydrate system (Woolsey Mound northern Gulf of Mexico). *Marine and Petroleum Geology* 59:491–504.

Macelloni, L., A. Simonetti, J. H. Knapp, C. C. Knapp, C. B. Lutken, and L. L. Lapham. 2012. Multiple resolution seismic imaging of a shallow hydrocarbon plumbing system, Woolsey Mound, northern Gulf of Mexico. *Marine and Petroleum Geology* 38:128–142. doi:10.1016/j.marpetgeo.2012.06.010.

Martens, C. S., and R. A. Berner. 1974. Methane production in the interstitial waters of sulfate-depleted marine sediments. *Science* 185:1167–1169.

———. 1977. Interstitial water chemistry of anoxic Long Island sediment. 1: Dissolved gases. *Limnology and Oceanography* 22:10–25.

Martens, C. S., J. P. Chanton, and C. K. Paull. 1991. Fossil biogenic methane at the Florida Escarpment. *Geology* 19:851–854.

Mead, R., and M. A. Goñi. 2006. A lipid molecular marker assessment of sediments from the northern Gulf of Mexico before and after the passage of Hurricane Lili. *Organic Geochemistry* 37:1115–1129.

Moncreiff, C. A., M. J. Sullivan, and A. E. Daehnick. 1992. Primary production dynamics in seagrass beds of Mississippi Sound: The contributions of seagrass, epiphytic algae, sand microflora, and phytoplankton. *Marine Ecology Progress Series* 25:161–171.

Morse, J. W., and M. J. Beazley. 2008. Organic matter in deep water sediments of the northern Gulf of Mexico and its relationship to the distribution of benthic organisms. *Deep Sea Research II: Tropical Studies in Oceanography* 55:2563–2571. doi:10.1016/j.dsr2.2008.07.004.

Morse, J. W., D. K. Gledhill, K. S. Sell, and R. S. Arvidson. 2002. Pyritization of iron in sediments from the continental slope of the northern Gulf of Mexico. *Aquatic Geochemistry* 8:3–13.

Morse, J. W., and G. T. Rowe. 1999. Benthic biogeochemistry beneath the Mississippi River Plume. *Estuaries* 22:206–214.

Mulder, T., and J. Alexander. 2001. The physical character of subaqueous sedimentary density flows and their deposits. *Sedimentology* 48:269–299.

Olsen, C. R., I. L. Larsen, P. D. Lowry, and N. H. Cutshall. 1986. Geochemistry and deposition of ^{7}Be in river-estuarine and coastal waters. *Journal of Geophysical Research* 91:896–908.

Orcutt, B., A. Boetius, M. Elvert, V. Samarkin, and S. B. Joye. 2005. Molecular biogeochemistry of sulfate reduction, methanogenesis and the anaerobic oxidation of methane at Gulf of Mexico cold seeps. *Geochimica et Cosmochimica Acta* 69:4267–4281.

Orcutt, B. N., S. B. Joye, S. Kleindienst, K. Knittel, A. Ramette, A. Reitz, and A. Boetius. 2010. Impact of natural oil and higher hydrocarbons on microbial diversity, distribution, and activity in Gulf of Mexico cold-seep sediments. *Deep Sea Research Part II: Tropical Studies in Oceanography* 57:2008–2021.

Paull, C. K., J. P. Chanton, C. S. Martens, P. D. Fullagar, A. C. Neumann, and J. A. Coston. 1991. Seawater circulation through the flank of the Florida platform: Evidence and implications. *Marine Geology* 102:265–279.

Paull, C. K., B. Hecker, R. Commeau, R. P. Freeman-Lynde, C. Neumann, W. P. Corso, and J. Curray. 1984. Biological communities at the Florida Escarpment resemble hydrothermal vent taxa. *Science* 226:965–967.

Qureshi, N. A. 1995. The role of fecal pellets in the flux of carbon to the sea floor on a river-influenced continental shelf subject to hypoxia. PhD diss., Louisiana State University. 255 pp.

Rabalais, N. N., R. E. Turner, and W. J. Wiseman Jr. 2002. Gulf of Mexico hypoxia, aka "the dead zone." *Annual Review of Ecology and Systematics* 33(1):235–263.

Redalje, D. G., S. E. Lohrenz, and G. L. Fahnenstiel. 1994. The relationship between primary production and the vertical export of particulate organic matter in a river impacted coastal ecosystem. *Estuaries* 17:829–838.

Reeburgh, W. S. 1980. Anaerobic methane oxidation: Rate depth distributions in Skan Bay sediments. *Earth and Planetary Science Letters* 47:345–352.

———. 2007. Oceanic methane biogeochemistry. *Chemical Reviews* 107(2):486–513.

Roberts, H. H. 2001. Fluid and gas expulsion on the northern Gulf of Mexico continental slope: Mud-prone to mineral-prone responses. *Natural Gas Hydrates: Occurrence, Distribution, and Detection* 124:145–161.

Roberts, H. H., and P. Aharon. 1994. Hydrocarbon-derived carbonate buildups of the northern Gulf of Mexico continental slope: A review of submersible investigations. *Geo-Marine Letters* 14:135–148.

Roberts, H. H., and R. S. Carney. 1997. Evidence of episodic fluid, gas, and sediment venting on the northern Gulf of Mexico continental slope. *Economic Geology* 92:863–879.

Roberts, H. H., D. Feng, and S. B. Joye. 2010. Cold-seep carbonates of the middle and lower continental slope, northern Gulf of Mexico. *Deep Sea Research Part II: Tropical Studies in Oceanography* 57:2040–2054.

Rowe, G. T., J. Morse, C. Nunnally, and G. S. Boland. 2008. Sediment community oxygen consumption in the deep Gulf of Mexico. *Deep Sea Research Part II: Tropical Studies in Oceanography* 55:2686–2691.

Sackett, W. M., J. M. Brooks, B. B. Bernard, C. R. Schwab, H. Chung, and R. A. Parker. 1979. A carbon inventory for Orca Basin brines and sediments. *Earth and Planetary Science Letters* 44:73–81.

Salvador, A. 1987. Late Triassic-Jurassic paleogeography and origin of Gulf of Mexico basin. *American Association of Petroleum Geologists Bulletin* 71:419–451.

———. 1991. Origin and development of the Gulf of Mexico basin. Pp. 389–444 in A. Salvador (ed.), *The Gulf of Mexico Basin*, vol. J: *The Geology of North America*. Boulder, CO: Geological Society of America. https://pubs.geoscienceworld.org/books/book/913/.

Sampere, T. P., T. S. Bianchi, S. G. Wakeham, and M. A. Allison. 2008. Sources of organic matter in surface sediments of the Louisiana continental margin: Effects of major depositional/transport pathways and Hurricane Ivan. *Continental Shelf Research* 28:2472–2487.

Santos, I. R., W. C. Burnett, J. Chanton, N. Dimova, and R. N. Peterson. 2009. Land or ocean? Assessing the driving forces of submarine groundwater discharge at coastal site in the Gulf of Mexico. *Journal of Geophysical Research: Oceans* 114:C04012.

Santos, I. R., W. C. Burnett, T. Dittmar, I. G. N. A. Suryaputra, and J. P. Chanton. 2009. Tidal pumping drives nutrient and dissolved organic matter dynamics in a Gulf of Mexico subterranean estuary. *Geochimica et Cosmochimica Acta* 73:1325–1339.

Santos, I. R., N. Dimova, R. N. Peterson, B. Mwashote, J. Chanton, and W. C. Burnett. 2009. Extended time series measurements of submarine groundwater discharge tracers (^{222}Rn and CH_4) at a coastal site in Florida. *Marine Chemistry* 113:137–147.

Sassen, R., S. L. Losh, L. Cathles, H. H. Roberts, J. K. Whelan, A. V. Milkov, and D. A. DeFreitas. 2001. Massive vein-filling gas hydrate: Relation to ongoing gas migration from the deep subsurface in the Gulf of Mexico. *Marine and Petroleum Geology* 18:551–560.

Sassen, R., H. H. Roberts, P. Aharon, J. Larkin, E. W. Chinn, and R. Carney. 1993. Chemosynthetic bacterial mats at cold hydrocarbon seeps, Gulf of Mexico continental slope. *Organic Geochemistry* 20:77–89.

Sassen, R., H. H. Roberts, R. Carney, A. V. Milkov, D. DeFreitas, B. Lanoil, and C. L. Zhang. 2004. Free hydrocarbon gas, gas hydrate, and authigenic minerals in chemosynthetic communities of the northern Gulf of Mexico continental slope: Relation to microbial processes. *Chemical Geology* 205:195–217.

Schroeder, W. W., and W. J. Wiseman. 1999. Geology and hydrodynamic of Gulf of Mexico estuaries. Pp. 3–28 in T. S. Bianchi, J. R. Pennock, and R. R. Twilley (eds.), *Biogeochemistry of Gulf of Mexico Estuaries*. New York: John Wiley and Sons.

Simonetti, A., J. H. Knapp, K. Sleeper, C. B. Lutken, L. Macelloni, and C. C. Knapp. 2013. Spatial distribution of gas hydrates from high-resolution seismic and core data, Woolsey Mound, northern Gulf of Mexico. *Marine and Petroleum Geology* 44:21–33. doi:10.1016/j.marpetge0.2013.04.004.

Sommerfield, C. K., C. A. Nittrouer, and C. R. Alexander. 1999. ^{7}Be as a tracer of flood sedimentation on the northern California continental margin. *Continental Shelf Research* 19:335–361.

Turner, R. E., N. N. Rabalais, and D. Justic. 2008. Gulf of Mexico hypoxia: Alternate states and a legacy. *Environmental Science & Technology* 42:2323–2327.

Valentine, D. L., and W. S. Reeburgh. 2000. New perspectives on anaerobic methane oxidation. *Environmental Microbiology* 2:477–484.

Vargo, G. A., K. L. Carder, W. Gregg, E. Shanley, C. Heil, K. A. Steidinger, and K. D. Haddad. 1987. The potential contribution of primary production by red tides to the west Florida shelf ecosystem. *Limnology and Oceanography* 32:762–767.

Walsh, J., D. R. Corbett, D. Mallinson, M. A. Goñi, and M. B. Dail. 2006. Mississippi Delta mudflow activity and 2005 Gulf hurricanes. *Eos, Transactions, American Geophysical Union* 87:477–478.

Waterson, E. J., and E. A. Canuel. 2008. Sources of sedimentary organic matter in the Mississippi. *Organic Geochemistry* 39:422–439.

Wilson, R. M., L. Macelloni, A. Simonetti, C. Lutken, K. Sleeper, M. Eemindio, M. Pizzi, L. Lapham, J. Knapp, and J. Chanton. 2014. Subsurface methane sources and migration pathways within a gas hydrate mound system, Gulf of Mexico. *Geochemistry, Geophysics, Geosystems* 15(1):89–107. doi:10.1002/2013GC004888.

Wysocki, L. A., T. S. Bianchi, R. T. Powell, and N. Reuss. 2006. Spatial variability in the coupling of organic carbon, nutrients, and phytoplankton pigments in surface waters and sediments of the Mississippi River plume. *Estuarine, Coastal and Shelf Science* 69:47–63.

Yan, B., U. Passow, J. Chanton, E. Nöthig, V. Asper, J. Sweet, M. Pitiranggon, A. Diercks, and D. Pak. 2016. Natural marine particles mediated sedimentation of oil and drilling mud components from the DWH spill. *Proceedings of the National Academy of Science* 113 (24):E3332–E3340.

Zieman, J. C., J. W. Fourqurean, and R. L. Iverson. 1989. Distribution, abundance and productivity of seagrasses and macroalgae in Florida Bay. *Bulletin of Marine Science* 44:292–311.

Zieman, J. C., and R. T. Zieman. 1989. *The Ecology of the Seagrass Meadows of the West Coast of Florida: A Community Profile*. Biological Report 85(7.25). Washington, DC: Department of the Interior, Minerals Management Service and Fish and Wildlife Service. 155 pp.

8

Organic Chemical Pollutants and Contamination in the Gulf of Mexico

José L. Sericano, Terry L. Wade, Victor Manuel Vidal Martinez, Gerardo Gold-Bouchot, and Anthony H. Knap

Introduction

Most of the physical energy enters the Gulf of Mexico (GOM) through the Yucatan Channel at up to 30 sverdrups (SV). The Gulf basin is approximately 1.6×10^6 km^2, with almost half of this area representing the shallow continental shelf, and contains a volume of water roughly estimated at 2.5×10^6 km^3. The Gulf receives large volumes of freshwater through the Mississippi River and its major tributaries, the Missouri, Illinois, and Ohio Rivers (Pereira et al. 1992) and the Rio Grande, between Mexico and the United States; together, these rivers drain an area representing more than 50% of the surface of the United States. The Usumacinta and Grijalva Rivers compose the largest fluvial system in Mexico. Together, they drain an area of ~1.1×10^5 km^2 into the GOM with an annual discharge of ~2678 $m^3 s^{-1}$ (Muñoz-Salinas and Castillo 2015). In the United States and Mexico, these rivers drain extensive agricultural regions, where large amounts of agricultural chemicals have been used for many years, and run through large population and industrial centers that also contribute chemicals to the GOM and organic contaminants that may accumulate in coastal environments. The extensive oil and gas activities onshore as well as the GOM shelf and slope also have the potential to add contaminants that cause additional risks to areas of high biological productivity and affect estuarine environments where fish and shrimp reproduce.

This chapter reviews and summarizes the data available to date on the distribution and effect of organic chemical pollutants in the GOM. These chemicals, many of them shown in Figure 8.1, are discussed in 2 groups: (1) persistent organic pollutants (POPs), which have raised awareness about their environmental and health impacts; and (2) petroleum hydrocarbons, as the GOM is one of the most oil productive regions in the world.

Emphasis was focused on polycyclic aromatic hydrocarbons (PAHs) because many of them have been linked to reproductive problems in animals and are also possible human carcinogens. Unless indicated otherwise, concentrations reported for solid matrices (i.e., sediments and biota) in this chapter are shown in ng g^{-1} or µg g^{-1} dry weight.

Persistent Organic Pollutants

POPs are highly toxic, persistent chemicals that are or were widely used as pesticides and/or were derived from industrial processes. The persistence and mobility of POPs are governed by different processes occurring in the environment, including sorption/desorption kinetics to particles, volatilization, chemical and/or biological degradation, and uptake by organisms. Due to their hydrophobic characteristics (i.e., their lipophilic nature), POPs can accumulate in tissue lipids and pass from one species to the next through the food web, resulting in potentially serious adverse effects on human and environmental health. Because of their persistence, POPs can be widely distributed, through the atmosphere and hydrosphere, within an environmental compartment and, in turn, be

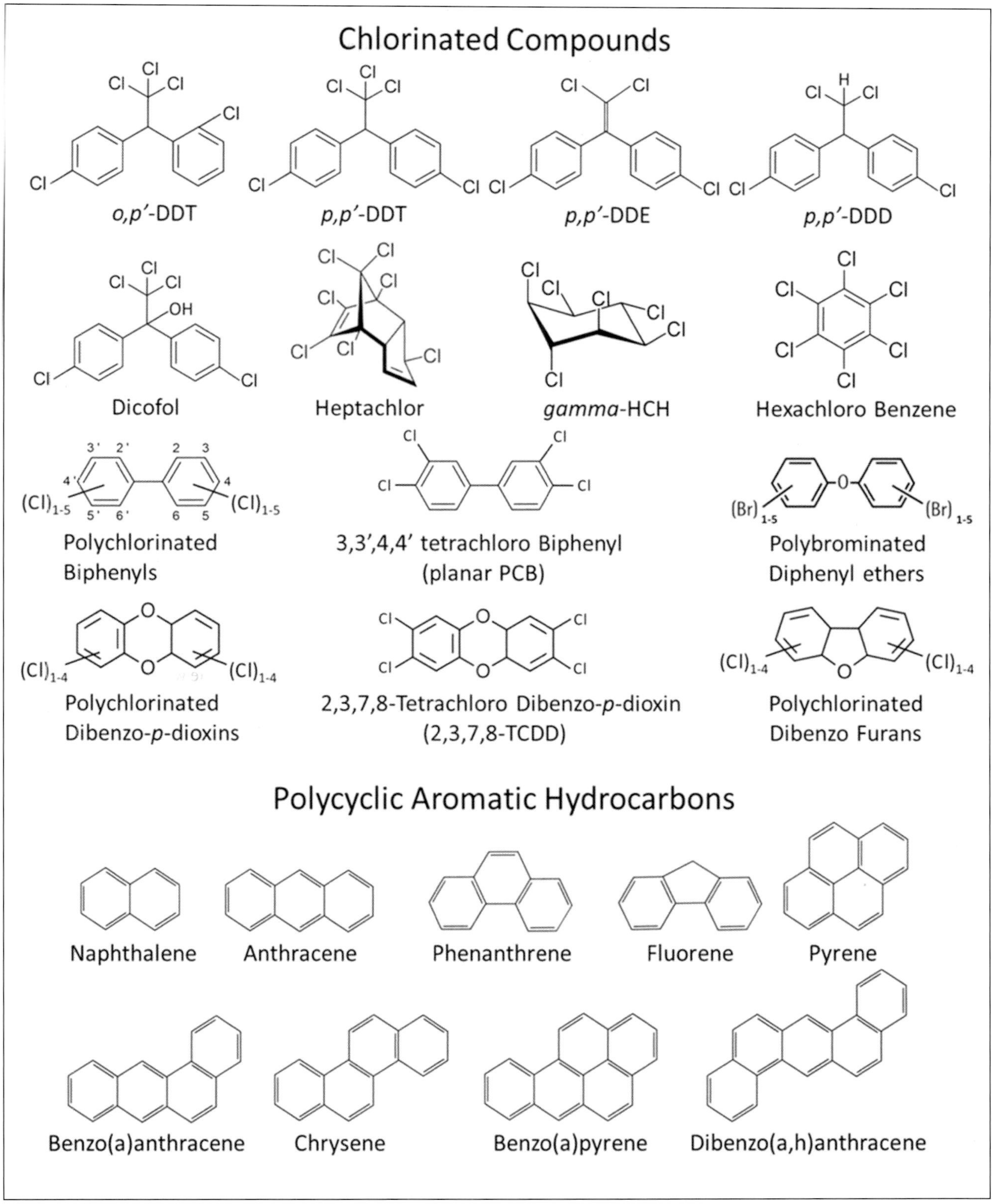

Figure 8.1. Structures of selected organic chemical pollutants reported in the Gulf of Mexico environments.

Table 8.1. Persistent organic pollutants under the Stockholm Convention, May 2013

Name	CAS #	Exemptions
Aldrin[a, *]	309-0-2	Production none; use as a local ectoparasiticide and insecticide
Chlordane[a, *]	57-74-9	Production by registered parties; use as a local ectoparasiticide, insecticide, termiticide, and as an additive in plywood adhesives
Dieldrin[a, *]	60-57-1	Production none; use in agricultural operations
Endrin[a, *]	72-20-8	None
Heptachlor[a, *]	76-44-8	Production none; use as a termiticide (including in the structure of houses and underground), for organic treatment, and in underground cable boxes
Hexachlorobenzene[a, b, c, *]	118-74-1	Production by registered parties; use as a chemical intermediate and a solvent for pesticides
Mirex[a, *]	2385-85-5	Production by registered parties; use as a termiticide
Toxaphene[a, *]	8001-35-2	None
Polychlorinated biphenyls[b, c, *]	Various	Production none
DDT[a, *]	50-29-3	Disease vector control; production and use as an intermediate in the production of dicofol and other compounds
Polychlorinated dibenzo-*p*-dioxins and dibenzofurans[c, *] Polychlorinated dibenzofurans[c, *]	Various	
α-Hexachlorocyclohexane[a]	319-84-6	None
β-Hexachlorocyclohexane[a]	319-85-7	None
Chlordecone[a]	143-50-0	None
Hexabromobiphenyl[b]	36355-01-8	None
Hexabromodiphenyl ether[b] Heptabromodiphenyl ether[b]	Various	Production none; use in recycling and reuse of articles containing these compounds
γ-Hexachlorocyclohexane[a] (lindane)	58-89-9	Production none; use in human health pharmaceuticals for control of head lice and scabies as second-line treatment
Pentachlorobenzene[a, b, c]	608-93-5	None
Tetrabromodiphenyl ether[b] Pentabromodiphenyl ether[b]	Various	Production none; use in recycling and reuse of articles containing these compounds
Perfluorooctanesulfonic acid (PFOS) and its salts[b] Perfluorooctanesulfonyl fluoride (PFOSF)[b]	Various	Production for permitted uses
Technical endosulfan[a]	115-29-7 959-98-8 33213-65-9	Production as allowed for the parties listed in the Register of specific exemptions; use in crop-pest complexes
Hexabromocyclododecane[b]	25637-99-4 3194-55-6	Production as allowed for the parties listed in the Register of Specific Exemptions; use in expanded polystyrene and extruded polystyrene in buildings
Hexachlorobutadiene[b]	87-68-3	None
Polychlorinated naphthalenes[c]	70776-03-3	Use in the production of polyfluorinated naphthalenes, including octafluoronaphthalene
Pentachlorophenol, its salts and esters[a]	Various	Use in utility poles and cross-arms

Source: See UNEP Stockholm Convention, http://chm.pops.int/TheConvention/ThePOPs/ListingofPOPs/tabid/2509/Default.aspx.

[a] Pesticide.

[b] Industrial chemical.

[c] Unintentional production.

[*] Dirty dozen.

transported globally, thus posing a serious risk in areas as distant from the original sources as the polar regions (e.g., Grotti et al. 2016; Corsolini and Sarà 2017). To mitigate the effect of POPs, the international community has developed joint strategies aimed to eliminate or restrict the production and use of POPs. In 1995, the United Nations Environment Programme (UNEP) addressed this concern with a global action plan to eliminate POPs; as a result, the first listing of dangerous and persistent substances, known as "the dirty dozen" (Table 8.1), was defined by the Intergovernmental Forum on Chemical Safety (IFCS) and the International Programme on Chemical Safety (IPCS). After five meetings of delegates between June 1998 and December 2000, actions regarding POPs were suggested. The negotiations were finally completed on May 23, 2001, in Stockholm and the results made publicly available. The final document, known as the Stockholm Convention, was made effective on May 17, 2004, after ratification by 128 parties and 151 signatories (UNEP 2010). In the process, 9 of the chemicals listed in the original dirty dozen were banned, the use of dichloro-diphenyl-trichloroethane (DDT) was limited to control the primary vector for malaria and dengue fever (*Anopheles* mosquitoes), and a reduction of the accidental production of dioxins and furans was recommended. This list is a "living" document to which new persistent toxic compounds that meet certain criteria for "persistence and global threat" can be added (UNEP 2017). During a conference held in Geneva on May 8, 2009, an agreement was made to add the first set of new POPs. The Stockholm Convention on POPs adds an important global dimension to international efforts to control these chemicals. As of March 2016, 179 parties and the European Union have ratified the convention (UNEP 2016). The United States, an original signatory in 2001, has not yet ratified the convention; however, the document has played a prominent role in the control of harmful chemicals at both national and global levels.

With the implementation of the Stockholm Convention on POPs, many of the chemicals listed are no longer produced or their production has been severely reduced worldwide. Even though reduced, many of these substances can still be detected in the marine environment because of their persistence in the environment, local unintentional production during industrial processes, or transboundary transport from developing nations where the full implementation of the Stockholm Convention has only recently been adopted. Nevertheless, global environmental health seems to have improved in many areas of the world, including the GOM, likely a result of these actions. This is particularly true for the group of "legacy" pesticides, defined as those chemicals that remain in the environment long after they were first introduced, included in the original Stockholm Convention's list. There are new generations of contaminants that are not yet regulated to the same extent, such as polybrominated diphenyl ethers (PBDEs), a flame-retardant mix that was until recently introduced into the environment in large amounts, mimicking PCBs (polychlorinated biphenyls) in the environment. Similar in structure and physicochemical properties (e.g., aqueous solubility, the log of the n-octanol/water-partition coefficient [log K_{ow}], and Henry's law constant) to PCBs, PBDEs are highly resistant to environmental degradation with high affinity for particles and partition into the lipids of organisms. Bioaccumulation of these contaminants is a function of both hydrophobicity, which is positively related to the chemical K_{ow}, and their molecular size. Less brominated congeners (i.e., tetra-, penta- and hexa-BDEs [brominated diphenyl ethers]) can accumulate in the bodies of animals (Yogui and Sericano 2009), including humans (Siddiqi et al. 2003; Johnson-Restrepo and Kannan 2009), while the uptake of large chemicals, although highly hydrophobic, is more difficult due to their unfavorable steric configurations (Sericano 1993).

Legacy Pesticides

The production of effective, inexpensive synthetic pesticides began to accelerate in the 1940s (Unsworth 2010) with the discovery of the effectiveness of some chlorinated compounds (e.g., DDT, aldrin, dieldrin, endrin, chlordane); DDT's broad-spectrum activity made it the most popular. During the mid-1900s, most pesticides were not recognized as harmful to the environment and were produced and used worldwide. These chemicals and/or their metabolites are still found in the environment decades after their use, mainly associated with soil or sediment particles by absorption, adsorption, or a combination of these processes, and can remain sorbed onto these particles for years because of their recalcitrance to degradation.

Polychlorinated Biphenyls

PCBs are a group of chlorinated aromatic hydrocarbons obtained by direct chlorination of biphenyls (Erickson 1986). PCBs are human-made, highly unreactive, and resistant to degradation. PCBs were used in many industrial and commercial applications, including adhesives; carbonless copy paper; coolants and lubricants in electrical equipment (transformers and voltage regulators); fire retardants; floor finishes; fluorescent lighting fixtures; insulation (fiberglass, foam, cork); oil in motors and transformers; pigments and dyes; plasticizers in paints, plastics, and rubber products; and surface coatings (e.g., De Voogt and Brinkman 1989; USEPA 1999). Many of these PCB-containing materials continue to be a worldwide problem today (e.g., Liebl et al. 2004; Ruus et al. 2006; Heinzow et al. 2007; Hwang et al. 2008; Jartun et al. 2009; Hu and Hornbuckle 2010; Frederiksen et al. 2012). PCBs consist of 2 connected phenyl rings with the potential of holding up to 5 chlorine atoms per ring (see Fig. 8.1). Depending on the position and number (1–10) of those chlorine atom substituents, there are 209 possible PCB congeners arranged in 10 homolog groups according to the number of chlorine atoms present. Only those PCB congeners with no chlorine substitution in *ortho* positions or having only 1 chlorine substitution in *ortho* positions (non-*ortho* and mono-*ortho* PCBs, respectively) are of environmental concern and considered harmful (e.g., Mihats et al. 2015; Shin et al. 2016; Francisco et al. 2017; Manning et al. 2017; see Fig. 8.1). These PCB congeners, also known as "dioxin-like" PCBs by the World Health Organization (WHO), can attain planar structures and exhibit biochemical and toxicological properties similar to those of the 17 toxic dioxins/furans (Srogi 2008; Wimmerová et al. 2016).

Commercial production of PCB mixtures, containing nearly 100 congeners with various degrees of chlorination, began in 1929 and continued for almost 40 years under different names (e.g., Aroclor in the United States, Clophen in Germany, Phenoclor and Pyralene in Italy, Kanechlor in Japan, Fenclor in France; Erickson 1986). Typically, there is a number associated with these names that indicates the degree or chlorination of the mixture. For example, Aroclor was commercialized by Monsanto Industrial Chemicals Company as Aroclor 1221, 1232, 1242, 1248, 1254, 1260, 1262, 1268, 1270, and 1016 (prepared by the fractional distillation of Aroclor 1242), where the last 2 digits indicate the percentage of chlorine by mass in the mixture. In the late 1970s, PCBs were reported as a new chemical hazard and were either banned or severely restricted worldwide because of concerns regarding the risks to human health and the environment (Jensen 1966). PCBs are present in marine plants and animals, such as birds, turtles, and marine mammals, and bioaccumulate through the food web with the potential to reach humans (Srogi 2008).

Dioxins/Furans

Polychlorinated dibenzo-*p*-dioxins (PCDDs, dioxins) and polychlorinated dibenzofurans (PCDFs, furans) are structurally and chemically related halogenated aromatic hydrocarbons that include 75 PCDD and 135 PCDF congeners (see Fig. 8.1). From these theoretically possible 210 congeners, only 17 are considered of toxicological significance in the environment (Safe 1990). These 17 congeners, with 4 to 6 chlorine atoms, share the common chlorine substitution in positions 2, 3, 7, and 8. The highest acutely toxic congener is 2,3,7,8-tetrachlorodibenzo-*p*-dioxin (2,3,7,8-TCDD).

Dioxins and furans are unintentional by-products in chlorine-containing manufacturing processes or formed during incomplete waste incineration or improperly designed combustion processes (Abad et al. 2006; Nzihou et al. 2012). Recently, uncontrolled burning of electronic waste (e-waste) has been identified as a major source of dioxins and furans to the environment (WHO 2010; Fu et al. 2012; Chakraborty et al. 2018). Dioxins and furans are extremely stable in the environment, where they are primarily associated with particulate and organic matter with limited leaching or volatilization potential. Like PCBs, dioxins and furans are highly resistant to degradation and are ubiquitous GOM contaminants (Wade et al. 2014).

Polybrominated Diphenyl Ethers

PBDEs are a group of chemicals added to a variety of products (e.g., electrical equipment, furniture and furnishing foam, rigid plastics, and textiles) as flame retardants (see de Wit 2002, and references therein). PBDEs are similar in structure to PCBs, with the difference that the 2 phenyl rings are connected by an ether bridge (see Fig. 8.1). PBDEs and PCBs are expected to exhibit similar toxic properties, ubiquitousness, and environmental persistence. Although PBDEs also have the potential of 209 different congeners depending on the position of the bromine atoms, commercial PBDE technical mixtures (particularly penta- and octa-BDEs) used as flame retardants contain only about 10 to 20 of these congeners (de Wit 2002). PBDEs enter the environment either accidentally during their production, through manufacture of products that contain these flame retardants, or by degradation/disposal of these products. PBDEs were first detected in the environment in the 1970s (Andersson and Blomkvist 1981; DeCarlo 1979), and since then, the environmental occurrence of PBDEs has been of increasing concern to scientists and policy makers. The source of PBDE to the environment can be any of the 3 commonly used technical mixtures or any combination of them. Although each mixture has different predominant congeners (e.g., 47 and 99 in penta-BDEs, 183 in octa-BDEs, and almost exclusively 209 in deca-BDEs), the congeners detected in the environmental samples are relatively consistent (e.g., 47, 99, 100, 153, and 154). Some of these PBDE congeners are considered toxic and capable of being accumulated by biota (Alonso et al. 2010; Díaz-Jaramillo et al. 2016; Pereira et al. 2017; Zhao et al. 2016). For these reasons, the penta- and octa-BDE mixtures have been recently added to the Stockholm Convention's list of restricted or banned POPs (see Table 8.1). PBDEs have been demonstrated to be ubiquitous contaminants in the North American environments (Bidleman et al. 1990; Knap and Binkley 1991; Hale et al. 2003; Yogui and Sericano 2009). Presently, the use of different PBDE mixtures is severely restricted or banned worldwide because of their environmental risks.

Persistent Organic Pollutants in the Gulf of Mexico

Several studies have reported the concentrations of legacy (i.e., formerly used) organochlorine pesticides in different compartments in the GOM since a series of pioneering studies (e.g., Baird et al. 1975; Butler and Schutzmann 1978; Fay and Newland 1972; Giam et al. 1972; Giam et al. 1973; Giam et al. 1974; Giam et al. 1976; Giam et al. 1977; Giam, Chan, and Neff 1978; Giam, Chan, Neff, et al. 1978; Giam et al. 1980; Petrocelli et al. 1974; Rosales et al. 1979) as discussed in Kennicutt et al. (1988). The following sections explore more recent studies on legacy pesticides and PCBs together with the occurrence of more contemporary POPs, such as PBDEs, in the atmosphere, water, sediments and, particularly, organisms from the GOM.

Atmospheric Transport and Deposition

One of the principal pathways for POPs to cycle in the environment is atmospheric transport, where dry or wet deposition can be a significant source of these chemicals to drainage basins and surface waters. While a significant number of recent studies describe the atmospheric concentrations and deposition of legacy pesticides in the Great Lakes and the US Mid-Atlantic regions (e.g., Gioia et al. 2005; Shunthirasingham et al. 2016; Yuan et al. 2010), the studies conducted in the GOM are scarce. One of the first reports of DDT and PCB concentrations in the GOM atmosphere was published by Giam et al. (1980). Samples collected near the Texas and Louisiana coastlines averaged 140 ± 94.3 (range = 38–224) and 720 ± 65.5 (range = 660–790) pg m^{-3} for DDTs (*p, p'*-DDT + *p, p'*-DDE) and PCBs, respectively. These concentrations decreased to 81.0 ± 54.4 (range = 41–143) and 317 ± 75.7 (range = 230–350) pg m^{-3} to 42.0 ± 21.2 (range = 26–73) and 163 ± 22.2 (range = 130–180) pg m^{-3}, respectively, as the sampling progressed to more open waters in the GOM.

Park et al. (2001) reported the atmospheric concentrations, deposition rates, and fluxes of chlorinated pesticides and PCBs for Galveston Bay, Texas, resulting from a nearly 2-year monitoring study (February 1995–August 1996).

The average pesticide concentrations reported in the air were generally in the order of HCH (hexachlorocyclohexane) isomers > chlordane-related compounds > DDT, including the degradation products DDD and DDE, isomers. Concentrations of total HCHs ranged from 47.8 to 674 pg m^{-3} and from 82.5 to 7170 pg L^{-1} in air vapor and rain-dissolved phases, respectively. The sum of these 2 HCH isomers accounted for 85% and 90% of the total HCH concentrations measured in those fractions, respectively. The sources of these 2 HCH isomers can be the past use of technical HCH mixtures, which include a number of stereoisomers (e.g., β, δ, and ε in addition to α and γ) in highly variable relative proportions, or the application of a refined γ-HCH product (lindane). The α to γ isomers in technical HCH varies from 3 to 7 (Willet et al. 1998); therefore, the relatively low α- to γ-HCH ratios (i.e., <2) detected in air and rain samples were inferred as the contribution of local/regional sources of lindane to the observed HCH contamination. Concentrations of total DDTs (i.e., the sum of *o,p'*- and *p,p'*-isomers of DDT, DDD, and DDE) ranged from not detected (n.d.) to 74 pg m^{-3} and from 1 to 1560 pg L^{-1} in air vapor and rain (dissolved) phases, respectively. Park et al. (2001) postulated that the relatively high proportions of *p,p'*-DDT to the total DDT concentrations encountered in these samples (25% and 40%, respectively) were the result of fresh atmospheric inputs of DDT from either local or distant sources, although they stressed the need of further studies to test this hypothesis. The range of total DDT concentrations in air reported by Park et al. (2001) were lower than the concentrations reported 20 years earlier in a sample collected in the GOM near Galveston Bay (158 pg m^{-3}) and similar to the levels encountered in samples collected in the open Gulf (26–73 pg m^{-3}) (Giam et al. 1980). Targeted chlordane-related compounds (i.e., heptachlor, heptachlor epoxide, oxychlordane, α-chlordane, γ-chlordane, *cis*-nonachlor, and *trans*-nonachlor) in air vapor and rain phases ranged from 31 to 291 pg m^{-3} and from n.d. to 402 pg L^{-1}, respectively. Dieldrin was the only cyclodiene compound detected in almost all the air and rain samples, while aldrin and endrin were rarely detected. Other pesticides reported included mirex and hexachlorobenzene (HCB). The concentrations were temporarily variable although no apparent seasonal trend was observed.

The annual dry plus wet deposition rates of these organochlorine pesticides to Galveston Bay were in the order of DDT isomers > HCH isomers > cyclodienes > chlordane-related compounds > HCB > mirex with 2.77, 2.48, 1.13, 1.08, 0.20, and 0.05 kg yr^{-1}, respectively. Chlordane-related compounds, cyclodiene pesticides, and DDTs were found to be deposited predominantly associated with particulate material via wet or dry deposition, while HCH isomers were primarily deposited associated with rainwater (Park et al. 2001). Total PCB concentrations ranged from 208 to 4780 pg m^{-3}, 9.10 to 141 pg m^{-3}, and 80.2 to 3340 pg L^{-1} in the vapor, particulate, and rain samples, respectively. The PCB levels detected in air samples were significantly higher than the concentrations reported in a sample collected just outside Galveston Bay (660 pg m^{-3}) by Giam et al. (1980). The annual dry plus wet deposition of PCB congeners to Galveston Bay was estimated to be 9.15 kg yr^{-1}. Although a definite temporal trend was not detected, the atmospheric concentrations of PCBs were observed to be lower when winds from the GOM were predominant. Interestingly, the annual flux of total PCBs from Galveston Bay water to air (71.2 μg m^{-2} yr^{-1}) was greater than the direct deposition rate (6.40 μg m^{-2} yr^{-1}), which potentially indicates that Galveston Bay is a net source of PCB congeners to the atmosphere, particularly during summertime.

In a year-long study (August 1998–September 1999), air and rain samples were also collected from a location in Corpus Christi Bay, Texas, and analyzed for chlorinated pesticides (Park et al. 2002). The concentrations detected during this study were much lower than those measured during February 1995–August 1996 in Galveston Bay's atmosphere (Park et al. 2001). Average concentrations in air (i.e., vapor phase) samples for α- and γ-HCH, heptachlor, α- and γ-chlordane, *trans*-nonachlor, dieldrin, and *p,p'*-DDE, the only DDT metabolite detected, were found to be between 2.0 and 6.0 pg m^{-3}. As discussed earlier, the restrictions on the production and use of these chemicals as a result of the implementation of the Stockholm Convention were likely responsible for the low concentrations encountered and the close-to-zero total fluxes reported. Other compounds detected in the Corpus Christi Bay area reflected the use of alternative pesticides (e.g., air and rain concentrations of chlorpyrifos were, on average, 32 pg m^{-3} and 170 pg L^{-1}, respectively). Chlorpyrifos, a chemical known to be highly toxic to marine biota (Maryoung et al. 2014; Palma et al. 2009; Wang et al. 2011) and virtually unrestricted at the time of the study, has been under restriction in the United States since 2000 (EPA 2018). Concentrations of total PCBs ranged from 9.0 to 115 pg m^{-3}, n.d. to 36 pg m^{-3}, and 464 to 8070 pg L^{-1} in the vapor, particulate, and rain samples, respectively. Similarly to Galveston Bay, the estimation of fluxes in Corpus Christi Bay indicated that the bay also acts as a net source of PCBs to the atmosphere. There are no reports known to us on PBDE and dioxins/furan concentrations in the GOM atmosphere.

Water

To our knowledge, the most recent report of chlorinated pesticide and PCB concentrations in microlayer and surface-seawater samples, collected during 4 surveys in 1985 and 1987 from the west-central part of the GOM off Texas, was published by Sauer et al. (1989). Previous studies on legacy pesticide and PCB concentrations in microlayer and surface-seawater samples are significantly older. None of the selected chlorinated pesticides and PCB congeners was detected above the method detection limits (MDLs = 100 to 1000 pg L^{-1}) in any microlayer samples. Similarly, most pesticides and PCB congeners were not detected in open-Gulf seawater samples, even at MDLs between 0.5 and 3.0 pg L^{-1} achieved by processing nearly 900 L of seawater. Only lindane and chlordane were detected in nearly every seawater filtrate sample analyzed at concentrations in the 10–150 and 4–34 pg L^{-1} ranges, respectively. Heptachlor epoxide (8–10 pg L^{-1}) and dieldrin (4–24 pg L^{-1})

were detected less frequently, while none of the samples analyzed revealed the presence of any of the DDT, DDD, or DDE isomers. As target chlorinated compounds were detected at very low concentrations in a few samples, the study concluded that pesticide contents in open water of the GOM were insignificant. However, amounts of suspended sediment, particle sizes, and organic carbon associated with those particles play an important role in the distribution of persistent organic pollutants within the aquatic ecosystems. Legacy pesticides and PCB congeners are, in general, very hydrophobic and become strongly associated with suspended particles once in the aquatic environment. This affinity for particles by hydrophobic substances is well described by their log K_{ow}, which are useful in predicting the fate of POPs in the environment. Chemicals with high log K_{ow} values are not usually encountered in water samples where the suspended load is small, as in open Gulf waters, because of their very low solubility.

Very few water samples collected seasonally over a 3-year period (1992–94) following the discharge plume of the Mississippi and Atchafalaya Rivers in the northwestern GOM showed the presence of target legacy pesticides and PCB congeners, since most remain sorbed onto particles (McMullin and Means 1996). The study reported the presence of HCB (230–7100 pg L^{-1}), α-chlordane (260–2600 pg L^{-1}), and *trans*-nonachlor (270–900 pg L^{-1}) in 14%, 8%, and 4%, respectively, of the 266 particulate samples analyzed. Interestingly, *p,p'*-DDE was detected in 9% of the dissolved-phase samples analyzed (range = 2000–8300 pg L^{-1}) and 5% of the colloidal-phase samples studied (range = 1900–13,000 pg L^{-1}) but not detected adsorbed onto particles. PCB congeners, in contrast, were detected only in the particulate-phase samples, with trichlorobiphenyls detected more frequently (41%, range = 370–946,000 pg L^{-1}), followed by tetrachlorobiphenyls (38%, 111–523,000 pg L^{-1}).

Carvalho et al. (2009) reported the concentrations of several chlorinated pesticides in filtered-water samples collected in 2000 from Laguna de Términos, Campeche, located in the southern portion of the GOM. DDT isomers were the most abundant chemicals in the dissolved phase (279 pg L^{-1}; range = 73–605 pg L^{-1}). Other chlorinated compounds, reported in the study at concentrations comparable to those found in open GOM waters, included lindane (average = 21 pg L^{-1}; range = 8–64 pg L^{-1}), total chlordane (18 pg L^{-1}; range = 3–33 pg L^{-1}), and dieldrin (8.4 pg L^{-1}; range = <3–13 pg L^{-1}). With an average value of 1180 pg L^{-1}, PCB congeners were detected at the highest concentrations (range = 70–3410 pg L^{-1}). There are no reports known to us on PBDE and dioxins/furan concentrations in GOM water.

Surface Sediments

The apparent fate of hydrophobic POPs, sorbed onto settling particles or carried down by dead biota, is to be removed to the underlying sediments, particularly in bays and/or estuaries, which are more likely than areas farther offshore to be affected by activities on land. The most extensive study of POPs in sediments from coastal areas of the northern GOM was conducted only during the first 2 years (1986 and 1987) of the National Oceanic and Atmospheric Administration's Status and Trends Program (NOAA's NS&TP). More than 300 samples were analyzed and reported for legacy pesticides and their concentrations (Sericano, Atlas, et al. 1990). Although nearly 50% and 57% of the data produced during 1986 and 1987, respectively, were below the MDL (typically below 0.25 ng g^{-1} depending on the mass of sample analyzed), and only 4% and 8% of the data, respectively, were above 1.00 ng g^{-1}, the first 2 years of NOAA's NS&TP permitted visualization of the geographic distribution of legacy pesticide concentrations along the northern coast of the GOM. *Trans*-nonachlor and α-chlordane, together with dieldrin in 1986, represented the major fractions of non-DDT chemicals, while most others were measurable in less than 20% of the samples. The highest concentrations were generally detected in samples from the eastern section of the northern GOM coast (e.g., Choctawhatchee, St. Andrew, Tampa, and Naples Bays), while the lowest were found along the western portion. With the exception of sediment samples from Matagorda Bay, a similar scenario was observed for DDT-related compounds. On average, nearly 70% of the total DDT concentrations corresponded to the sum of the *o,p'* and *p,p'* isomers of DDD and DDE, indicating the active degradation of DDT in the media or the input of already-degraded DDT to the northern GOM from inland sources. The reported median concentration for total DDTs encountered in 1986 and 1987 for the entire study area (0.87 and 0.89 ng g^{-1}, respectively) were within the range of concentrations reported earlier for NOAA's Benthic Surveillance Study (0.1–7.0 ng g^{-1}; NOAA 1987). Sediment samples were also collected in the GOM during the following years as new locations were added to the program, and in 2007 several of the sites

Table 8.2. Legacy-persistent organic pollutant concentrations in northern Gulf of Mexico coastal sediments, 1986 and 2007

Compound	Median	Avg ± 1 Std	Range	Median	Avg ± 1 Std	Range	Change
	1986 (n = 24)			2007 (n = 24)			(%)
HCB	0.09	0.20 ± 0.28	<MDL–0.97	0.03	0.11 ± 0.27	n.d.–1.30	-45.0
Lindane	<MDL	0.03 ± 0.06	n.d.–0.26	n.d.	<MDL	n.d.–0.04	
Heptachlor	0.02	0.02 ± 0.02	n.d.–0.08	n.d.	n.d.	n.d.	
Heptachlor epoxide	n.d.	0.03 ± 0.05	n.d.–0.16	n.d.	<MDL	n.d.–0.09	
α-Chlordane	0.07	0.14 ± 0.18	n.d.–0.79	n.d.	0.02 ± 0.05	n.d.–0.20	-85.7
trans-Nonachlor	0.04	0.08 ± 0.11	n.d.–0.47	n.d.	<MDL	n.d.–0.03	
Aldrin	n.d.	<MDL	n.d.–0.04	n.d.	0.02 ± 0.05	n.d.–0.13	
Dieldrin	0.18	0.21 ± 0.18	0.04–0.63	n.d.	<MDL	n.d.–0.06	
Mirex	0.03	0.04 ± 0.05	n.d.–0.18	n.d.	n.d.	n.d.	
o-p′-DDE	n.d.	<MDL	n.d.–0.14	n.d.	<MDL	n.d.–0.02	
p-p′-DDE	0.30	0.53 ± 0.79	0.04–3.78	0.07	0.08 ± 0.08	n.d.–0.26	-84.9
o-p′-DDD	0.09	0.19 ± 0.53	n.d.–2.65	0.03	0.04 ± 0.07	n.d.–0.31	-78.9
p-p′-DDD	0.42	0.95 ± 1.87	0.07–8.29	<MDL	0.03 ± 0.03	n.d.–0.11	-96.8
o-p′-DDT	n.d.	0.04 ± 0.09	n.d.–0.40	n.d.	n.d.	n.d.–0.03	
p-p′-DDT	0.09	0.93 ± 3.71	<MDL–18.3	0.02	0.03 ± 0.05	n.d.–0.20	-96.8
ΣPCBs	3.88	8.30 ± 13.6	1.00–66.7	0.71	0.93 ± 0.96	0.03–3.98	-8.8

Note: Concentrations of ng g^{-1}, dry wt. Locations (collected at the same locations each year): Apalachicola Bay (Cat Point Bar and Dry Bar); Aransas Bay (Long Reef); Atchafalaya Bay (Oyster Bayou); Barataria Bay (Middle Bank and Bayou Saint Denis); Caillou Lake (Caillou Lake); Calcasieu Lake (St. Johns Island); Cedar Key (Black Point); Charlotte Harbor (Bird Island); Copano Bay (Copano Reef); Corpus Christi (Nueces Bay); Galveston Bay (Confederate Reef, Hanna Reef, Todd's Dump, and Yacht Club); Joseph Harbor Bayou (Joseph Harbor Bayou); Matagorda Bay (East Matagorda, Gallinipper Point, Lavaca River Mouth, and Tres Palacios Bay); Mesquite Bay (Ayres Reef); and Sabine Lake (Blue Buck Point).

sampled during the first 2 years were resampled. Table 8.2 compares the averages, medians, and ranges of POP concentrations measured in 2007 with those encountered in 1986. After 20 years, POP average concentrations in sediments have substantially decreased (by 45.0% to 96.8% compared to concentrations reported in 1986), with several analytes mostly undetected in 2007.

Total chlorinated pesticide concentrations reported for sediment samples collected in Sarasota Bay (n.d.–61.4 ng g^{-1}; Sherblom et al. 1995) exhibited a range similar to those reported for samples from Tampa Bay during NOAA's NS&TP almost 10 years earlier in 1986–87 (Sericano, Atlas, et al. 1990). Hemming et al. (2005) reported lower total DDT concentrations in sediment samples from Choctawhatchee Bay (2.69–43.2 ng g^{-1}) than those found in 1986–87 (up to 175 ng g^{-1}; Sericano, Atlas et al. 1990).

The limited data available from the southern coast of the GOM indicate that total DDT concentrations in sediments are even lower in these areas. Concentrations ranging from 0.008 to 0.630 ng g^{-1} (average = 0.148 ng g^{-1}) were reported for samples collected during 2000 in Laguna de Términos (Carvalho et al. 2009). The sum of the isomers of DDE and DDD accounted for 70% to 90% of the total DDT loads. Other reported POPs were total HCHs (n.d.–0.005 ng g^{-1}) and total chlordane (n.d.–0.023 ng g^{-1}). The predominance of γ-HCH over α- and β-HCH suggested that a refined HCH product (γ-HCH = lindane) has been used in Campeche instead of technical HCH, which generally contained higher percentages of α-HCH among other HCH isomers.

Coastal sediments are a sink for POPs, including PBDEs (Palm et al. 2002); however, limited data are available regarding the concentrations of these chemicals in

Table 8.3. PBDE concentrations in northern Gulf of Mexico sediments and native oysters

Matrix	Period	n	Median	Avg ± 1 Std	Range
Sediment	2006–7	35	0.10	0.15 ± 0.31	n.d.–1.80
Oysters	1996	25	3.60	8.41 ± 13.6	n.d.–48.8
Oysters	2006–7	50	1.35	4.44 ± 7.19	n.d.–27.8

Note: Concentrations are the sum of congeners 47, 99, 100, 153, and 154 in ng g-1, dry wt.

sediments from the GOM. In 2006, PBDEs were added to the list of target chemicals for NOAA's NS&TP, and sediments from 35 locations in the GOM were sampled and analyzed for these flame retardants between 2006 and 2007 (Table 8.3). As discussed earlier, congeners 47, 99, 100, 153, and 154 are the PBDEs most likely to bioaccumulate in plants and organisms; therefore, for comparative purposes, the concentrations shown are the sum of these congeners. In line with what has been observed for most POPs, PBDE concentrations in sediment samples are considerably lower than those measured in oysters. The highest concentration, corresponding mainly to equal amounts of congeners 47 and 99, was detected in sediments from the vicinity of the Yacht Club in Galveston Bay (1.80 ng g^{-1}).

Data in the GOM sediments are equally scarce for dioxins and furans, and to our knowledge limited only to data reported for Galveston Bay during NOAA's NS&TP. Five locations within the bay revealed an average total dioxin/furan concentration (i.e., sum of the seventeen 2-, 3-, 7-, 8-substituted dioxin and furan congeners) of 120 ± 99.3 pg g^{-1} (median = 82.8; range = 46.7–287 pg g^{-1}). Dioxin concentrations were the most prominent (average = 102 ± 85.5; median = 74.9; range = 27.6–240 pg g^{-1}).

Biota

Organisms such as bivalve mollusks have been the preferred tool to monitor POP concentration trends in the aquatic environment (e.g., Besada et al. 2014; Braune et al. 2005; Schöne and Krause 2016; Rigét et al. 2016; Sericano et al. 2014), bioaccumulation (e.g., Coat et al. 2011; Luna-Acosta et al. 2015; Magnusson et al. 2006; Taffi et al. 2015), and associated human health risks (e.g., Ameur et al. 2013; Cheung et al. 2007). Since the introduction of the Mussel Watch concept by Edward Goldberg (1975), bivalve mollusks have become one of the most frequently used organisms to investigate environmental concentrations and trends of POPs. Farrington et al. (2016) have summarized the Mussel Watch concept: Bivalves are cosmopolitan, sedentary species able to concentrate chemicals from their surrounding seawater, with limited ability to metabolize organic contaminants, making it possible for researchers to assess bioavailability of POPs in the marine environment. Bivalves have generally stable local populations, are robust to moderate levels of pollution, and can be transplanted for monitoring purposes. Additionally, many bivalves have an important commercial value. Forty years after Goldberg's proposal for Mussel Watch, the concept continues to be used in monitoring studies worldwide (e.g., Carlton et al. 2010; Nakata et al. 2012; Regoli et al. 2014; Sparks et al. 2014).

In 1986, NOAA initiated the longest, most comprehensive, continuous chemical contaminant monitoring program involving native mussels and oysters to assess the effects of human activity on US coastal and estuarine areas (Lauenstein and Daskalakis 1998). Nationwide, NOAA's NS&TP evolved from sampling 145 locations every winter, within a 1-week window at any given location, to a biennial sampling in more than 300 sites around the United States divided into 2 groups, each sampled in alternative years. In the GOM, sampling locations increased from 51 to 85 sites, and the findings were reflected in many publications over the years (e.g., Apeti et al. 2012; Ellis et al. 1993; Jackson et al. 1994; Kim et al. 2008; Sericano, Atlas, et al. 1990; Sericano et al. 1993; Sericano et al. 2014; Wade et al. 1988; Wade et al. 1998). Sericano, Atlas, et al. (1990) published a report summarizing NOAA's NS&TP legacy pesticide data in the GOM, with nearly 300 pooled samples analyzed during the first 2 years of the program. Concentrations, occurrences, and spatial-distribution patterns of non-DDT pesticides were very similar for the 1986 and 1987 sampling periods. Dieldrin, α-chlordane, and *trans*-nonachlor were detected in more than 95% of the samples, with concentrations ranging over 2 orders of magnitude, followed by heptachlor epoxide, lindane, and mirex with over 85%, 80%, and 38% of the samples, respectively. The highest non-DDT pesticide concentrations were measured in samples from Galveston, Choctawhatchee, St. Andrew, Tampa, and

Table 8.4. Legacy-persistent organic pollutant concentrations in northern Gulf of Mexico native oysters, 1986 and 2012

Compound	Median	Avg ± 1 Std	Range	Median	Avg ± 1 Std	Range	Change
	1986 (n = 11)			2012 (n = 11)			(%)
HCB	<MDL	0.22 ± 0.33	n.d.–1.06	n.d.	n.d.	n.d.	
Lindane	0.77	0.77 ± 0.46	n.d.–1.38	n.d.	n.d.	n.d.	
Heptachlor	<MDL	0.28 ± 0.29	n.d.–0.94	n.d.	n.d.	n.d.	
Heptachlor epoxide	1.61	2.40 ± 2.12	0.66–8.35	n.d.	0.42 ± 1.39	n.d.–4.61	-82.5
α-Chlordane	9.73	11.8 ± 9.11	3.25–31.4	3.70	4.13 ± 5.87	n.d.–20.0	-65.0
trans-Nonachlor	8.21	12.7 ± 12.51	2.29–40.6	1.63	2.57 ± 3.53	n.d.–10.1	-79.8
Aldrin	<MDL	<MDL	n.d.–0.34	n.d.	n.d.	n.d.	
Dieldrin	5.44	8.43 ± 11.4	2.16–42.1	n.d.	2.06 ± 6.84	n.d.–22.7	-75.6
Mirex	1.70	3.24 ± 4.42	n.d.–11.8	n.d.	n.d.	n.d.	
o-p′-DDE	0.75	5.30 ± 14.8	n.d.–49.7	n.d.	2.18 ± 6.16	n.d.–20.6	-58.9
p-p′-DDE	12.1	16.7 ± 12.0	3.84–40.1	4.78	9.59 ± 11.6	n.d.–34.6	-42.6
o-p′-DDD	2.88	5.39 ± 7.88	0.57–28.3	n.d.	0.77 ± 2.56	n.d.–8.50	-85.7
p-p′-DDD	9.84	11.4 ± 12.4	n.d.–42.0	1.15	3.52 ± 5.76	n.d.–19.1	-69.1
o-p′-DDT	0.92	2.06 ± 4.26	n.d.–14.5	n.d.	<MDL	n.d.–0.32	
p-p′-DDT	0.97	1.16 ± 1.47	n.d.–5.23	n.d.	0.78 ± 2.46	n.d.–8.19	-32.8
ΣPCBs	102	148 ± 104	45.8–356	14.1	40.3 ± 72.8	4.33–255	-90.5

Note: Concentrations of ng g^{-1}, dry wt. Locations (collected at the same locations each year): Breton Sound (Bay Gardene and Sable Island); Lake Borgne (Malheureux Point); Mississippi Sound (Pascagoula Bay and Pass Christian); Mobile Bay (Cedar Point Reef); Tampa Bay (Cockroach Bay, Hillsborough Bay, Mullet Key Bayou, and Papys Bayou); and Terrebonne Bay (Lake Barre).

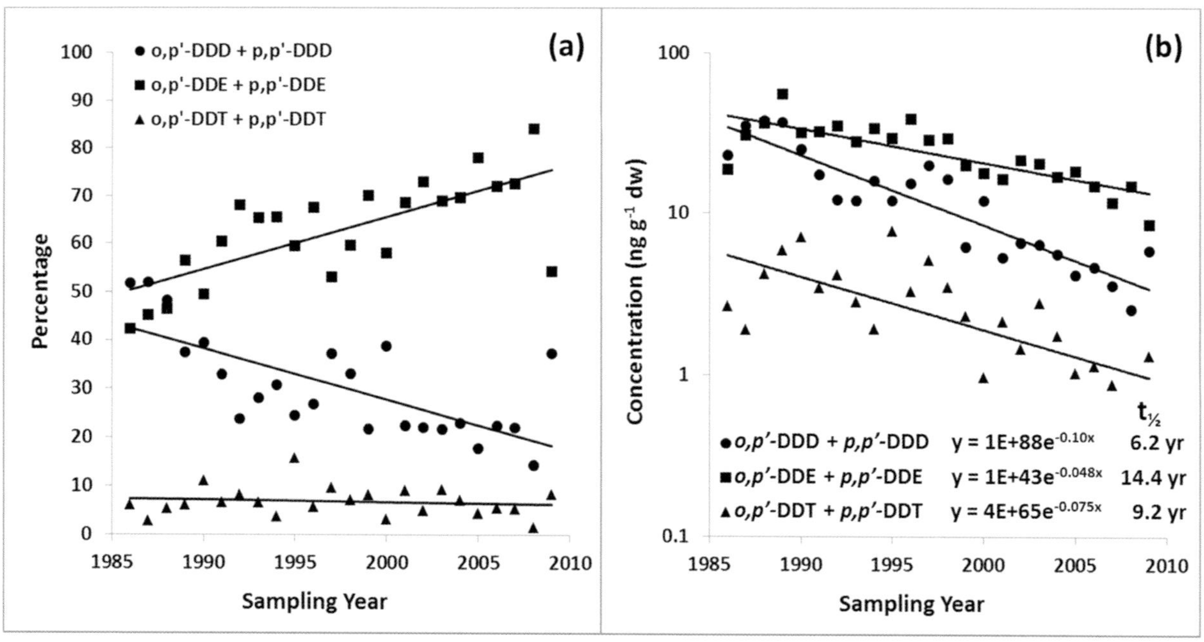

Figure 8.2. Relative proportions of isomers of DDT, DDE, and DDD (*a*) and their dissipation half-life (*b*).

Naples Bays and Mississippi and Breton Sounds. Table 8.4 compares the averages, medians, and ranges of POP concentrations in oyster tissues measured in 2012 with those encountered in 1986 at the same locations. Similarly to sediments, POP average concentrations have shown important decreases after 27 years (range = -32.8% to -90.5%).

Most DDT-related compounds in oyster tissues were detected as the degradation products of *o,p'*- and *p,p'*-DDT (*o,p'*- and *p,p'*-DDE and DDD, respectively). The contributions of these 2 metabolites (sum of their isomers) to the total DDT concentrations in oysters were nearly equal (i.e., 44% and 50%, on average, respectively; Sericano, Wade, et al. 1990). The relative proportions of DDEs to DDDs changed over time (Fig. 8.2a) probably due to their different dissipation rates exhibited from 1986 to 2009 with estimated half-lives of 14.4 and 6.9 years, respectively (Fig. 8.2b) or the input of DDEs from a different source (i.e., dicofol, an insecticide closely related to DDT, introduced after the ban of technical DDT in 1972), including the in situ degradation of DDTs to DDEs.

Dicofol is manufactured from technical DDT, and in the early years the final product contained substantial amounts of DDTs, DDDs, and DDEs. Technical dicofol was reported to contain, on average, 1.5% of DDT impurities (i.e., 8.3% on dicofol content basis; Di Muccio et al. 1988). The most abundant impurities were *o,p'*-DDE (22.6%), *o-p'*-DDT (13.0%), *p,p'*-DDE (10.3%), *p,p'*-DDD (4.11%), and *p,p'*-DDT (2.74%). For this reason, the use of dicofol formulations was temporarily canceled in the United States in 1986 by the EPA. Technology evolved to produce technical-grade dicofol that contains less than 0.1% DDT, and, in 1998, the EPA allowed the use of dicofol with some restrictions. The United States was responsible for the greatest dicofol usage in North and Central America between 2000 and 2012, although annual applications decreased from 323 t, in 2000, to 32.7 t, in 2012 (Li et al. 2015). The amount of *p,p'*-DDE + *o,p'*-DDE present in technical dicofol (nearly 33% of DDT impurities) in use until 1986 in the United States very well could explain the uptrend observed in the relative proportions of DDE shown in Figure 8.2a.

To put NOAA's NS&TP DDT data (1986–2009) in perspective, it can be compared to a previous coastal-wide study (1965–72) in the northern GOM, keeping in mind the substantial changes in analytical methods over the years (Butler 1973; Duinker et al. 1988; Sericano, Wade, et al. 1990). Although the concentrations reported in these studies overlap, it is clear that concentrations, as recorded in oysters, were higher and declined more rapidly during 1965–72, after a peak in 1968, than in the 1986–2009

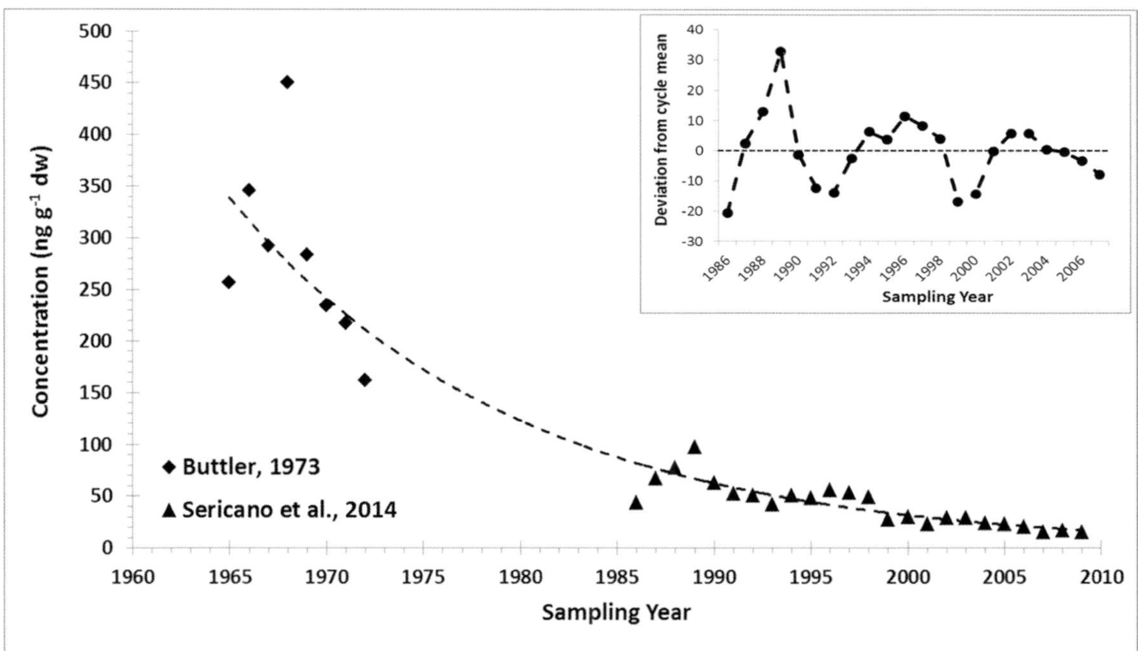

Figure 8.3. Environmental dissipation of total DDT reflected by bivalves from the Gulf of Mexico, 1965–2009. Insert shows percentage deviation of total DDT concentrations in bivalves from the average concentration within each cycle of 6 to 8 years observed between 1986 and 2009 (Butler 1973; Sericano et al. 2014).

period (Fig. 8.3). This decline rate over time is consistent with the exponential decline of many organic contaminants in the environment.

Archived mussel and oyster samples of a Mussel Watch Program of the 1970s were analyzed by using a more current analytical methodology and then compared to NOAA's NS&T data generated in 1992. POPs, in general, and DDTs and metabolites, in particular, exhibited significant decreases between 1970 and 1992 (Lauenstein 1995). A study of the influence of climate cycles on the annual variations in contaminant concentrations in GOM oyster tissues indicated that, in spite of a strong local component, a relationship between body burden and a wider climate influence (i.e., El Niño/Southern Oscillation) may exist (Kim et al. 1999). The 6- to 8-year oscillations in the GOM, observed in NOAA's NS&T total DDT concentrations (Fig. 8.3 insert; Sericano, unpublished data), seem to confirm the hypothesis by Kim et al. (1999).

The ratio of (*p,p'*-DDE + *p,p'*-DDD)/*p,p'*-DDT is commonly employed in the literature to distinguish between old (>1) and recent (<1) inputs of fresh DDT (e.g., Qiu et al. 2004; Ma et al. 2016). This ratio in oyster samples from the GOM ranged from 6 to 334 (average = 46) in 1986 and from 1.4 to 180 (average = 36) in 2009 (Sericano, unpublished data). The lowest ratio detected in 2009 corresponded to a sample collected in Galveston Bay. These ratios in Butler's study (1965–72) were lower and ranged from 5.1 to 22 (average = 9.5), indicating fresher, although somewhat degraded, inputs of DDT to the northern GOM (Butler and Schutzmann 1978). The half-life of total DDT in the GOM, as recorded in oysters and calculated from merging Butler's study (1965–72) and NOAA's NS&TP (1986–2009), was 10.2 years. The half-life of total DDT in the GOM was shorter than estimates for those on the east and west coasts (13.3 and 13.9 years, respectively; Sericano et al. 2014). According to these calculations and in absence of new inputs, it is expected that total DDT concentrations will be below 10% of today's levels by 2050 (Sericano et al. 2014).

Sericano and Wade (2011) discussed in detail the concentrations and annual trends of POPs in oysters from 2 important bays in the GOM, Galveston and Tampa Bays. In both areas, the higher concentrations were found in oysters from the inner portion of the bays, with levels decreasing outward. POP distributions in oysters from Galveston Bay sites sampled during the NOAA's NS&TP revealed concentrations high enough for potential biological effects to be observed in oysters (Jackson et al. 1998).

Fish species have also been subjects of study to assess POP concentrations (e.g., Lewis et al. 2002; Gallo et al. 2011; Gallo et al. 2015), especially if quantifying POP concentrations in fish provides a good surrogate model for human exposure to these contaminants from seafood consumption (e.g., Rose et al. 2015; Miranda and Yogui 2016; Sun et al. 2018). Although the potential for these contaminants to enter bodies of fish exist in all cells/tissues in contact with water and food, the ones with the highest and more efficient uptake are respiratory (i.e., gills) and gastrointestinal tissues. Final POP concentrations in tissues, however, depend on many factors, including species, gender, lipid contents, and age. Concentrations of organic chemicals in fish continue to be elevated in some areas of the GOM. Even after discontinuing production and use of DDTs in December 1972 (USEPA 1972), this POP continues to be a threat to the health of the fish, birds, and humans in the vicinity of a former manufacturing facility on the Tombigbee River, downstream of the Mobile River, and into Mobile Bay. DDT concentrations in fish were up to 2 orders of magnitude (range = 419–49,800 ng g^{-1} wet weight [w.w.]) greater than those in fish from nearby locations (range = 24–113 ng g^{-1} w.w.), while average concentrations (9045 ng g^{-1} w.w.) were greater than average levels (range = 10–2233 ng g^{-1} w.w.) reported at any site of the US Geological Survey's Biomonitoring of Environmental Status and Trends (BEST) Program (Hinck et al. 2009).

Marine mammals at the top of the marine food web, particularly dolphins, provide information on the bioavailability and bioconcentration of POPs over large geographic areas and time spans. Concentrations of organochlorine and organobromine compounds were measured in bottlenose dolphins (*Tursiops truncatus*) after an unusual mortality event in the GOM in 1990 (Kuehl and Haebler 1995). The study reports that average contaminant concentrations, in μg g^{-1} on lipid basis, all increased in the order of fetus, mature female, and adult male (Table 8.5). In a similar study, PCB and total DDT concentrations in bottlenose dolphins stranded on the Texas and Florida coastlines were also reported as higher in adult males than in mature females (Salata et al. 1995; Table 8.5). Mature female dolphins are able to unload portions of their POP burden to their fetuses through placental and lactation transfers (Kuehl and Haebler 1995; Ridgway and Reddy 1995; Salata et al. 1995; Yordy et al. 2010). Fukushima and Kawai (1981) estimated that 70–90% of the female striped dolphins' contaminant load is offloaded during lactation and 5–10% of the total POP content is transferred during gestation. Similar figures were estimated for transplacental transfer of total DDTs (5.1%) and PCBs (3.7%) in bottlenose dolphins from the GOM (Salata et al. 1995). Contaminant concentrations in immature females are typically higher than those in mature females, reflecting the maternal transfer of contaminants through lactation and, in most cases, are comparable to those in males of a similar age (Tuerk et al. 2005; Yordy et al. 2010). Lactation is the most effective mother-fetus transfer but also seems to change the contaminant profile in the female after the first reproductive event. Congener-specific blubber/milk partition coefficients showed that the lower-molecular-weight POPs were selectively transferred via lactation (Yordy et al. 2010).

A recent study suggests that POPs, found to be in the lower half of the range of previously reported concentrations in dolphins from the northern GOM, were likely not the primary contributors to the poor health conditions and increased mortality observed in bottlenose dolphins following the Deepwater Horizon oil spill (Balmer et al. 2015).

As indicated earlier, PBDEs were added in 2006 to the list of chemicals of interest for NOAA's NS&TP. Between 2006 and 2007, sediment and oysters were sampled from 35 and 50 locations in the GOM, respectively, and analyzed for PBDEs. Archived oysters from a subset of locations sampled in 1996 were also analyzed for PBDEs. The total concentrations of selected tetra- to hexa-DBEs (i.e., congeners 47, 99, 100, 153, and 154) encountered during these sampling activities are shown in Table 8.3. As discussed previously, these PBDE congeners have the highest potential to bioaccumulate. In 1996, the highest concentrations were measured in samples from Lake Pontchartrain (New Orleans, 48.8 ng g^{-1}), Tampa Bay (Peter O. Knight Airport, 40.4 ng g^{-1}), and Mobile Bay (Dog River, 35.8 ng g^{-1} and Hollingers Island, 24.6 ng g^{-1}). In 2006–7, the highest values were found in samples from Tampa Bay (Peter O. Knight Airport, 27.8 ng g^{-1}), Mississippi Sound (Biloxi Bay, 25.6 ng g^{-1} and Pass Christian, 22.6 ng g^{-1}), and Galveston Bay (Ship Channel, 22.3 ng g^{-1}). With 65% to 70% contributions to the total concentrations, PBDE 47 was the most predominant congener detected in both sampling periods, followed by congeners 100 and 99. Concentrations decreased by a factor of 2 when the average concentration detected in 2006–7 is compared to the average value calculated for 1996 (see Table 8.3). Concentrations of PBDEs were also measured in blubber samples of the rough-toothed dolphin (*Steno bredanensis*) collected from a single mass stranding event near Apalachicola, Florida, in 1997 (Tuerk et al. 2005). The study revealed higher average concentrations in juvenile male and female dolphins (1367 ng g^{-1} lipid wt.) than in mature female dolphins (518 ng g^{-1} lipid wt.). Despite the differences that might exist between bottlenose and rough-toothed dolphins and the significance of the correction for water content in blubber samples (e.g., ~30% in adult bottlenose dolphins; Dunkin et al. 2005), the concentrations in mature female dolphins stranded in 1997 (518 ng g^{-1} lipid wt.) are nearly 3 times higher than those reported by Kuehl and Haebler (1995) in dolphins stranded in 1990 (190 ng g^{-1} lipid wt.).

Dioxin and furan concentrations were detected at very low levels (pg g^{-1}) in bivalves collected every other

Table 8.5. Average and range of concentrations of chemical contaminants in blubber of stranded bottlenose dolphins (*Tursiops truncatus*) on the northern Gulf of Mexico coastline

Contaminant	Fetus	Mature female	Adult male	Reference
PCBs (mg g^{-1})	4.0 (2.2–5.8)	7.2 (1.5–18)	93 (64–187)	Kuehl and Haebler 1995
p-p'-DDE (mg g^{-1})	2.8 (0.6–4.9)	3.7 (0.6–14)	37 (15–78)	
p-p'-DDT (ng g^{-1})	39 (n.d.–76)	9 (n.d.–15)	907 (234–1990)	
Heptachlor epoxide(ng g^{-1})	288 (n.d.–424)	520 (90–1580)	5020 (2310–7280)	
Mirex (ng g^{-1})	62 (10–113)	94 (23–275)	502 (271–810)	
Oxychlordane (ng g^{-1})	97 (n.d.–177)	108 (n.d.–369)	714 (407–1120)	
PBDEs (ng g^{-1})	115 (52–177)	190 (36–432)	3110 (126–16,300)	
PCBs (mg g^{-1})	n.a.	27 (4.2–63)	63 (8.6–253)	Salata et al. 1995
Total DDTs (mg g^{-1})	n.a.	8.7 (0.4–21.7)	22 (1.4–75)	

Note: Concentrations in ng g^{-1} or mg g^{-1} lipid wt. as indicated.

year between 1995 and 1998 in US coastal areas as part of NOAA's NS&TP (Wade et al. 2014). In the GOM, the average total dioxin/furan concentration detected was 6.88 ± 10.8 pg g^{-1} (median = n.d.; range = n.d.–41.2 pg g^{-1}) with nearly equal contributions of dioxins (average = 3.85 ± 7.44; median = n.d.; range = n.d.–28.5 pg g^{-1}) and furans (average = 3.03 ± 4.66; median = n.d.; range = n.d.–16.1 pg g^{-1}).

Petroleum Hydrocarbons in the Gulf of Mexico

Composition of Petroleum

There are many references on the composition of petroleum that can be consulted for further detail (e.g., Overton et al. 2016; Pampanin and Sydnes 2013). Petroleum is a complex mixture of thousands of individual compounds, the majority of which (>80%) consist of the elements carbon and hydrogen; other petroleum compounds also have 1 or more nitrogen, sulfur, or oxygen atoms. In addition, petroleum can contain incorporated metals, predominantly nickel and vanadium. In fact, this nickel/vanadium ratio is used as a marker of specific oils (Yakubov et al. 2016). Petroleum can contain compounds ranging in molecular weight from 16 to >1000 Da (Dalton, unified atomic weight unit). Methane (16 Da) is the predominant gas found in petroleum. Natural gas (1 to 5 carbon atoms) can also contain the hydrocarbon gases ethane, propane, butanes, and pentanes as well as traces of carbon dioxide and nitrogen. The next range of petroleum components, the condensates, comprise hydrocarbons, with 6 or more carbon atoms, in a liquid form when temperature and pressure are reduced. Other components of oils are considered semivolatile and are extractable into organic solvents such as dichloromethane. High-molecular-weight components of petroleum (i.e., asphaltenes and resins) are not amenable to typical analyses (Overton et al. 2016) and are not commonly measured during oil spill responses other than as tar balls (Knap et al. 1980; Wade and Quinn 1975) so are not discussed further here. The complex composition of petroleum requires many different methods for resolution during an oil spill.

Petroleum Sources to the Gulf of Mexico

Petroleum enters the coastal zone in association with runoff from roads and parking lots, sewage discharge, boating activities, industrial activities, inappropriate disposal of used crankcase oil, spills, natural seepage, and atmospheric deposition (NRC 2003). In offshore areas, the sources of petroleum hydrocarbons include oil seeps, spills, offshore-produced water discharges, transport from nearshore areas, and atmospheric deposition. The GOM has a significant number of natural oil seeps that can be documented from satellites (MacDonald et al. 2015). Natural oil seeps can indicate whether producible oil reservoirs are present, which is a tool used for the exploration of petroleum. The GOM is a highly productive petroleum area, one of the most productive basins in the world. However, with that production come the issues of disposal of drilling muds and produced waters and risk of catastrophic events. The Deepwater Horizon (DWH) oil spill, in which 11 people died and 17 were injured in April 2010, is an example of this risk. Reported concentrations of total petroleum and aliphatic and polycyclic aromatic hydrocarbons in the GOM before and after the DWH (Macondo) incident are shown in Tables 8.6 and 8.7.

Most of the total petroleum hydrocarbon (TPH) concentrations, measured by gas chromatography with a flame ionization detector (GC-FID), reported for samples collected before April 20, 2010, were at a concentration of <1 μg L^{-1} (Calder 1977; Macko et al. 1982). After the accidental release, TPH concentrations ranged from n.d. to 7,270,000 μg L^{-1} (Wade et al. 2016). The higher concentrations were measured in samples collected around the leaking wellhead or from surface oil slicks (Reddy et al. 2011; Wade et al. 2016). After the DWH oil spill, 70 beach-water samples and 70 sediment samples were collected before oil could have made landfall to establish background concentrations (Nowell et al. 2013). After oil had made landfall, 48 of these sites were resampled. Nowell et al. (2013) reported that only 1 oil component, toluene, had significantly higher concentrations in post-landfall water samples. None of the water samples exceeded any human-health benchmark, and only 1 of the 118 samples (a post-landfall sample) exceeded the aquatic-life benchmark. PAHs frequently exceeded the aquatic-life benchmarks for sediment samples collected before and after the landfall, which indicates that many coastal sediments were contaminated with PAHs before the DWH oil spill. Only 5 sites (10.4%) had even higher post-landfall concentrations and, based on geochemical data, may have been from DWH oil.

Table 8.6. Hydrocarbon concentrations reported for Gulf of Mexico water before April 20, 2010 (pre-spill data)

Year	Area	n	Depth (m)	Total petroleum hydrocarbons		Aliphatic hydrocarbons		Aromatic hydrocarbons		Instrumentation	Reference
				Avg ($\mu g\ L^{-1}$)	Range ($\mu g\ L^{-1}$)	Avg ($\mu g\ L^{-1}$)	Range ($\mu g\ L^{-1}$)	Avg ($\mu g\ L^{-1}$)	Range ($\mu g\ L^{-1}$)		
1973	Florida Straits	6	1–500	47	trace–75					Gravimetric	Iliffe and Calder 1974
	Mid-Gulf	5	1–500	12	n.d.–18						
	Yucatan Channel	5	1–500	12	trace–24						
1975–76	Northeastern GOM	45	10	0.4	0.1–1.31	0.2	0.01–1.14	0.21	0.03–1.01	GC-FID	Calder 1977
1976	Seadock (TX)	11 10	5 15					2.4 ± 0.4 1.3 ± 0.5		Fluorescence spectroscopy	Hiltabrand 1978
	Loop (LA)	8 11	5 15					1.9 ± 0.5 0.9 ± 0.5			
	Ameraport (AL)	11 10	5 15					1.7 ± 0.2 0.6 ± 0.4			
1979	Western GOM	5	3–10			0.285	0.005–0.657			GC-FID	Macko et al. 1982
1979	Vicinity of Buccaneer oilfield	76	1–20			5.46	0.42.8			GC-FID and GC/MS	Middleditch et al. 1979
	Northern GOM	39						4.7			
	Mississippi outflow	24						4.3			
1980	GOM								0.1–0.2	UV fluorescence (chrysene equivalents)	Levy et al. 1981
1982	GOM								1–10	UV fluorescence (chrysene equivalents)	Atwood, Burton, et al. 1987
1988	Green Canyon 233/234	1	570					0.028		GC/MS	Wade et al. 1989
1987–88	Florida Straits	119						3.1 ± 4.3			Botello et al. 1997
1999	GOM	1	Surface					<0.001		GC/MS	Mitra and Bianchi 2003
		1	Bottom					0.085			

Table 8.7. Hydrocarbon concentrations reported for Gulf of Mexico water after April 20, 2010 (post-spill data)

Year	Area	n	Depth (m)	Total petroleum hydrocarbons		Aliphatic hydrocarbons		Aromatic hydrocarbons		Instrumentation	Reference
				Avg ($\mu g\ L^{-1}$)	Range ($\mu g\ L^{-1}$)	Avg ($\mu g\ L^{-1}$)	Range ($\mu g\ L^{-1}$)	Avg ($\mu g\ L^{-1}$)	Range ($\mu g\ L^{-1}$)		
2010	GOM	45	Varies					47 ± 187	0.0–1231	GC/MS	Sammarco et al. 2013
		63	Varies	202,000 ± 1,424,000	n.d.–11,400,000						
2010	Eastern GOM	282	Varies		0.70–5.04					Total scanning fluorescence	Wade, Sweet, Walpert, et al. 2011
		48	Varies						0.020–0.160	GC/MS	
2010	GOM	12	Surface/ subsurface		2–442				0.01–59	GC-FID and GC/ MS	Wade, Sweet, Sericano, et al. 2011
2010	GOM		Subsurface						29.4–189	GC/MS	Diercks et al. 2010

Hydrocarbons Released

Gaseous and higher-molecular-weight hydrocarbons were released during the DWH oil spill. The gaseous components of oil (e.g. methane, ethane, propane, and butane) were found in subsurface (1100 m depth) plumes (Camilli et al. 2010). There were reductions of dissolved-oxygen concentrations at depth in areas where DWH oil occurred (Camilli et al. 2010; Joye et al. 2011). The reduced oxygen resulted from blooms of methanotrophic bacteria that degraded the released methane back to low levels after 120 days (Kessler et al. 2011). Monoaromatic components (e.g., benzene, toluene, and xylenes) were also found in the subsurface plumes (Camilli et al. 2010). These components are toxic but due to their low boiling point and high solubility were not often detected in water samples during the DWH oil spill. We focus instead on the most toxic components of petroleum.

The Toxic Component of Petroleum: Polycyclic Aromatic Hydrocarbons

PAHs are some of the more toxic components of petroleum, and selected PAH components are on the EPA's priority pollutant list as suspected carcinogens (Overton et al. 2016). Due to concerns about their toxicity, PAHs have been preferentially studied in the marine environment of the OM. PAHs are aromatic compounds containing 2 to 8 conjugated ring systems (see Fig. 8.1.). Some of these rings can contain nitrogen, sulfur, and oxygen in addition to hydrogen and carbon. Although PAHs are moderately persistent in the environment, many of these compounds have characteristics similar to those of POPs. In 2005, the Stockholm Convention's list was expanded to include, among others, several PAHs (UNEP 2005).

PAHs present in the marine environment have 2 predominant sources, petrogenic and pyrogenic (Overton et al. 2016), while a third source, biogenic, is relatively minor. Petrogenic PAHs derive from oil and oil products, while pyrogenic PAHs are formed by the incomplete combustion of organic materials (biomass burning, forest fires, coal burning, carbonization of coal, refineries, etc.). Pyrogenic PAHs are composed predominantly of compounds having more rings (e.g., 5–6) than petrogenic PAHs (e.g., 2–3). Petrogenic PAHs are found in petroleum and petroleum products and have a predominance of lower-molecular-weight and alkylated components (e.g., the addition of a methyl group to the ring). The sources of petrogenic PAHs to the coastal zone and offshore GOM include all of those described previously for petroleum. The pyrogenic PAH sources also include all of those indicated for petroleum except natural seepage and oil spills. In addition, pyrogenic PAH can be produced by flaring of gas on offshore platforms as well as the in situ burning of surface oil as part of oil-spill response efforts.

Water Column: The Deepwater Horizon Oil Spill

There are many historical studies of oil in the water column derived from spills such as the Ixtoc I spill (Boehm et al. 1982). We concentrate our review on the DWH oil spill. A recent study (Wade et al. 2016) reviewed the PAH concentrations measured in offshore water samples from the GOM before and after the DWH oil spill on April 20, 2010. PAH concentrations from reports prior to the spill ranged from 0.028 to 10.6 μg L^{-1} in samples collected from the Gulf surface to a depth of 570 m. In the reviewed studies, some high concentrations were determined by fluorescence techniques, and some were reported as chrysene equivalents. Two past studies determined PAHs by gas chromatography-mass spectrometry (Wade et al. 1989; GC/MS; Mitra and Bianchi 2003). PAH concentrations in unfiltered 20–30 L water samples collected in the vicinity of a known seep had a mean concentration of 0.028 μg L^{-1} (Wade et al. 1989). This concentration was below the median field blank reported in the Gulf Science Data (Wade et al. 2016). The second pre-DWH study reported dissolved PAH from filtered samples but excluded 2-ring aromatics (Mitra and Bianchi 2003). PAH concentrations were low, <0.001 to 0.085 μg L^{-1}, and in the range of field blanks reported in the Gulf Science Data. These concentrations could have been possibly higher if the samples had not been filtered and the determination had included the 2-ring PAHs.

WATER BACKGROUND POLYCYCLIC AROMATIC HYDROCARBON CONCENTRATIONS

PAH concentrations from the Gulf Science Data were measured in 16,557 samples, of which 42% were collected between April 20 and August 3, 2010, which is the reported date of the last potentially recoverable surface oil (OSAT 2011), with 58% of the samples collected after that date (Wade et al. 2016). Field blanks, consisting of water bottles filled with organic free water, taken to the field, opened on shipboard, and returned to the lab for analyses, were also included. The total PAH concentrations in field blanks had a mean concentration of 0.22 μg L^{-1} and a median concentration of 0.056 μg L^{-1}. Based on these field blanks, concentration for the open GOM below 0.056

μg L^{-1} were defined as background (Wade et al. 2016). In samples collected in 2012 and 2013 at the DWH spill site, PAH concentrations ranged from 0.024 to 0.058 μg L^{-1} in 20 L filtered-water samples (Adhikari et al. 2015). Liu et al. (2016) reported that rapid weathering quickly reduced hydrocarbon concentrations in coastal surface waters of the northern GOM by August 2010 and showed results generally in the range of those of the field blanks (Adhikari et al. 2015; Liu et al. 2016; Wade et al. 2016). To our knowledge, there are no prespill PAH concentration measurements reported in the absence of a known spill or natural seep.

DISTRIBUTION OF POLYCYCLIC AROMATIC HYDROCARBONS DURING AND AFTER THE DEEPWATER HORIZON OIL SPILL

Of the 16,557 water samples measured before and after August 3, 2010, 78% had PAH concentrations below the value of the defined background PAH concentration for the GOM (0.056 μg L^{-1}; Wade et al. 2016). The highest PAH concentration in subsurface waters reported during the spill was 154,054 μg L^{-1} and was collected by a remotely operated vehicle (ROV) at the riser jet at the blowout area. Generally, most samples analyzed were orders of magnitude below this concentration. Eighteen samples had concentrations above 100 μg L^{-1}. Highest PAH concentrations were found in water samples collected during the spill near the surface (0–1 m) or those collected from a submersible directly in the spill jet (Wade et al. 2016).

The study of PAH concentrations at various water depths and distances from the wellhead, before and after August 3, 2012, showed that samples with higher PAH concentrations were found at the wellhead (1524 m depth); in subsurface intrusion layers, characterized by very small oil droplets and dissolved hydrocarbons (800–1200 m); or at the surface from the large droplets that rise quickly to the surface (Wade et al. 2016). PAH concentrations above background were found throughout the water column, but mostly at the surface or in the intrusion layer of oil treated with chemical dispersant in a plume within 25 km off the wellhead to the southwest, partially due to the targeted sampling of these areas.

After the spill but before August 3, 2010, most of the samples with high PAH concentrations were located in depth ranges between 1–10 m and 800–1400 m. After August 3, 2010, both of these areas had only a few samples above background level. Surface samples (0–1 m) with concentrations above background were likely due to targeted sampling of areas with surface slicks present where petroleum hydrocarbons are known to be concentrated (Burns and Knap 1986; Duce et al. 1972; Knap et al. 1986). The source of the hydrocarbons may be from the DWH oil spill but could be from other sources (e.g., natural seeps, ship operations, riverine discharge, and atmospheric deposition). It has been estimated that on average 1.0 (NRC 2003) to 3.5 (MacDonald et al. 2015) million barrels of oil enter the GOM from natural seeps each year. Based on these estimates, the amount of oil released during the DWH oil spill (4.1 million barrels) would enter the GOM in 1.2 to 4.1 years. In addition, an estimated 0.074 million barrels of oil enter annually from platform production operations, transportation accidents, and unburned fuel from engines (NRC 2003). There are ubiquitous sources of PAHs in coastal environments (Iqbal et al. 2008). Atmospheric deposition is another direct source of PAHs to open ocean and coastal areas and an indirect source through runoff from drainage basins (Park et al. 2001, 2002).

Tarballs

When oil undergoes weathering, it can create buoyant solids or tarballs, which are commonly found on GOM beaches. Tarballs from Texas beaches were likely used by Karankawa Indians to waterproof pottery (TSHA 2010). These tarballs are also found in open waters of the GOM (Atwood, Dinkel-Mckay, et al. 1987; Atwood, Burton, et al. 1987). Regulations regarding washing oil tankers with seawater have greatly reduced the amount of tarballs; however, they are still an unwelcome occurrence on GOM beaches. The source is likely natural seepages as well as unreported small spills.

Sediments

Oil components are only sparingly water soluble and therefore associate with particles in the water column. When these particles become denser than seawater, they sink to the sediments. This process leads to sediments being sinks for oil and other hydrophobic contaminants. During the DWH oil spill profuse flocs of microbial-produced, mucus-abundant materials were observed in surface water (Passow et al. 2012). Once it combines with other material (e.g., particulates, detritus), it is operationally defined as "marine snow." The production of these exopolymers or macroscopic aggregates by planktonic microbes has been reported as early as in the 1980s (Knauer et al. 1982; Alldredge and Youngbluth 1985; Alldredge and Silver 1988); however, not until after the

DWH oil spill did the production of marine snow become the focus of attention to study their significance in the fate of oil and chemical dispersants in the ocean. It has been hypothesized that the production of these exopolymers by planktonic microbes significantly impacted the distribution of oil through aggregation and subsequent gravitational settling to the seafloor (Passow et al. 2012; Passow 2016; Quigg et al. 2016). This process could be important for the removal of oil by sedimentation. It has been estimated that as much as 14% of the total oil released was transported to the ocean floor, by the formation and incorporation of oil, via sedimentation of marine snow (Daly et al. 2016). The formation and rapid sedimentation of oil-rich marine snow associated with the long residence times of oil on the seafloor observed after the spill can have adverse effects on benthic communities, including corals (Hsing et al. 2013; DeLeo et al. 2016; Fisher et al. 2016)

To assess the ecological impact of spilled oil to the environment, it is important to distinguish the spilled-oil signature from background hydrocarbons. The discussion of oil fingerprinting is beyond the scope of this book, but readers can consult the review by Overton et al. (2016) for further details. PAHs are extensively found in soils, sea-bottom sediments, and petroleum products (Blumer 1976). The GOM is one of the few coastal ecosystems in the world where both natural petroleum seeps (Kennicutt 1988; MacDonald et al. 2015) and anthropogenic sources occur (Mitra and Bianchi 2003). PAHs in sediments may retain their toxic nature for years (Wolfe et al. 1994; Douglas et al. 1996; Hostettler et al. 1999; Reed et al. 1999; Culbertson et al. 2007; Culbertson et al. 2008; Viñas et al. 2010). Perhaps one of the longest-studied spills in the United States was the West Falmouth oil spill, off the Massachusetts coast, in 1969. Recent work has shown that a substantial amount of moderately degraded petroleum still existed 8–20 cm below the sediment surface 30–40 years after the spill (Reddy et al. 2002; Culbertson et al. 2007).

Source Identification

Source-identification studies are routinely used in oil-spill investigations to distinguish oil sources (Wang et al. 1994; Boehm et al. 1997; Peters et al. 2004). Petroleum biomarkers such as hopanes and steranes, which are recalcitrant to degradation, are the most commonly studied biomarkers because of their high resistance to weathering (Wang et al. 1995; Douglas et al. 1996; Sauer et al. 1998; Wang et al. 1999; Peters et al. 2004; Wang et al. 2005; Malmquist et al. 2007; Yim et al. 2011). Many studies of the DWH oil spill have been focused on distribution of oil in salt marshes (Lin and Mendelssohn 2012; Silliman et al. 2012) and in sediments (Sammarco et al. 2013). Unfortunately, little work has been focused on the sources and levels of weathering of oil residues in sediments, despite their importance in understanding how extensively marsh sediments remained contaminated by DWH oil. One and a half years after the DWH incident, a set of subtidal and intertidal marsh sediment cores were collected from 5 stations in Barataria Bay, Louisiana, and analyzed to determine the spatial and vertical distributions and source of hydrocarbon residues based on their chemical composition (Kirman et al. 2016). An archived core, collected before the DWH oil spill from the same area, was also analyzed to assess the prespill hydrocarbon distribution in the area. Analyses of aliphatic hydrocarbons, PAHs, and stable carbon isotopes showed that the distribution of petroleum hydrocarbons in Barataria Bay was patchy and limited in areal extent. High TPH and PAH concentrations (77,399 $\mu g\ g^{-1}$ and 219,065 $ng\ g^{-1}$, respectively) were detected in the surface sediments of 1 core from the surface to a depth of 9 cm. Based on a sedimentation rate of 0.39 $cm\ yr^{-1}$, determined using ^{137}Cs, the anthropogenic hydrocarbons in sections of these sediment cores were deposited ~50–60 years ago (Kirman et al. 2016). The historical hydrocarbon concentrations from deeper core sections increased significantly toward the sediment surface and were attributed to recent inputs. Although the oil present in the bay's sediments has undergone moderate weathering, biomarker analyses performed on samples from 1 core indicated the presence of hydrocarbons from the DWH oil spill (see Kirman et al. 2016; Overton et al. 2016). The effects of oiling events on Barataria Bay and other marsh ecosystems in this region remain uncertain, as oil undergoes weathering changes over time (Kirman et al. 2016). These estuarine sediments indicate nonpoint source contamination by oil and combustion hydrocarbons as well as a recent input from DWH oil.

Biota

PAHs are metabolized by vertebrates (e.g., fish, humans, etc.); therefore, PAHs are not usually detected in appreciable concentrations in these organisms unless they were recently exposed. However, PAHs may be converted by these organisms into more active and potent carcinogenic metabolites (e.g., Bonner et al. 2005; Da Silva et al. 2006; Johnson-Restrepo et al. 2008; Beg et al. 2015). Invertebrates, particularly bivalves, exposed to PAHs in the environment, however, tend to accumulate these compounds with little modification, if any, in their tissues to reach equilibrium with surrounding concentrations (Sericano 1993); therefore, these chemicals were also monitored in bivalves by the NOAA's NS&TP.

Table 8.8 compares the PAH concentrations measured in native oysters collected at the same locations 23 years apart (1989 and 2012). With the exception of a large, unexplained increase observed for acenaphthene (from 24.7 ± 10.4 ng g^{-1} to 6.42 ± 2.36 ng g^{-1}), PAH concentrations in the northern GOM coastal areas decreased by 11.2% and 92.7% (average = 56.9 ± 20.4) between 1989 and 2012. These decreases are somewhat lower than those observed in mean legacy OC pesticide concentrations (range = 32.8% to 90.5%, average 68.3 ± 18.9%; see Table 8.4).

A recent long-term monitoring study of PAHs in ambient air of Porto, Portugal, concluded that the observed decreasing trend in PAH concentrations could be due to a decline in the consumption of fossil fuels and identified automobile traffic emissions as the major contributors of PAH levels in air (Albuquerque et al. 2016). If human activity is the dominant source of PAHs to the environment through the combustion of various fuels, a decrease in PAH concentrations would be expected as cleaner sources of energy are introduced; however,

Table 8.8. PAH concentrations in northern Gulf of Mexico native oysters, 1989 and 2012

Compound	Median	Avg ± 1 Std	Range	Median	Avg ± 1 Std	Range	Change
	1989 (n = 14)			2012 (n = 14)			%
Acenaphthene	6.11	6.42 ± 2.36	3.06–9.75	22.0	24.7 ± 10.4	12.5–52.6	285
Acenaphthylene	6.86	7.61 ± 3.06	3.56–13.3	2.16	3.50 ± 3.01	0.73–9.69	-54.0
Anthracene	13.9	16.5 ± 12.8	4.76–49.0	5.77	6.73 ± 5.06	1.28–17.3	-59.2
Benz[a]anthracene	9.35	15.0 ± 15.1	4.85–62.3	3.06	5.49 ± 5.30	0.37–17.9	-63.4
Benzo[a]pyrene	3.80	4.31 ± 2.98	0.55–10.2	3.44	4.63 ± 4.01	<MDL–12.1	7.42
Benzo[b]fluoranthene	10.4	13.3 ± 13.7	2.32–52.0	4.18	8.13 ± 8.42	0.57–25.0	-38.9
Benzo[g,h,i]perylene	5.96	7.08 ± 6.24	2.21–25.8	1.86	2.95 ± 2.78	0.31–8.32	-58.3
Benzo[k]fluoranthene	5.59	11.0 ± 13.1	1.77–52.1	1.15	2.12 ± 2.18	<MDL–7.02	-80.7
Chrysene	27.2	34.9 ± 32.8	5.92–106	10.4	13.8 ± 13.1	1.07–47.0	-60.5
Dibenzo[a,h]anthracene	6.42	6.32 ± 3.20	1.78–14.4	<MDL	0.46 ± 0.59	n.d.–1.86	-92.7
Fluoranthene	67.4	68.8 ± 53.8	9.24–162	15.0	37.2 ± 42.2	4.14–134	-45.9
Fluorene	10.6	11.0 ± 3.01	6.74–17.4	3.42	4.74 ± 2.80	1.64–11.5	-56.9
Indeno[1,2,3-c,d]pyrene	8.12	8.53 ± 4.40	2.48–15.7	0.83	1.43 ± 1.42	0.28–4.82	-83.2
Naphthalene	31.1	29.4 ± 14.7	5.17–52.0	24.0	26.1 ± 7.40	16.7–41.5	-11.2
Phenanthrene	22.6	29.3 ± 10.9	17.5–48.9	13.5	16.7 ± 10.9	7.39–44.3	-43.0
Pyrene	43.2	58.4 ± 57.6	7.49–221	20.4	29.8 ± 30.0	4.93–100	-49.0

Note: Concentrations of ng g^{-1}, dry wt. Locations (collected at the same locations each year): Breton Sound (Bay Gardene and Sable Island); Lake Borgne (Malheureux Point); Mississippi Sound (Pascagoula Bay and Pass Christian); Mobile Bay (Cedar Point Reef and Hollingers Island Channel); Tampa Bay (Cockroach Bay and Peter O. Knight Airport, Mullet Key Bayou, Navarez Park, Old Tampa Bay, and Papys Bayou); and Terrebonne Bay (Lake Barre).

the presence of PAHs in the GOM from other sources, minor oil spills and seepage, for example, might explain the larger range detected. Noreña-Barroso et al. (1999) reported the concentrations of PAHs in oyster tissues, collected in 1991 from the Términos Lagoon, one of the most productive shrimp-fishing and oyster-farming areas in the GOM. The total PAH concentrations reported in that study (range = 785–3220 ng g^{-1}) slightly overlapped with the range of concentrations encountered, as part of the NOAA's NS&TP, in 1989 along the northern GOM (range = 79.4–912 ng g^{-1}) for the same suite of PAHs (see Table 8.8).

Impact of Organic Chemical Pollutants

Bioaccumulation, Bioconcentration, and Biomagnification

Bioaccumulation is the uptake and retention of bioavailable chemicals from external sources: air, food, sediments, and water (Neff 2007). This process includes the increase in chemical concentration in an organism from both environmental exposure and consumption of food items containing the chemical (Newman and Unger 2003). Clearly, for bioaccumulation to occur, the uptake rate of the chemical should be higher than the loss rate from the tissues of the organism (Neff 2007; Amiard and Amiard-Triquet 2015). In the absence of or very low metabolic activity, bioaccumulation of POPs is directly related to the compound's lipophilicity (Neff 2007). As a result of bioconcentration (dietary accumulation), POPs can also move up throughout the food web (van der Oost et al. 2003). POPs are persistent in the environment with the potential to increase in concentrations with progression up the food web (biomagnification) and eventually be consumed by humans (Nikinmaa 2014). The most relevant effects of POPs on humans include cancer, developmental impairments, damage to the nervous system, endocrine disruption, reproductive alterations, and damage to the immune system (Dewailly and Knap 2006). Consequently, after confirmation of bioaccumulation and biomagnification along the food web, the next question is whether the concentrations of these chemicals constitute a health risk for biota and humans. In the literature, this question has been addressed by using a relatively large number of effect indicators, such as bioconcentration factors (BCFs), toxic equivalent factors (TEF, relative toxic potency factors), toxic equivalency quotient (TEQ, the measured concentration of a chemical divided by the corresponding TEF), chemical-specific cancer risks, or hazard quotients (HQs), among others (van den Berg 2006).

Effects of Chemical Organic Pollutants on Marine Biota

There is a large amount of information on the presence and concentrations of chemical organic pollutants in coastal areas, both American and Mexican, bordering the GOM, particularly those included in the original dirty dozen list in the Stockholm Convention (see Table 8.1). These chemicals have the potential to cause negative effects in biota once they reach a threshold concentration, including implication for human health. The effects of pollutants on marine organisms and ecosystems have been well documented; however, relatively few articles were published on the effects of pollutants in the GOM, particularly those listed in Table 8.1. Now research funded by the Gulf of Mexico Research Initiative (GOMRI) has resulted in many articles on the effects of petroleum, in different forms, alone or mixed with dispersants.

Persistent Organic Pollutants

A study involving oysters from 5 Florida bays indicated that relatively high concentrations of certain xenobiotics (including some pesticides and PCBs) in tissues are correlated with the oyster defense-related activities (Fisher et al. 2000; Oliver et al. 2001). Similarly, exposure of *Crassostrea virginica* to water-soluble fractions, generated from contaminated sediments containing variable amounts of PAHs, increased the oyster's susceptibility to infection caused by *Perkinsus marinus* in laboratory experiments (Chu 1999). *P. marinus* is one of the primary factors that adversely impacts the abundance and productivity of *C. virginica* at various coastal locations on the GOM coast (Burreson et al. 1994; Ray 1996; Aguirre-Macedo et al. 2007; Powell et al. 2008). It is worthy to note, however, that abundance and distribution of *P. marinus* in the field are also positively correlated with warm temperatures and high salinities (Andrews and Ray 1988). In contrast, negative correlations between parasite abundance and contaminant concentrations in GOM oysters suggest that some contaminants might affect parasites more than their hosts. The study concluded that "as parasites are potentially more sensitive than hosts to environmental disturbance, the possibility that contaminants may reduce the abundance of some parasites is deserving of attention" (Kim et al. 2008, 122).

Significant negative associations were also reported between the concentrations of some POPs (e.g., aldrin, dieldrin, endrin, DDT) and the community metrics of the helminth parasites infecting 2 fish species (*Sphoeroides testudineus* and *Eucinostomus gula*) and the crab (*Uca thayeri*) (Vidal Martínez and Aguirre-Macedo 2014). Watanabe et al. (2003) reported the risk of cancer through the consumption of contaminated fish from the lower Mississippi River. Nine species (blue catfish, carp, channel catfish, cobia, crayfish, flathead catfish, red drum, spotted gar, and striped bass) were identified as potentially dangerous for human consumption due to elevated concentrations of aldrin, followed by dieldrin, α-HCH, γ-HCH, and heptachlor epoxide. Exposure to POPs was associated with DNA damage in the grey mullet (*Mugil cephalus*) and 4 species of fish (i.e., *Centropomus parallelus*, *Mugil cephalus*, *Oreochromis* sp., and *Ariopsis felis*) from Coatzacoalcos, Veracruz, Mexico (González-Mille et al. 2010).

The high concentrations of legacy pesticides measured in eggs from a female alligator (*Alligator mississippiensis*) dosed experimentally with *p,p'*-DDE, toxaphene, chlordane, and dieldrin were associated with an increased embryonic mortality at contaminated sites in Florida, as eggs from dosed females exhibited a decreased viability (13–22%) compared to that for control females (20–45%) (Rauschenberger et al. 2004). Also, in female Morelet's crocodiles from Campeche, Mexico, embryonic mortality seem to be increased because of high PCB and HCH isomer concentrations (González-Jauregui et al. 2012). Nearly half of the biota sampled from 5 coastal lagoons of Yucatan, Mexico, had HCH isomer concentrations that exceeded the NOAA's probable effect level (Gold-Bouchot et al. 2014) above which negative effects are possible (Buchman 1999).

TEF and TEQ are used extensively to express the toxicity of dioxins, furans, and PCBs as equivalents to the most toxic form of dioxin (2,3,7,8-TCDD; van den Berg et al. 2006). The TEQs derived from PCB concentrations in dolphins and pygmy whales from the eastern GOM were elevated enough to induce perturbation of the immune response and cytochrome P450 (Watanabe et al. 2000). High PCB concentrations measured in dolphins (*T. truncatus*) from Florida were nearly at or above the threshold concentrations of 17 $\mu g\ g^{-1}$ lipid proposed for adverse effects in marine mammals (Kannan et al. 2000), according to which most dolphins sampled in Biscayne Bay would be at risk of reproductive failure and decreased immune function (Litz et al. 2007). There is also a potential association between PCB concentrations and the infertility rates in a Florida shark (*Sphyrna tiburo*; Gelsleichter et al. 2005). The TEQs derived from PCB congeners detected in the blue crab (*C. sapidus*) and 2 catfish species (*Ariopsis felis* and *Ictalurus furcatus*) from the Houston Ship Channel, Texas, were high (0.21–1.6 ng TEQ g^{-1} lipid and 0.18–1.2 ng TEQ g^{-1} lipid, for crab and fish, respectively; Howell et al. 2011). This was related to the high concentration of PCB 126, a planar congener that has the potential to mimic the effects of 2,3,7,8-TCDD. It must be noted that the Houston Ship Channel area in Galveston Bay is under a fish-consumption advisory guidelines recommending that fish consumption should not exceed a specified maximum amount because the levels of PCDD/Fs in both catfish and blue crabs have not declined since the 1990s (Dean et al. 2009).

Petroleum Hydrocarbons

Changes in community structure of different types of marine organisms have been used to assess the impact of pollutants, particularly in sediments. In the southern GOM, negative changes in free-living nematodes' community structure related to total hydrocarbon concentrations in sediments were reported by Gold-Bouchot and Herrera-Rodríguez (1996) in the area close to the offshore oil-producing zone. Hernández-Arana et al. (2005) reported changes in the macrobenthos community structure in sediment samples collected along 4 transects crossing the oil-producing zone in the southern Gulf, while Kuk-Dzul et al. (2012) reported changes in microbenthic community structure in coastal lagoons along the northern coast of the Yucatan Peninsula due to environmental factors (e.g., water salinity, pH, and temperature) and pollutants (e.g., pentachlorobenzene, trichlorobenzene, and low-molecular-weight PAHs), which explained 52% of the variance in abundance and species composition among sites. Brown et al. (2000) also reported the effects of pollutants in sediments and other environmental factors as driving forces for changes in the structure of microbenthic communities in estuaries in the northern Gulf. Several studies have reported the impact of oil and dispersants on the growth rates, histology, and community structure of deep corals and other benthic communities following the DWH oil spill (DeLeo et al. 2016; Fisher et al. 2014; Fisher et al. 2016; Hsing et al. 2013; Prouty et al. 2016; Silva et al. 2016; White et al. 2012).

In the case of particular groups of benthic organisms, the effects of oil on meiofauna (free-living nematodes and copepods) diversity were reported by Landers, Nichols,

Barron, et al. (2014), and the effects of metals on meiofauna were reported by Landers, Nichols, Schimmer, et al. (2014). Effects of oil on salt marsh crabs (Zengel, Pennings, et al. 2016) and on larval settlement and condition of blue crabs (Grey et al. 2015) have been reported. Also, the effects of oil on crab transcriptome have been recently published (Yednock et al. 2015). The decline of benthic foraminifera was one of the effects found after the DWH oil spill (Schwing et al. 2015). Furthermore, oil and chemically dispersed oil has affected oyster larvae (Laramore et al. 2014) as well as oyster physiology (Soniat et al. 2011). The effects of oil spill associated with the DWH accident on periwinkles have also been recently reported (Zengel, Weaver, et al. 2016).

A number of researchers studied oil and dispersant effects on zooplankton at different levels. Almeda, Wambaugh, Chai, et al. (2013) reported the effects of Louisiana light sweet crude oil exposure on the survival of different zooplankton at larval and adult stages and indicated that gelatinous zooplankton, at larval stages, were more sensitive to crude oil than adult stages. Some of the most toxic PAHs of crude oil can be bioaccumulated by the studied zooplankton and potentially transferred up the food web to reach apex predators. In a companion study, these authors also indicated that the negative impact of oil spills on mesozooplankton may be increased by the use of chemical dispersant and UV radiation, although the harmful effects of crude-oil exposure in copepods and the transfer of PAHs to higher trophic levels can be attenuated by microbial food-web interactions (Almeda, Wambaugh, Wang, et al. 2013). These interactions may play an important role in the fate of PAHs in marine environments. Almeda, Baca, et al. (2014); Almeda, Bona, et al. (2014); and Almeda, Hyatt, et al. (2014) examined the effect of oil and chemically dispersed oil in marine planktonic copepods (*Acartia tonsa*, *Temora turbinate*, and *Parvocalanus crassirostris*), meroplanktonic barnacle nauplii (*Amphibalanus improvises*), and marine microzooplankton (oligotrich ciliates, tintinnids, and heterotrophic dinoflagellates). The 3 species of copepods studied were able to ingest crude-oil droplets, and it was observed that exposure to oil alone or dispersant (Corexit 9500A) alone caused a reduction of the production rates of egg and fecal pellets and egg hatching. However, exposure to dispersant-treated crude oil was significantly more toxic than crude oil alone (Almeda, Baca, et al. 2014). Also, the application of chemical dispersants seem to increase the impact of crude-oil spills on meroplanktonic larvae with a potential effect on the population dynamics of marine benthic invertebrates (Almeda, Bona, et al. 2014). The transfer of energy from lower to higher trophic levels might be negatively impacted by dispersant-treated crude oil on microzooplankton and also might affect the dynamics of marine planktonic communities (Almeda, Hyatt, et al. 2014). The lethal and sublethal toxicity of dispersant-treated crude oil to marine planktonic copepods seems to be increased by natural ultraviolet B radiation (Almeda et al. 2016). Carassou et al. (2014) investigated the response of mesozooplankton community structure to the DWH oil spill, comparing data compiled during the oil spill to data from a long-term plankton survey off Alabama. Significant variations observed in zooplankton assemblage structure in May and June 2010 were possibly due to the oil spill; however, these changes were no longer observed by July 2010, leading to the conclusion that "although changes in zooplankton community composition were observed during the oil spill, variations were weak and recovery was rapid."

The effect of oil contamination on phytoplankton has been less studied. The general toxicity and mutagenicity of upper-water-column samples along the northeastern GOM and Florida shelf were studied at the time of the DWH oil spill and during several cruises thereafter (Paul et al. 2013). In 2010, 21% of the water samples were toxic to bacteria (Microtox assay), 34% were toxic to phytoplankton (QwikLite assay), and 43% showed DNA-damaging activity (λ-Microscreen Prophage Induction assay). The first 2 analyses correlate toxicity with petroleum hydrocarbon concentrations. By December 2010, Microtox toxicity was nearly absent while QwikLite assay still tested positive at multiple locations. The Microscreen mutagenicity assay indicated that certain water-column samples from the Florida shelf were mutagenic at least 1.5 years after capping the Macondo well. A comparison of the available baseline phytoplankton data from the Louisiana shelf, with samples collected monthly at the same locations during and after the oil spill, showed significantly lower phytoplankton abundance and a shifting of the phytoplankton community from ciliates and phytoflagellates to diatoms and cyanobacteria (Parsons et al. 2015). A series of toxicological experiments conducted with 2 microalgae species (*Isochrysis galbana and Chaetoceros* sp.) to better understand the potential impacts of the oil spill and dispersant on marine microalgae revealed that exposure to oil did not affect growth and motility; however, both parameters were rapidly inhibited after the addition of dispersant (Garr et al. 2014). As reported by Ozhan et al. (2014), a literature survey indicates that oil in the marine environment can promote the growth of some phytoplankton while inhibiting the growth of others, which will have an effect on the phytoplankton assemblage.

Fish have been extensively studied after the DWH oil spill. For example, Xu et al. (2016) reported the loss of cell viability at early developmental stages in mahi-mahi embryos and larvae, which followed the induction of differentially expressed cardiac-associated genes at later stages. In addition, the authors observed a perturbation in metabolism, steroid biosynthesis, and cytochrome P450 genes, suggesting that other targets, besides the heart, may be involved in the development of toxicity. Fodrie et al. (2011) and Able et al. (2015) conducted studies on the effects on fish assemblage, following the spill in coastal areas of the GOM. Fodrie et al. (2011) concluded that "immediate, catastrophic losses of 2010 cohorts were largely avoided, and that no shifts in species composition occurred following the spill." However, they warned that "the potential long-term impacts facing fishes as a result of chronic exposure and delayed, indirect effects now require attention." Similar findings were reported by Able et al. (2015) while evaluating the juvenile and adult marsh-fish response to the spill and observing the marsh-fish assemblage structure across several subhabitats, such as marsh edge, creeks, ponds, and depressions. After comparing representative oiled and unoiled sites in Louisiana marshes (Caminada, Terrebonne, and Barataria Bays), they reported the absence of consistent differences in species composition, abundance, and size as a function of oiling 2–3 years after the spill reached these environments. Fodrie et al. (2014) reported that during the 4 years following the spill, the responses of estuarine fishes to oil pollution have been extensively studied at the organismal and population levels. However, they noted discrepancies between the negative impact detected among individual organisms and the absence of the same impacts among populations.

Based on the lessons learned from the DWH and previous oil spills, the authors indicated that factors obscuring negative population impacts, despite known organismal responses and factors dampening population-levels, might be responsible for the observed mismatch. This highlights critical knowledge gaps to be considered to assess ecosystem responses at the basin scale of disturbance. In 2011 and 2012, Murawski et al. (2014) surveyed offshore fish populations in the GOM following reports of abnormal skin lesions after the spill. Their study indicated that, in 2011, the incidence of skin lesions was more frequent in some bottom-dwelling species captured north of the DWH location; however, the overall frequency and severity of lesions declined by the following year. The elevated incidence of skin lesions detected in some species during 2011 was positively correlated with the high concentrations of PAH metabolites measured in bile and possibly related to PAH exposure from the spill. A study of the gene expression in the salt marsh minnow (*Fundulus grandis*) from the northern coast of the GOM revealed a broad, complex genomic response to DWH oil exposure that could compromise fish health (García et al. 2012). In similar study, Dubansky et al. (2013) reported that *F. grandis* collected from Grande Terre, Louisiana, had divergent gene expressions in the liver and gill tissue coincident with the arrival of contaminating oil for 1 year following peak landfall.

A novel approach has been to use the community structure of parasites in fish and other organisms such as shrimp to evaluate the impact of pollutants. The parasites and symbionts of the pink shrimp (*Penaeus duorarum*) were used in the Campeche Sound, as well as the helminth parasite communities of the Mexican flounder (*Cyclopsetta chittendeni*) to describe correlative statistical associations with PAH concentrations observed in sediments following an accidental oil spill (Centeno-Chalé et al. 2015). The study revealed statistically significant negative correlations between hydrocarbon concentrations and the abundances of helminth parasites and concluded that the oil spill had a strong negative effect on helminth communities immediately after the spill. Despite this, the authors observed that the impacted area was repopulated by the flatfish and its helminth parasite after 5 to 8 months. Vidal-Martínez et al. (2015) used the low- to high-molecular-weight PAH ratios (PAHL/PAHH) to understand the environmental and anthropogenic factors influencing the probability of occurrence of marine parasitic species. If this ratio was <1, the most likely origin of the PAHs was pyrolytic, whereas values >1 indicated a petrogenic origin. These authors found an increase in the probability of occurrence of the cestode *Oncomegas wageneri* infecting the flatfish *Syacium gunteri* for sites with ratio values up to 1.89, while sites with ratio values above 1.90 could be considered harmful for the transmission stages of *O. wageneri* and its hosts (copepod, shrimp, and shoal flounder). In the coastal lagoons of the northern Yucatan Peninsula, Pech et al. (2009) used the checkered puffer (*Spheroides testudineus*) and its parasites as environmental indicators of the potential impact of chemical pollution.

Histopathology has also been used as a tool to evaluate the effects of pollutants. Gold-Bouchot et al. (1995), for example, used the histopathology of the American oyster (*Crassostrea virginica*) to assess the impact of the Mexican oil industry in 3 coastal lagoons in the southern Gulf. Strong dose-response curves were reported for cadmium and the unresolved complex mixture (UCM) of hydrocarbons in the oyster soft tissue.

Biomarkers at different levels of biological organization have also been used as tools for the evaluation of pollution effects. Zapata-Pérez et al. (2005) used ethoxycoumarin O-deethylase (ECOD, an enzyme closely related to EROD [7-ethoxyresorufin-O-deethylase]) activity and vitellogenin induction in 2 species of shrimp (*Peneaus duorarum* and *P. setiferus*) to evaluate the status of shrimp populations exposed to pollutants in the Campeche Sound in the southern GOM. The study failed to demonstrate significant differences between shrimp collected near an oil-production area and those from a control location; however, the authors reported a significant correlation between ECOD activity and metabolites of pyrene (1-OH-pyrene) and phenanthrene (1-OH-phenanthrene). Gold-Bouchot et al. (2007) also used biomarkers in the American oyster (*C. virginica*) in Laguna de Términos, a protected area in the southern Gulf. Although the contaminant body burden and biomarker levels were moderately high, edema, which was positively correlated with most target analytes (e.g., chlordanes, pentachloroanisol, aldrin/dieldrin/endrin, PCBs, and PAHs) in the gills, was the only histopathological lesion observed. Zapata-Pérez et al. (2007) used the catfish (*Ariopsis felis*) to evaluate the effects of pollutants in coastal lagoons from the Yucatan Peninsula. The study revealed that vitellogenin gene expressions, associated with some endocrine disruptors detected in the Terminos Lagoon, were clearly elevated and related to the concentrations of total DDTs and PCBs.

With the perspective given by time since the blowout of the Macondo well in April 2010, a few review studies have been published summarizing the effects of the spill on different biological communities. Murawski et al. (2016) concluded that impacts on plankton (phytoplankton, zooplankton, and ichthyoplankton) were short-lived and that these communities returned to previous conditions within a year. Impact on soft-bottom benthic communities seems to be minimal except in areas of high deposition of oil. A similar conclusion was reached by Centeno-Chalé et al. (2015) for the parasite communities of *Cyclopsetta chittendeni* in the southern Gulf. However, impact on hard-bottom communities had longer, lasting impacts. The authors point out that long-term effects on communities and fisheries are not well understood and may pose a risk given the long persistence of oil in some environments. A study reported by Vignier et al. (2017) suggests that *Crassostrea virginica* gametes, released during the DWH oil spill, were likely exposed to a cytotoxic oil/dispersant mix, negatively affecting the reproduction success and early development of oysters. Similarly, Buskey et al. (2016) found that, under laboratory conditions, early life stages in fish are more sensitive to oil than adults. Field studies found that chemical dispersants may be more toxic than oil alone. Joye et al. (2016) showed that carbon from petroleum was incorporated into marine food webs, but the mechanism is not clear.

As mentioned earlier, the induction-equivalence factors are used to compare the toxicity of environmental contaminants (Safe 1990). The contribution of PAHs, PCBs, and PCDDs/PCDFs to the total induction equivalents, assuming an additive mode of action in mollusks from US coastal environments, indicated the importance of PAH concentrations on the induction of cytochrome P450-dependent EROD enzyme (Gardinali and Wade 1998). Although PAHs have, in general, induction equivalency factors (IEFs) several orders of magnitude lower than those of dioxins/furans and planar/mono-*ortho* PCBs, the relative importance of PAHs is related to their much higher concentrations in mollusk samples from the GOM.

Conclusion

The concentrations of legacy pesticides and related persistent organic pollutants (i.e., dioxins/furans, PCBs, and PBDEs) in the GOM are lower than concentrations reported before the advent of imposed regulations. The decreasing PAH concentrations could be a result of regulations on combustion sources, proper disposal of used motor oil, and replacement of traditional sources of energy by more environmentally friendly ones. However, marine organisms are still contaminated with these toxic organic chemicals, and their physiology and health are likely being affected. Even though some reports indicate that parasites may be potentially more sensitive than hosts to environmental contamination, contaminated shellfish, for example, are more susceptible to infections that adversely affect their abundance and productivity (Chu 1999; Kim 2008; Powell et al. 2008; Vignier et al. 2017). Similarly, the consumption of contaminated fish carries potentially harmful human health implications. Although the health of the GOM has improved after the global implementation of the Stockholm Convention and changes in the United States have been progressing in the right direction, there is more to do regarding the ratification of the Stockholm Convention by the United States to further reduce the concentrations of these chemicals in the GOM.

References

Abad, E., K. Martínez, J. Caixach, and J. Rivera. 2006. Polychlorinated dibenzo-p-dioxins, dibenzofurans and "dioxin-like" PCBs in flue gas emissions from municipal waste management plants. *Chemosphere* 63:570–580.

Able, K. W., P. C. Lopez-Duarte, F. J. Fodrie, O. Jensen, C. W. Martin, and B. J. Roberts. 2015. Fish assemblages in Louisiana salt marshes: Effects of the Macondo oil spill. *Estuaries and Coasts* 385:1385–1398.

Adhikari, P. L., K. Maiti, and E. B. Overton. 2015. Vertical fluxes of polycyclic aromatic hydrocarbons in the northern Gulf of Mexico. *Marine Chemistry* 168:60–68.

Aguirre-Macedo, M. L., R. A. Simá-Álvarez, M. K. Román-Magaña, and J. I. Güemez-Ricalde. 2007. Parasitic survey of the American oyster (*Crassostrea virginica*) in coastal lagoons of the southern Gulf of Mexico. *Journal of Aquatic Animal Health* 19:270–279.

Albuquerque, M., M. Coutinho, and C. Borrego. 2016. Long-term monitoring and seasonal analysis of polycyclic aromatic hydrocarbons (PAHs) measured over a decade in the ambient air of Porto, Portugal. *Science of the Total Environment* 543:439–448

Alldredge, A. L., and M. W. Silver. 1988. Characteristics, dynamics and significance of marine snow. *Progress in Oceanography* 20:41–82.

Alldredge, A. L., and M. J. Youngbluth. 1985. The significance of macroscopic aggregates (marine snow) as sites for heterotrophic bacterial production in the mesopelagic zone of the subtropical Atlantic. *Deep Sea Research* 12:1445–1456.

Almeda, R., S. Baca, C. Hyatt, and E. J. Buskey. 2014. Ingestion and sublethal effects of physically and chemically dispersed crude oil on marine planktonic copepods. *Ecotoxicology* 23:988–1003.

Almeda, R., S. Bona, C. R. Foster, and E. J. Buskey. 2014. Dispersant Corexit 9500A and chemically dispersed crude oil decreases the growth rates of meroplanktonic barnacle nauplii (*Amphibalanus improvisus*) and tornaria larvae (*Schizocardium* sp.). *Marine Environmental Research* 99:212–217.

Almeda, R., T. E. Harvey, T. L. Connelly, S. Baca, and E. J. Buskey. 2016. Influence of UVB radiation on the lethal and sublethal toxicity of dispersed crude oil to planktonic copepod nauplii. *Chemosphere* 152:446–458.

Almeda, R., C. Hyatt, and E. J. Buskey, 2014. Toxicity of dispersant Corexit 9500A and crude oil to marine microzooplankton. *Ecotoxicology and Environmental Safety* 106:76–85.

Almeda, R., Z. Wambaugh, C. Chai, Z. Wang, Z. Liu, and E. J. Buskey. 2013. Effects of crude oil exposure on bioaccumulation of polycyclic aromatic hydrocarbons and survival of adult and larval stages of gelatinous zooplankton. *PLOS ONE* 8:e74476.

Almeda, R., A. Wambaugh, Z. Wang, C. Hyatt, Z. Liu, and E. J. Buskey. 2013. Interactions between zooplankton and crude oil: Toxic effects and bioaccumulation of polycyclic aromatic hydrocarbons. *PLOS ONE* 8:e67212.

Alonso, V., V. Linares, M. Bellés, M. L. Albina, A. Pujol, J. L. Domingo, and D. J. Sánchez. 2010. Effects of BDE-99 on hormone homeostasis and biochemical parameters in adult male rats. *Food and Chemical Toxicology* 48:2206–2211.

Ameur, W. B., Y. El Megdiche, E. Eljarrat, S. B. Hassine, B. Badreddine, T. Souad, H. Bèchir, D. Barceló, and M. R. Driss. 2013. Organochlorine and organobromine compounds in a benthic fish (*Solea solea*) from Bizerte Lagoon (northern Tunisia): Implications for human exposure. *Ecotoxicology and Environmental Safety* 88:55–64.

Amiard, J.-C., and C. Amiard-Triquet. 2015. Conventional risk assessment of environmental contaminants. Pp. 25–49 in C. Amiard-Triquet, J. C. Amiard, and C. Mouneyrac (eds.), *Aquatic Ecotoxicology: Advancing Tools for Dealing with Emerging Risks*. New York: Elsevier Science.

Andersson, O., and G. Blomkvist. 1981. Polybrominated aromatic pollutants found in fish in Sweden. *Chemosphere* 10:1051–1060.

Andrews, J. D., and S. M. Ray. 1988. Management strategies to control the disease caused by *Perkinsus marinus*. *American Fisheries Society* 18:206–224.

Apeti, D. A., G. G. Lauenstein, and D. W. Evans. 2012. Recent status of total mercury and methyl mercury in the coastal waters of the northern Gulf of Mexico using oysters and sediments from NOAA's Mussel Watch Program. *Marine Pollution Bulletin* 64:2399–2408.

Atwood, D. K., F. J. Burton, J. E. Corredor, G. R. Harvey, A. J. Mata-Jimenez, A. Vasquez-Botello, and B. A. Wade. 1987. Results of the CARIPOL Petroleum Pollution

Monitoring Project in the wider Caribbean. *Marine Pollution Bulletin* 18:540–548.

Atwood, D. K., S. Dinkel-Mckay, G. C. Remaro, and E. Van Vleet. 1987. Floating tar and dissolved/dispersed hydrocarbons in the northern Gulf of Mexico and the Straits of Florida. *Caribbean Journal of Science* 23:73–76.

Baird, R. C., N. P. Thompson, T. L. Hopkins, and W. R. Weiss. 1975. Chlorinated hydrocarbons in mesopelagic fishes of the eastern Gulf of Mexico. *Bulletin of Marine Science* 24:473–481.

Balmer, B. C., G. M. Ylitalo, L. E. McGeorge, K. A. Baugh, D. Boyd, K. D. Mullin, P. E. Rosel, et al. 2015. Persistent organic pollutants (POPs) in blubber of common bottlenose dolphins (*Tursiops truncatus*) along the northern Gulf of Mexico coast, USA. *Science of the Total Environment* 527–528:306–312.

Beg, M. U., S. N. Al-Subiai, N. Al-Jandal, S. A. Butt, K. R. Beg, and M. Al-Husaini. 2015. Seasonal effect on biomarkers of exposure to petroleum hydrocarbons in fish from Kuwait's marine area. *Marine Pollution Bulletin* 100:673–680.

Besada, V., J. L. Sericano, and F. Schultze. 2014. An assessment of two decades of trace metals monitoring in wild mussels from the Northwest Atlantic and Cantabrian coastal areas of Spain, 1991–2011. *Environment International* 71:1–12.

Bidleman, T., E. L. Atlas, R. Atkinson, B. Bonsang, K. Burns, W. C. Keene, A. H. Knap, J. Miller, J. Rudolph, and S. Tanabe. 1990. The long-range transport of organic compounds. Pp. 259–302 in A. H. Knap (ed.), *Proceedings of a NATO ASI Series C: Mathematical and Physical Series*, vol. 297. Dordrecht, Netherlands: Academic Publishers.

Blumer, M. 1976. Polycyclic aromatic compounds in nature. *Science* 234:35–45.

Boehm, P. D., G. S. Douglas, W. A. Burns, P. J. Mankiewicz, D. S. Page, and A. E. Bence. 1997. Application of petroleum hydrocarbon chemical fingerprinting and allocation techniques after the *Exxon Valdez* oil spill. *Marine Pollution Bulletin* 34:599–613.

Boehm, P. D., D. L. Fiest, D. Mackay, and S. Paterson. 1982. Physical-chemistry weathering of petroleum hydrocarbons from the Ixtoc I blowout: Chemical measurements and weathering model. *Environmental Science & Technology* 16:498–505.

Bonner, M. R., D. Han, J. Nie, P. Rogerson, J. E. Vena, P. Muti, M. Trevisan, S. B. Edge, and J. L. Freudenheim. 2005. Breast cancer risk and exposure in early life to polycyclic aromatic hydrocarbons using total suspended particulates as a proxy measure. *Cancer Epidemiology, Biomarkers & Prevention* 14:53–60.

Braune, B. M., P. M. Outridge, A. T. Fisk, D. C. G. Muir, P. A. Helm, K. Hobbs, P. F. Hoekstra, et al. 2005. Persistent organic pollutants and mercury in marine biota of the Canadian Arctic: An overview of spatial and temporal trends. *Science of the Total Environment* 351–352:4–56.

Brown, S. S., G. R. Gaston, C. F. Rakocinski, and R. W. Heard. 2000. Effects of sediment contaminants and environmental gradients on macrobenthic community trophic structure in Gulf of Mexico estuaries. *Estuaries* 23:411–424.

Buchman, M. F., 1999. *NOAA Screening Quick Reference Tables.* NOAA HAZMAT Report 99-1. Seattle, WA: Coastal Protection and Restoration Division, National Oceanic and Atmospheric Administration. 12 pp.

Burns, K. A., and A. H. Knap. 1986. The Bahia Las Minas oil spill: Hydrocarbon uptake by reef-building corals. *Marine Pollution Bulletin* 20:391–398.

Burreson, E. M., R. Sima Alvarez, V. M. Vidal Martinez, and M. L. Aguirre Macedo. 1994. *Perkinsus marinus* (Apicomplexa) as a potential source of oyster *Crassostrea virginica* mortality in coastal lagoons of Tabasco, Mexico. *Diseases of Aquatic Organisms* 20:77–82.

Buskey, E. J., H. K. White, and A. J. Esbaugh. 2016. Impact of oil spills on marine life in the Gulf of Mexico. *Oceanography* 29:174–181.

Butler, P. A. 1973. Residues in fish, wildlife, and estuaries: Organochlorine residues in estuarine mollusks, 1965–72. National Pesticide Monitoring Program. *Pesticides Monitoring Journal* 6:238–368.

Butler, P. A., and R. L. Schutzmann. 1978. Residues of pesticides and PCBs in estuarine fish, 1972–1976. National Pesticide Monitoring Program. *Pesticides Monitoring Journal* 12:51–59.

Calder, J. M., 1977. *Hydrocarbons in the Water Column of the MAFLA Lease Area 1976.* Final Report BLM Contract No. 08550-CTS-30. Seattle, WA: Elsevier. 498 pp.

Camilli, R., C. M. Reddy, D. R. Yoerger, B. A. S. Van Mooy, M. V. Jakuba, J. C. Kinsey, S. P. McIntre, and J. V. Maloney. 2010. Tracking hydrocarbon plume transport and biodegradation at Deepwater Horizon. *Science* 330:201–204.

Carassou, L., F. J. Hernandez, and W. M. Graham. 2014. Change and recovery of coastal mesozooplankton community structure during the Deepwater Horizon oil spill. *Environmental Research Letters* 9:124003.

Carlton, D., C. D. Hunt, and E. Slone. 2010. Long-term monitoring using resident and caged mussels in Boston Harbor yield similar spatial and temporal trends in chemical contamination. *Marine Environmental Research* 70:343–357.

Carvalho, F. P., J. P. Villeneuve, C. Cattini, J. Rendón, and J. Mota de Oliveira. 2009. Pesticide and PCB residues in the aquatic ecosystems of Laguna de Terminos, a protected area of the coast of Campeche, Mexico. *Chemosphere* 74:988–995.

Centeno-Chalé, O. A., M. L. Aguirre-Macedo, G. Gold-Bouchot, and V. M. Vidal-Martínez. 2015. Effects of oil spill related chemical pollution in Mexican flounder *Cyclopsetta chittendeni* from the Campeche Sound, Gulf of Mexico. *Ecotoxicology and Environmental Safety* 119:162–169.

Chakraborty, P., S. Selvaraj, M. Nakamura, B. Prithiviraj, A. Cincinelli, and J. J. Bang. 2018. PCBs and PCDD/Fs in soil from informal e-waste recycling sites and open dumpsites in India: Levels, congener profiles and health risk assessment. *Science of the Total Environment* 621:930–938.

Cheung, K. C., H. M. Leung, K. Y. Kong, and M. H. Wong. 2007. Residual levels of DDTs and PAHs in freshwater and marine fish from Hong Kong markets and their health risk assessment. *Chemosphere* 3:460–468.

Chu, F-L. E. 1999. Effects of field-contaminated sediments and related water soluble components on haemocyte function and *Perkinsus marinus* susceptibility and expression in oysters. *Biomarkers* 4:537–548.

Coat, S., D. Monti, P. Legendre, C. Bouchon, F. Massat, and G. Lepoint. 2011. Organochlorine pollution in tropical rivers (Guadeloupe): Role of ecological factors in food web bioaccumulation. *Environmental Pollution* 159:1692–1701.

Corsolini, S., and G. Sarà. 2017. The trophic transfer of persistent pollutants (HCB, DDTs, PCBs) within polar marine food webs. *Chemosphere* 177:189–199.

Culbertson, J. B., I. Valiela, Y. S. Olsen, and C. M. Reddy. 2008. Effect of field exposure to 38-year-old residual petroleum hydrocarbons on growth, condition index, and filtration rate of the ribbed mussel, *Geukensia demissa*. *Environmental Pollution* 154:312–319.

Culbertson, J. B., I. Valiela, E. E. Peacock, C. M. Reddy, A. Carter, and R. VanderKruik. 2007. Long-term biological effects of petroleum residues on fiddler crabs in salt marshes. *Marine Pollution Bulletin* 54:955–962.

Da Silva, D., J. Buzitis, M. M. Krahn, M. C. Bicego, and A. M. S. Pires-Vanin. 2006. Metabolites in bile of fish from Sao Sebastiao Channel, Sao Paulo, Brazil as biomarkers of exposure to petrogenic polycyclic aromatic compounds. *Marine Pollution Bulletin* 52:175–183.

Daly, K. L., U. Passow, J. Chanton, and D. Hollander. 2016. Assessing the impacts of oil-associated marine snow formation and sedimentation during and after the Deepwater Horizon oil spill. *Anthropocene* 13:18–33.

Dean, K. E., M. P. Suarez, H. S. Rifai, R. M. Palachek, and L. Koenig. 2009. Bioaccumulation of polychlorinated dibenzodioxins and dibenzofurans in catfish and crabs along an estuarine salinity and contamination gradient. *Environmental Toxicology and Chemistry*. 28:2307–2317.

DeCarlo, V. J., 1979. Studies on brominated chemicals in the environment. *Annals of the New York Academy of Sciences* 320:678–681.

DeLeo, D. M., D. V. Ruiz-Ramos, I. B. Baums, and E. E. Cordes. 2016. Response of deep-water corals to oil and chemical dispersant exposure. *Deep Sea Research Part II: Tropical Studies in Oceanography* 129:137–147.

De Voogt, P., and U. A. T. Brinkman. 1989. Production, properties and usage of polychlorinated biphenyls. Pp. 3–45 in R. D. Kimbrough and A. A. Jensen (eds.), *Halogenated Biphenyls, Terphenyls, Naphthalenes, Dibenzodioxins, and Related Products*. New York: Elsevier Science.

Dewailly, E., and A. H. Knap. 2006. Food from the oceans and human health: Balancing risks and benefits. *Oceanography* 19:84–87.

de Wit, C. A. 2002. An overview of brominated flame retardants in the environment. *Chemosphere* 46:583–624.

Díaz-Jaramillo, M., K. S. B. Miglioranza, M. Gonzalez, E. Barón, J. M. Monserrat, E. Eljarrat, and D. Barceló. 2016. Uptake, metabolism and sub-lethal effects of BDE-47 in two estuarine invertebrates with different trophic positions. *Environmental Pollution* 213:608–617.

Di Muccio, A., I. Camoni, P. Citti, and D. Pontecorvo. 1988. Survey of DDT-like compounds in dicofol formulations. *Ecotoxicology and Environmental Safety* 16:129–132.

Douglas, G. S., A. E. Bence, R. C. Prince, S. J. McMillen, and E. L. Butler. 1996. Environmental stability of selected petroleum hydrocarbon source and weathering ratios. *Environmental Science & Technology* 30:2332–2339.

Dubansky, B., A. Whitehead, J. Miller, C. D. Rice, and F. Galvez. 2013. Multi-tissue molecular, genomic, and developmental effects of the Deepwater Horizon oil spill on resident Gulf killifish (*Fundulus grandis*). *Environmental Science & Technology* 47:5074–5082.

Duce, R. A., J. G. Quinn, C. P. Olney, S. R. Piotrowicz, B. J. Ray, and T. L. Wade. 1972. Enrichment of heavy metals and organic compounds in the surface microlayer on Narragansett Bay, Rhode Island. *Science* 176:161–163.

Duinker, J. C., A. H. Knap, K. S. Binkley, G. H. Van Dam, A. Darrell-Rew, and M. T. J. Hillebrand. 1988. Method to represent the qualitative and quantitative characteristics of PCB mixtures: Marine mammal tissues and commercial mixtures as examples. *Marine Pollution Bulletin* 19:74–79.

Dunkin, R. C., W. A. McLellan, J. E. Blum, and D. A. Pabst. 2005. The ontogenetic changes in the thermal properties of blubber from Atlantic bottlenose dolphin *Tursiops truncatus*. *Journal of Experimental Biology* 208:1469–1480.

Ellis, M. S., K-S. Choi, T. L. Wade, E. N. Powell, T. J. Jackson, and D. H. Lewis. 1993. Sources of local variation in polynuclear aromatic hydrocarbons and pesticide body burden in oysters (*Crassostrea virginica*) from Galveston Bay, Texas. *Comparative Biochemistry and Physiology Part C: Toxicology and Pharmacology* 106C:689–698.

EPA (Environmental Protection Agency). 2018. Chlorpyrifos. https://www.epa.gov/ingredients-used-pesticide-products/chlorpyrifos.

Erickson, M. D. 1986. *Analytical Chemistry of PCBs*. Stoneham, MA: Butterworth Publishers.

Farrington, J. W., B. W. Tripp, S. Tanabe, A. Subramanian, J. L. Sericano, T. L. Wade, and A. H. Knap. 2016. Edward D. Goldberg's proposal of "the Mussel Watch": Reflections after 40 years. *Marine Pollution Bulletin* 110:501–510.

Fay, R. R., and L. O. W. Newland. 1972. Organochlorine insecticide residue in Aransas Bay, Texas. *Pesticides Monitoring Journal* 6:97–102.

Fisher, C., P. Y. Hsing, C. Kaiser, D. Yoerger, H. H. Roberts, W. Shedd, E. E. Cordes, et al. 2014. Footprint of Deepwater Horizon blowout impact to deep-water coral communities. *Proceedings of the National Academy of Sciences of the United States of America* 111:11744–11749.

Fisher, C., P. Montagna, and T. Sutton. 2016. How did the Deepwater Horizon oil spill impact deep-sea ecosystems? *Oceanography* 29:182–195.

Fisher, W. S., L. M. Oliver, J. T. Winstead, and E. R. Long. 2000. A survey of oysters *Crassostrea virginica* from Tampa Bay, Florida: Associations of internal defense measurements with contaminant burdens. *Aquatic Toxicology* 51:115–138.

Fodrie, F. J., K. W. Able, F. Galvez, K. L. Heck, O. P. Jensen, P. C. Lopez-Duarte, C. W. Martin, R. E. Turner, and A. Whitehead. 2014. Integrating organismal and population responses of estuarine fishes in Macondo spill research. *BioScience* 64:778–788.

Fodrie, F. J., K. L. Heck, and S. J. Bograd, 2011. Response of coastal fishes to the Gulf of Mexico oil disaster. *PLOS ONE* 6:e21609.

Francisco, A. P., A. C. Nardocci, M. Y. Tominaga, C. Rodrigues da Silva, and J. V. de Assunção. 2017. Spatial and seasonal trends of polychlorinated dioxins, furans and dioxin-like polychlorinated biphenyls in air using passive and active samplers and inhalation risk assessment. *Atmospheric Pollution Research* 8:979–987.

Frederiksen, M., H. W. Meyer, N. E. Ebbehøj, and L. Gunnarsen. 2012. Polychlorinated biphenyls (PCBs) in indoor air originating from sealants in contaminated and uncontaminated apartments within the same housing estate. *Chemosphere* 89:473–479.

Fu, J., T. Wang, P. Wang, G. Qu, Y. Wang, Q. Zhang, A. Zhang, and G. Jiang. 2012. Temporal trends (2005–2009) of PCDD/Fs, PCBs, PBDEs in rice hulls from an e-waste dismantling area after stricter environmental regulations. *Chemosphere* 88:330–335.

Fukushima, M., and S. Kawai. 1981. Variation of organochlorine residue concentrations and burden in striped dolphin with growth. Pp. 97–117 in T. Fujiyama (ed.), *Studies on the Levels of Organochlorine Compounds and Heavy Metals in Marine Organisms*. Okinawa: University of the Ryukyus Press.

Gallo, M. V., G. D. Deane, A. P. DeCaprio, and L. M. Schell. 2015. Changes in persistent organic pollutant levels from adolescence to young adulthood. *Environmental Research* 140:214–224.

Gallo, M. V., L. M. Schell, A. P. DeCaprio, and A. Jacobs. 2011. Levels of persistent organic pollutant and their predictors among young adults. *Chemosphere* 83:1374–1382.

Garcia, T. I., Y. Shen, D. Crawford, M. F. Oleksiak, A. Whitehead, and R. B. Walter. 2012. RNA-Seq reveals complex genetic response to Deepwater Horizon oil release in *Fundulus grandis*. *BMC Genomics* 13:474. https://doi.org/10.1186/1471-2164-13-474.

Gardinali, P. R., and T. L. Wade. 1998. Contribution of PAHs, PCBs, and PCDD/PCDFs to the total induction equivalents (sigma IEs) in mollusks. *Marine Pollution Bulletin* 37:27–31.

Garr, A., S. Laramore, and W. Krebs. 2014. Toxic effects of oil and dispersant on marine microalgae. *Bulletin of Environmental Contamination and Toxicology* 93:654–659.

Gelsleichter, J., C. A. Manire, N. J. Szabo, E. Cortés, J. Carlson, and L. Lombardi-Carlson. 2005. Organochlorine concentrations in bonnethead sharks (*Sphyrna tiburo*) from four Florida estuaries. *Archives of Environmental Contamination and Toxicology* 48:474–483.

Giam, C. S., E. L. Atlas, H. S. Chan, and G. S. Neff. 1977. Estimation of fluxes of organic pollutants to the marine environment-phthalate plasticizer concentration and fluxes. *Revue Internationale d'Océanographie Médicale* 47:79–84.

———. 1980. Phthalate esters, PCB and DDT residues in the Gulf of Mexico atmosphere. *Atmospheric Environment* 14:65–69.

Giam, C. S., H. S. Chan, and G. S. Neff. 1976. Distribution of phthalates, DDTs and PCBs in the Gulf of Mexico. Pp.

375–385 in H. L. Windom and R. A. Duce (eds.), *Marine Pollutant Transfer*. Lexington, MA: D. C. Heath.

———. 1978. Phthalate ester plasticizers, DDT, DDE, and polychlorinated biphenyls in biota from the Gulf of Mexico. *Marine Pollution Bulletin* 9:249–252.

Giam, C. S., H. S. Chan, G. S. Neff, and E. L. Atlas. 1978. Phthalate ester plasticizers: A new class of marine pollutant. *Science* 199:419–421.

Giam, C. S., A. R. Hanks, R. L. Richardson, W. M. Sackett, and W. K. Wong. 1972. DDT, DDE and polychlorinated biphenyls in biota from the Gulf of Mexico and Caribbean Sea—1971. *Pesticides Monitoring Journal* 6:139–143.

Giam, C. S., R. L. Richardson, D. Taylor, and M. K. Wong. 1974. DDT, DDE and PCBs in the tissues of reef dwelling groupers (*Serranidae*) in the Gulf of Mexico and the Grand Bahamas. *Bulletin of Environmental Contamination and Toxicology* 11:189–192.

Giam, C. S., R. L. Richardson, M. K. Wong, A. R. Hanks, and W. M. Sackett. 1973. Chlorinated hydrocarbons in plankton from the Gulf of Mexico and northern Caribbean. *Bulletin of Environmental Contamination and Toxicology* 9:376–382.

Gioia, R., J. H. Offenberg, C. L. Gigliotti, L. A. Totten, S. Du, and S. J. Eisenreich. 2005. Atmospheric concentrations and deposition of organochlorine pesticides in the US Mid-Atlantic region. *Atmospheric Environment* 39:2309–2322.

Goldberg, E. D. 1975. The Mussel Watch: A first step in global marine monitoring. *Marine Pollution Bulletin* 6:111–114.

Gold-Bouchot, G., and M. Herrera-Rodríguez. 1996. Efectos de los hidrocarburos sobre la comunidad de nemátodos bénticos de la sonda de Campeche, México [Effects of oil on benthic community nematodes probe Campeche, Mexico]. Pp. 279–296 in A. Vázquez-Botello, F. Vázquez, E. Benítez, and D. Zárate-Lomelí (eds.), *Golfo de México, Contaminación e Impacto Ambiental: Diagnóstico y Tendencias* [Gulf of Mexico, pollution and environmental impact: Diagnosis and trends). EPOMEX Serie Científica. Campeche: Universidad Autónoma de Campeche.

Gold-Bouchot, G., J. P. Rodas Ortiz, J. A. Domínguez Maldonado, and V. Ceja Moreno. 2014. Contaminantes orgánicos persistentes (COP) en sedimentos de lagunas y zona costera [Persistent organic pollutants (POPs) in lagoon sediments and coastal zone]. Pp. 43–54 in J. I. Euán Ávila, A. García de Fuentes, M. Á. Liceaga Correa, and A. Munguía Gil (eds.), *La costa del estado de Yucatán, un espacio de reflexión sobre la relación sociedad-naturaleza, en el contexto de su ordenamiento ecológico territorial* [The coast of the state of Yucatan, a space for reflection on the relationship between society and nature, in the context of its territorial ecological ordering], vol. 2. Mexico City: Plaza y Valdés.

Gold-Bouchot, G., R. Simá, O. Zapata, and J. Guemez. 1995. Histopathological effects of petroleum hydrocarbons and heavy metals on the American oyster, *Crassostrea virginica*, from Tabasco, Mexico. *Marine Pollution Bulletin* 31:439–445.

Gold-Bouchot, G., O. Zapata-Pérez, V. Ceja-Moreno, G. Rodríguez-Fuentes, R. Simá-Álvarez, M. L. Aguirre-Macedo, and V. Vidal-Martínez. 2007. Biological effects of environmental pollutants in the American oyster, *Crassostrea virginica*: A field study in Laguna de Terminos, Mexico. *International Journal of Environment and Health* 1:171–184.

González-Jauregui, M., C. Valdespino, A. Salame-Mendez, G. Aguirre-Leó, and J. Rendón-von Osten. 2012. Persistent organic contaminants and steroid hormones levels in morelet's crocodiles from the southern Gulf of Mexico. Archives of Environmental Contamination and Toxicology 62:445–454.

González-Mille, D. J., C. A. Ilizaliturri-Hernandez, G. Espinosa-Reyes, R. Costilla-Salazar, F. Díaz-Barriga, I. Ize-Lema, and J. Mejía-Saavedra. 2010. Exposure to persistent organic pollutants (POPs) and DNA damage as an indicator of environmental stress in fish of different feeding habits of Coatzacoalcos, Veracruz, Mexico. *Ecotoxicology* 19:1238–1248.

Grey, E. K., S. C. Chiasson, H. G. Williams, V. J. Troeger, C. M. Taylor, and H. M. Patterson. 2015. Evaluation of blue crab, *Callinectes sapidus*, megalopal settlement and condition during the Deepwater Horizon oil spill. *PLOS ONE* 10:e0135791.

Grotti, M., S. Pizzini, M. L. Abelmoschi, G. Cozzi, R. Piazza, and F. Soggia. 2016. Retrospective biomonitoring of chemical contamination in the marine coastal environment of Terra Nova Bay (Ross Sea, Antarctica) by environmental specimen banking. *Chemosphere* 165:418–426.

Hale, R. C., M. Alaee, J. B. Manchester-Neesvig, H. M. Stapleton, and M. G. Ikonomou. 2003. Polybrominated diphenyl ether flame retardants in the North American environment. *Environment International* 29:771–779.

Heinzow, B., S. Mohr, G. Ostendorp, M. Kerst, and W. Körner. 2007. PCB and dioxin-like PCB in indoor air of public buildings contaminated with different PCB sources—deriving toxicity equivalent concentrations from standard PCB congeners. *Chemosphere* 67:1746–1753.

Hemming, J. M., J. M. Brown, M. Brim, and R. Jarvis. 2005. Sediment quality survey of the Choctawhatchee Bay system in the Florida Panhandle. *Marine Pollution Bulletin* 50:889–903.

Hernández-Arana, H. A., R. M. Warwick, M. J. Attrill, A. A. Rowden, and G. Gold-Bouchot. 2005. Impact of oil related activities on benthic macroinfauna assemblages of the Campeche shelf, southern Gulf of Mexico. *Marine Ecology Progress Series* 289:89–107.

Hinck, J. E., R. J. Norstrom, C. E. Orazio, C. J. Schmitt, and D. E. Tillit. 2009. Persistence of organochlorine chemical residues from the Tombigbee River (Alabama, USA): Continuing risk to wildlife from a former DDT manufacturing facility. *Environmental Pollution* 157:582–591.

Hostettler, F. D., W. E. Pereira, K. A. Kvenvolden, A. van Geen, S. N. Luoma, C. C. Fuller, and R. Anima. 1999. A record of hydrocarbon input to San Francisco Bay as traced by biomarker profiles in surface sediment and sediment cores. *Marine Chemistry* 64:115–127.

Howell, N. L., H. S. Rifai, and L. Ko. 2011. Comparative distribution, sourcing, and chemical behavior of PCDD/Fs and PCBs in an estuary environment. *Chemosphere* 83:873–881.

Hsing, P. Y., B. Fu, E. A. Larcom, S. P. Berlet, T. M. Shank, A. F. Govindarajan, A. J. Lukasiewicz, P. M. Dixon, and C. R. Fisher. 2013. Evidence of lasting impact of the Deepwater Horizon oil spill on a deep Gulf of Mexico coral community. *Elementa: Science of the Anthropocene* 1:000012. http://doi.org/10.12952/journal.elementa.000012.

Hu, D., and K. Hornbuckle. 2010. Inadvertent polychlorinated biphenyls in commercial paint pigments. *Environmental Science & Technology* 44:2822–2827.

Hwang, H.-W., E.-K. Park, T. M. Young, and B. D. Hammock. 2008. Occurrence of endocrine-disrupting chemicals in indoor dust. *Science of the Total Environment* 404:26–35.

Iqbal, J., E. B. Overton, and D. Gisclair. 2008. Polycyclic aromatic hydrocarbons in Louisiana river and coastal environments: Source fingerprint and forensic analysis. *Environmental Forensics* 9:63–74.

Jackson, T. J., T. L. Wade, T. J. McDonald, D. L. Wilkinson, and J. M Brooks. 1994. Polynuclear aromatic hydrocarbon contaminants in oysters from the Gulf of Mexico (1986–1990). *Environmental Pollution* 83:291–298.

Jackson, T. J., T. L. Wade, J. L. Sericano, J. M. Brooks, J. M. Wong, B. Garcia-Romero, and T. J. McDonald. 1998. Galveston Bay: Temporal changes in the concentrations of trace organic contaminants in National Status and Trends oysters (1986–1994). *Estuaries* 21:718–730.

Jartun, M., R. T. Ottesen, E. Steinnes, and T. Volden. 2009. Painted surfaces—important sources of polychlorinated biphenyls (PCBs) contamination to the urban and marine environment. *Environmental Pollution* 157:295–302.

Jensen, S. 1966. Report of a new chemical hazard. *New Scientist* 32:612.

Johnson-Restrepo, B., and K. Kannan. 2009. An assessment of sources and pathways of human exposure to polybrominated diphenyl ethers in the United States. *Chemosphere* 76:542–548.

Johnson-Restrepo, B., J. Olivero-Verbel, S. Lu, J. Guette-Fernández, R. Baldiris-Avila, I. O'Byrne-Hoyos, K. M. Aldous, R. Addink, and K. Kannan. 2008. Polycyclic aromatic hydrocarbons and their hydroxylated metabolites in fish bile and sediments from coastal waters of Colombia. *Environmental Pollution* 151:452–459.

Joye, S. B., A. Bracco, T. M. Ozgokmen, J. P. Chanton, M. Grosell, I. R. MacDonald, E. E. Cordes, J. P. Montoya, and U. Passow. 2016. The Gulf of Mexico ecosystem, six years after the Macondo oil well blowout. *Deep Sea Research Part II: Tropical Studies in Oceanography* 129:4–19.

Joye, S. B., I. R. MacDonald, I. Leifer, and V. Asper. 2011. Magnitude and oxidation potential of hydrocarbon gases released from the BP oil well blowout. *Nature Geosciences* 4:160–164.

Kannan, K., A. L. Blankenship, P. D. Jones, and J. P. Giesy. 2000. Toxicity reference values for the toxic effects of polychlorinated biphenyls to aquatic mammals. *Human and Ecological Risk Assessment* 6:181–201.

Kennicutt, M. C., II. 1988. The effect of biodegradation on crude oil bulk and molecular composition. *Oil and Chemical Pollution* 4:89–112.

Kennicutt, M. C., II, J. M. Brooks, E. L. Atlas, and C. S. Giam. 1988. Organic compounds of environmental concern in the Gulf of Mexico: A review. *Aquatic Toxicology* 11:191–212.

Kessler, J. D., D. L. Valentine, M. C. Redmond, M. Du, E. W. Chan, S. D. Mendes, E. W. Quiroz, et al. 2011. A persistent oxygen anomaly reveals the fate of spilled methane in the deep Gulf of Mexico. *Science* 331:312–315.

Kim, Y., E. N. Powell, T. L. Wade, and B. J. Presley. 2008. Relationship of parasites and pathologies to contaminant body burden in sentinel bivalves: NOAA Status and Trends "Mussel Watch" Program. *Marine Environmental Research* 65:101–127.

Kim, Y., E. N. Powell, T. L. Wade, B. J. Presley, and J. M. Brooks. 1999. Influence of climate change on interannual variation in contaminant body burden in Gulf of Mexico oysters. *Marine Environmental Research* 48:459–488.

Kirman, Z. D., J. L. Sericano, T. L. Wade, T. S. Bianchi, F. Marcantonio, and A. S. Kolker. 2016. Composition and depth distribution of hydrocarbons in Barataria Bay marsh sediments after the Deepwater Horizon oil spill. *Environmental Pollution* 214:101–113.

Knap, A. H., and K. S. Binkley. 1991. The occurrence of organic contaminants in the troposphere of the North Atlantic Ocean as measured by aircraft. *Atmospheric Environment* 25A:1507–1516.

Knap, A. H., K. A. Burns, R. Dawson, M. Ehrhardt, and K. Palmork. 1986. Dissolved/dispersed hydrocarbons, tar balls and the surface microlayer: Experiences from an IOC/UNEP Workshop in Bermuda. *Marine Pollution Bulletin* 17:313–319.

Knap, A. H., T. M. Iliffe, and J. N. Butler. 1980. Has the amount of tar in the open ocean changed in the past decade? *Marine Pollution Bulletin* 11:161–164.

Knauer, G. A., D. Hebel, and F. Cipriano. 1982. Marine snow: Major site of primary production in coastal waters. *Nature* 300:630–631.

Kuehl, D. W., and R. Haebler. 1995. Organochlorine, organobromine, metals, and selenium residues in bottlenose dolphins (*Tursiops truncatus*) collected during an unusual mortality event in the Gulf of Mexico, 1990. *Archives of Environmental Contamination and Toxicology* 28:494–499.

Kuk-Dzul, G., G. Gold-Bouchot, and P. L. Ardisson. 2012. Benthic fauna variability in relation to environmental factors and organic pollutants in tropical lagoons from northern Yucatan Peninsula. *Marine Pollution Bulletin* 64:2725–2733.

Landers, S. C., A. C. Nichols, N. K. Barron, C. A. Schimmer, R. Tao, K. Yu, P. M. Stewart, and E. Ólafsson. 2014. Nematode and copepod diversity from Louisiana near the Deepwater Horizon oil spill. *Proceedings of the Biological Society of Washington* 127:47–57.

Landers, S. C., A. C. Nichols, C. A. Schimmer, P. M. Stewart, S. Ramroop, D. A. Steffy, and F. A. Romano III. 2014. Meiofauna and trace metals from sediment collections in Florida following the Deepwater Horizon oil spill. *Gulf of Mexico Science* 32:1–10.

Laramore, S., W. Krebs, and A. Garr. 2014. Effects of Macondo Canyon 252 oil (naturally and chemically dispersed) on larval *Crassostrea virginica* (Gmelin, 1791). *Journal of Shellfish Research* 33:709–718.

Lauenstein, G. G. 1995. Comparison of organic contaminants found in mussels and oysters from a current mussel watch project with those from archived mollusc samples of the 1970s. *Marine Pollution Bulletin* 30:826–833.

Lauenstein, G. G., and K. D. Daskalakis. 1998. U.S. long-term coastal contaminant temporal trends determined from mollusk monitoring programs, 1965–1993. *Marine Pollution Bulletin* 37:6–13.

Lewis, M. A., G. I. Scott, D. W. Bearden, R. L. Quarles, J. Moore, E. D. Strozier, S. K. Sivertsen, A. R. Dias, and M. Sanders. 2002. Fish tissue quality in near-coastal areas of the Gulf of Mexico receiving point source discharges. *Science of the Total Environment* 284:249–261.

Li, L., J. Liu, and J. Hu. 2015. Global inventory, long-range transport and environmental distribution of dicofol. *Environmental Science & Technology* 49:212–222.

Liebl, B., T. Schettgen, G. Kerscher, H. Broding, A. Otto, J. Angerer, and H. Drexler. 2004. Evidence for increased internal exposure to lower chlorinated polychlorinated biphenyls (PCB) in pupils attending a contaminated school. *International Journal of Hygiene and Environmental Health* 207:315–324.

Lin, Q., and I. A. Mendelssohn. 2012. Impacts and recovery of the Deepwater Horizon oil spill on vegetation structure and function of coastal salt marshes in the northern Gulf of Mexico. *Environmental Science & Technology* 46:3737–3743.

Lipscomb, Carol A. 2010. Karankawa Indians. https://tshaonline.org/handbook/online/articles/bmk05. Uploaded on June 15, 2010. Modified on May 10, 2016. Published by the Texas State Historical Association.

Litz, J. A., L. P. Garrison, L. A. Fieber, A. Martinez, J. P. Contillo, and J. R. Kucklick. 2007. Fine-scale spatial variation of persistent organic pollutants in bottlenose dolphins (*Tursiops truncatus*) in Biscayne Bay, Florida. *Environmental Science & Technology* 41:7222–7228.

Liu, Z., J. Liu, W. S. Gardner, G. C. Shank, and N. E. Ostrom. 2016. The impact of the Deepwater Horizon oil spill on petroleum hydrocarbons in surface waters of the northern Gulf of Mexico. *Deep Sea Research Part II: Tropical Studies in Oceanography* 129:292–300.

Luna-Acosta, A., H. Budzinski, K. Le Menach, H. Thomas-Guyon, and P. Bustamante. 2015. Persistent organic pollutants in a marine bivalve on the Marennes-Oléron Bay and the Gironde Estuary (French Atlantic Coast). Part 1: Bioaccumulation. *Science of the Total Environment* 514:500–510.

Ma, J., J. Pan, X. Yang, X. Liu, S. Tao, L. Zhao, X. Qin, Z. Sun, H. Hou, and Y. Zhou. 2016. DDT, DDD, and DDE in soil of Xiangfen County, China: Residues, sources, spatial distribution, and health risks. *Chemosphere* 163:578–583.

MacDonald, I. R., O. Garcia-Peneda, A. Beet, S. Daneshgar Asl, L. Feng, G. Graettinger, D. French-McCay, et al. 2015. Natural and unnatural oil slicks in the Gulf of Mexico. *Journal of Geophysical Research: Oceans* 120:8364–8380.

Macko, S. A., J. K. Winters, and P. L. Parker. 1982. Gulf of Mexico dissolved hydrocarbons associated with the IXTOC I mousse. *Marine Pollution Bulletin* 13:174–177.

Magnusson, K., R. Ekelund, R. Grabic, and P-A. Bergqvist. 2006. Bioaccumulation of PCB congeners in marine benthic infauna. *Marine Environmental Research* 61:379–395.

Malmquist L. M. V., R. R. Olsen, A. B. Hansen, O. Andersen, and J. H. Christensen. 2007. Assessment of oil weathering by gas chromatography–mass spectrometry, time warping and principal component analysis. *Journal of Chromatography A* 1164:262–270.

Manning, T. M., A. C. Roach, K. J. Edge, and D. J. Ferrell. 2017. Levels of PCDD/Fs and dioxin-like PCBs in seafood from Sydney Harbour, Australia. *Environmental Pollution* 224:590–596.

Maryoung, L. A., R. Lavado, and D. Schlenk. 2014. Impacts of hypersaline acclimation on the acute toxicity of the organophosphate chlorpyrifos to salmonids. *Aquatic Toxicology* 22:284–290.

McMullin, D. J., and J. C. Means. 1996. Spatial and temporal trends of pesticide residues in water and particulates in the Mississippi River plume and the northwestern Gulf of Mexico. *Journal of Chromatography A* 754:169–185.

Mihats, D., W. Moche, M. Prean, and E. Rauscher-Gabernig. 2015. Dietary exposure to non-dioxin-like PCBs of different population groups in Austria. *Chemosphere* 126:53–59.

Miranda, D. A., and G. T. Yogui. 2016. Polychlorinated biphenyls and chlorinated pesticides in king mackerel caught off the coast of Pernambuco, northeastern Brazil: Occurrence, contaminant profile, biological parameters and human intake. *Science of the Total Environment* 569–570:1510–1516.

Mitra, S., and T. S. Bianchi. 2003. A preliminary assessment of polycyclic aromatic hydrocarbon distributions in the lower Mississippi River and Gulf of Mexico. *Marine Chemistry* 82:273–288.

Muñoz-Salinas, E., and M. Castillo. 2015. Streamflow and sediment load assessment from 1950 to 2006 in the Usumacinta and Grijalva Rivers (southern Mexico) and the influence of ENSO. *Catena* 127:270–278.

Murawski, S. A., J. W. Fleeger, W. F. Patterson III, H. Chuanmin, K. Daly, I. Romero, and G. A. Toro-Farmer. 2016. How did the Deepwater Horizon oil spill affect coastal and continental shelf ecosystems of the Gulf of Mexico? *Oceanography* 29:160–173.

Murawksi, S., W. Hogarth, E. B. Peebles, and L. Barbeiri. 2014. Prevalence of external skin lesions on PAH concentrations in Gulf of Mexico fishes, post-Deepwater Horizon. *Transactions of the American Fisheries Society* 143:1084–1097.

Nakata, H., R-I. Shinohara, Y. Nakazawa, T. Isobe, A. Sudaryanto, A. Subramanian, S. Tanabe, et al. 2012. Asia-Pacific mussel watch for emerging pollutants: Distribution of synthetic musks and benzotriazole UV stabilizers in Asian and US coastal waters. *Marine Pollution Bulletin* 64:2211–2218.

National Centers for Coastal Ocean Science. 2010. National Status and Trends: Mussel Watch Program [July 15, 2010–August 15, 2010]. NOAA National Centers for Environmental Information. https://inport.nmfs.noaa.gov/inport/item/39400.

Neff, J. M. 2007. *Bioaccumulation in Marine Organisms: Effect of Contaminants from Oil Well Produced Water*. Amsterdam: Elsevier. 468 pp.

Newman, M. C., and M. A. Unger. 2003. Bioaccumulation from food and trophic transfer. Pp. 95–110 in M. C. Newman and M. A. Unger (eds.), *Fundamentals of Ecotoxicology*, 2nd ed. Boca Raton, Fl: Lewis Publishers.

Nikinmaa, M. 2014. *An Introduction to Aquatic Toxicology*. Waltham, MA: Academic Press. 240 pp.

NOAA (National Oceanic and Atmospheric Administration). 1987. *National Status and Trends Program—a Preliminary Assessment of Findings of the Benthic Surveillance Project—1984*. Rockville, MD: NOAA Office of Oceanography and Marine Assessment.

Noreña-Barroso, E., G. Gold-Bouchot, O. Zapata-Perez, and J. L. Sericano. 1999. Polynuclear aromatic hydrocarbons in American oyster *Crassostrea virginica* from the Terminos Lagoon, Campeche, Mexico. *Marine Pollution Bulletin* 38:637–645.

Nowell, L. H., A. S. Ludtke, D. K. Mueller, and J. C. Scott. 2013. *Organic Contaminants, Trace and Major Elements, and Nutrients in Water and Sediment Sampled in Response to the Deepwater Horizon Oil Spill*. US Geological Survey Scientific Investigations Report 2012–5228. https://pubs.er.usgs.gov/publication/sir20125228.

NRC (National Research Council). 2003. *Oil in the Sea III: Inputs, Fates and Effects*. Washington DC: National Academies Press. 280 pp.

Nzihou, A., N. J. Themelis, M. Kemiha, and Y. Benhamou. 2012. Dioxin emissions from municipal solid waste incinerators (MSWIs) in France. *Waste Management* 32:2273–2277.

Oliver, L. M., W. S. Fisher, J. T. Winstead, B. L. Hemmer, and E. R. Long. 2001. Relationships between tissue contaminants and defense-related characteristics of oysters (*Crassostrea virginica*) from five Florida bays. *Aquatic Toxicology* 55:203–222.

OSAT (Operation Science Advisory Team). 2011. *Unified Area Command Summary Report for Sub-surface Oil and Dispersant Detection: Ecotoxicity Addendum*. https://www.restorethegulf.gov/sites/default/files/u306/FINAL%20OSAT%20Ecotox%20Addendum.pdf.

Overton, E. B., T. L. Wade, J. R. Radovis, B. M. Meyer, M. S. Miles, and S. R. Later. 2016. Chemical composition of Macondo and other crude oils and compositional alterations during oil spills. *Oceanography* 29:50–63.

Ozhan, K., M. L. Parsons, and S. Bargu. 2014. How were phytoplankton affected by the Deepwater Horizon oil spill? *BioScience* 64:829–836.

Palm, A., I. T. Cousins, D. Mackay, M. Tysklind, C. Metcalfe, and M. Alaee. 2002. Assessing the environmental fate of chemicals of emerging concern: A case study of the polybrominated diphenyl ethers. *Environmental Pollution* 117:195–213.

Palma, P., V. L. Palma, R. M. Fernandez, A. Bohn, A. M. V. M. Soares, and I. R. Barbosa. 2009. Embryo-toxic effects of environmental concentrations of chlorpyrifos on the crustacean *Daphnia magna*. *Ecotoxicology and Environmental Safety* 72:1714–1718.

Pampanin, D. M., and M. O. Sydnes. 2013. Polycyclic aromatic hydrocarbons a constituent of petroleum: Presence and influence in the aquatic environment. In V. Kutcherov and A. Kolesnikov (eds.), *Hydrocarbon*. IntechOpen. doi:10.5772/48176.

Park, J-S., T. L. Wade, and S. Sweet. 2001. Atmospheric distribution of polycyclic aromatic hydrocarbons and deposition to Galveston Bay, Texas. *Atmospheric Environment* 35:3241–3249.

———. 2002. Atmospheric deposition of PAHs, PCBs, and organochlorine pesticides to Corpus Christi Bay, Texas. *Atmospheric Environment* 36:1707–1720.

Parsons, M. L., W. Morrison, N. N. Rabalais, R. E. Turner, and K. N. Tyre. 2015. Phytoplankton and the Macondo oil spill: A comparison of the 2010 phytoplankton assemblage to baseline conditions on the Louisiana shelf. *Environmental Pollution* 207:152–160.

Passow, U. 2016. Formation and rapidly-sinking, oil associated marine snow. *Deep Sea Research II: Tropical Studies in Oceanography* 129:232–240.

Passow, U., K. Ziervogel, V. Asper, and A. Diercks. 2012. Marine snow formation in the aftermath of the Deepwater Horizon oil spill in the Gulf of Mexico. *Environmental Research Letters* 7:035301.

Paul, J. H., D. Hollander, P. Coble, K. L. Daly, S. Murasko, D. English, J. Basso, J. Delaney, L. McDaniel, and C. W. Kovach. 2013. Toxicity and mutagenicity of Gulf of Mexico waters during and after the Deepwater Horizon oil spill. *Environmental Science & Technology* 47:9651–9659.

Pech, D., V. Vidal-Martínez, L. Aguirre-Macedo, G. Gold-Bouchot, J. Herrera-Silveira, O. Zapata-Pérez, and D. Marcogliese. 2009. The checkered puffer (*Spheroides testudineus*) and its helminths as bioindicators of chemical pollution in Yucatan coastal lagoons. *Science of the Total Environment* 407:2315–2324.

Pereira, W. E., C. E. Rostad, and T. J. Leiker. 1992. Synthetic organic agrochemicals in the lower Mississippi River and its major tributaries: Distribution, transport and fate. *Journal of Contaminant Hydrology* 9:175–188.

Pereira, L. C., F. Valente Duarte, A. T. Inácio Ferreira Varela, A. Pinto Rolo, C. M. Marques Palmeira, and D. Junqueira Dorta. 2017. Exposure to BDE-153 induces autophagy in HepG2 cells. *Toxicology in Vitro* 42:61–68.

Peters, K. E., C. C. Walters, and J. M. Moldowan. 2004. *The Biomarker Guide*, vol. 1: *Biomarkers and Isotopes in the Environment and Human History*. Cambridge: Cambridge University Press. 471 pp.

Petrocelli, S. R., J. W. Anderson, and A. R. Hanks. 1974. DDT and dieldrin residues in selected biota from San Antonio Bay, Texas, 1972. *Pesticides Monitoring Journal* 8:167–172.

Powell, E. N., K. A. Ashton-Alcox, J. N. Kraeuter, S. E. Ford, and D. Bushek. 2008. Long-term trends in oyster population dynamics in Delaware Bay: Regime shifts and response to disease. *Journal of Shellfish Research* 27:729–755.

Prouty, N. G., C. R. Fisher, A. W. J. Demopoulos, and E. R. M. Druffel. 2016. Growth rates and ages of deep-sea corals impacted by the Deepwater Horizon oil spill. *Deep Sea Research Part II: Tropical Studies in Oceanography* 129:196–212.

Qiu, X., T. Zhu, J. Li, H. Pan, Q. Li, G. Miao, and J. Gong. 2004. Organochlorine pesticides in the air around the Taihu Lake, China. *Environmental Science & Technology* 39:1368–1374.

Quigg, A., U. Passow, W. Chin, C. Xu, S. Doyle, L. Bretherton, M. Kamalanathan, et al. 2016. The role of microbial exopolymers in determining the fate of oil and chemical dispersants in the ocean. *Limnology and Oceanography Letters* 1:3–26.

Rauschenberger, R. H., J. J. Wiebe, J. E. Buckland, J. T. Smith, M. S. Sepulveda, and T. S. Gross. 2004. Achieving environmentally relevant organochlorine pesticide concentrations in eggs through maternal exposure in *Alligator mississippiensis*. *Marine Environmental Research* 58:851–856.

Ray, S. M., 1996. Historical perspective on *Perkinsus marinus* disease of oysters in the Gulf of Mexico. *Journal of Shellfish Research* 15:9–11.

Reddy, C. M., J. S. Arey, J. S. Seewald, S. P. Sylva, K. L. Lemkau, R. K. Nelson, C. A. Carmichael, et al. 2011. Composition and fate of gas and oil released to the water column during the Deepwater Horizon oil spill. *Proceedings of the National Academy of Sciences of the United States of America* 109:20229–20234.

Reddy, C. M., T. I. Eglinton, A. H. Hounshell, H. K. White, L. Xu, R. B. Gaines, and G. S. Frysinger. 2002. The West Falmouth oil spill after thirty years: The persistence of petroleum hydrocarbons in marsh sediments. *Environmental Science & Technology* 36:4754–4760.

Reed, M., Ø. Johansen, P. J. Brandvik, P. Daling, A. Lewis, R. Fiocco, D. Mackay, and R. Prentki. 1999. Oil spill modeling towards the close of the 20th century: Overview of the state of the art. *Spill Science Technology Bulletin* 5:3–16.

Regoli, F., D. Pellegrini, A. M. Cicero, M. Nigro, M. Benedetti, S. Gorbi, D. Fattorini, et al. 2014. A multidisciplinary weight of evidence approach for environmental risk assessment at the Costa Concordia wreck: Integrative indices from Mussel Watch. *Marine Environmental Research* 96:92–104.

Ridgway, S., and M. Reddy. 1995. Residue levels of several organochlorines in *Tursiops truncatus* milk collected at varied stages of lactation. *Marine Pollution Bulletin* 30:609–614.

Rigét, F., K. Vorkamp, R. Bossi, C. Sonne, R. J. Letcher, and R. Dietz. 2016. Twenty years of monitoring of persistent organic pollutants in Greenland biota: A review. *Environmental Pollution* 217:114–123.

Rosales, M. T. L., A. V. Botello, H. Bravo, and E. F. Mandelli. 1979. PCBs and organochlorine insecticides in oysters from coastal lagoons of the Gulf of Mexico, Mexico. *Bulletin of Environmental Contamination and Toxicology* 21:652–656.

Rose, M., A. Fernandes, D. Mortimer, and C. Baskaran. 2015. Contamination of fish in UK fresh water systems: Risk assessment for human consumption. *Chemosphere* 122:183–189.

Ruus, A., N. W. Green, A. Maage, and J. Skei. 2006. PCB-containing paint and plaster caused extreme PCB-concentrations in biota from the Sørfjord (western Norway): A case study. *Marine Pollution Bulletin* 52:100–103.

Safe, S. 1990. Polychlorinated biphenyls (PCBs), dibenzo-p-dioxins (PCDDs), di-benzofurans (PCDFs), and related compounds: Environmental and mechanistic considerations which support the development of toxic equivalency factors (TEFs). *Critical Reviews in Toxicology* 21:51–88.

Salata, G. G., T. L. Wade, J. L. Sericano, J. W. Davis, and J. M. Brooks. 1995. Analysis of Gulf of Mexico bottlenose dolphins for organochlorine pesticides and PCBs. *Environmental Pollution* 88:167–175.

Sammarco, P. W., S. R. Kolian, R. A. F. Warby, J. L. Bouldin, W. A. Subra, and S. A. Porter. 2013. Distribution and concentrations of petroleum hydrocarbons associated with the BP/Deepwater Horizon oil spill, Gulf of Mexico. *Marine Pollution Bulletin* 73:129–143.

Sauer, T. C., G. S. Durell, J. S. Brown, D. Redford, and P. D. Boehm. 1989. Concentrations of chlorinated pesticides and PCBs in microlayer and seawater samples collected in open-ocean waters off the U.S. East Coast and in the Gulf of Mexico. *Marine Chemistry* 27:235–257.

Sauer, T. C., J. Michel, M. O. Hayes, and D. V. Aurand. 1998. Hydrocarbon characterization and weathering of oiled intertidal sediments along the Saudi Arabian coast two years after the Gulf War oil spill. *Environment International* 24:43–60.

Schöne, B. R., and R. A. Krause Jr. 2016. Retrospective environmental biomonitoring—Mussel Watch expanded. *Global and Planetary Change* 144:228–251.

Schwing, P. T., I. C. Romero, G. R. Brooks, D. W. Hastings, R. A. Larson, and D. J. Hollander. 2015. A decline in benthic foraminifera following the Deepwater Horizon event in the northeastern Gulf of Mexico. *PLOS ONE* 10:e0120565.

Sericano, J. L. 1993. The American oyster (*Crassostrea virginica*) as a bioindicator of trace organic contamination. PhD diss., Texas A&M University. 242 pp.

Sericano, J. L., E. L. Atlas, T. L. Wade, and J. M. Brooks. 1990. NOAA's Status and Trends Mussel Watch Program: Chlorinated pesticides and PCBs in oysters (*Crassostrea virginica*) and sediments from the Gulf of Mexico, 1986–1987. *Marine Environmental Research* 29:161–203.

Sericano, J. L., and T. L. Wade. 2011. Contamination profiles and temporal trends of persistent organic pollutants on oysters from the Gulf of Mexico. In B. G. Loganathan and P. K. S. Lam (eds.), *Global Contamination Trends of Persistent Organic Chemicals*. Boca Raton, FL: CRC Press.

Sericano, J. L., T. L. Wade, E. L. Atlas, and J. M. Brooks. 1990. Historical perspective on the environmental bioavailability of DDT and its derivatives to Gulf of Mexico oysters. *Environmental Science & Technology* 24:1541–1548.

Sericano, J. L., T. L. Wade, J. M. Brooks, E. L. Atlas, R. R. Fay, and D. L. Wilkinson. 1993. National Status and Trends Mussel Watch Program: Chlordane-related compounds in Gulf of Mexico oysters, 1986–90. *Environmental Pollution* 82:23–32.

Sericano, J. L., T. L. Wade, S. T. Sweet, J. Ramirez, and G. G. Lauenstein. 2014. Temporal trends and spatial distribution of DDT in bivalves from the coastal marine environments of the continental United States, 1986–2009. *Marine Pollution Bulletin* 81:303–316.

Sherblom, P. M., D. Kelly, and R. H. Pierce. 1995. Baseline survey of pesticide and PAH concentrations from Sarasota Bay, Florida, USA. *Marine Pollution Bulletin* 30:568–573.

Shin, E., J. Kim, S.-D. Choi, Y.-W. Kang, and Y.-S. Chang. 2016. Estimated dietary intake and risk assessment of polychlorinated dibenzo-p-dioxins and dibenzofurans and dioxin-like polychlorinated biphenyls from fish consumption in the Korean general population. *Chemosphere* 146:419–425.

Shunthirasingham, C., A. Gawor, H. Hung, K. A. Brice, K. Su, N. Alexandrou, H. Dryfhout-Clark, et al. 2016. Atmospheric concentrations and loadings of organochlorine pesticides and polychlorinated biphenyls in the Canadian Great Lakes Basin (GLB): Spatial and temporal analysis (1992–2012). *Environmental Pollution* 217:124–133.

Siddiqi, M. A., R. H. Laessig, and K. D. Reed. 2003. Polybrominated diphenyl ethers (PBDEs): New pollutants-old diseases. *Clinical Medicine and Research* 1:281–290.

Silliman, B. R., J. van de Koppel, M. W. McCoy, J. Diller, G. N. Kasozi, K. Earl, P. N. Adams, and A. R. Zimmerman. 2012. Degradation and resilience in Louisiana salt marshes after the BP-Deepwater Horizon oil spill. *Proceedings of the National Academy of Sciences* 109:11234–11239.

Silva, M., P. J. Etnoyer, and I. R. MacDonald. 2016. Coral injuries observed at mesophotic reefs after the Deepwater Horizon oil discharge. *Deep Sea Research Part II: Tropical Studies in Oceanography* 129:96–107.

Soniat, T. M., S. M. King, M. A. Tarr, and M. A. Thorne. 2011. Chemical and physiological measures on oysters (*Crassostrea virginica*) from oil-exposed sites in Louisiana. *Journal of Shellfish Research* 30:713–717.

Sparks, C., J. Odendaal, and R. Snyman. 2014. An analysis of historical Mussel Watch Programme data from the west coast of the Cape Peninsula, Cape Town. *Marine Pollution Bulletin* 87:374–380.

Srogi, K. 2008. Levels and congener distributions of PCDDs, PCDFs and dioxin-like PCBs in environmental and human samples: A review. *Environmental Chemistry Letters* 6:1–28.

Sun, R., X. Luo, Q. X. Li, T. Wang, X. Zheng, P. Peng, and B. Mai. 2018. Legacy and emerging organohalogenated contaminants in wild edible aquatic organisms: Implications for bioaccumulation and human exposure. *Science of the Total Environment* 616–617:38–45.

Taffi, M., N. Paoletti, P. Liò, S. Pucciarelli, and M. Marini. 2015. Bioaccumulation modelling and sensitivity analysis for discovering key players in contaminated food webs: The case study of PCBs in the Adriatic Sea. *Ecological Modelling* 24:205–215.

Tuerk, K. J. S., J. H. Kucklick, P. R. Becker, H. M. Stapleton, and J. E. Baker. 2005. Persistent organic pollutants in two dolphin species with focus on toxaphene and polybrominated diphenyl ethers. *Environmental Science & Technology* 29:692–698.

UNEP. 2005. *Stockholm Convention: "New POPs" Screening Additional POPs Candidates*. https://www.wwf.or.jp/activities/lib/pdf_toxic/chemical/newpopsfinal.pdf. 37 pp.

———. 2010. *Stockholm Convention on Persistent Organic Pollutants (POPs), as Amended in 2009: Text and Annexes*. Geneva: United Nations Environment Programme. https://www.wipo.int/edocs/trtdocs/en/unep-pop/trt_unep_pop_2.pdf. 52 pp.

———. 2016. Stockholm Convention: Status of ratification. http://chm.pops.int/Countries/StatusofRatifications/PartiesandSignatoires/tabid/4500/Default.aspx.

———. 2017. *The 16 New POPs*. chm.pops.int/Portals/0/download.aspx?d=UNEP-POPS-PUB-Brochure-16NewPOPs-201706.English.pdf. 25 pp.

Unsworth, J. 2010. History of pesticide use. IUPAC (International Union of Pure and Applied Chemistry). http://agrochemicals.iupac.org/index.php?option=com_sobi2&sobi2Task=sobi2Details&catid=3&sobi2Id=31.

USEPA (US Environmental Protection Agency). 1972. DDT ban takes effect. https://archive.epa.gov/epa/aboutepa/ddt-ban-takes-effect.html.

———. 1986. Dicofol: Intent to cancel registrations of pesticide products containing dicofol. *Federal Register* 51:19508–19525.

———. 1999. Use authorization for and distribution in commerce of non-liquid polychlorinated biphenyls. *Federal Register* 64:237.

van den Berg, M., L. S. Birnbaum, M. Denison, M. De Vito, W. Farland, F. Feeley, H. Fiedler, et al. 2006. The 2005 World Health Organization reevaluation of human and mammalian toxic equivalency factors for dioxins and dioxin-like compounds. *Toxicological Sciences* 93:223–241.

van der Oost, R., J. Beyer, and N. P. E. Vermeulen. 2003. Fish bioaccumulation and biomarkers in environmental risk assessment: A review. *Environmental Toxicology and Pharmacology* 13:57–149.

Vidal-Martínez, V. M., and M. L. Aguirre-Macedo. 2014. Helmintos parásitos de organismos acuáticos como indicadores de impacto ambiental [Helminth parasites: Potential bioindicators of environmental pollution in the Neotropics]. In J. I. Euán Ávila, A. García de Fuentes, M. Á. Liceaga Correa, and A. Munguía Gil (eds.), *La costa del estado de Yucatán, un espacio de reflexión sobre la relación sociedad-naturaleza, en el contexto de su ordenamiento ecológico territorial* [The coast of the state of Yucatan, a space for reflection on the relationship between society and nature, in the context of its territorial ecological order], vol. 2. Mexico City: Plaza y Valdés.

Vidal-Martinez, V., E. Torres-Irineo, D. Romero, G. Gold-Bouchot, E. Martinez-Meyer, D. Valdés-Lozano, and L. Aguirre-Macedo. 2015. Environmental and anthropogenic factors affecting the probability of occurrence of *Oncomegas wageneri* (Cestoda: Trypanorhynca) in the southern Gulf of Mexico. *Parasites and Vectors* 8:609.

Vignier, J., A. K. Volety, A. Rolton, N. Le Goic, F. L. E. Chu, and R. Robert. 2017. Sensitivity of eastern oyster (*Crassostrea virginica*) spermatozoa and oocytes to dispersed oil:

Cellular responses and impacts on fertilization and embryogenesis. *Environmental Pollution* 225:270–282.

Viñas, L., M. A. Franco, J. A. Soriano, J. J. González, J. Pon, and J. Albaigés. 2010. Sources and distribution of polycyclic aromatic hydrocarbons in sediments from the Spanish northern continental shelf: Assessment of spatial and temporal trends. *Environmental Pollution* 158:1551–1560.

Wade, T. L., E. L. Atlas, J. M. Brooks, M. C. Kennicutt II, R. G. Fox, J. Sericano, B. Garcia, and D. DeFreitas. 1988. NOAA Gulf of Mexico Status and Trends Program: Trace organic contaminant distribution in sediment and oysters. *Estuaries* 11:171–179.

Wade, T. L., M. C. Kennicutt II, and J. M. Brooks. 1989. Gulf of Mexico hydrocarbon seep communities. III: Aromatic hydrocarbon burdens of organisms from oil seep ecosystems. *Marine Environmental Research* 27:19–30.

Wade, T. L., and J. G. Quinn. 1975. Hydrocarbon in the Sargasso Sea surface microlayer. *Marine Pollution Bulletin* 6:54–57.

Wade, T. L., J. L. Sericano, P. R. Gardinali, G. Wolff, and L. Chambers. 1998. NOAA's "Mussel Watch" project: Current use organic compounds in bivalves. *Marine Pollution Bulletin* 37:20–26.

Wade, T. L., J. L. Sericano, S. T. Sweet, A. H. Knap, and N. L. Guinasso Jr. 2016. Spatial and temporal distribution of water column total polycyclic aromatic hydrocarbons (PAH) and total petroleum hydrocarbons (TPH) from the Deepwater Horizon (Macondo) incident. *Marine Pollution Bulletin* 103:286–293.

Wade, T. L., S. T. Sweet, J. L. Sericano, D. A. DeFreitas, and G. G. Lauenstein. 2014. Polychlorinated dibenzo-p-dioxins and dibenzofurans detected in bivalve samples from the NOAA National Status and Trends Project. *Marine Pollution Bulletin* 81:317–324.

Wang, X., J. Li, H. Xing, and S. Xu. 2011. Review of toxicology of atrazine and chlorpyrifos on fish. *Journal of Northeast Agricultural University* [English ed.] 18:88–92.

Wang, Z., M. Fingas, and G. Sergy. 1994. Study of 22-year-old Arrow oil samples using biomarker compounds by GC/MS. *Environmental Science & Technology* 28:1733–1746.

———. 1995. Chemical characterization of crude oil residues from an Arctic beach by GC/MS and GC/FID. *Environmental Science & Technology* 29:2622–2631.

Wang, Z., M. Fingas, Y. Shu, L. Sigouin, M. Landriault, P. Lambert, R. Turpin, P. Campagna, and J. Mullin. 1999. Quantitative characterization of PAHs in burn residue and soot samples and differentiation of pyrogenic PAHs from petrogenic PAHs—the 1994 mobile burn study. *Environmental Science & Technology* 33:3100–3109.

Wang, Z., C. Yang, M. Fingas, B. Hollebone, X. Peng, A. B. Hansen, and J. H. Christensen. 2005. Characterization, weathering, and application of sesquiterpanes to source identification of spilled lighter petroleum products. *Environmental Science & Technology* 39:8700–8707.

Watanabe, K. H., F. W. Desimone, A. Thiyagarajaha, W. R. Hartley, and A. E. Hindrichs. 2003. Fish tissue quality in the lower Mississippi River and health risks from fish consumption. *Science of the Total Environment* 302:109–126.

Watanabe, M., K. Kannan, A. Takahashi, B. G. Loganathan, D. K. Odell, S. Tanabe, and J. P. Giesy. 2000. Polychlorinated biphenyls, organochlorine pesticides, tris(4-chlorophenyl) methane, and tris(4-chlorophenyl)methanol in livers of small cetaceans stranded along Florida coastal waters, USA. *Environmental Toxicology and Chemistry* 19:1566–1574.

White, H. K., P-Y. Hsing, W. Cho, T. Shank, E. Cordes, R. Quattrini, R. K. Nelson, et al. 2012. Impact of the Deepwater Horizon oil spill on a deep-water coral community in the Gulf of Mexico. *Proceedings of the National Academy of Sciences of the United States of America* 109:20303–20308.

WHO (World Health Organization). 2010. Dioxins and their effects on human health. http://www.who.int/mediacentre/factsheets/fs225/en/.

Willet, K. L., E. M. Ulrich, and R. A. Hites. 1998. Differential toxicity and environmental fates of hexachlorocyclohexane isomers. *Environmental Science & Technology* 32:2197–2207.

Wimmerová, S., M. van den Berg, J. Chovancová, H. Patayová, T. A. Jusko, M. B. M. van Duursen, Ľ. Palkovičová Murínová, R. F. Canton, K. I. van Ede, and T. Trnovec. 2016. Relative effect potency estimates of dioxin-like activity for dioxins, furans, and dioxin-like PCBs in adults based on cytochrome P450 *1A1* and *1B1* gene expression in blood. *Environment International* 96:24–33.

Wolfe, D. A., M. J. Hameedi, J. A. Galt, G. Watabayashi, J. Short, C. O'Claire, S. Rice, et al. 1994. The fate of the oil spilled from the *Exxon Valdez*. *Environmental Science & Technology* 28:560A–568A.

Xu, E. G., E. M. Mager, M. Grosell, C. Pasparakis, L. S. Schlenker, J. D. Stieglitz, D. Benetti, et al. 2016. Time- and oil-dependent transcriptomic and physiological responses to Deepwater Horizon oil in mahi-mahi (*Coryphaena hippurus*) embryos and larvae. *Environmental Science & Technology* 50:7842–7851.

Yakubov, M. R., D. V. Milordov, S. B. Yakubova, D. N. Borisov, D. V. Ivanor, and T. Sinyashin. 2016. Concentrations of vanadium and nickel and their ratio in heavy oil asphaltenes. *Petroleum Chemistry* 56:16–20.

Yednock, B. K., T. J. Sullivan, and J. E. Neigel. 2015. De novo assembly of a transcriptome from juvenile blue crabs (*Callinectes sapidus*) following exposure to surrogate Macondo crude oil. *BMC Genomics* 16:521. https://doi.org/10.1186/s12864-015-1739-2.

Yim, U. H., S. Y. Ha, J. G. An, J. H. Won, G. M. Han, S. H. Hong, M. Kim, J-H. Jung, and W. J. Shim. 2011. Fingerprint and weathering characteristics of stranded oils after the Hebei Spirit oil spill. *Journal of Hazardous Materials* 197:60–69.

Yogui, G. T., and J. L. Sericano. 2009. Polybrominated diphenyl ether flame retardants in the U.S. marine environment: A review. *Environment International* 35:655–666.

Yordy, J. E., R. S. Wells, B. C. Balmer, L. H. Schwacke, T. K. Rowles, and J. R. Kucklick. 2010. Life story as a source of variation for persistent organic pollutant (POP) patterns in a community of common bottlenose dolphins (*Tursiops truncates*) resident to Sarasota bay, FL. *Science of The Total Environment* 408:2163–2172.

Yuan, Y., T. Harner, K. Su, K. A. Brice, P. Blanchard, and J. Ma. 2010. A captured episode of γ-hexachlorocyclohexane air pollution in the Toronto area after the Canadian lindane ban. *Atmospheric Pollution Research* 1:168–176.

Zapata-Pérez, O, V. Ceja-Moreno, R. Olmos Pérez, M. del Río-García, M. Yarto, A. Mendoza-Cantú, I. Ize-Lema, A. Gavilán-García, A. Sanchez-Teyer, and G. Gold-Bouchot. 2007. Ecotoxicological effects of POPs on ariidae *Ariopsis felis* (Linnaeus, 1766) from three coastal ecosystems in the southern Gulf of Mexico and Yucatan Peninsula. *Journal of Environmental Science and Health Part A* 42:1511–1518.

Zapata-Pérez, O., M. Del-Río, J. Domínguez, R. Chan, V. Ceja, and G. Gold-Bouchot. 2005. Preliminary studies of biochemical changes (ethoxycoumarin O-deethylase activities and vitellogenin induction) in two species of shrimp (*Farfantapeneaus duorarum* and *Litopenaeus setiferus*) from the Gulf of Mexico. *Ecotoxicology and Environmental Safety* 61:98–104.

Zengel, S., S. C. Pennings, B. Silliman, C. Montague, J. Weaver, D. R. Deis, M. O. Krasnec, N. Rutherford, and Z. Nixon. 2016. Deepwater Horizon oil spill impacts on salt marsh fiddler crabs (*Uca* spp.). *Estuaries and Coasts* 39:1154–1163.

Zengel, S., J. Weaver, S. C. Pennings, B. Silliman, D. R. Deis, C. L. Montague, N. Rutherford, Z. Nixon, and A. R. Zimmerman. 2016. Five years of Deepwater Horizon oil spill effects on marsh periwinkles *Littoraria irrorata*. *Marine Ecology Progress Series* 576:1–10.

Zhao, X. X., B. Ren, Z. Luo, and R. Zhu. 2016. Life-cycle exposure to BDE-47 results in thyroid endocrine disruption to adults and offsprings of zebrafish (*Danio rerio*). *Environmental Toxicology and Pharmacology* 48:157–167.

Contributors

James W. Ammerman is the science coordinator for the Long Island Sound Study, the EPA National Estuary Program for Long Island Sound, and also an adjunct professor at Stony Brook University. He has also held research or faculty appointments at Columbia, Texas A&M, and Rutgers Universities. He has a PhD from Scripps Institution of Oceanography and is a Fellow of the American Association for the Advancement of Science. His scientific interests focus on nutrients and hypoxia, and he has served both as New York Sea Grant director and as a program manager at the National Science Foundation.

Thomas S. Bianchi is full professor and holder of the Jon and Beverly Thompson Endowed Chair in Geological Sciences at the University of Florida (UF), Gainesville, Florida. Before joining UF, he held full professor positions at Tulane University and Texas A&M University. His general areas of expertise are organic geochemistry, chemical oceanography, and global carbon cycling in aquatic ecosystems. He is the recipient of 2 Fulbright Research Awards and is a Fellow of the American Association for the Advancement of Science, the Geochemical Society, the European Association of Geochemistry, and the Association for the Sciences of Limnology and Oceanography.

Sumit Chakraborty is a staff scientist at Mote Marine Laboratory. His research focuses on understanding fine-scale phytoplankton dynamics, primary production, and harmful algal blooms in coastal oceans. He is also interested in coastal biogeochemistry and carbon cycles. He takes an interdisciplinary approach in his research, using combinations of marine bio-optics, remote sensing, and measurements from field and laboratory based experiments.

Jeff Chanton, a Gulf Coast native, received his AB from New College and his PhD from the University of North Carolina at Chapel Hill. He joined Florida State University (FSU) in 1989. He is a Lawton Professor, an AGU Fellow, and a Florida Wildlife Conservation communicator. He teaches graduate and undergraduate classes and directs FSU's Environmental Science Graduate and Undergraduate Programs. He speaks on the science of climate change to churches, civic groups, and government organizations, including a presentation to the 2014 Florida gubernatorial candidates.

Reide Corbett is a professor in the Department of Coastal Studies at East Carolina University and a senior scientist at the Coastal Studies Institute located on the Outer Banks of North Carolina. He holds a PhD and MS in chemical oceanography from Florida State University. His primary scientific interest is the cycling of biologically active constituents (e.g., carbon, nitrogen, phosphorus) and sediment processes in coastal and open-ocean environments.

Kendra L. Daly is a professor in the College of Marine Science at the University of South Florida. Her research focuses on the ecology of the lower trophic marine food web and its role in biogeochemical cycles. She is a Fellow of the American Association for the Advancement of Science.

Gerardo Gold-Bouchot is full professor in the Oceanography Department and the Geochemical and Environmental Research Group (GERG) at Texas A&M University (TAMU). Before joining TAMU, he held a full professor position at the Center for Research and Advanced Studies (CINVESTAV) located at Merida, Yucatan, Mexico. His general areas of expertise are marine pollution, ecotoxicology, and chemical oceanography. Lately he has been interested in the CDOM and FDOM dynamic in coastal ecosystems. He has been vice president and president of the Society of Environmental Toxicology and Chemistry (SETAC) Latin America section and a national researcher in Mexico.

Laodong Guo received his PhD in chemical oceanography from Texas A&M University and is currently a professor at University of Wisconsin–Milwaukee. His research interests focus on the biogeochemical cycling of carbon, nutrients, and trace elements/radionuclides in aquatic systems with an emphasis on the role of colloids and natural organic matter. He has published more than 180 peer-reviewed articles in earth and environmental journals.

Dennis A. Hansell is professor of ocean sciences at the University of Miami. His research addresses the cycling of carbon through organic matter dissolved in the global ocean.

Christopher T. Hayes is an assistant professor of marine science at the University of Southern Mississippi, based at the Stennis Space Center. He received a bachelor's degree in chemistry from McGill University and master's and doctoral degrees in earth and environmental science from Columbia University. He began his current faculty position after a postdoctoral fellowship at the Massachusetts Institute of Technology. His research is in marine biogeochemistry and paleoceanography.

Anna Hayward graduated from Florida State University in 2017 with a major in biochemistry and worked with Angela Knapp measuring nitrogen in marine samples.

Ruoying He is a Goodnight Innovation Distinguished Professor in the Department of Marine, Earth, and Atmospheric Sciences at North Carolina State University. His research spans from coastal circulation dynamics, to air-sea interaction, to biophysical interactions. As the director of the Ocean Observing and Modeling Group, he conducts coastal ocean observations and remote-sensing data analyses and leads the development of predictive models of ocean circulation, air-sea interactions, and physical-biogeochemical couplings.

Jorge Herrera-Silveira is titular professor and director of the Primary Production Laboratory of the Center for Research and Advanced Studies (CINVESTAV) located at Merida, Yucatan, Mexico. He holds a PhD in biology (aquatic ecology) from the University of Barcelona, Spain. He is an expert and collaborator with international organizations such as United Nations Educational, Scientific and Cultural Organization; Global Environment Facility; Ramsar Convention; International Union for Conservation of Nature; and the Coastal Conservation Association. The research topics addressed are eutrophication and harmful algal blooms; water quality; health status of coastal habitats; restoration of mangroves; "blue carbon"; vulnerability to, adaptation to, and mitigation for the effects of climate change; and evaluation of services ecosystems of tropical coastal environments and their resilience. He has received recognition from the Mexican government and nongovernmental organizations for his contribution to the investigation of the coastal wetlands of Mexico and his leadership in the conservation of the coastal natural heritage.

Gary L. Hitchcock is an associate professor emeritus in the Department of Marine Biology and Ecology at the University of Miami. His research focus was phytoplankton ecology and primary productivity. Much of his research concentrated on the application of Lagrangian techniques to plankton dynamics in the Atlantic and Indian Oceans, as well as harmful algal blooms in the Gulf of Mexico.

David Hollander is a professor at the University of South Florida's College of Marine Science. He is a co–principal investigator and theme leader for the C-IMAGE consortium funded by the Gulf of Mexico Research Initiative to study long-term impacts of the 2010 Deepwater Horizon oil disaster. He received his PhD from the Swiss Federal Institute of Technology, ETH-Zurich.

Karen H. Johannesson is the Cochran Family Professor of Earth and Environmental Sciences at Tulane University in New Orleans. She holds a BS in geology from the University of New Hampshire, an MS in geology and geophysics from Boston College, and a PhD in hydrology/hydrogeology from the University of Nevada. Her research is focused chiefly on trace-element speciation and biogeochemical cycling in natural waters.

Samantha Joye received her BSc, MSc, and PhD at the University of North Carolina at Chapel Hill. She joined the faculty at the University of Georgia in 1998 after 2 years at Texas A&M University, College Station. She holds the Athletic Association Professorship in Arts Sciences and is a Fellow of the American Geophysical Union, the American Association for the Advancement of Science, Association for the Sciences of Limnology and Oceanography, and the American Academy of Microbiology. She teaches graduate and undergraduate classes in geobiology, marine sciences, and science communication. She is an outspoken ocean and environmental advocate.

Anthony H. Knap is director of the Geochemical and Environmental Research Group, full professor of oceanography, and holder of the James R. Whatley Endowed Chair of Geosciences at Texas A&M University. He has been president, director, and senior scientist at the Bermuda Institute of Ocean Sciences, St. Georges, Bermuda. His general areas of expertise are organic geochemistry, chemical oceanography, environmental sciences, and global climate change. Knap was the founder and principal investigator of the Bermuda Atlantic Time-series Study (BATS) for 25 years. He has served as the chairman of the Health of the Ocean Panel and cochairman of the Coastal Panel for the Global Ocean Observing System and recently stepped down from the National Academy of Sciences Gulf Research Program Advisory after 3 years.

Angela Knapp is an assistant professor of chemical oceanography in the Earth, Ocean, and Atmospheric Sciences Department at Florida State University. She received her BA in biology-geology from Whitman College and her PhD in geosciences from Princeton University. Her research focuses on the marine nitrogen cycle, especially using isotope geochemistry to identify and quantify fluxes of nitrogen to the ocean.

Jeffrey W. Krause is a senior marine scientist at the Dauphin Island Sea Lab and assistant professor at the University of South Alabama. He received a BS in marine science from Southampton College of Long Island University and a PhD in oceanography from Oregon State University. His research focuses on silicon biogeochemistry and phytoplankton physiology/ecology in coastal, open-ocean, and high-latitude systems.

Laura Lapham is an associate professor at the Chesapeake Biological Laboratory, University of Maryland Center for Environmental Science. She received her BS from Florida State University and her PhD from University of North Carolina at Chapel Hill. She is a biogeochemist who focuses on microbial transformations of organic carbon in the sediments and has worked on gas-hydrate seep systems since 2000.

Chih-Ping Lee is a PhD candidate in the Chemical Division of the Institute of Oceanography, National Taiwan University. He received his bachelor's degree in geosciences and chemistry and his master's in geosciences at National Taiwan University. His major research interest is trace-element speciation in aquatic systems, especially metalloids.

Andrew R. Margolin received his PhD from the University of Miami and was awarded a National Science Foundation Graduate Research Fellowship. Recently, he completed a postdoctoral research associate appointment at the Virginia Institute of Marine Sciences. He studies the carbon biogeochemistry in marginal seas with a focus on the carbonate system.

Frank J. Millero is a professor emeritus at the Rosenstiel School of Marine and Atmospheric Science, University of Miami. He is well known as an expert on the physical chemistry of natural waters. He received his PhD from Carnegie Mellon University. Over the years he has received numerous awards for his research accomplishments.

Elise S. Morrison received her PhD from the University of Florida and her BS from the University of California, Davis. She is currently a postdoctoral research associate at the University of Florida's Department of Geological Sciences. Her research focuses on linking biogeochemical cycles with microbial ecology.

Christopher L. Osburn is an associate professor of marine biogeochemistry at North Carolina State University. He received his PhD from Lehigh University and is a recent Fulbright Specialist. His research focuses on the transport and cycling of dissolved and particulate organic matter in rivers, estuaries, and coastal waters.

Donald G. Redalje is a professor of marine science at the University of Southern Mississippi. He received his BA in environmental biology from the University of California, Santa Barbara, and his PhD in oceanography from the University of Hawaii at Manoa. His research has focused on the physiological ecology of marine phytoplankton with a particular focus on using phytoplankton pigment taxonomy to help understand the health of estuarine, coastal, and open-ocean ecosystems as well as marine ecosystem management and policy. He has also focused on the relationship between primary production and the flux of organic matter to the subeuphotic zone ocean. Recently his focus has been directed toward the use of marine phytoplankton to produce algal biofuels, algal protein supplements, and other high-value products.

Delores M. Robinson is a professor in the Department of Geological Sciences and Center for Sedimentary Basin Studies at the University of Alabama. She earned her PhD from the University of Arizona and MS from Vanderbilt University. Her areas of expertise are tectonics and structural geology. She is a Fellow of the Geological Society of America.

Kelsey Rogers earned a BS in geology at University of North Carolina at Chapel Hill and, in 2018, her PhD in oceanography from Florida State University. She is a Gulf of Mexico Research Institute Scholar and has studied at Woods Hole Oceanographic Institute, University of California, Irvine, and Laurence Livermore National Labs. She is currently working on a postdoctorate at the University of Copenhagen, Denmark.

Brad E. Rosenheim is an associate professor of geological oceanography at the College of Marine Science of the University of South Florida in Saint Petersburg. He earned a BS in environmental science from the University of Vermont and a PhD in marine geology and geophysics from the Rosenstiel School of Marine and Atmospheric Sciences at the University of Miami. His research revolves around the development of isotopic-measurement techniques and their application to problems concerning climate change and carbon cycling.

Peter H. Santschi is director of the Laboratory for Oceanographic and Environmental Research, as well as of the Coastal Zone Laboratory, and is a Regents Professor in the Marine Sciences Department at Texas A&M University at Galveston. He has a PhD and MS equivalent in chemistry from the University of Bern, Switzerland. His research focuses on marine and environmental biogeochemical and radiochemical processes.

José L. Sericano is a research scientist with the Geochemical and Environmental Research Group at Texas A&M University. He received his PhD in chemical oceanography from Texas A&M University and has been involved in assessing and characterizing persistent organic pollutants (POPs) in environmental samples for more than 35 years. Since 1991, he has been regularly invited to conduct training courses and provide technical support to scientists in Latin America, southern Asia, and the Middle East as part of cooperative programs founded by the International Atomic Energy Agency, the Korea Ocean Research and Development Institute, the Organization of American States, and the United Nations Environment Programme, among many others. His general research interest focuses on the determination, impact, and fate of trace organic contaminants.

David Valdes is a professor in the Department of Marine Resources at the Center for Research and Advanced Studies (CINVESTAV) located at Merida, Yucatan, Mexico. He holds a PhD and MS in chemical oceanography from National Mexico University (UNAM). His primary scientific interest is the cycling of biologically active constituents (e.g., carbon, oxygen, nitrogen, phosphorus) and sediment processes in coastal and open-ocean environments.

Victor Manuel Vidal Martínez is a professor in the Sea Resources Department at the Center for Research and Advanced Studies (CINVESTAV) located at Merida, Yucatan, Mexico. He holds a PhD from Exeter University, United Kingdom. His primary scientific interest is the use of parasites of aquatic organisms as bioindicators of environmental impact in freshwater, coastal, and open-ocean environments.

Terry L. Wade is deputy director of the Geochemical and Environmental Research Group (GERG) and research professor of oceanography and interdisciplinary faculty of toxicology at Texas A&M University. Wade received his PhD in 1978 in chemical oceanography from the University of Rhode Island. He has held a soft-money research position at GERG since 1984, where he has managed many large environmental programs on the fate and effects of contaminants.

John P. Walsh is director of the Coastal Resources Center and a professor in the Graduate School of Oceanography at the University of Rhode Island. He has a PhD in oceanography from the University of Washington, an MS in marine environmental science from Stony Brook University, and a BA in geology from Colgate University. His research focuses on coastal and ocean sedimentary processes.

Jennifer Warrillow is the laboratory manager of the Ocean Observing and Modeling Group in the Department of Marine, Earth, and Atmospheric Sciences at North Carolina State University. She has a BA in aquatic biology from the University of California, Santa Barbara, and an MS in natural resources from Cornell University.

Liang-Saw Wen is an associate professor at the Institute of Oceanography, National Taiwan University. He holds a PhD in chemical oceanography from Texas A&M University. He was a senior research fellow with the Environmental Technology Institute at the National Science and Technology Board of Singapore and the deputy director of the Taiwan Ocean Research Institute. His research is focused mainly on trace-metal speciation and reactions in aquatic environments, especially reactions with colloids.

Index